AERIAL ENGAGEMENTS

AND THE USE OF AIR POWER FROM THE GREAT WAR TO THE PERSIAN GULF WARS

RODNEY MACLEAN

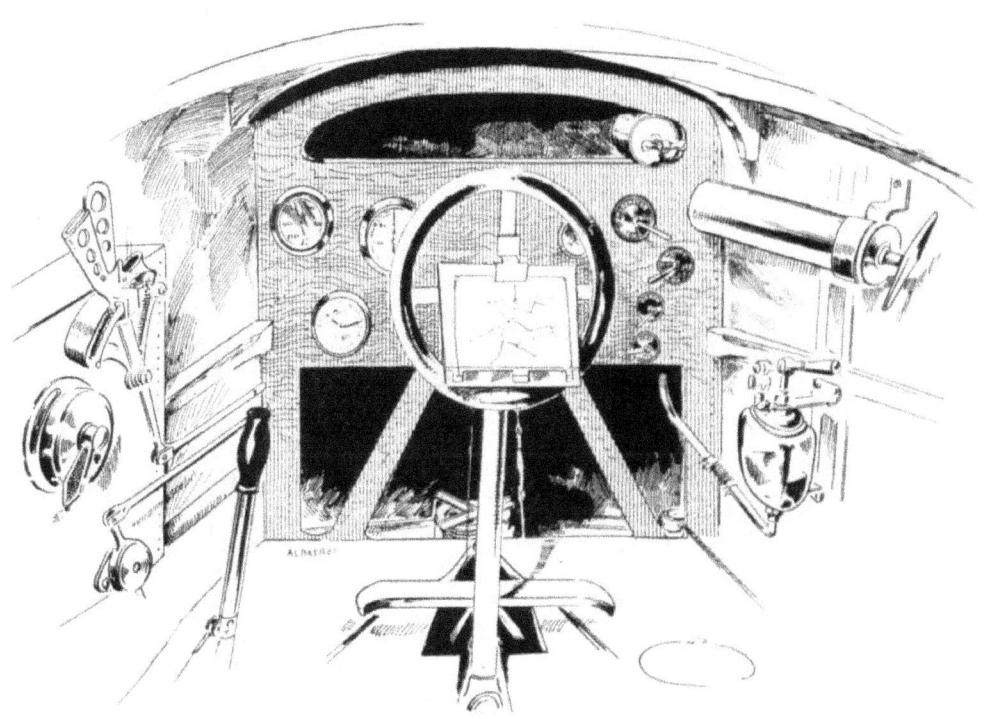

Sketch showing the instruments carried on the captured German Albatros C-1

ARCHIVES
d'Histoire et Militaire

ADELAIDE
2017

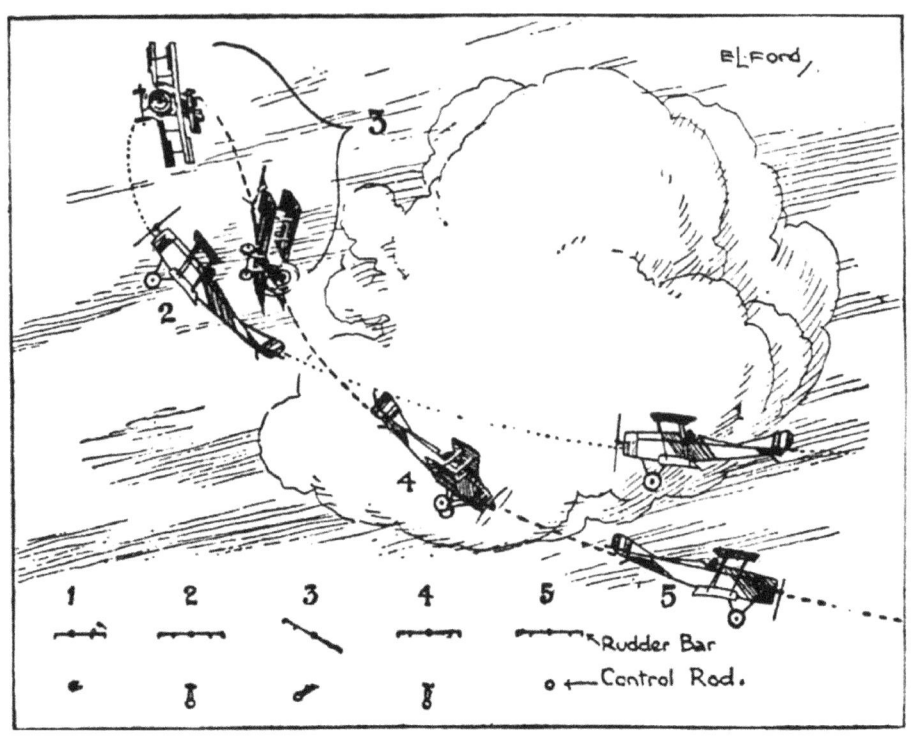

The Immelmann Turn

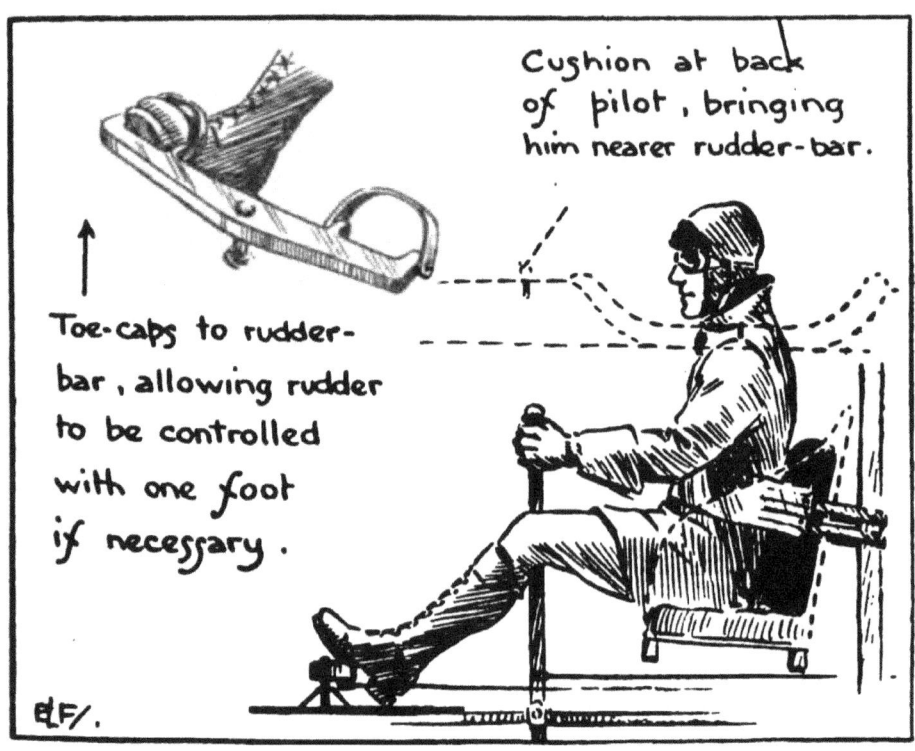

Copyright © 2017 Rodney MacLean
All rights reserved.
ISBN-13 : 978-0994371409
ISBN-10 : 0994371403
BISAC HIS 027140
History ~ Military ~ Aviation ~ Aeronautics ~ Flight ~ Naval Aviation ~ Fighter Pilots ~ Tactics

To Elena

Arabella and Harald

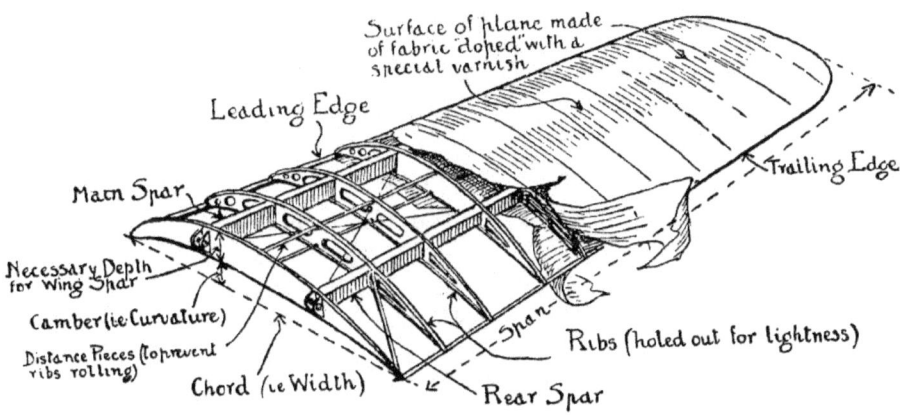

CONTENTS

PREFACE

CHAPTER I
THE GREAT AIR WAR BEGINS ~ 1

CHAPTER II
THE ZEPPELIN CAMPAIGN ~ 31

CHAPTER III
WINNING THE GREAT WAR ~ 63

CHAPTER IV
NAVAL AVIATION 1914-1918 ~ 87

CHAPTER V
THE SPANISH CIVIL WAR ~ 109

CHAPTER VI
BLITZKRIEG TO THE BLITZ ~ 119

CHAPTER VII
THE EUROPEAN THEATRE ~ 145

CHAPTER VIII
THE PACIFIC THEATRE ~ 169

BEWARE OF THE HUN IN THE SUN (RAF TRAINING POSTER 14.06.1918 ~ CWM 19850452-075)
A CUNNING FIGHTER PILOT WILL ALWAYS TRY TO USE THE SUN TO TRY TO BLIND HIS OPPONENT. IN A SURPRISE ATTACK, THE ENEMY MAY COME OUT OF THE SUN WHERE THERE IS A DIFFICULTY IN SEEING HIM, REMEMBER TO LOOK FOR THIS ESPECIALLY WHEN ABOUT TO ENGAGE A MACHINE THAT MAY WELL PROVE TO BE A DECOY.

Contents

CHAPTER IX
KOREAN WAR ~ 221

CHAPTER X
VIETNAM WAR ~ 241

CHAPTER XI
MIDDLE-EASTERN WARS ~ 261

CHAPTER XII
THE FALKLANDS WAR ~ 275

CHAPTER XIII
DESERT SHIELD & DESERT STORM ~ 285

EPILOGUE

Appendix I
APPLICATIONS OF ANTIQUE AIR POWER

GLOSSARY OF TERMS & ABBREVIATIONS

BIBLIOGRAPHY

INDEX

F-4 Phantom ~ courtesy of *Fighter Sweep*.com

LIST OF MAPS

MAP ILLUSTRATING RFC AERIAL RECONNAISANCE AREAS OF AUGUST 1914

MOVEMENTS OF ROYAL FLYING CORPS FROM AUGUST TO OCTOBER 1914

MAP ILLUSTRATING AERIAL RECONNAISANCE AREA SEPTEMBER 1914 ~ 6

BATTLE OF BRITAIN RAF FIGHTER COMMAND PLOTTING ROOMS ~ 128

RAF FIGHTER COMMAND GROUPS AND SECTORS OF SPRING 1941 ~ 134

RAF COASTAL COMMAND HQ OPERATIONS ROOM WALL CHART ~ 142

THE MALTA CONVOY ROUTES OF 14-16th JUNE 1942 ~ 152

RADIUS OF ACTION OF AIRCRAFT FROM MALTA ~ 152

NEARNESS OF MAIN RAF BOMBING OBJECTIVES ~ 160

ALLIED ESCORT FIGHTER RANGES ~ 167

ROUTE OF JAPANESE NAVAL FORCE TO PEARL HARBOUR ~ 168

JAPANESE PLAN TO CAPTURE AREAS OF AUSTRALIA ~ 183

JAPANESE FU-GO BALLOON INCIDENT LOCATIONS ~ 217

B-29 TARGET CITIES IN JAPAN ~ 219

WESTERN PACIFIC B-52 ROUTES TO NORTH VIETNAM ~ 254

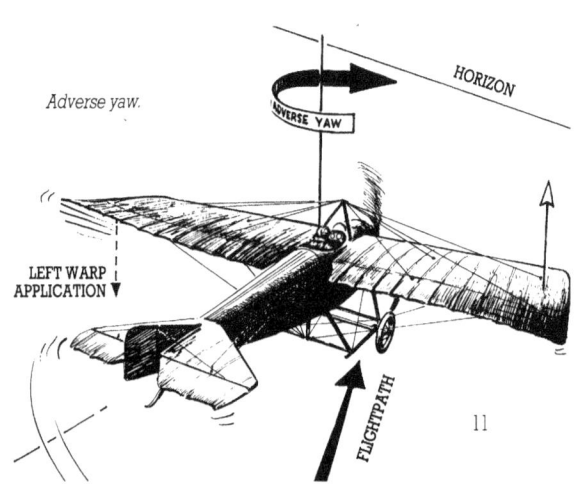

Adverse yaw.

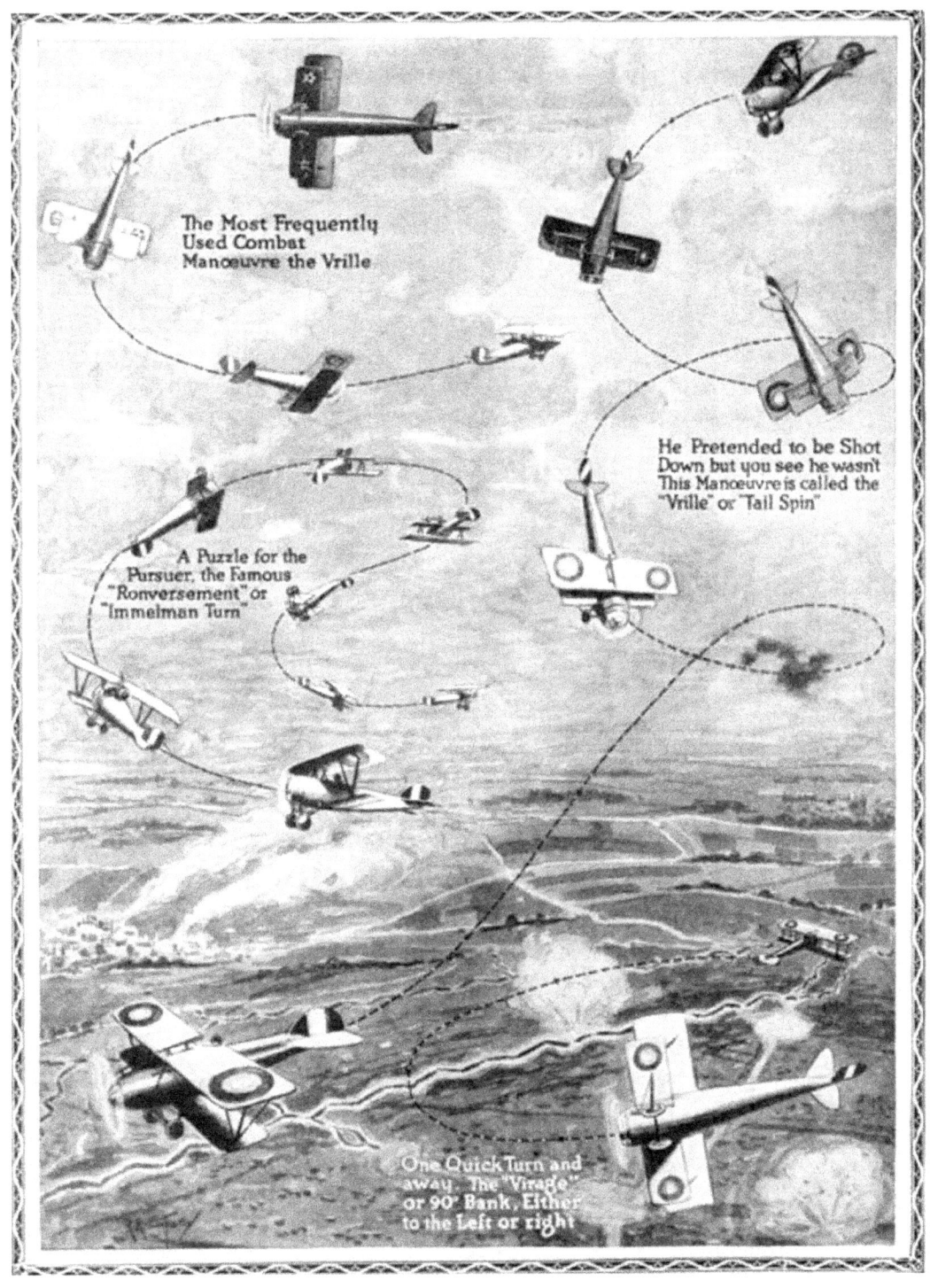

Centre left : Immelmann Turn ~ Climbing as if to loop-the-loop but instead twists the machine about and straightens out at a higher level.

Right : The same manouevre repeated in a series of loops, mimicking the action of a machine which is out of control.

Top left : a most useful manouevre ~ the Vrille ~ the pilot banks sharply and dives, then straightens out.

In the lower right-hand corner ~ The Virage ~ the quickest way to turn an æroplane.

-

PREFACE

In some ways we live in a harsh world, where old-fashioned history and truth still continue to have significant value and virtue. History can teach us to understand, empathize and sympathize with people who are different from us ~ either because they're from different cultures or of a different era. By evoking the layers of history as a vivid tapestry with endless threads woven through time to illuminate the present and the future, we can see how particular ideas and events developed in earlier periods, and acquire a strong understanding of how they might develop and emerge into the future.

Whilst history never repeats, forewarned is forearmed, since it often follows familiar patterns. The task then is to learn how to distinguish those aspects and tactics of war that are transient or unique in time and place, from those which are likely to recur elsewhere in the coming future. When looking at the principles that underlie the conduct of military and naval aviation through the rich stream of history, the greater the range of experience that can be absorbed, the more balanced ones own judgement becomes ~ as does our depth of understanding. And so the quest unfolds, to find the tipping point in a campaign, and to tell the harrowing history of military and naval warriors. Of how fresh ideas in policy, equipment and tactics acquired a winning advantage for one side in their aerial engagements, even when the future of their cause looked bleak.

The compiled wisdom in this book transposes the winning ways of military and naval aviators from previous wars into the awareness of the present, to fill the need of a practical book that reveals an archive of aerial adventure in times of war. In aerial warfare as in everything else, a glimpse of the future can often be grasped through the medium of a reflected past. However, the further back we go, the more the conditions of material and social outlook differ from those of the present, and it becomes ever more challenging to stand in the shoes of those whose deeds we probe. Past people were not the same as us *only dressed in petticoats and top hats*. They had quite different ideas from our own. Therefore, they need to be understood on their own terms, not ours. This book delves into their world, from the perspective of those who mastered their aircraft and conquered the skies. They are radiant glimpses that reach out and make your heart beat faster as they thrill on the edge of life's fragility.

Here, those winning aspects that enabled aviators to defeat their foes are revealed to the reader. Often it is not the technology that conveys a winning edge in warfare, but its acquired mastery. To survive in the past, the best aces were frequently the most calculating of all. Those who balanced risk taking with a clear understanding of their aircraft and the capabilities of their opponents, and the position at which they found themselves at the beginning of an engagement. Now a well-trained pilot and his aircrew are still the best part of any aircraft, and the perspective of quality over quantity continues to win. However, an armed aircraft is not an object in the usual sense. For the pilot to divide him or herself from it, even in thought is one of the principal roots of error. It is hoped that the knowledge within, will interest readers of history, and assist aviators to avoid the mistakes so often made by others in the past. The recovery of the past or looking backward, to see how to go forward becomes helpful at times. By connecting with those of the past who made worthy contributions to civilization while they describe their role in their own words, adds depth to our impression of how they advanced the adventure of flight with clear vision and devotion. This compilation traces the history of successful aerial engagements from the origin of human flight to recent Middle East actions.

Rodney MacLean   2017

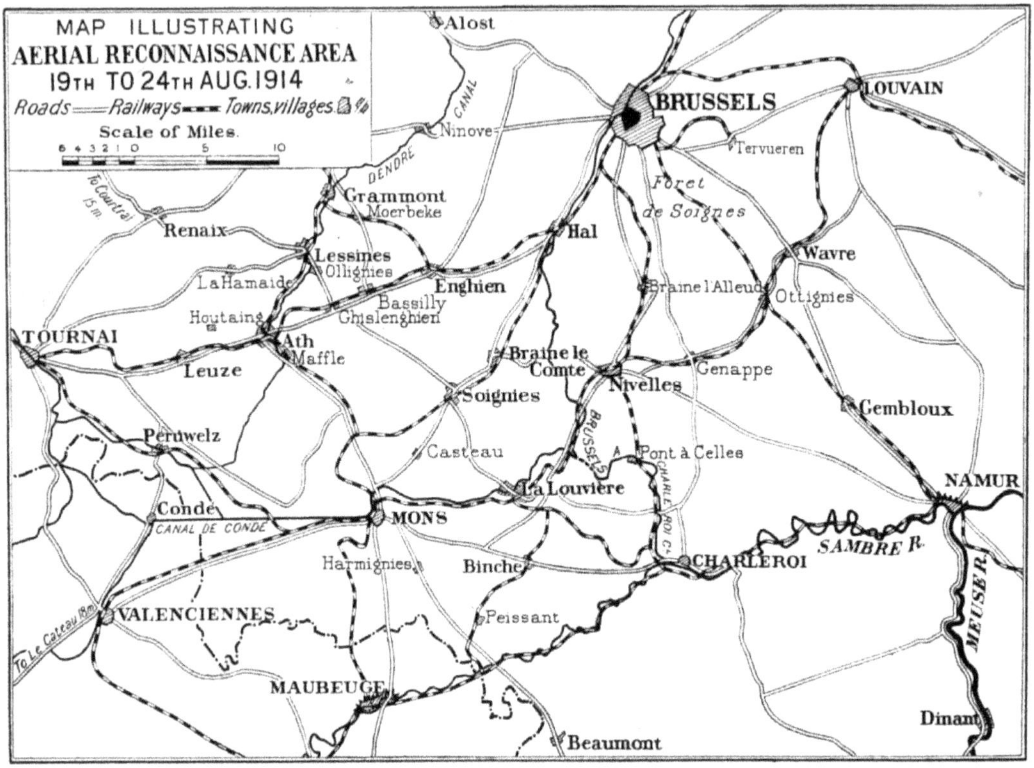

MAP ILLUSTRATING
AERIAL RECONNAISSANCE AREA
19TH TO 24TH AUG. 1914
Roads ——— Railways ==== Towns, villages

MOVEMENTS OF THE ROYAL FLYING CORPS FROM AUG. 16TH TO OCT. 12TH 1914.

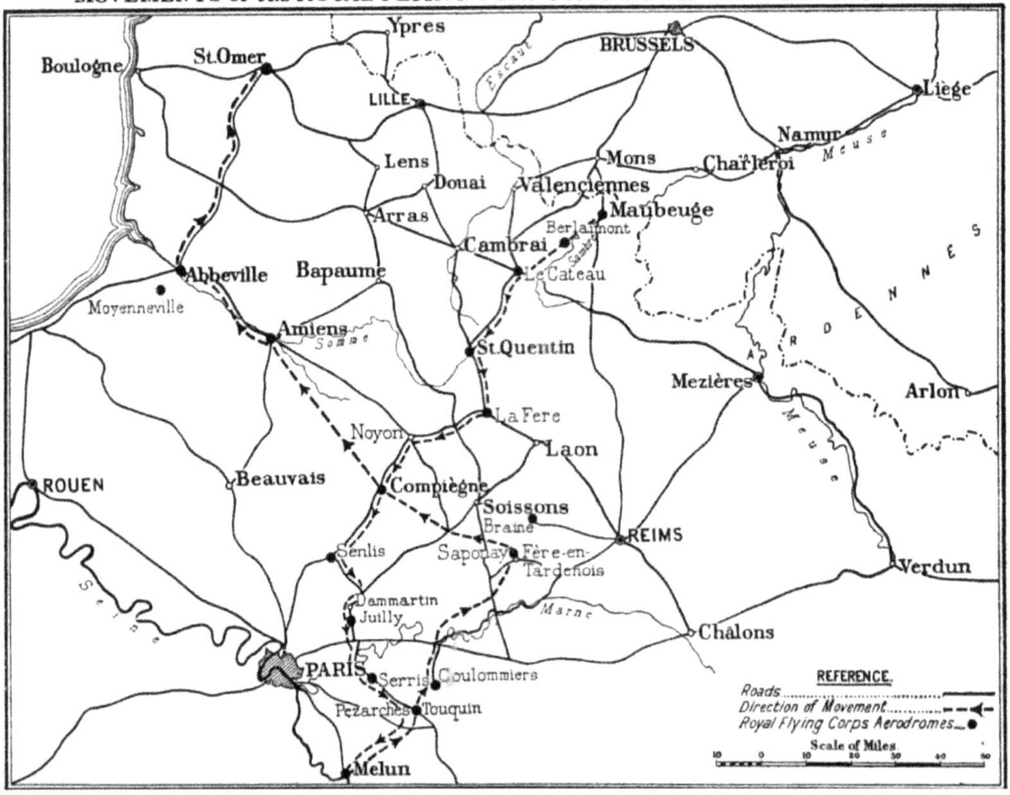

REFERENCE.
Roads ....
Direction of Movement .... ---◄
Royal Flying Corps Aerodromes ... ●

CHAPTER I

THE GREAT AIR WAR BEGINS

Prelude ~ The Prussian military caste were consumed with a desire to create a German Empire, and prepared for action. In 1862, Kaiser Wilhelm I appointed Otto von Bismarck his Foreign Minister to exploit rather than initiate events to acquire the German states, then part of Denmark and Austria. Prussia first fought Denmark in 1864, to obtain the disputed Danish duchy of Schleswig-Holstein. This was used to trigger Seven Weeks War, in which the vanquished Austrians were compelled to give up Hanover, Hesse, Nassau, and Frankfurt. Then the cunning von Bismarck seized upon Prince Leopold Hohenzollern's candidacy for the vacant Spanish throne to dupe France into declaring war on them. After humbling the French into yielding Alsace-Lorraine, the unification of Imperial Germany was proclaimed at Versailles. Kaiser Wilhelm I's grandson, Wilhelm II intended that his Imperial Germany would have *a place in the sun,* and seized a handful of colonies in the Pacific and on the coast of Africa. Wilhelm II took a more aggressive stance in 1899 when he defied the British and continued supplying weapons to Afrikaner rebels during the South African War. In 1905 he sailed into the international Moroccan port of Tangier to make a speech in favour of Moroccan independence, and in 1911 sent the gunboat *Panther* into the Moroccan port of Agadir. The French interpreted the move as a German move to turn Agadir into a naval base on the Atlantic. The German ambassador in Paris instead levered from the French their African colony of Middle Congo, which became the dual German colonies of Kameroon and Togoland as *compensation* for recognizing the French position in Morocco. Kaiser Wilhelm II next visited Damascus, where he paid for the rebuilding of the tomb of Saladin, declared himself a sympathizer of Islam, and commissioned the *Berlin-Baghdad Railway* to range Imperial Germany on the side of the Ottoman Empire and project German arms into the Middle East.

* * *

After the assassination of the Archduke Ferdinand and his wife Sophie on the streets of Sarajevo, the Great War started on Sunday 4th August 1914, when Germany declared war on France and invaded neutral Belgium and headed toward Paris. The German Army continued to thrust deeper into Belgium, which Britain was morally obliged to support. The British Government sent an ultimatum that, if German forces had not withdrawn by midnight a state of war would exist between Britain and Germany. As no response came, Britain as the guarantor of Belgian neutrality was officially at war. The British Expeditionary Force under Field Marshal Sir John French embarked for France six days later to take up positions between Maubeuge and Le Cateau, while the RFC prepared to fly to France.

Starting in mid-August, 63 RFC aeroplanes took off from Dover to fly to their French airfields. Lieutenant Louis Strange of No. 5 Squadron RFC piloting a Henri Farman, crossed a gusty English Channel from Dover to the Amiens aerodrome on 16th August, with as all pilots were issued; maps of Belgium and France, a revolver, field-glasses, a spare pair of flying goggles, an inner-tube for a life preserver, a roll of tools, two water bottles, a small stove, plus a haversack brimming full of field provisions; such as chocolate bars, biscuits, sandwiches, cold chicken, pork pies, and canned bully beef ~ along with orders, *to ram on sight any enemy Zeppelin airships encountered.* Lt. Louis Strange's scout aeroplane was the only machine fitted with an improvised Lewis machine-gun mounting in the observer's forward cockpit, *I fixed a safety strap to the leading edge of top wing so as to enable a passenger to stand up and fire all round over the top of plane and behind. I took Lieutenant Rabagliati as my passenger on the trip. Great success. Increases range of fire greatly, and I hear that these belts are to be fitted to all machines.*

When the airmen of the Royal Flying Corps touched down at Amiens Aerodrome to refuel, crowds of welcoming villagers offered them bottles of wine, bouquets of flowers, fruit, and eggs. From Amiens they flew to Maubeuge, where the French provided blankets and straw for their bedding. For the first few days in the war zone, they saw nothing of the British army till one evening. British troops marched through Maubeuge on their way to Mons. Captain Joubert de la Ferté wrote: *We were rather sorry they had come, because up till that moment we had only been fired on by the French whenever we flew. Now we were fired on by French, and English... to this day I can remember the roar of musketry that greeted two of our machines as they left the aerodrome and crossed the main Maubeuge-Mons road, along which a British column was proceeding.* To avert further episodes of this, British airmen worked all night painting Union Jacks in the form of a shield, on the lower-planes of all Royal Flying Corps aircraft.

THE ROYAL FLYING CORP'S FIRST WAR MISSION

In the gathering storm of war, the RFC flew its first war mission on 19th August 1914, when two aircraft from No. 3 Squadron were ordered over the Nivelle-Genappe area, to report if any Belgian forces were in the vicinity. On the 22nd August, Captain Joubert de la Ferté, with Lt. Allen in the observer cockpit of his Bleriot monoplane, took off to report on the fighting around Charleroi. Joubert and Allen flying at 2,000 feet over Charleroi seeing heavy fighting also observed that the French were being pushed back. That evening, the First Division of the small British Expeditionary Force, marching north towards Mons to launch an offensive against the Germans, was unaware that the French on their right flank were falling back — as Joubert and Allen had reported to GHQ only to be disbelieved.

Scouting above the British Expeditionary Force's left flank near Mons the following day, Joubert and Allen observed roads covered with *grey streams of men where we knew that there were no Allied troops*. In one of the finest reconnaissance's ever made, Joubert had discovered the fourfold larger German force of General von Kluck's 2nd Corps, sweeping westward on the Brussels-Ninove road, and the advancing British First Division was marching straight into a trap. After a bullet rattled off the steel plate under his seat and another pierced the side of his fuel tank, Joubert plugged the hole with his finger and made for home. His report warned Sir John French at Le Cateau and was supported by the maps of two other airmen showing long black lines of German troops on every road. HQ called off the offensive and issued orders to withdraw ; *We move in four hours!* Barely enough time to support the struggling French force. The Royal Flying Corps prevented Mons from becoming a disaster and gave British forces just enough time to prepare for the orderly retreat that started on August 24th and continued for several weeks. As the squadrons of the RFC retreated, subsequent positions of their ground parties were easily located from the air, as much of their recently requisitioned road transport was highly decorative. One squadron's vehicles consisted of copious furniture vans, while another squadron's ammunition lorry having belonged to a famous sauce supplier was painted in vivid scarlet with the word's *The World's Appetizer* in golden letters along the sides. Captain Louis Strange of No. 5 Squadron RFC on August 28th tried his idea of using petrol as a weapon. Putting together some simple petrol bombs, he took them with him on a flight he made in his Henri Farman that afternoon. Spotting some German trucks trundling along a dirty road ~ his first two bombs missed their mark, but his third hit a truck that swerved into a ditch and caught fire. Exploding petrol splashed over the following German truck and that caught fire too. Strange flew home, well pleased with his adventure.

Scout aircraft, competing with the enemy in the same airspace would increasingly be required to fight for their information, to fly fighting patrols and obtain *elbow room* (later termed air superiority) for bombing and reconnaissance two-seater aircraft. Aerial observers, who were usually cavalry officers with experience in traditional scouting, became prime targets as rival pilots and aircrew took pistols aloft. In the first months of the war, two-seater pushers had the tactical advantage as the observer could handle a gun while the pilot controlled the plane. On scouting flights in two-seater aeroplanes it was standard for an observer to call for a turn by tapping his pilot on the right or left shoulder. A prod between the shoulder blades meant straight ahead. A rap on his helmet to descend. A wave with the flat of the hand: an enemy aeroplane. A clenched fist: anti-aircraft fire. Some scouting aircraft were armed with machine guns and the Lewis gun was the preferred option for Allied aircraft. It was light, reliable, and was fed by a magazine-drum on top that carried 47 rounds. The problem was that there was an acute shortage of these handy weapons after the factory in Liege was captured, and production at the Birmingham Small Arms factory remained sluggish. Demand for the Lewis gun became so intense after the weapon proved itself handy in trench warfare that it was only possible at first, to issue a few to each frontline squadron.

Aerial reconnaissance had its limitations as well as its advantages, and the danger of relying on negative reports was not at first apparent. Several sightings of troops on the move turned out to be long patches of tar on roadways, and one report of enemy tents proved to be shadows cast by gravestones in a churchyard. General von Kluck failed to keep in touch with and outflank the British during the retreat from Mons, because he misinterpreted an observer's report, that all roads through the Floret de Mormal were clear of troops. He presumed that this meant that there were no troops in the area, however one forest contained a considerable British force that, hearing the approach of any aircraft, simply moved off the roads and under the cover of trees, so that nothing was seen of them. During the retreat, RFC aerial scouts plotted the daily movements of the German forces showing Sir John French how von Kluck's troops were constantly trying to envelop his British force, and made it possible for him to keep his army, continually just out of reach.

* * *

PUSHERS ~ THE HENRI FARMAN & F.E.2d & F.E.2b NIGHT BOMBERS

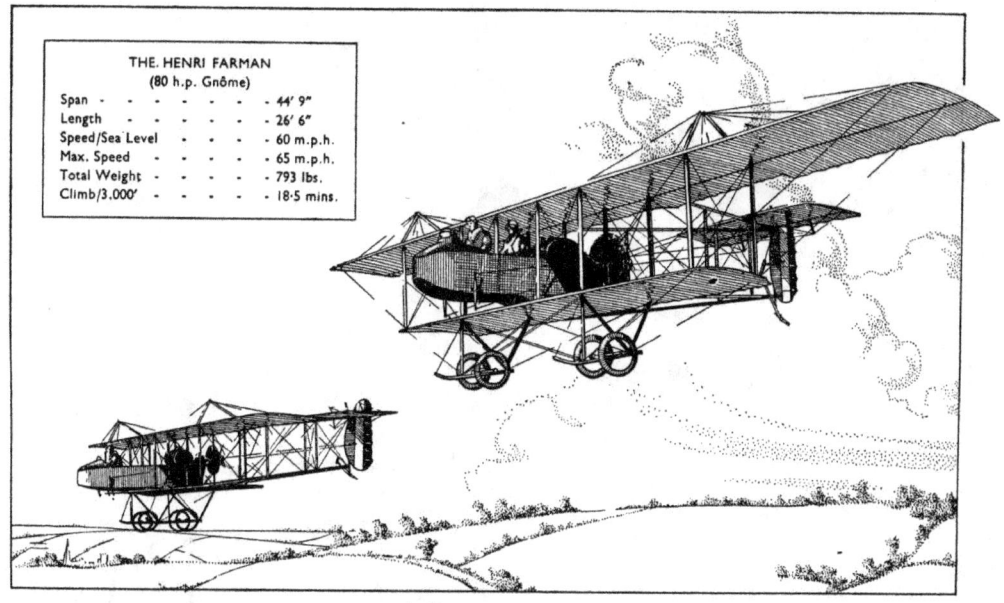

THE HENRI FARMAN
(80 h.p. Gnôme)
- Span — 44' 9"
- Length — 26' 6"
- Speed/Sea Level — 60 m.p.h.
- Max. Speed — 65 m.p.h.
- Total Weight — 793 lbs.
- Climb/3,000' — 18·5 mins.

THE F.E.2d NIGHT BOMBER (Top)
(250 h.p. Rolls-Royce Eagle VII)
- Span — 47' 9"
- Length — 33' 2"
- Loaded Weight — 3,549 lbs.
- Max. Speed — 100 m.p.h.

Three movable Lewis guns

and THE F.E.2b (Bottom)
(120–160 h.p. Beardmore engine)

FRENCH AVIATORS OF THE GREAT WAR 1914 -1918

Capitaine René Roeckel~ Commandant l'Escadrille MF-7

Charles Nungesser ~ 45 victories

Roland Garros ~ 4 victories

René Fonck ~ 75 victories

FRENCH AVIATORS OF THE GREAT WAR

Gervais Raoul Lufbery ~ 16 victories

Georges Guynemer ~ 53 victories

Adolphe Pégoud ~ 6 victories

Jean Navarre ~ 14 victories

Védrines, flying a Bleriot 160hp Blindé known popularly known as *la vache* (French : the cow) that had been designed as a Zeppelin destroyer claimed the downing of a Taube over Suippe on the 2nd of September 1914, but, the victory was not confirmed as infantry fire was also directed at the enemy aircraft. Brindejonc des Moulinais noted in his diary on 2nd September. *At 0750 a Taube dropped a bomb 60m from our camp, Védrines took off at once but the Taube was shot down by infantry and landed 10km from our camp, radiator holed. Pilot and observer were able to get back to German lines after 2 days of pursuit in woods.* Védrines' forward gunner, René Vicaire said ~ *There is no doubt that my shots downed the Taube but we got no credit because the CO disliked Védrines.*

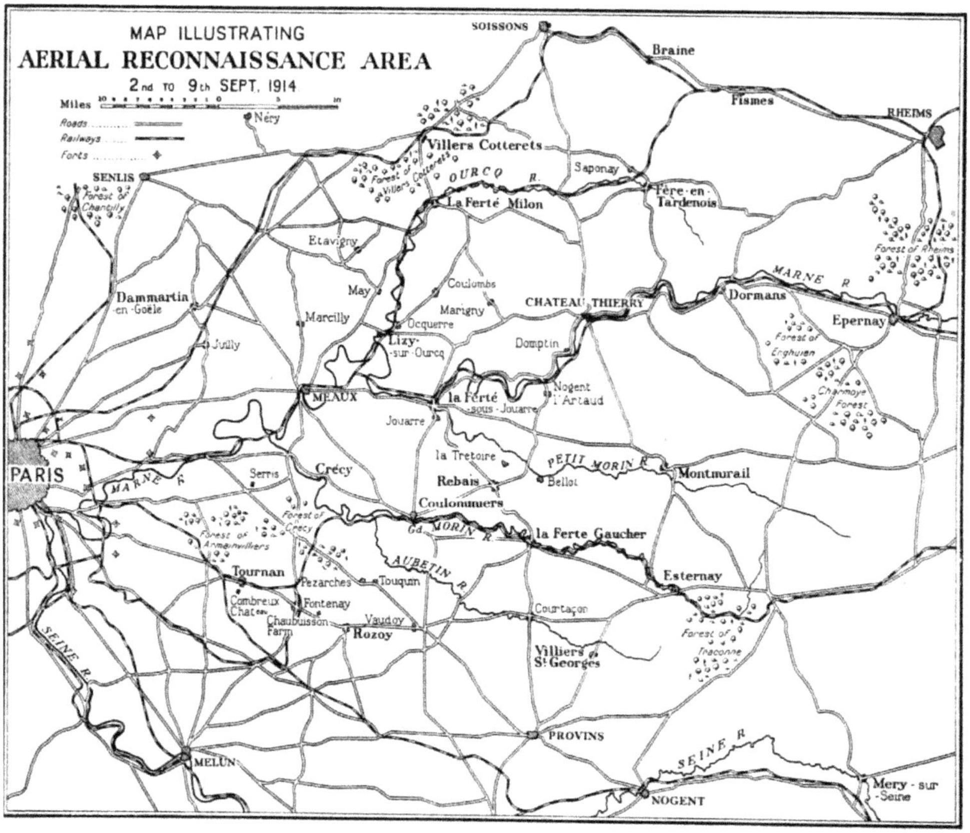

RECONNAISANCE MISSIONS ~ THE MARNE & TANNENBERG

The Battle of Tannenberg ~ In the early months of the war German AEG biplanes routinely flew reconnaissance flights over French positions, and often brought back useful reports to their HQ, but on the Eastern Front they delivered a triumph. On a reconnaissance over East Prussia on 22nd August, German pilot *Leutnant* Ernst Canter and his observer *Leutnant* Merten, in a Rumpler Taube discovered in the distance, what proved to be the unexpected advance of the Russian 1st Corps kicking up long columns of dust. In frantic haste and landing where they could, they continued by bicycle, cart and commandeered motorcar to deliver their report on the position and strength of the Russian forces, personally to General von Prittwitz in his forward field HQ who initiated the Battle of Tannenberg. In all subsequent German movements and actions, von Prittwitz was further informed by the inexplicably naive Russian staff practice of sending radio messages in un-encoded form rather than enciphered, which enabled the Germans to successfully respond to every Russian move. Their Russian opponents, who although formidable individually were organized inadequately and poorly equipped. Even after August 26th when the Russians used cipher, the Austrian cryptographers soon cracked it. On the night of the 31st the victorious Paul von Hindenberg signaled Kaiser Wilhelm II claiming 92,000 prisoners, *and their 300 guns still in the forest, will be collected!* The Kaiser promoted von Hindenburg to Field Marshal and decorated him with the *Pour le Mérite*. Hindenburg, who attributed his supposed brilliance to a refusal to read romances and poetry, wrote in his memoirs *Ohne Flieger, kein Tannenberg* or without airmen there would be no Tannenberg.

The Battle of the Marne ~ At 2000 hours on September 3rd, General Joseph Gallieni, defending Paris received a pilots report confirmed by a RFC report from Joubert and Allen that the right wing of von Kluck's German 2nd Corps in their great south westward wheel towards Paris, had changed direction gliding from west to east, exposing themselves to a flank attack. Gallieni pounced on the opportunity. *We will fall on the back of their neck!* General Joseph Joffre declared; *Gentlemen we will fight on the Marne.* When an observer flight over the German western flank precisely reported a serious gap, the Allies attacked with everything they could scrape together, and were just able to stem the German advance.

The French pilot, *Capitaine* René Roeckel and his observer Chatelain bombed the railway tunnel at Soissons to block German ammunition trains from getting to the front to prevent an enemy artillery barrage. They then spent ten hours artillery spotting near Vaubecourt that silenced half of the artillery of the German 16th Corps. In the intervening time, German aerial observers found the British units pressing into the breach. Alerted, the Germans were able to close ranks, but once halted they were driven back to the Aisne. Paris had been saved, but the tactical pattern was changing. As machine guns forced troops underground, barbed wire entanglements thickened, and room for maneuver on the ground was lost; the grim and bloody affair of trench warfare replaced these early fluid battles.

The first air-to-air kill ~ On 5th October 1914, *Sergent* Joseph Frantz flying a Voisin III light bomber of *L'Aviation Militaire, Escadrille* VB24 with his observer *Sapeur* Louis Quenault in the forward cockpit armed with a Hotchkiss 8mm machine gun were somewhere over the Vesle near Jonchery. At 1015 Frantz spotted a German Aviatik B-type two-seater scout returning to the German lines. Joseph Frantz then dived to cut off the retreat of the Aviatik scout, manœuvring to place himself behind the German. He swung the Voison into a headlong attack, to a point where they could clearly see the movements of the German pilot and his passenger. In a flash, the German observer in the front seat shouldered a repeating carbine and opened fire on the Voisin, but the German Aviatik's tail fin disturbed his aim, masking the French bomber. Quenault opened fire, peppering the Germans with a burst from his Hotchkiss: *tat tat tatter tat* ... then his gun jammed. Quenault with commendable composure started to strip the guns receiver to clear the jam. The Aviatik lurched before their eyes. Began to dive, turned on its back, and nose-dived into the ground in a cloud of black smoke. The Germans had crashed in a copse of firs beside a small pond in marshy ground, where the motor had almost completely buried itself. Frantz and Quenault descended nearby and walked over to see their victims. The pilot *Vizefeldwebel* Wilhelm Schlichting lay ten feet away from the twisted fuselage, among the scattered fragments of the Aviatik scouts shattered wings. The observer *Leutnant* Fritz von Zangen was in his final convulsions as Frantz and Quenault arrived, his hands clawing at the earth... An automobile arrived, containing General Franchet d'Esperey, a robust man, who had come to congratulate them.
I will give you the Medaille Militaire, d'Esperey declared to Frantz ~ *I already have it, mon Général!*
Then I will make you Chevalier a la Legion d'Honneur, and award the Medaille Militaire to your observer!

BRITISH EMPIRE AVIATORS OF THE GREAT WAR

Albert Ball ~ 44 victories

Lieutenant Louis Arbon Strange (1914)

William Avery *Billy* Bishop ~ 72 victories

Edward Mannock ~ 61 victories

BRITISH EMPIRE AVIATORS OF THE GREAT WAR

Rodrick Dallas ~ 32 victories

Arthur Rhys-Davids ~ 27 victories (including Werner Voss)

James McCudden ~ 57 victories

Lanoe Hawker ~ 7 victories

AERIAL OBSERVERVATION

WIRELESS OPERATORS BADGE   OBSERVERS BADGE

All reconnaissance observers must carry out a simultaneous search of the earth for movement and the sky for foes, and in addition keep their guns ready for instant use. As regards observation, keep your arms free of the machine to avoid unnecessary vibration when you train your field glasses on train stations and the rolling stock, don't forget the precise direction of trains and motor transport; don't forget the railways and roads on every side; don't forget the canals ~ and for the Lord's and everybody else's sake don't be surprised by Hun aircraft. Recco

The primary task of the fighters, was to keep the skies clear, so that their bombers and observation aircraft could operate without being shot down, because the vital sorties flown by aerial observers so often required by ground commanders formed such an important part of the overall strategic picture. Allied reconnaissance aircraft were also needed to swoop just above the German trenches before any infantry attack to see whether their artillery had actually severed the German barbed wire defences.

Major Lanoe Hawker, CO of No. 24 Squadron RFC detected camouflaged enemy weapons in daring fashion. *I fly low and draw their fire, then mark it down on the map!* After one such sortie, Hawker's aircraft had 50 bullet holes in it, and one in his leg. He thought up new concepts of aerial gunnery, designing the ring sight that was adopted by all, plus an upward firing gun sight: while useless in a dogfight, it worked brilliantly against enemy two-seater scouts, which could be approached from far behind, and unaware of their assailant, be shot down from a range of 1,000 feet. Hawker trained his pilots to do tight turns without losing height, flying upward spirals as an evasive manœuvre, and pinned his directive on No. 24 Squadron's notice board ~ *Attack Everything!* Scouts flying over suspicious forests would circle a few times, then swoop down and hurl out a bomb. Detonations often proved effective, as enemy horses stampeded to reveal a hidden formation. Some of Hawker's pilots mounted twin Lewis guns, giving them double the firepower or a spare if one gun ever jammed in the biting cold of high altitude. Lanoe Hawker also created the 94-round increased capacity *double drum* (one welded on top of another) for the Lewis gun, which was issued to both the RFC and the RNAS.

An observer must be something of a sleuth, particularly when using smoke as a clue. In the early morning a thin layer of smoke above a wood may be a bivouac. But a few miles behind the lines, it can evidence heavy artillery. A narrow stream of smoke near a railway will make an observer scan the line closely for a stationary train, as the Hun engineer drivers usually try to avoid detection by shutting off steam. The Hun has many other dodges to avoid publicity. When Allied aircraft appear, motor and horse transport will remain immobile at the roadside or under trees. Artillery are parked under cover: the Germans very rarely move artillery in the daytime, preferring the night or early morning, when there are no troublesome eyes in the air. Recco

On the Western Front as it were the Allies who were more often on the offensive, most aerial engagements occurred over German held territory. The first peril for Allied airmen was the German anti-aircraft artillery. Pilots dreaded high explosive shells and shrapnel, since an explosion within 100 yards could destroy an aircraft's engine, petrol tank, or wings. Allied casualties compounded as their damaged aircraft were more likely to come down on the Kaiser's side of the fighting line. The Germans had less need for artillery spotting aircraft, as they had wisely dug in on higher ground during their 1914 retreat that ever after provided them with an expansive view over the plains along most of the front. In many places forward observation officers were sufficient to spot the fall of shot. The Germans still required spotter aircraft and observation balloons, but to a lesser extent.

After landing back at the aerodrome, the pilots proceeded to tea and a bath, while we, the unfortunate observers, copied our notes into a detailed report, elaborated the sketches of the new Hun aerodromes, and drove in our unkempt state to headquarters, there to discuss the reconnaissance with spotlessly neat staff officers. At the end of the report one must give the height at which the job was done, and say whether the conditions were favourable or otherwise for observation. Recco

ROLAND GARROS AND THE CONCEPTION OF THE FIGHTER

Before the war, French pilot Roland Garros achieved fame as the first airman to fly across the Mediterranean, from the French coastal town of St. Raphael, south over the islands of Corsica and Sardinia to land on the North African coast at Bizerte near Tunis. He also held the world altitude record of 18,000 feet. In a quirk of fate, Roland Garros happened to be in Berlin to take part in an air show when war broke out in August 1914. However, during the night of 3rd August, Garros escaped, flew to France, and the next morning enlisted for service. *Capitaine* Garros was posted to Escadrille 23 of *L'Aviation Militaire* assigned to the aerial defence of Paris. Later, Roland Garros felt an inventory revelation, as he glimpsed at an old photograph of the poet Paul Verlaine through the blades of a fan. *There are some glances that pass through the blades of the fan, and some glances that do not.* Visualizing machine gun bullets deflecting off the spinning blades of his propeller arc, Garros with the help of aeroplane designer Raymond Saulnier shaped a streamlined steel deflector and fastened it to the propeller blades of a Morane-Saulnier monoplane. When they test-fired it, seven or eight orange winks of deflected bullets appeared as they skimmed away. It worked! Next time Garros attacked, his Hotchkiss machine gun would fire most of its rounds through the blades of his propeller.

Roland Garros armed with the world's first true fighter, firing forward in the pilot's line-of-sight, took off from St. Pol, near Dunkerque on 1st April 1915 to bomb the railroad station at Ostend with two modified 155mm artillery shells. At a point several kilometres away, he spotted a German two-seater scouting above the French trenches drawing the fire of the French anti-aircraft batteries. Garros closed in on his prey. From a distance of 100 feet, he fired a 25-round Hotchkiss clip of 8mm bullets into his victim. Garros quickly reloaded, fired off another clip from behind his armoured propeller. The German dived away with Garros clinging to his targets tail. As they descended Garros closed and triggered a short burst from his third clip. His prey fell, dead at the controls.

The chase became more and more chaotic ~ we were now no higher than 300m. At that moment an immense flame burst out of the German motor and spread instantly. The machine went into an immense spiral and dashed into the ground horribly in a great cloud of smoke. I gazed below me for a long time to convince myself that it was not a nightmare. I went by car to see the wreck. Those first on the scene had pilfered souvenirs: side arms, insignia, and the like. I took energetic steps to retrieve them. The two corpses were in a horrible state, naked and bloody. The observer had been shot through the head. The pilot was too horribly mutilated to be examined. The remains of the aeroplane were pierced everywhere with bullet holes.   *Capitaine* Roland Garros

With this invention, a pilot was now able to aim his entire plane as a weapon and shoot through the propeller to destroy an opponent. Roland Garros had shaped the future of the tactics of the air. On 15th April, Roland Garros shot down his second German in a head on attack, and on the 18th he got his third victory. Late on 19th April, Garros accompanied by *Capitaine* de Malherbe took off on a mission, to bomb the railway station at Courtrai. Both dived down from 2000m to 60m to release their bombs, when a rifle bullet fired by a German sentinel hit and fractured the petrol pipe on his monoplane, forcing him down over enemy territory. German troops closed in on the scene of the crash. Although Garros successfully evaded capture for a several hours, he had not sufficiently set fire to his aircraft to destroy its secret propeller. General Ferdinand Foch described Roland Garros, *As modest as brilliant, who has never ceased to give the example of the most admirable spirit*. During the last week, of April of 1914, the newspaper of the German Fourth Army published the following account of the capture of Garros.

On 19th April 1915, at about seven o'clock in the region of Sainte-Catherine and Landelede, two aeroplanes suddenly appeared flying at a great height. One of them disappeared in the direction of Menin, pursued by our anti-aircraft fire. The other, which was piloted by Roland Garros, headed toward Landelede. Precisely at this moment a train was passing on the Inglemunster-Courtrai line, arriving from the north. Immediately on perceiving the train, Garros made a descent at an angle of some sixty degrees, from 2000m to about 40, executing a series of tight turns over the train. Garros dropped one bomb which fell upon the rails, digging a crater one metre deep and two across. Some sentinels opened fire on him at a range of 100m. The aviator dropped a second bomb and climbed to 700m. Suddenly his motor stopped. The aeroplane wavered and descended in a glide in the direction of Hulste. Garros set fire to his machine on touching the ground and took refuge in a peasant home. The soldiers who were pursuing him finally discovered him crouching in a ditch behind a thick hedge. The soldiers asked him if he did not have a companion. Garros gave them his word of honour that he had been alone in the machine, the engine of but 80-hp and able consequently, to thus carry but a single passenger. Garros explained that at 700m his motor had been hit during the shooting and that this had forced him to land. Subsequently the soldiers of the Landsturm who affected his capture have been awarded a bonus of 100 marks.

Kriegszeitung

IMPERIAL GERMAN AVIATORS OF THE GREAT WAR

Max Immelmann ~ 15 victories

Werner Voss ~ 48 victories   Baron Walter von Bülow ~ 28 victories

IMPERIAL GERMAN AVIATORS OF THE GREAT WAR

Kapitainleutant Heinrich Mathy ~ Commander Zeppelin L 31

Zeppelin Flotten Kapitan Peter Strasser

Oberleutnant Rudolf Berthold ~ 44 victories

Baron Manfred von Richthofen ~ 80 victories

THE FOKKER MONOPLANE

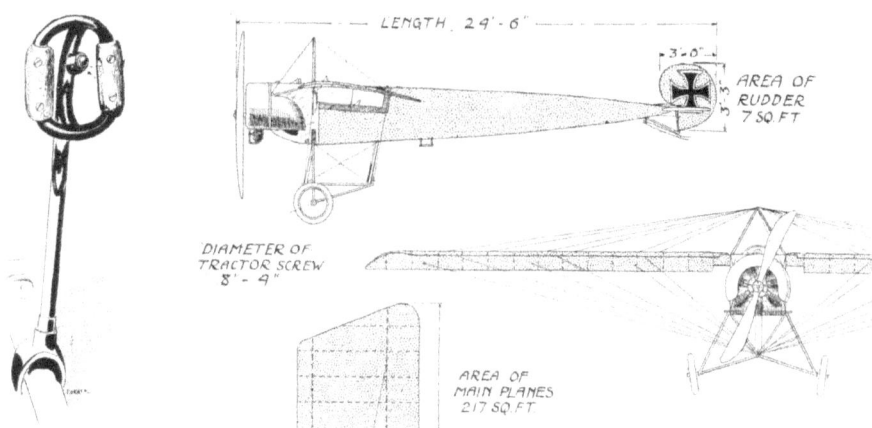

From Flight 1914
PLAN OF THE CAPTURED FOKKER MONOPLANE ~ FRONT, AND SIDE ELEVATIONS TO SCALE WITH CONTROL LEVER

The monoplane wreckage and its secret propeller were salvaged and transported to Isegheim for inspection. A few weeks later a team of engineers working for aircraft designer Antony Fokker, invented an alternative system that used interrupter gear studs on the propeller shaft to activate a cam and push rod system to prevent the machine-gun hammer from falling. No shots could then be fired while the propeller blades passed by the machine gun muzzle. With its Spandau machine gun trigger fitted to the joystick, German pilots were able to fly through the air at 130 kph and shoot with the same hand, rather than having to use separate hands for shooting and flying, as in all other aircraft. The wings of the Fokker monoplane were well strengthened, to give it superior steadiness in a dive. Although nimble, its primary advantage was that it was extremely difficult to see; having such thin wings that head-on, the Fokker Eindecker seemed to appear only when it was too late to take evasive action. The main evasive manœuvre was the hard turn, and Fokker pilots usually attempted to follow. The aerial opponents would then try to out-turn each other, to acquire a more effective firing position. In single combat, a prolonged *waltz* or circling contest might ensue or the engagement be broken off.

From May 1915, Eindecker monoplanes fitted with interrupter gear began shooting down great numbers of Allied aircraft with little loss to themselves. The first Fokker Eindecker was issued to the finest German pilot, a 24-year old Saxon *Hauptman* Oswald Boelcke who ardently wrote to his parents; *I believe in the saying that, the strong man is mightiest alone ~ I have attained my ideal with this single-seater, now I can be pilot, observer and fighter all in one.* The second Eindecker went to *Leutnant* Max Immelmann from Dresden, and together they collaborated to exploit speed, height, and concealment to develop early fighter tactics. It was Oswald Boelcke who first devised the basic two aircraft formation, so that while one pilot engaged an opponent ~ the other, his *wingman* protected him from attack. It was also Oswald Boelcke who pioneered the hawk-like technique, of gaining altitude and then with the sun behind him, diving down at his prey thru a daring angle while sighting his machine gun on his victim below. While still hurtling down, Oswald Boelcke would gently moved his feet on the rudder bar to cause his Fokker to slightly wag its tail and fire a short burst at close range; raking his targets fuselage. Next he either ; continued diving until he was well out of range or used the Eindecker's climbing power to regain altitude and wait for another opportunity to attack. Oswald Boelcke's tactic of aerial attack was not only masterly in itself, but demanded astute flying in the defence and the requisite manœuvre created by his wingman Max Immelmann: *The Immelmann Turn* enabled fast diving Fokker pilots to strike again and again with little loss of time. A diving Fokker pilot was rarely able to follow a targets evasive hard turn. But, by pulling up into a steep climb (as if to loop) and aileron turn sideways over the vertical, coming out in the opposite direction, height was gained as direction was changed and an attacking pilot was poised again for another swooping attack dive. On 1st August, *Leutnant* Max Immelmann was awarded the Iron Cross First Class, and shot down another;

First Boelcke had him in his sights, then I, and finally both of us did, and we came within 80m of our foe. Boelcke's machine gun jammed. I fired 300 rounds. Then I suddenly saw ~ I could hardly believe my eyes ~ the enemy flyer threw up both of his arms. His crash helmet fell out and went down in wide circles, and a moment later his machine plunged downward from a height of 2000m. A pillar of dust showed where he hit the ground.
Max Immelmann

Hauptmann Boelcke und Oberleutnant Immelmann im Kreise ihrer Kameraden.

Oswald Boelcke and his ground crew in front of his Fokker E.IV ~ note the air-intake slots in the engine cowling of this model.

MAX IMMELMANN & OSWALD BOELCKE

One must not wait till they come across, but seek them out and hunt them down. I always go off to the lines in the evening with Oberleutnant Immelmann to hunt the French. They treat my single-seater with a holy respect... they bolt so quickly that it is really glorious. If a Frenchman comes along, I pounce on him like a hawk and give him a good hammering with our machine gun ~ I have whacked about a dozen Frenchmen in this way. It is great fun for me.
Oswald Boelcke

The German High Command, anxious that the secrets of their Eindecker not be revealed, ordered the pilots of these monoplanes, from September 1915 to fly only within their own lines so none fell into Allied hands. Oswald Boelcke considered aerobatics useful only when up against a skilful foe that refused to hold still for a good shot. *When a fellow is in such a funk and is going into turns, he can never hit anything. I on the other hand always wait for the favourable moment and put in a few well-directed shots.*

Because of Boelcke's tactic of flying aircraft in pairs with a wingman to look out for danger while the other attacked the tail of his target, Eindecker pilots were now able to concentrate on their attacks and still keep a good lookout. Oswald Boelcke also taught his pilots to attack two-seaters from below and behind. If an enemy two-seater spotted the attack and rolled to the side to bring his observer's gun to bear, Boelcke turned in the opposite direction, then turned once more and came in again from below, advising all his loyal pilots; *Everything depends on sticking together when the Staffel goes into battle. It does not matter who actually scores the victory as long as the Staffel wins.* The Germans in their new Fokker monoplanes shot their victims down with impunity and took control of the skies between the late summer of 1915 and the summer of 1916. It was during this time Boelcke wrote his fundamental precept of tactical aviation : *A single aircraft is vulnerable to attack whether on the offensive or defensive and the minimum grouping of fighter aircraft should be two. In a pair, the leader selects the target and closes in for the kill as his wingman covers him.* Boelcke's further principles, the *Dikta Boelcke* were :

Fight as a single unit and obey the formation leader's commands. The best position in aerial combat is that where one can shoot at the enemy from close range without his being able to reply. Therefore, climb before the attack and dive from the rear. Altitude imparts speed in a dive and widens the patrol area. Use natural cover, clouds, and the glare of the sun. Attack when the enemy is unsuspecting or preoccupied with other tasks. Hold fire till the enemy is within range and squarely in your sights. Turn tighter than your opponent, coming into position on his tail. Never turn your back from the enemy, face him with your guns. To parry an attack from ahead, turn directly towards the opponent and present a small and fast target possible. To parry an attack from behind, enter and maintain as tight a turn as possible to make it difficult for the enemy to stay on your tail. When over enemy lines, never forget your line of retreat. Foolish acts of bravery are fatal.

While Allied pilots flew with shakily mounted Lewis guns fed from 47-shot drums and had to change heavy ammunition drums during combat in the wind stream, with scarcely enough space to carry four spare drums ~ German fighters had a fixed Model 08/15 Spandau machine gun fed from a continuous 500-round belt. The flaw of the Eindecker was that its 80-hp Oberursel rotary engine (copied from the French *Le Rhone*) had atmospheric inlet valves causing its performance to diminish at high altitude and in hot weather, which made the Eindecker in turn vulnerable to more powerful Allied fighters in such circumstances. Another German anxiety with the Oberursel rotary was its protracted production time; as the same time-consuming craftsman methods used before the war, were still the norm, so that in November and December of 1915, only 32 Eindeckers were delivered. The winning feature of the Eindecker was neither its speed nor its power, so much as its maneuverability and climbing ability that enabled its pilots to quickly gain position for repeated, rapid diving attacks. The Fokker Eindeckers shot down over 1,000 Allied aircraft, claiming so many British B.E.2c aircraft, that their pilots became known as *Fokker fodder*.

During the time of the Fokker menace, some Allied pilots advocated the evasive tactic of turning in under the Fokker as it dived to attack to get out of the way. Other two-seater crews preferred to hold a steady course so that their observer had a better prospect of firing from a steady gun platform. Others improvised various devices to gain some advantage. Captain Lanoe Hawker fixed a Lewis gun to the side of his aeroplane, with just enough of a downward angle to miss his scout's propeller. He received the Victoria Cross, perhaps as much in recognition of his innovations, as for his well-known bravery. Fortunately for Allied airmen, new British aircraft were on the drawing board that could fire forward without gun synchronization and in less than a year these new craft would enter production. The French meanwhile developed their Nieuport XI *Le Bebe* with a machine gun mounted high on the top wing above the arc of the propeller, which would enter service with *Escadrille N.3* in January 1916.

MAX IMMELMANN & OSWALD BOELCKE

Max Immelmann who only ever went on leave once, not to miss any action in the air, took part in a flying display over Leipzig on 28th November 1915 to raise funds for Christmas presents for the German Air Service. When he returned, two weeks before Christmas, Immelmann sent his mother 100 marks and his sister 50 to buy herself presents, as he preferred not to spend the money in France. After Christmas, Immelmann claimed his seventh victim, forcing Lieutenant Darley RFC down behind the German lines. Max Immelmann landed beside his prisoner and seeing his thumb almost severed, took out his knife and completed the amputation, bandaging Darley's hand and seeing him off to hospital, and prison. With such anecdotes, Immelmann was regularly invited to private lunches and intimate dinners with members of the German royal houses. After dinner with the Crown Prince of Bavaria and the King of Saxony, Immelmann wrote: *It was quite a small affair, we were only seven men.* After Max Immelmann and Oswald Boelcke both scored their eighth victories on 12th January 1916, Kaiser Wilhelm II decorated them with Prussia's highest military honour, the coveted *Pour le Mérite*.

The greatest surprise came in the evening. We were just at dinner when I was called to the phone. At the other end was the Commander-in-Chief's Adjutant, who congratulated me for receiving the Pour le Mérite. I thought he was joking. But he told me that Immelmann and I had both received this honor at the telegraphic order of the Kaiser. My surprise and joy were great. I went in and said nothing, but sent Captain K. to the phone, and he received the news and broke it to all. First, everyone was surprised, then highly pleased. That evening I received several messages of congratulation, and the next day, January 13th had nothing to do all day but receive other such messages. Everybody seemed elated. One chap would not let me go, and I didn't escape till I promised to visit him. From all comers I received messages: by telephone and telegraph. The King of Bavaria, who happened to be in Lille with the Bavarian Crown Prince invited me to dinner for the 14th of January.

Now comes the best of all. On the 14th that is, yesterday, it was ideal weather for flying. So I went up at nine o'clock to look around. As it was getting cloudy near Lille, I changed my course to take me south of Arras. I was up hardly an hour, when I saw the smoke of bursting bombs near P. I flew in that direction, but the Englishman who was dropping the bombs saw me and started for home. I soon overtook him. When he saw I intended to attack him, he suddenly turned and attacked me. Now started the hardest fight I have as yet been in. The Englishman continually tried to attack me from behind, and I tried to do the same to him. We circled round and round each other. I took my experience of December 28th to heart (the time I used up all my ammunition) so I only fired when I could get my sights on him. In this way, we circled around, I often not firing a shot for several minutes. This merry-go-round was immaterial to me, since we were over our lines. But I watched him, for I felt that sooner or later he would make a dash for home. I noticed that while circling around he continually tried to edge over toward his own lines, which were not far away. I waited my chance, and was able to get at him in real style, shooting his engine to pieces. He glided toward his own lines, leaving a tail of smoke behind him. I had to stop him in his attempt to reach safety ~ so, in spite of his wrecked motor, I had to attack him again. About 200 metres inside our positions I overtook him, and fired both my guns at him at close range (I no longer needed to save my cartridges). The moment I caught up to him, we passed over our trenches then I turned back as I had little fuel left. I landed near the village of F. Here I was received by Division Staff and told what had become of the Englishman. To my joy, I learned that after I had left him, he had come to earth near the English positions. The trenches are only a hundred meters apart at this place. One of the passengers, the pilot, it seems, jumped out and ran to the English trenches. He seems to have escaped, in spite of the fact that our infantry fired at him. Our field artillery quickly opened fire on his machine, and among the first shots one hit it and set it afire. The other aviator, probably the pilot, who was either dead or severely wounded, was burned up with the machine. Nothing but the skeleton of the airplane remains. I rode in the Division automobile, because I had to be with the King of Bavaria at 5:30 in Lille. The King and Crown Prince both conversed with me for quite a while, and they were especially pleased at my most recent success. Once home, I began to see the black side of being a hero. Everyone congratulates you. All ask you questions. My ninth success, followed close on the Pour le Mérite.

Writing to congratulate Immelmann on his 12th victory, *The Eagle of Lille* scored his 13th and a thrilled Kaiser Wilhelm II exclaimed: *One cannot write as fast as Immelmann shoots!* Max Immelmann after scoring 15 kills, suffered a synchronizer malfunction on 18th June, and crashed on the German side of *No Man's Land*. His wreckage clearly showed that one propeller blade had been shot off precisely at the line of his machine gun fire. By August 1916 Germany could only put into the air 250 operational machines, about half of the Allied number they expected to meet in the contested skies. over the Western Front ~ this is reflected in this Imperial German Air Service Order of October 1916; *The present system of aerial warfare has shown the inferiority of the isolated fighting aeroplane; dispersal of forces and a continuance of fights carried out when in a minority must be avoided by flying in large formations up to a Jagdstaffel. Fighting Staffels must be trained most carefully to operate in close formation as a single tactical unit, which is the manner in which they must carry out attacks.*

Hauptmann Oswald Boelcke ~ NPG Postcard

This order was issued following Boelcke's idea of concentrating fighters to overwhelm the Allies. Oswald Boelcke had convinced Major General Herman von der Lieth-Thomsen, of the validity of his ideas and the two men decided on the formation of specialized *Jagdstaffeln* or hunting echelons, known to their pilots as *Jastas*. *Jasta 1* was formed on 23rd August under *Hauptman* Martin Zander. When Oswald Boelcke's score reached 16, he was asked to form *Jasta 2* at Lagnicourt aerodrome in the province of Artois in northern France. On a grand tour of all theatres of war, Boelcke handpicked the men he wanted to fly with him. Boelcke's *Jasta 2* were equipped with the fast new Albatros D.II fighter, light with a ceiling of 17,060 feet, powered by a 160-hp Mercedes it was fast, manœuvrable, and very well armed. Being the first aircraft to be fitted with twin synchronized Spandau 08/15 machine guns firing through the propeller, it had at least double the firepower of most Allied fighters, with 600 rounds of continuous belt-fed ammunition thereafter carried for each machine gun.

Flying at first in pairs, Boelcke next instructed his pilots to operate in flights of four, both in the V and the diamond formations with the flight leader positioned in front where his view was clear and his signals could be seen. Then they practiced full *Jasta* take offs and patrols. The operational debut for *Jasta 2* was the bright clear morning of Sunday, 17th September 1916. Once airborne, Boelcke quickly spotted a two-flight formation of British aeroplanes crossing the lines: F.E.2b bombers with a B.E.2c fighter escort. Boelcke immediately began to climb to put himself and *Jasta 2* above them in the glare of the sun. They were heading into German territory, so he would pick his own time to jump them. He speculated that when they reached their target they would all be looking down - the time to attack. Oswald Boelcke waggled his wings and gave the hand signal and every pilot picked his own target. *Jasta 2* knifed down and cut into the enemy formation. Very few British airmen survived that attack.

Between friends and foe, Oswald Boelcke enjoyed a superb reputation for bravery and chivalry. In January 1916, after he rescued a French schoolboy from drowning in a canal, the French awarded him their Life Saving Medal. Oswald Boelcke devoted many hours to visiting wounded men he had brought down. Then on 28th October, just as Boelcke was attacking ~ he collided with Erwin Bohme. The left wing of Bohme's Albatros sliced through Boelcke's right wing, sending him out of control in a steepening glide with the stricken wing whistling to his death. Field Marshall von Mackensen wrote to Oswald Boelcke's parents, *I have never gazed into a finer pair of gleaming blue eyes. I encountered the eyes of a man who was absolutely fearless, a rare hero.* The mother of Manfred von Richthofen wrote in her diary; *Crashed and dead. Inconceivable. He lived in the entire Volk. Everyone knew him ~ even if they never saw him. He was the first great Kampfflieger in the grand style… with him, one of the immortals of war has passed away.*

WERNER VOSS

Whatever Oswald Boelcke told us was taken as gospel. When one has shot down one's first, second or third opponent, then one begins to find out how the trick is done. I started shooting when I was much too far away. That was merely a trick of mine. I did not mean so much as to hit him as to frighten him. He began flying curves and this enabled me to draw near, and I succeeded in catching him. The aggressive spirit, the offensive, is the chief thing everywhere in war and the air is no exception. Fight on and fly on to the last drop of blood and the last drop of fuel, to the last beat of the heart. Baron Manfred von Richthofen

Werner Voss began 1914, as a 17 year-old son of a Lutheran tailor who was expected to follow in the family tradition. However, on 16th November he falsified his age to enlist in the 11th Westphalian Hussar Regiment. Within six months Werner Voss had been promoted to *Unteroffizier* and awarded the Iron Cross. Weary of the filth at the front, he smartly transferred to the German Air Service and volunteered for flight training. Werner Voss was such a gifted student that he was requested to remain as an instructor until he was posted to serve as an observer, before training as a war pilot.

By the time he received his pilot's badge in May 1916, Werner Voss was the sole surviving airman of those with whom he had begun operations with. The tragedy of this gave Voss a heartfelt sympathy for the allied aircrews of two-seater scouts ; *Poor devils. I know how they felt. I have flown such a type. They must be destroyed because they spy out our secrets, but I would prefer to shoot down their fighters.*

Voss knew that British and French scout aircraft flew regular patrols to continually renew the mosaic of photographs that completed the allied photographic maps of *Hun-land*, as the British called the German-held territory beyond *No-Man's-Land*. Voss was a crack shot and tried to sight his guns first on his targets engine to give them some hope of survival. Nevertheless, this gentleman of the air who evoked the admiration of all, could not excuse an enemy, who was prepared to fight back.

Commissioned *Leutnant* Werner Voss on 9th September 1916, he was posted to Oswald Boelcke's *Jasta 2* to fly as wingman to Baron Manfred von Richthofen who respected him as a close friend. Scoring his first two victories at the age of 18 on 27th November, Werner Voss became the *teenage ace*, the mercurial one with the mysterious gift of instinctive awareness, knowing where all aircraft were. Werner Voss shot down 28 allied aeroplanes with *Jasta 2* and was awarded the *Pour le Mérite* to become the most feared German ace at this time. Whenever one of his victims survived on the ground, Voss would find them to hand out quality cigars and offer them an autographed postcard, which were welcomed, and prized. Before flying, he always dressed in his best uniform, just in case he happened to land and meet some ladies on the ground. As pilots had to have their kills confirmed to count as victories toward their tally, Voss once needing proof to confirm his claim for an unseen kill, landed his Albatros behind enemy lines, climbed out running to souvenir his fallen victim's machine gun. Dashing back into his bird, Voss took off just in time to wave good-bye, to an advancing British patrol.

Werner Voss was given command of *Jasta* 10 and achieved great fame throughout Germany. However, when they came to visit his airfield they often found Voss in the hangar in worn oily overalls tuning his Albatros or his 1914 *Wanderer* motorcycle, which he rode around the airfield. Voss in the German fashion began to decorate his aircraft; painting his light green with a white ring, then painting the under surfaces of his wings and wheel discs pale-blue, and finally highlighting the bird's nose in the style of Japanese mask with two threatening eyes, lurid eyebrows, and a bold moustache.

Back in the air Werner Voss, the teenage ace was on the tail of his next victim. Spotting a Nieuport Scout at 16,000 feet, Voss darted at it, firing a short accurate burst that splinters the instrument board of its horrified pilot, Keith Caldwell. More hits. Caldwell's engine splutters and he seems doomed. Without the use of his engine, Caldwell dives to 10,000 feet losing height fast. He has not escaped.

Werner Voss on his tail, fires more bursts into the fabric of the New Zealander aviator's Scout. Caldwell by theatrical deception, tries to elude the formidable Voss. Throwing up his arms as if hit by a bullet and convincingly letting go of the controls he droops his head over the side of the Scouts cockpit feigning death. Voss follows him down, till a few thousand feet off the ground, and sees the *corpse of Caldwell come to life!* Spinning his engine into life, he dives for the lines, landing his stricken Scout in a shell hole behind the allied trench line. Exasperated, the duped Voss turns back to claim the airman who deceived him, strafing Caldwell on the ground, struggling to free himself from his pranged Scout. Voss misses ~ and *Killer* Keith Caldwell will survive the war.

TRENCHARD'S RESPONSE

Until the Royal Flying Corps are in possession of a machine as good or better than a German Fokker, it seems that a change in the tactics employed becomes necessary. It must be laid down as a hard and fast rule that a machine proceeding on reconnaissance must be escorted by at least three other fighting machines. These machines must fly on close formation and a reconnaissance should not be continued if any of the machines become detached. This should apply to both short and distant reconnaissance. From recent experience, it seems that the German are now employing their aeroplanes in groups of three or four, and our aeroplanes frequently encounter these numbers. Flying in close formation, must be practiced by all pilots.

Carry the war into enemy territory and keep it there, to raid prominent enemy aerodromes and attack any hostile machine that offers combat, leaning forward over German territory constantly.   Gen. Hugh Trenchard
Commander RFC France

These two orders set the agenda for the RFC over the Western Front in 1916. To counter the Fokker superiority, Trenchard introduced formation flying and formed special fighter squadrons to patrol deep into German airspace and intercept any German aircraft before they could reach the battlefield. Two-seater scouts were to make reconnaissance flights deep into German-held territory with a strong fighter escort. Then the fighter squadrons of the RFC would fly other missions over the battlefield to support ground forces with light bombs and machine gun fire. Trenchard's plan of flying offensive patrols, with fighters assigned to special squadrons behind enemy lines, eventually led to a period where reconnaissance aircraft could work safer. As Trenchard had intended, Allied morale stiffened by using aggressive tactics, and better aircraft arriving in late May heralded an undisputed mastery of the skies. Trenchard was able to report that in one Autumn week ;

Only 14 hostile machines crossed the line of the Fourth Army area, whereas something like two or three thousand of our machines crossed their lines, destroying among other targets, 521 heavy guns.

This air supremacy can be attributed to a combination of many factors. Most important was the early development in the RFC of formation fighting patrols, as a departure from the principle of the single aircraft operating independently. The British tactical unit became the flight of five or six aeroplanes, handled as one body under one leader.

The accuracy of British aerial gunnery improved following the advent of tracer ammunition. Although some early work had been done before the Great War, the first military issue tracer rounds designated as SPK Mark IV T or *Sparklet* for the RFC, first came into service in July of 1916 and were more effective with every seventh round fired being visible tracer. Now, a fighter pilot maneuvering to get on his opponent's six o'clock or tail could see his stream of fire and adjust his aim accordingly. Another significant change was the decision that aircraft on contact patrols should operate at low rather than at medium altitude, as the struggling infantry dreaded letting off flares to indicate their forward positions. On the ground, forward infantry officers were extremely concerned that if the fired a flare skyward ~ it could almost at once bring down upon them a firestorm of German artillery

ENEMY OBSERVATION KITE BALLOONS

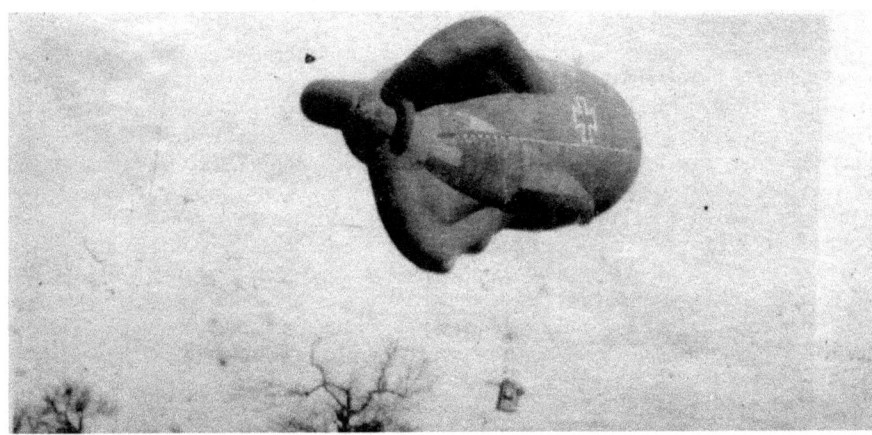

It was found to be the best tactic: to dive steeply to a point about half a mile from the balloon and on a level with it, to flatten out and go straight at the balloon with all the added velocity gained in the dive. At 200 yard's range one took a sighting shot and at 50 yards opened fire with the Lewis gun. One carried straight on to within about 20 yards of the balloon, firing all the time, hopped over it and zoomed away.   Sholto Douglas RFC

RFC pilots received the following instructions to tackle enemy balloons:
The sausage-shaped Drachen or German observation balloons are mostly tethered five miles inside German lines and hovering at heights up to 5,000 feet. Their role is to direct counter-battery artillery fire, the fall of shot corrected by observers in these captive observation balloons. It is the fighter pilot's task to shoot down enemy aircraft and these balloons spotting for the artillery. Before delivering an attack, the air should be scanned for hostile aircraft, since kite balloons are defended by enemy aircraft specially detailed for that purpose, who will keep some distance away until they see their opponent intent upon his mission. The Drachens are sometimes fitted with a kite, flown from a strong wire cable fixed to the top of the envelope that is difficult to see, and this is a danger to an attacking pilot. It is better to attack such balloons from below.

Attacking aircraft need to fire a mixture of tracer bullets to ignite these highly flammable gasbags; this will burst its envelope into flame and disintegrate in 20 seconds or less. As soon as the Drachen has been set on fire, the attacking pilot should either execute a steep turn or fly back along the ground, or climb to a higher altitude before returning and use the burning balloon as cover from hostile anti-aircraft defences. They are tenaciously defended by anti-aircraft guns and one Drachen will now count as a kill towards a pilot's tally of aerial victories. The Germans are now hauling their balloons down than by winch. Whenever an Allied aircraft is spotted, they pass the balloon cable under a pulley and hitch it to a truck, which is ready to drive away at speed, pulling the balloon very rapidly down to the ground. An attacking aircraft is therefore forced to descend very low and come under heavy machine-gun fire. The present tactic to counter to this consists of a hedge-hopping fighter attack at ground level. If there is rain about or the air is moist, it is almost impossible to ignite an observation balloon.

OVER THE VERDUN FRONT 1916

In the interim, the Germans in the spring of 1916 began to focus their forces for a killer blow against Verdun, 140 miles east of Paris, with the intention ~ *to bleed France white*. The French countered by increasing the number of their escadrilles in the Verdun sector from four to 16, including six fighter escadrilles, their aim being to seize air supremacy from the Germans, and hold Verdun. At this time, the French High Command released military communiqués detailing the exploits of their aerial aces, or those who had achieved five confirmed victories in the air. French newspapers were keen to publish these thrilling reports as their readers admired their flying heroes. When a pilot became an ace he was offered a posting to one of France's four top *Escadrilles de Chasse* defending Verdun. Known as *Les Cigognes* or The Storks, they flew into action with a different depiction of a stork, the heraldic emblem of the occupied province of Alsace and the French sign of good luck painted on their fuselages. Their commander, Major Tricornot de Rose, directed the pilots of France's elite Stork escadrilles to, *Seek the enemy in order to engage and destroy him.* Three legendary Stork aces were;

Jean Navarre a fearless aviator since 1914, who once took off after a Zeppelin armed only with a kitchen knife. By the time the Battle of Verdun had begun on 21st February 1916, Navarre had only achieved two victories, but this reckless French sub-lieutenant of 20 years soon became the French hero of the air war over Verdun. The infantry called him, *la Sentinelle de Verdun,* as he spent up to 10 hours a day in the air descending only to refuel his plane, reload his guns and revive his energies with light food and wine. Navarre painted his agile Nieuport XI *Le Bebe* red and embellished it with the skull and crossbones to challenge and intimidate the Germans over the skies of Verdun. Flying in a combination of headlong attacks and masterful aerobatics he shot down 10 German aircraft in less than three months. By the end of May his tally had risen to 12 official kills and he was ranked above all other allied fighter aces. His preferred tactic was to fly up to within a few feet, slightly below and astern of a German aircraft, stand up in his cockpit, sight his machine gun, and dispatch his opponent with a fatal burst. Navarre dressed to kill in eccentric style, in a shaggy bearskin coat, leather breeches, fur lined boots, gloves of leather lined with lambs wool for warmth that did not diminish his touch, and the *au courant* silk stocking of a lady friend wrapped around his head for his flying helmet, a most pleasant, practical and warm headdress to comfort him against the chills of high altitudes.

My method consists in attacking almost point blank. It is more risky, but everything lies in manœuvring so as to remain in the dead angle of fire. Except as a last means I never take recourse to aerobatics. I owe myself to my country ~ and death is the risk of the profession.   *Capitaine* Georges Guynemer

Capitaine Georges Guynemer flew in an all white SPAD VII aeroplane, famous for its speed and strength. Before his first year's service had expired he was decorated and promoted for gallantry in rushing to the aid of a comrade attacked by five enemy machines. He entered the combat at the height of 10,000 feet, and inside two minutes had dropped two opponents. The others fled, and he hotly pursued. But just then, Guynemer was brought down by a near miss from a 75mm shell that went through the water reservoir in front of his cockpit, sprayed wreckage in all directions, and stripped the fabric from his port upper main wing. His SPAD dropped in a tailspin and hit the ground, shearing off the landing gear, then the wings folded up and tore off. The SPAD cast off various pieces and slewed through 45° and planted itself in the ground. Guynemer was concussed and suffered a gash on the knee from his magneto, but had survived. Recovering in hospital he wrote to his Father; *The SPAD is solid! She loops wonderfully! I've read while in bed that the crowd gave me an ovation in Paris. It's the result of ubiquity. Modern science really does wonderful things! As long as one has not given everything, one has given nothing!* He preferred flying alone, far behind the German lines. Revered by his French countrymen as the *Winged Sword of France* he seemed doomed and felt fatalistic himself. Asked by an admiring lady, *what new decoration is there for you to earn?* He answered ~ *the wooden cross!* By August 1917, Guynemer with 54 confirmed kills became the leading Allied ace. On the morning of 11th September, under an overcast sky and in drizzling rain, *Capitaine* Georges Guynemer took off, and during an aerial engagement, mysteriously vanished into a large white cloud, never to be seen again.

Capitaine René Fonck was known for his tactical positioning, precise flying and for advocating the three-V fighter formation of a leader covered by two wingmen. He taught his new pilots that he found 15 rounds enough for a kill and that that if a bullet ever struck him he would transfer to the trenches since wounds prove a lack of skill. Whenever he came upon an enemy formation, Fonck would first attempt to shoot down the leader and in the subsequent upset aim for a second. He flew the limited production SPAD XII *moteur-canon* fighter with a hand-loaded 37mm Puteaux cannon that fired through the propeller shaft with deadly results from precise hands. He called it his *avion magique*.

OVER THE VERDUN FRONT 1916

Finding myself only 600 feet away from one of those numerous two-seater planes whose mission it was to drop into the trenches propaganda leaflets designed to discourage our Poilus, I had to aim 60 feet ahead of him in order to shoot him down. That cost the Boches dearly, for coming at them, in a few seconds I sent them to join their papers, while my brave comrades below were advancing magnificently. My bursts are from three to eight. Besides the advantage of economizing on ammunition, this method has also the advantage of facilitating my aim and reducing the chance of jamming or damaging my gun. To obtain good results, you must also know how to control your nerves, how to have absolute self-mastery, and how to think coolly in difficult situations. I always believed that it is indispensable to maintain absolute confidence in ultimate success, along with the most complete disdain for danger. These are the necessary qualities for a fighter pilot. Capitaine René Fonck

When aerial operations over Verdun began, the French had few aeroplanes on the spot. A rapid concentration was made and a vigorous offensive policy was adopted. The eventual result after the arrival of the elite aces of the Storks, was that superiority in the air was obtained and scout aeroplanes detailed for artillery co-operation and photography were enabled to carry out their work on most days in safety. Then the French introduced their responsive Nieuport II fighter, nicknamed *Le Bébé* because they were so small to the Verdun Front in 1916, French pilots began to vanquish the Germans in the air. After this the Germans acquired a very defensive outlook and failed to push home their decisive advantage. From the accounts of prisoners the French learnt that German airmen had received orders, restricting their aircraft to flying only over German held territory although exceptions could be made to fly over the front, if the day was very cloudy and a surprise attack could be made.

This led directly to a severe intelligence lapse by the Germans, who having no observers in the air failed to spot their chance to strike a colossal blow. During a two week period, when the French were hard at it reinforcing Verdun with 190,000 men and 22,500 tons of munitions, they had 8,000 trucks moving day and night along a 36-mile road, which the German planes could have strafed into oblivion. As no German aircraft appeared, they remained unaware of the transfer of supplies. This put new heart into the desperate French soldiers who called their road, *la Voie Sacree* or the Sacred Way.

In August of 1914, Charles Nungesser had initially enlisted with the French 2nd Hussar Regiment. Out on patrol on 20th September to gather information on the advancing Germans, a small scouting group of himself, his officer, and two infantrymen found themselves cut off during a German thrust. Hearing an enemy car approaching… his wounded officer and the others hid beside a roadside ditch. Nungesser with lightning initiative and leadership leapt out to slam shut a level-crossing gate across the road blocking a German staff car. Nungesser shot all four occupants and drove the Mors sedan with his countrymen at high speed back to the 2nd Hussars HQ, whose commander was so impressed with his bravery that he was promoted, given the bullet holed car with the moniker, *l'Hussar de Mor* and recommended for a medal. During the ceremony, Nungesser asked the General who decorated him with the *Medaille Militaire*, for a transfer to *L'Aviation Militaire*, to train as a pilot.

Charles Nungesser shot down his first plane, after flying without permission. For this enterprise he was awarded the *Croix de Guerre* and given eight days house arrest. In November of 1915, he was given a Nieuport *Bébé* and Charles Nungesser enjoyed himself decorating his fighter with a black heart framing a coffin, two candlesticks, and a skull and crossbones calling himself the *Knight of Death*. After flying over a town at ten metres on full throttle he earned another eight days arrest, suspended. In the first week of April 1916, Charles Nungesser was posted to Escadrille N 65 flying over Verdun. Already other pilots had received the splendid Nieuport Scout XVII *Superbébé* armed with a synchronized Vickers machine gun firing through the propeller. Fitted with a strong 110-hp *Le Rhone* engine, one French pilot said, *It climbs like a witch!* ~ to 10,000 feet in nine minutes. The version flown by the ace Charles Nungesser was fitted with the even more powerful 130-hp Pierre Clerget aero-engine. In this aeroplane, he shot down six aircraft and a balloon in quick succession, and achieved rapid promotion to *Capitaine* Nungesser after claiming ten confirmed victories over the Verdun Front.

On 12th May, a lone Albatros flew over Nungesser's aerodrome, dropping a streamered note challenging him to single combat. *To my worthy opponent, Monsieur Skull and Bones, Meet me at 4 o'clock.* Other members of his Escadrille warned him of a trap as ate his lunch, and then he took off at 0230. Over the German airfield of Douai… six Albatros fighters waiting for him, now whined toward him. Unfazed by the odds of the ruse, the *Knight of Death* flamed their leader in a head on attack and shot another down. The remainder of the cowardly pack scattered. This was not the only known occasion when the enemy offered single combat, conniving a trap for others to spring.

ESCADRILLE LAFAYETTE

One of Charles Nungesser's drinking partners was Jean Navarre and they enjoyed the plethora of attractions in Paris as often as they were on leave, whether it was with Colette's courtesans or touring the many boulevards, café and bars with his American friends from the *Escadrille Lafayette*. After a spot of leave, Charles Nungesser often fronted for morning patrols still in his tuxedo with a beauty on his arm. One time, while driving from the front to Paris through sluggish traffic, he spotted his black-hearted Nieuport heading the same way ~ it was Jean Navarre! ~ who borrowed Nungesser's morbid mount after his own had been shot up and declared, *I had forgotten what a woman looked like!* Charles Nungesser certainly had not. He was enjoying the charms of Margaretha Zelle or Mata Hari, the ethereal beauty and rumoured German spy. Aware of her rumoured espionage, he amused himself with the tactic of plying the exotically attired and highly desirable dancer with alluring tales of a new French super aeroplane powered by eight, top-secret engines to keep her enthralled. Berlin replied to her report in a coded message that she should, *lay off the liquor!* Mata Hari with the confirmed covert German codename of H-21 was arrested in Paris, charged with spying and unable to explain herself, was dramatically executed unbound and unblindfolded by a firing squad of ten French Zouaves.

Bomber leader ~ Capitaine Felix Happe, commanding officer of *Escadrille MF 29* commenced using the *three-V* formation in August 1916, after observing a flock of wild geese migrating. He then ordered his Maurice Farman bomber pilots to fly in a close *V* while above he flew a roving commission to thwart attackers. From a higher altitude, Happe also plotted the position of each bombs impact. Generally the weight of the defensive fire from his French gunners made it too risky for the lightly built Fokkers. However, after a three-machine raid on Rothweil from which only Happe returned, he introduced larger formations, always taking his entire escadrille of French bombers out flying multiples of the *three-V* which gave each bomber such effective covering fire, flying in the V formation that it became the standard defensive bomber formation in all air forces for decades. After Happe's unit bombed the poison gas plant at Dornach, the Aviatik aircraft factory at Freiburg and the Mauser factory at Oberndorf, the Germans placed a price of 25,000 marks on his head. Happe dropped them a note, *Splendid! You will know my machine by its red wheels. Don't bother to shoot anyone else.* Felix Happe soon after ceased daylight raids and trained *Escadrille* MF 29 for night operations. The French airmen flying Happe's Maurice Farman biplane bombers found them risky aircraft to fly as they were slow when fully laden and their pusher layout presented a blind spot that enemy pilots made the most of, but with few German night-fighters this was of little concern. The preferred French bomber was the sturdy steel-framed Voison III that could carry 300kg of bombs. When it was realized how robust the Voison III was, some were armed with the quick-firing Hotchkiss cannon, a fearsome piece of aerial ordinance, although the machine gun remained the standard armament on most fighting aeroplanes.

Prior to the entry of America into the Great War, when the US Congress granted President Wilson's request for a declaration of war against Imperial Germany on 4th April 1917, many Americans had already sailed to Britain to join the Royal Flying Corps. Others enlisted in France's *L'Aviation Militaire*. Of the many intrepid Americans who had volunteered to fight on the Allied side, French-born Raoul Lufbery was the first. At the beginning of 1914, Lufbery who had already been serving in the French Foreign Legion transferred to *L'Aviation Militaire's Escadrille MS 23* as a mechanic and volunteered for pilot training. He began learning how to fly in May and qualified in July. When the Great War began he flew regular sorties at the controls of a Voison III light-bomber.

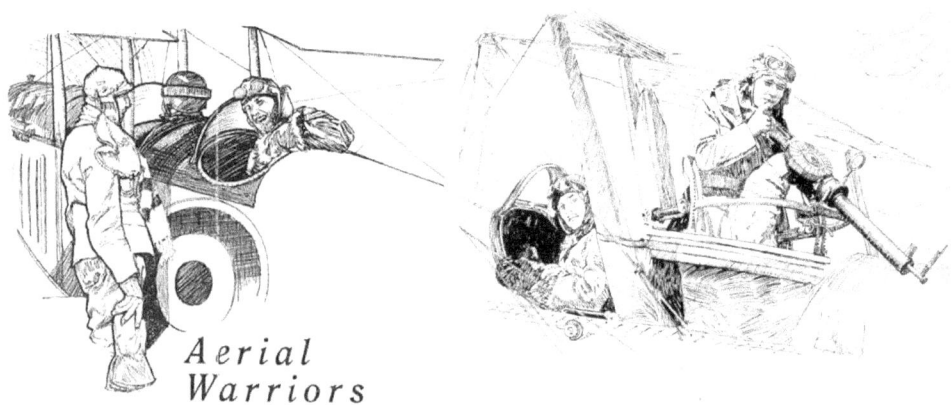

Aerial Warriors

THE LAFAYETTE FLYING CORPS

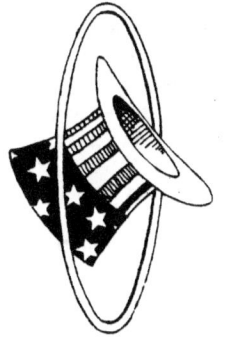

HAT-IN-THE-RING 94th AERO SQUADRON

ESCADRILLE LAFAYETTE N. 124

Norman Prince, a trainee aviator from Pride's Crossing, Massachusetts travelled to France soon after the war began in 1914, with the purpose of forming an American volunteer squadron. There, he met another American, Dr Edmund L. Gros, director of the American Field Ambulance Service. Together they set up a committee, comprising Dr Gros and five other Americans, one being the millionaire William K. Vanderbilt, who provided the finance. Dr Gros as a doctor and commander of an ambulance unit remained a non-combatant, while Norman Prince sought recruits from among the ranks of those Americans who had already joined the French forces. Then Prince intended to join the unit and fly. At first, the French opposed Prince's plan, but after their troops became stalemated in the squalid muddy trenches during a dreary winter of inactivity, the notion of volunteer airmen from overseas was warmly welcomed. William Thaw, another Foreign Legion volunteer who came from a wealthy Pittsburgh family was commissioned their First Lieutenant and others were enlisted as corporals, becoming sergeants when qualified as pilots. They received a new uniform every three months, 125 francs per month and for each confirmed victory there would be a bonus of 1,000 francs.

The American unit was formed and designated *Escadrille Lafayette N. 124* on 16th April 1916. They were placed under the command of *Capitaine* Georges Thénault and posted to the Luxeuil aerodrome in a quiet sector to learn to master their Nieuport II fighters. The seven foundation aviators were an adventurous lot. Three were financially comfortable, one was a medical student, and another was a graduate of Harvard. The symbol of *Escadrille Lafayette* painted boldly on the side of their aeroplanes, was the head of a Red Indian in a warrior chief's eagle feathered headdress, his jaw agape yelling a ferocious war cry. One month later Kiffin Rockwell claimed the escadrille's first victory when he shot down a German LVG. The next day, they were posted to Bar-le-Duc aerodrome on the Verdun Front.

The Lufbery Circle ~ During the aerial fighting over the battle of Verdun, Raoul Lufbery developed his tactic of the defensive circle, where any enemy aircraft attempting to get into a firing position behind an Allied fighter in defensive circle with the others following around could be engaged by the pilot of the next aircraft astern. Although it was still possible for an enemy pilot to get into the circle and shoot down an aircraft, it required perfect timing, great skill, and courage, consequently it was only rarely attempted. With the tail of every aircraft covered, height was gradually lost and sufficient fuel was needed to maintain formation against a prolonged attack. By circling tight, two-seaters under attack, benefited even more from the Lufbery Circle with most of their defensive guns able to target the enemy. If enemy fighters tried to attack from above, all rear gunners could engage them. If an enemy fighter attacked from below, the gunners on the far side of the defensive circle could fire at it. Another tactic that Lufbery taught his American brothers was the *flat-turn* a little known trick outside of the military flying fraternity of how to skid sideways by applying the rudder and opposite aileron.

Raoul Lufbery himself was a solitary man who led by example. Whenever the weather became too rough for flying he would take long walks in the woods with another airman to pick mushrooms for the cooks. Other times Lufbery would drive his Hispano-Suiza roadster, on loan to him from the car company (who made the SPAD's 150-hp aero engine) with his men to Paris for a good time, and it was from a circus there that he acquired the *Escadrille's* two lion cub mascots Whisky and Soda, who roamed freely around their aerodrome at Bar-le-Duc. On another road trip they abducted from a nearby regiment, a French chef who had formerly cooked at the Ritz in New York. In the *Escadrille Lafayette*, 38 Americans had served at various times during the 20 months before America entered the war, achieving 42 confirmed victories.

THE UNITED STATES AIR SERVICE

All of the Americans, who volunteered to fly with the French *Lafayette Flying Corps,* were awarded an individually numbered certificate of recognition, but fewer were privileged to be members of *Escadrille Lafayette*. After the U.S.A. entered the war, *Escadrille Lafayette* became *Escadrille Jean d'Arc* and French pilots replaced the Americans. Of the French fighter escadrilles chosen to escort the French bombers (such as Felix Happe's MF 29) prominent were the Lafayettes. On the ground, they also played baseball together with the bomber crews and nearby nautics of RNAS No. 4 Wing, and also indulged in some tremendous parties. Whiskey and Soda always gave the newcomers a bit of a start.

The American Lafayettes eventually all became members of the United States Air Service. The French, however deciding not to use their Nieuport 28 fighter because of its known dangerous design flaws, such as stress breaks in its copper fuel lines from excessive engine vibration that caused a few fires, and a tendency from flight stress to shed its top wing fabric leading to upper wing failure, loaned these horrors to American fighter squadrons until the SPAD XIII became available. And so the flawed Nieuport 28 was the first aircraft to equip the USAS. Captain Eddie Rickenbacker who became America's highest ranking fighter ace with 26 kills witnessed for himself this structural wing problem during a dogfight with three Albatros. After becoming separated from his wingman, Rickenbacker shot one to pieces and put his Nieuport into a high-speed dive to escape the two others. However, the fabric on the top right wing of his Nieuport tore off in the slipstream. Then his upper wing broke free. Spinning over the German lines, Rickenbacker idled his engine, lifted the Nieuport's nose up, and leveled into a shallow glide towards the French lines. Rickenbacker, who often said that everything he knew about flying, he learned from Raoul Lufberry, managed to return safely with just a lower wing. After this, the Americans counted the days till they were re-equipped with SPAD XIII fighters.

He drove himself to exhaustion. He'd fly the required patrol. Then he and I would come back to the field, have a cup of coffee, get into our second ships, and go hunting by ourselves. Rick always patrolled at just enough rpm's to prevent stalling. He saved the ship for the moment he needed it. When he fought, however, he called for maximum performance and drove the plane until it nearly fell apart. Most of the pilots he killed never knew what hit them. Out of the sun, a quick burst and gone. That was Rickenbacker.   Reed Chambers

Fighting in the air is not sport. It is scientific murder. The experienced fighter pilot does not take unnecessary risks. His business is to shoot down enemy planes, not to get shot down. His trained hand and eye and judgment are as much a part of his armament as his machine-gun, and a 50 : 50 chance is the worst he will take or should take except where the show is of the kind that justifies the sacrifice of plane or pilot.
Captain Eddie Rickenbacker

On 19th May 1918, Raoul Lufbery commanding the 94th *Hat-in-the-Ring* Squadron, took off in another pilot's Nieuport 28 to save a new pilot being targeted by the skilled crew of a German observation aircraft. Lufbery soared, closed on the pair, and fired a burst at the German aircraft. His gun jammed and he flew off to clear his machine gun. The Germans then turned on the vulnerable Lufbery, hitting him over the village of Maron seven miles southeast of the Toul aerodrome. Lufbery's Nieuport took a critical hit from a tracer bullet igniting his reserve fuel tank. The entire front of his aeroplane instantly erupted in flames. The bullet zipped on to carve off part of Lufbery's thumb. Diving down to 2,000 feet, Raoul Lufbery was seen reaching out of his Nieuport aiming his aeroplane for the Moselle River trying to save his life. Seconds later he was seen falling out of his Nieuport or jumping to his death rather than being roasted alive. Lufbery landed lengthwise on the picket fence fronting a mademoiselle's flower garden near the village of Maron. Within the hour Eddie Rickenbacker and the others had arrived to see that the French had taken Raoul Lufbery's corpse away to rest in the town square where it was covered with the flowers of the village. The following day, German pilots flew over to show their respect and to drop bouquets of flowers over the village gravesite of Raoul Lufbery. Captain Rickenbacker was appointed to command the 94th *Hat-in-the-Ring* Squadron, and his men became the highest scoring U.S. fighter squadron with 67 victories.

THE SPAD & THE NIEUPORT SCOUT

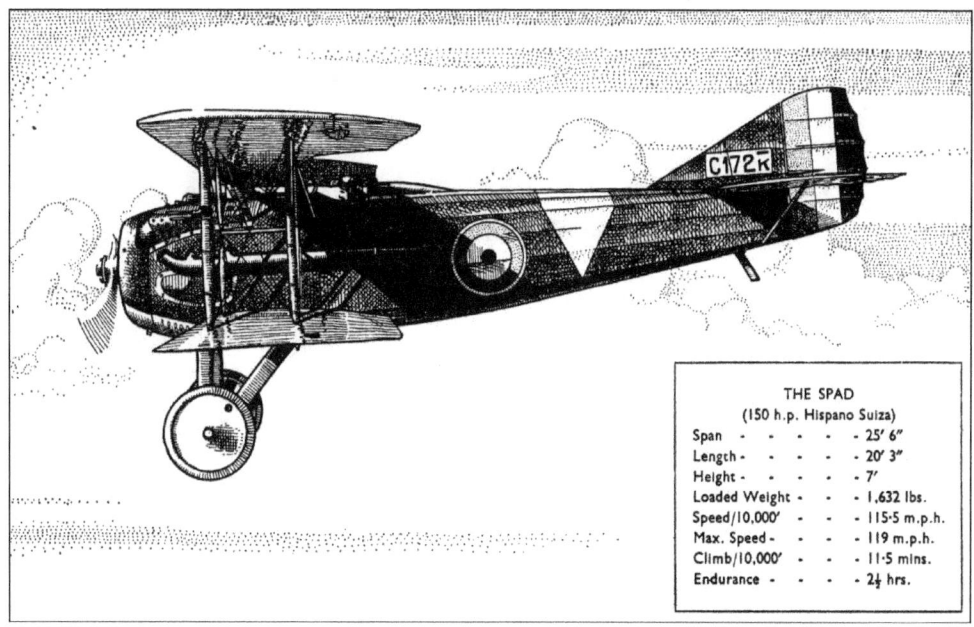

THE SPAD
(150 h.p. Hispano Suiza)
Span - - - - 25′ 6″
Length - - - - 20′ 3″
Height - - - - 7′
Loaded Weight - - 1,632 lbs.
Speed/10,000′ - - 115·5 m.p.h.
Max. Speed - - 119 m.p.h.
Climb/10,000′ - - 11·5 mins.
Endurance - - - 2½ hrs.

THE NIEUPORT SCOUT
(110 h.p. Le Rhône)
Span - - - - 26′
Length - - - - 19′
Height - - - - 7′
Loaded Weight - - 1,232 lbs.
Speed/10,000′ - - 101 m.p.h.
Max. Speed - - 107 m.p.h.
Service Ceiling - - 17,400′
Climb/10,000′ - - 9 mins.

THE LOOP & THE ROLL OR BARREL

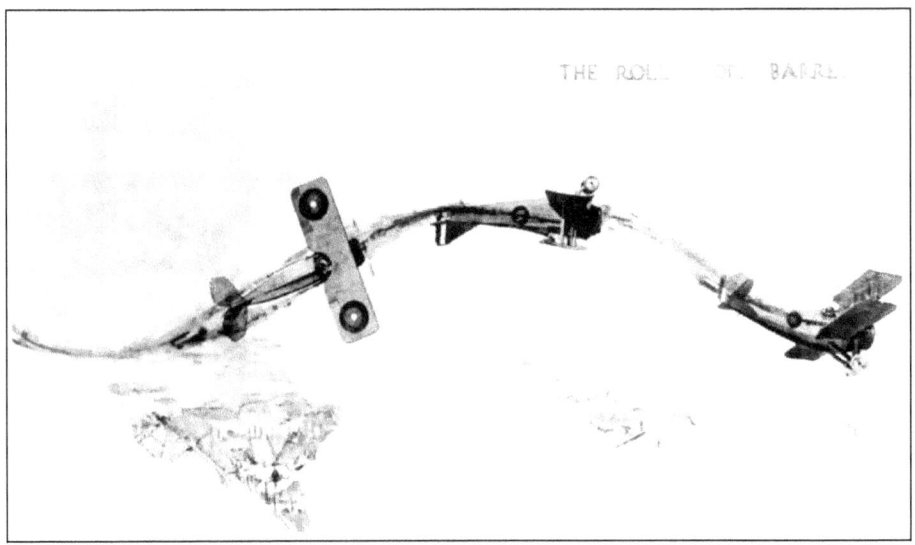

Lieutenant Albert Louis Deullin of *L'Aviation Militaire* described the tactical principles he had acquired over the Western front as a *chasse* aviator and flight leader in *Pursuit Work in a Single-Seater*. His expertise was absorbed by the United States Army Air Service as part of its *French System* of flight instruction that all American pilots received at the USAS training airfields near Issoudon in France

As a pilot, he should be, before everything else, skillful in maneuvering. He can never practice too much aerobatics; the short turn without change of height, climbing and descending spirals, nose spins, renversements, retournements, looping the loop, short climbs at a very steep angle, dives and so on are the beginning of his period of instruction. He will only be ready for the chasse when he can execute them in precision relative to an adversary who manœuvres likewise. Lieutenant Albert Deullin

The Renversement ~ The fastest *about face* turn if being pursued by an opponent. When flying along level, pull smartly on the stick lifting the nose, let the aeroplane slip over to one side, over onto its back with a quick, sharp kick on the right rudder, throttling the motor. Just as the aeroplane comes over on its back, restore the rudder to its neutral position and pull the stick back to your seat to bring the aeroplane out in the opposite direction in a glide. No altitude was lost during this manouvre.

THE CHANDELLE & THE SPIN OR VRILLE

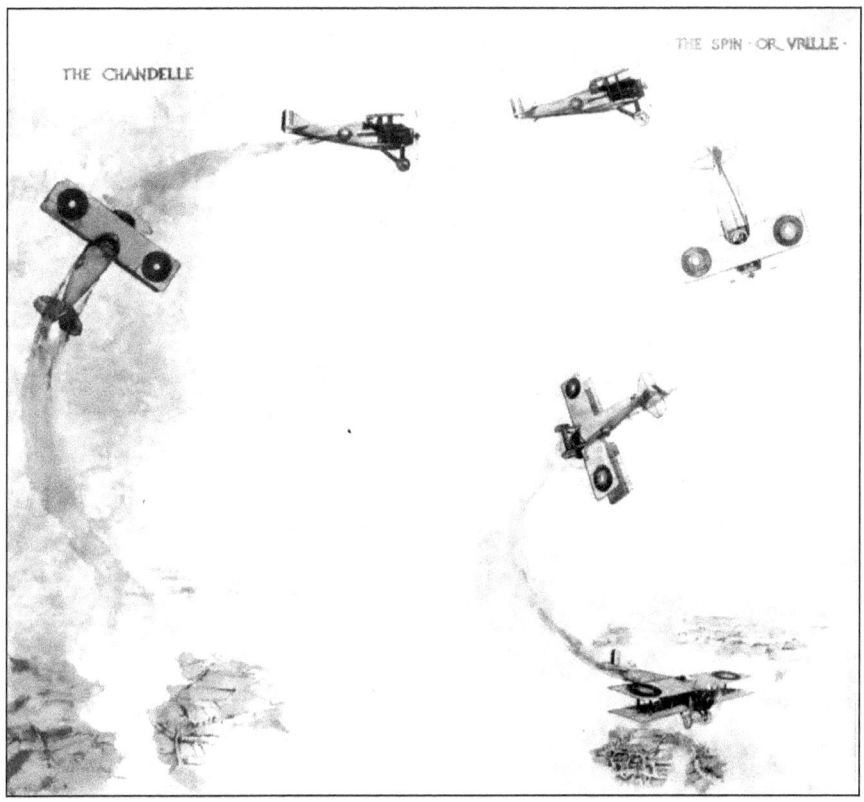

The Chandelle ~ is the manoeuvre, when an aeroplane has acquired excessive speed in a dive, by pulling the machine into an angle greater than the best climbing angle. By using full motor, a very steep ascent may be made until the machine begins to lose its speed, when a more normal angle is assumed. The height that can be acquired in a chandelle depends entirely upon the type of machine, and the margin of excess speed at the beginning of an ascent.

The Vrille ~ How to initiate, and recover from a spin. Fly level, throttle down the engine and hold the nose of the aeroplane level for a few seconds until its flying speed is almost lost. Pull the stick into one corner and kick the rudder over to the same side, causing the aircraft to rear up, go over on its back and fall immediately into a spinning nose dive, with its wings spinning around the fuselage axis. To recover from the spin, straighten the rudder to neutral, move the stick back to the centre and push it slightly forward. When the aeroplane stops spinning in a straight nose-dive, as flying speed increases the aeroplane can be slowly pulled up level, and the throttle opened. The perils for excited pilots were in over control by falling into a reverse spin, or else by pushing the joystick too far forward to turn a somersault, and coming out of a spin on their back, having insufficient height to react to any predicaments before falling too near the ground.

Wing Slips ~ In a wing slip, the aeroplane falls sideways and is controlled by a slight pressure on the stick and rudder. To initiate a wing slip, bank the aeroplane over slowly reducing throttle, and putting on reverse rudder to prevent the aeroplane from diving, at the same time push the stick slightly forward in to overcome any tendency to spiral. To recover from the wing slip, it is necessary to push the rudder down so as to cause the aeroplane to dive, and pull in the stick as though coming out of a spiral.

Lieutenant Frank Luke U.S.A.S. with his Blériot-built SPAD XIII at La Ferme de Rattentout, S.E of Verdun, 19th September 1918.

Charles Nungesser with his decorated mount...

CHAPTER II
THE ZEPPELIN CAMPAIGN

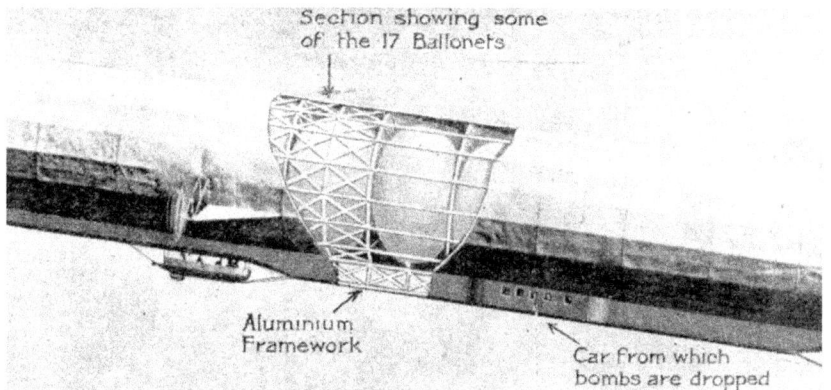

In early 1900, Count Ferdinand von Zeppelin was determined to push the potential of the airship. The first of his mighty airships was 420 feet long with two Daimler engines powering twin airscrews. In a few years he made enormous progress, due to his unique design of a long tubular skeletal framework of duralumin that enclosed the lifting gasbags of his airships. This design concept enabled his airships to be lengthened to extraordinary dimensions and achieve the lift required to fit larger motors and increasing fuel capacity. In 1907, von Zeppelin's third airship made a 211-mile flight from Friedrichshafen to Frankfurt in seven hours, and in 1909 he flew from Friedrichshafen to Berlin, descending in the presence of Kaiser Wilhelm II. With the evolving capacities of his Zeppelin airships and the growing confidence of German people, the Count established the world's first airline *Delag*, announced his plans for coming passenger flights, and commissioned six airships of the Parseval type.

On 4th March 1912, the *Viktoria Luise* made her first trip with 23 passengers on board flying from Friedrichshafen to Frankfurt. The *Viktoria Luise* completed her 400th trip on 26th November 1913, having carried a total of 8,551 passengers and travelled 29,430 miles, many of them flown over the sea between the main Zeppelin aerodromes of Berlin, Frankfurt, Berlin, Gotha, Heligoland, Copenhagen, and Gotha. On board, her passengers were served five-course dinners. On one flight, several British naval officers (including Sir John Jellicoe) on a pre-war visit to Berlin flew across Germany in the *Viktoria Luise*. On their return to London they reported on the military potential of German airships fearing that in times of war they could be used against Britain. The performance of this Zeppelin had been experienced at first hand by these British officers. They had seen the *Viktoria Luise* operate at high altitudes and easily navigate through clouds. Furthermore, Zeppelins could take advantage of clouds and remain concealed for considerable periods because of their stability and ability to hover stationary, and being able to hover virtually motionless over a target, meant greater potential bombing accuracy. Zeppelin airships also could climb away at great speed after releasing their ballast or bomb load. If the Germans used prevailing winds, they could launch their Zeppelins aloft in the afternoon, shut down the engines and let the prevailing wind carry them over to England on silent raids at night. However, the Zeppelins could be seen by moonlight at their supposed combat altitudes with the naked eye, while the loud drumming of their motors was magnified by the resonance of the giant hull covered with tightly stretched fabric, perhaps rendering the Zeppelins detectable by listening devices.

German airship seized in France ~ 1912 saw the construction of the first Zeppelin designed expressly for naval and military purposes: to carry a bomb load or stay aloft for weeks if necessary on extended maritime patrols. The trial flight of the fourth airship in this series, Zeppelin Z4 on 3rd April 1913, commanded by *Hauptman* Georg Fritz caused a sensation at a time when relations between Germany and France were near breaking point. After reaching the French frontier, it turned to traverse the entire line of French forts facing the German frontier from south to north. Trouble compounded when Z4 was blown off course, lost its way in the mist, exhausted its fuel supply, and descended among French cavalry on the parade ground of the French fortress of Lunéville. French troops surrounded and seized the stranded Germans. Soldiers were placed around the Zeppelin to keep back sightseers, and that evening while the Mayor of Lunéville entertained the stranded Germans at dinner, the Z4 was thoroughly inspected, measured, and photographed by French military technicians before being released. The French confirmed their suspicion that such airships could be transformed into bombers.

ILL WIND BLOWS FRANCE GOOD : SECRETS OF THE GERMAN Z4 REVEALED

1. Securing a permanent record for France of Germany's newest military dirigible : a French soldier photographs the Z4.
2. Officers of the German military air-ship, which fell into French hands at Lunéville : Capitan Glund and Lieut. Jacobi.
3. Bent on getting the fullest possible knowledge of Germany's best dirigible : French officers boarding the Z4 at Lunéville.
4. During the 24 hours' detention of the air-ship in France : soldiers guarding the German dirigible on Lunéville parade ground.
5. During the French investigation : General Hirschauer of France talking to the Officer-Commanding at Lunéville.
6. After the Z4's involuntary and descent with her secrets on to Lunéville parade-ground ~ the German dirigible in French hands.

It is very evident that when the newest German dirigible, Z4 made an involuntary descent at Lunéville, the French authorities did not hesitate to possess themselves of that air-ship's jealously guarded secrets. To quote the Times description of the Z4, which was minutely examined by General Hirschauer and other French military and aeronautical experts, together with the engineers of the Lebaudy and Clemen-Bayard air-ship firms : Z4 reproduces the main features of the Zeppelin type, but the internal arrangements of the car appear to be unusually complete.

A small aluminium ladder leading from the car reaches the interior of the vessel. The frame of the vessel is supported from the inside by a bewildering maze of slender aluminium girders and stays, and in the roof hangs several gas-bags. In the corridor, running the length of the hull are stored tools, cables, ropes, and spare parts, and opening off from it are the captain's cabin, with navigation instruments, a photographic dark-room, a wireless telegraphy cabin, and a lavatory. All the wire for the communications of orders from the car under the bows to the car astern run through this corridor…

THE SHEARING OF A GERMAN SAMSON! : THE Z4 POWERLESS IN FRANCE

KEENLY INTERESTED IN THEIR COUNTRY'S TREASURE-TROVE

FRENCH SOLDIERS EXAMINING THE GERMAN DIRIGIBLE Z4 DURING HER DETENTION AT LUNÉVILLE

The two cars below the hull are fitted with quick-firing guns, and on the very top of the air-ship there is a platform, surrounded by netting measuring twelve square feet, which is evidently intended for a quick-firer or some special gun. No gun ammunition, or any form of bomb-throwing apparatus was found on board. Some of the spectators had at first taken three of the silencers around the motors, which in shape resemble howitzers, for bomb-throwing machinery. All that can be said is that the air-ship was constructed for carrying guns, both in the car and on the platform above it.

The German authorities have thanked the French for their courtesies extended to the airmen of the Z4. The officers aboard the Z4 were Captain Glund of the Zeppelin Company : Captain Georg Fritz of the Berlin Experimental Aeronautical Section : Lieutenant Felix Jacobi of the Aeronautical Division of Metz ; and First Lieutenant Johann Brandeiss of the Berlin Aeronautical Section. The last three were on board as members of a commission to ascertain the capabilities of the air-ship before its purchase for the German Army.

THE RISE OF THE ZEPPELIN MENACE

In 1914 the Germans possessed an air fleet of eight Zeppelin airships. The four airships they built in 1914 were 518 feet long, filled with 794,000 cubic feet of hydrogen gas, and powered by four motors. The motors were fitted to the streamlined passenger and gunnery gondolas beneath the Zeppelin. These giant airships performance of 80 kph over a distance of 12,500km, at a height of 2,000m enabled them to stay in the air for flights of at least 30 hours, often days. The Germans had the option to deploy their Zeppelins for high-level reconnaissance or for long range bombing missions. But instead they squandered them on low-level assignments in broad daylight. The first to fall was Zeppelin *L VI* commanded by *Oberleutnant* Kleinschmidt, ordered on 6th August to bombard the Belgian fortress at Liege blocking von Moltke's dash across Belgium. After dropping its eight modified 420mm heavy artillery shells, when Zeppelin *L VI* might have been expected to fly well, low cloud and a lack of ceiling due to poor performance brought this Zeppelin down within the range of Belgian AAA guns and rifle fire, who at once shot so many holes in *L VI* that it was barely able to stagger back across the frontier. Unable to remain in the air *L VI* crash-landed in a forest near Bonn, wrecked and written off. Orders then went out for German aeroplanes to reconnoitre the fort of Liege. One observer ordered his pilot to land directly between two of the forts, then bravely scouted them at ground level and flew home to deliver his report. By mid-August Liege had been captured by the advancing Germans.

Zeppelins *L VII* and *L VIII* were then ordered to survey General Joffre's counter-attack in the Argonne. *L VII* was forced down by French gunners and on 16th August, *L VIII* was brought down by German ground troops at Badonvillers after misreading the Zeppelins flare-gun recognition signals. On the Eastern Front, Zeppelin *SL 2* flew 300 miles into Russian airspace on 22nd August 1914, from where it spotted and reported a Tsarist formation massing for an attack on the Austro-Hungarian fortress of Przemysl. Other long-range reconnaissance sorties were flown with success, but in their intended role of a bomber, the Zeppelins proved an early failure. During the Battle of Lemberg in September, the Germans repeated the same tactical error they had committed in the West, when they ordered Zeppelin *L V* over at low level during daylight to bomb the rail junction at Mlawa. It was immediately disabled, forced to descend and captured with its crew of 30. None of these Zeppelins, all severely punctured, and forced to land had ignited. The reason was, that their outer envelope enclosed smaller envelopes filled with hydrogen, and hydrogen in an enclosed container such as a balloon's envelope, was not flammable. Like petrol, it became explosive only when mixed with air in the correct ratio to form an explosive mixture. To set a Zeppelin ablaze, it was necessary to concentrate machine gun fire into a particular point of the Zeppelin, puncture the airship's outer envelope for the gas to escape and merge with the surrounding air, and then to ignite the gaseous mixture with tracer.

Less than four weeks fighting had halved the German strength of their new Zeppelin airship arm. These early losses persuaded the German High Command that their hydrogen filled airships were of little use on such tactical missions. The huge Zeppelin airships, which could be built for a fraction of the cost of a warship and in six weeks instead of two years for a cruiser, were redeployed on maritime missions to make use of their extraordinary range. The Zeppelin crews then waited for the outcome of a feasibility study into their planned use to attack London with incendiaries. Admiral von Tirpitz in the German HQ at Charleville on 18th November referred to, *the immensity of the effect if London could be set on fire in thirty places. All that flies and creeps should be concentrated on that city.* Tirpitz's proposed Fire Plan anticipated the burning of London, and demoralizing her civilian population enough, to pressure their government into seeking a peaceful resolution, and force the British Empire out of the war.

In the sky, no French or British aeroplane of the day could match the potential operational altitude of an oxygen-equipped Zeppelin of more than two miles. But if a fighter pilot ever spotted a Zeppelin at a reasonable height, he would need to think about his tactics. Unless there were clouds in the sky that could be used as cover to conceal his attack, it would be preferable to approach a Zeppelin from below to avoid being silhouetted against the sky. To get within firing range, an attacking pilot would need to manœuvre into a survivable attack position. Either from astern and beneath the airship or in order to brave the defending fire from the gondola's Parabellum MG14 machine guns and remain under enemy fire from the rooftop machine gun for as short a time as possible, it would be wisest to attack the Zeppelin from ahead. Owing to the incredible size of enemy airships, deflection shooting was irrelevant. A pilot only needed to maintain the accuracy of his aim when he fired off his entire Lewis gun magazine against a definite mark, to obtain a good firing group, and always attempt to return to the same point of attack ~ to ignite, and detonate the gas-filled Zeppelin.

THE FIRE PLAN AND WHAT BECAME OF IT

German Zeppelins flying too high to be attacked by British fighters began night bombing the coastal towns of southern England in January of 1915, while their navigators experimented with radio transmission cross bearings, and their bombardiers used optical bombsights to measure their air speed over the ground to make adjustments for wind-drift. Always the Zeppelins were at the mercy of the weather. On 13th May 1915, *Hauptman* Erich Linnarz, the commander of Zeppelin *LZ 38* came to Southend on the south coast of England, then flew over the southern shires dropping a ton of bombs and then wafting away. Following the visit of *L 38* to Southend, the tactics of her commander were reported by William Ledicott of the Special Police Reserve, in *The Southern Standard* of 14th May 1915.

Now the one I saw was a huge thing, about 300 feet long. At 0240 hours, four of us were standing at the corner with one of the ordinary police force ~ when we heard a whirring noise, the engines were running quite slowly. We all went into the middle of the road, and searched the sky, and saw coming from the direction of Hamlet Court Road straight across the Great Eastern yard, a large dim shape, which, as it came near, we made out to be a huge Zeppelin. Corporal Frost immediately went for the fire brigade, while the policemen on duty went to warn the authorities. I then stood in the middle of the roadway and took out my watch and noticed the Zeppelin's movements and timed them. She circled the Great Eastern Railway Station. Then she turned sharply round ~ suddenly the engines were shut off and everything was quiet ~ and she came to a standstill over the middle of the roadway at Cobweb Corner. She then stopped her engines, and the sound ceased.

The Zeppelin gradually dropped down and it was then that I had a good view of her. I should then estimate her height at from 600 feet. She was now distinct and remained motionless for four minutes ~ I timed her by my watch. I then walked from the centre of the road and the first thing I noticed was when a bomb fell within five yards of the special constables in front of the tramway centre, but fortunately it did not burst. It fell with terrible force and buried itself under the wooden paving. She then started her engines and proceeded straight up the London Road, dropping bombs all the way, from each side of her as fast as they could drop them. These bombs are dreadful things, and no sooner are they in a house than the building is a mass of flames. They are saturated with petrol and throw out huge volumes of smoke. They also smell very strongly of spirit. During this time the Zeppelin which had been going very slowly, started to circle ~ the intensity of her engines was so great that the noise she was making showed that she was putting on more speed. She gradually went out of sight, and I lost the sound of her engines at 0309. She appeared to go directly out to sea. No guns were fired at her while she was over the town. One thing that struck me was the cool way in which people behaved.

With a full moon to guide him on the night of 31st May, *Hauptman* Erich Linnarz commanding *LZ 38* set off for London, eager to be the first to wreak havoc on the capital of the British Empire. He crossed the English coastline over Margate at 2012 hours, and was observed heading for London. Shortly after, when all London was ringing with news and alarmist rumours, the Admiralty issued to the press the following very brief communiqué. *Zeppelins are reported to have been seen near Ramsgate, and in certain outlying parts of London. Many fires are reported, but these cannot absolutely be connected with the visit of airships. Further particulars will be issued as soon as they can be collated and collected.*

The Zeppelin aircrew spotted the illuminated metropolis and the vivid arcing of London's trams. The issue of the potential of the Fire Plan was practically decided on that first raid against London, when Erich Linnarz in the course of a ten-minute flight across north-east London from 2045 hours, deliberately dropped a series of 89 x 11.5kg incendiary bombs on civilian housing that only succeeded in starting three fires of serious importance. The casualties of this first air raid on London were seven killed and 35 wounded. When the Zeppelin flew off, one of her crew dropped a taunting note on a piece of card. *You English! ~ We have come and will come again. Kill or cure ~ German.* On its return journey after crossing the lines at Armentieres *LZ 38* was attacked by Lieutenant R.H. Mullock RFC, but his machine gun jammed at the critical moment. After a quick ascent *LZ 38* climbed out of reach.

Had the Zeppelin commanders possessed a more practical knowledge of London, regarding the whereabouts of flammable buildings, a blaze might have been started which would have been exceedingly difficult to control. The opportunity to achieve this was wasted during the autumn of 1915 by the German hesitancy to press the theory to the limit. Later, when the occasion did arise to strike a decisive blow at the city itself by igniting the vast wooden warehouses lining the river, which were most favourable to the rapid spread of big fires, the Zeppelin commanders inexplicably chose to mix their weapons with a high proportion of explosive bombs and their opportunity to implement Tirpitz's suggested Fire Plan was overlooked under exceptionally favourable conditions.

LATEST ZEPPELIN DEVICE FOR SECURE BOMB-DROPPING
ARMOURED CAR SUSPENDED BY THREE CABLES FROM A ZEPPELIN AIRSHIP

A means by which a Zeppelin airship takes cover behind low-flying clouds and at the same time approaches the object which the bomb-aimer seeks to target. A message was received from a Copenhagen correspondent that Zeppelins are employing suspended cars.

HOW THE RUSSIANS BROUGHT A ZEPPELIN TO EARTH

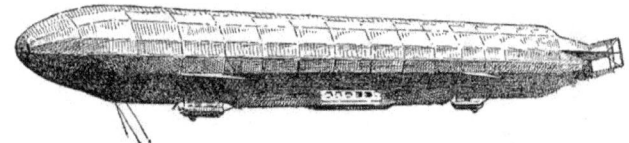

While a Russian cavalry brigade with horse artillery was moving from Mlawa in Poland, towards the Prussian border in late September of 1914, Zeppelin L. V appeared at great speed. The horse artillery immediately opened fire. The first shot fell short, and the second overshot the mark. The third damaged the balancing mechanism in the rudder. With her nose tilted in the air, the Zeppelin managed to move off, disappearing behind a wood, and dropping bombs. The battery of limbered up, galloped through the woods, unlimbered again, and then re-opened fire, with their first shot taking effect. The German airship sank slowly to earth and was captured.

R.A.J. WARNEFORD

Flight Lieutenant Reginald A.J. Warneford V.C.

The Gun platform of Zeppelin LZ 38 ~ Felix Schwormstaedt

Flying over Zeebrugge on the 17th May 1915, Flight Lieutenant Reginald A.J. Warneford of the RNAS sighted a U-boat near a small steamer, clearing the harbour and heading northwest. Warneford dived on the submarine and from 200 feet, hurled his grenades at it. The U-boat escaped undamaged, and so Warneford returned to Dunkerque. After his promising escapade, Warneford was issued an improved single-seat Morane-Saulnier Parasol monoplane, with Garros inspired metal deflector plates fitted to its propeller blades, plus a bomb rack fitted to his undercarriage to carry six 20-lb Hales incendiary bombs, which could be released by a lever mounted on the starboard side of the fuselage.

On the evening of 6th June 1915, No. 1 Wing RNAS based at Dunkerque received an airship alert from the Admiralty that three Zeppelins were on their way back to Belgium following an abortive attack on England. Squadron Commander Longmore ordered Warneford in his Morane Parasol to intercept the raiders over Ghent. Warneford strapped himself in and took off at 1305 hours. Nearing Dixmude, he spotted the faintly penciled outline of a Zeppelin to the north beyond Ostend and turned towards it. The Zeppelin in sight was the *LZ 37* commanded by *Oberleutnant* Otto von Hagen that had a top speed of 80 kph. Catching up with the Zeppelin beyond Bruges at 0105 hours, Warneford came under vigorous machine-gun fire from the gondolas and worse, from the forward rooftop gun crew firing their automatic Parabellum MG14 at 700 rounds a minute. As Warneford realized he needed to force his bomb-laden Morane still higher, he climbed away to gain altitude for another attack. However, the Zeppelin commander with supreme confidence turned *LZ 37* after him, ordered his Zeppelin to rise and pursue the English fighter ~ to keep the Morane under heavy automatic fire.

Warneford made off west, in a ruse to give the Zeppelin's commander the notion that he had given up his attack. As his Morane veered away, all machine gun fire ceased from the Zeppelin and Warneford manœuvred into a safer position far behind his quarry. At 0215 hours *LZ 37* approached Ghent and began its descent to dock at its mooring poles. This was Warneford's opportunity and he grasped it. Ten minutes later he was at his absolute ceiling of 10,500 feet and above the Zeppelin. He switched off his engine and swooped down towards the Zeppelin 4,500 feet below. At 7,000 feet Warneford made a fore to aft pass 150 feet above the Zeppelin's spine… in Warneford's own words ;

I put my ship in a shallow dive toward the tail of the giant, its crew firing every gun on board right at me. I could hear their bullets cracking into the wings and body of my ship. That bloody awful mammoth bag got huge. I was about 300 feet above them, safe from their guns for the moment, and I let loose all six of my little bombs.

R.A.J. WARNEFORD

As the sixth and final bomb cleared the rack there came a tremendous roar of explosions as one hydrogen gasbag after another exploded inside of *LZ 37* as it was almost ripped in two over Belgium. Savage eruptions tore through the Zeppelin in seconds and the blast of flaming hot air heaved Warneford's Parasol 200 feet straight up into the sky, overturning it, fracturing a petrol pipe, and damaging a fuel pump. Victor and victim went fluttering through the sky. For the first time an aircraft had brought down a Zeppelin raider and although Warneford was struggling to regain control of his Morane after the Zeppelin explosion, he was unable to restart his engine and glided down 35 miles into German held territory, landing in a small field beside a side of a small wooded hill at 0240 hours. In 15 minutes, he had his Morane repaired — but heard the sounds of German cavalry searching the wooded hill close by.... In a while they rode away, and he had his chance to become airborne again.

I pulled and pushed and bounced her along until I got her nose pointing downhill which was luckily pretty steep. Then I swung the prop. I kept on handling and pushing her ~ and she started to move slowly at first and then as she gathered speed and I knew she wouldn't stop I made a leap for the cockpit just as the Huns cavalry charged out of the wood firing their carbines in my direction.

Once airborne, Warneford flew on through mist and fog having to descend to near ground level a few times to get his bearings. As his engine starved of fuel, began to splutter and give up Warneford caught sight of the English Channel and made a bumpy landing on the beach at Cap Gris Néz. He was so weary that he fell fast asleep by the side of the machine, and was finally discovered by French soldiers only 20 yards from the cliffs of Cap Gris Néz. After a hero's welcome at St Pol on 7th June, Warneford reported to Squadron Commander Longmore's HQ at Dunkerque where he was handed a special telegram sent from King George V in London : *I most heartily congratulate you upon your splendid achievement of yesterday in which you single-handed destroyed an enemy Zeppelin. I have much pleasure in conferring upon you the Victoria Cross for this gallant act.* George RI.

Warneford was stunned. He was now the second airman to be decorated with the Victoria Cross, having with superb courage and skill, destroyed the first of the Zeppelins in flight. Squadron Commander Longmore ordered the new British air hero to keep out of trouble until he had visited Buckingham Palace where King George V excitedly decorated Warneford with the Victoria Cross. After his celebrated return to Dunkerque, the French requested his presence in Paris to confer upon him the *Knight's Cross of the Legion d'Honneur*.

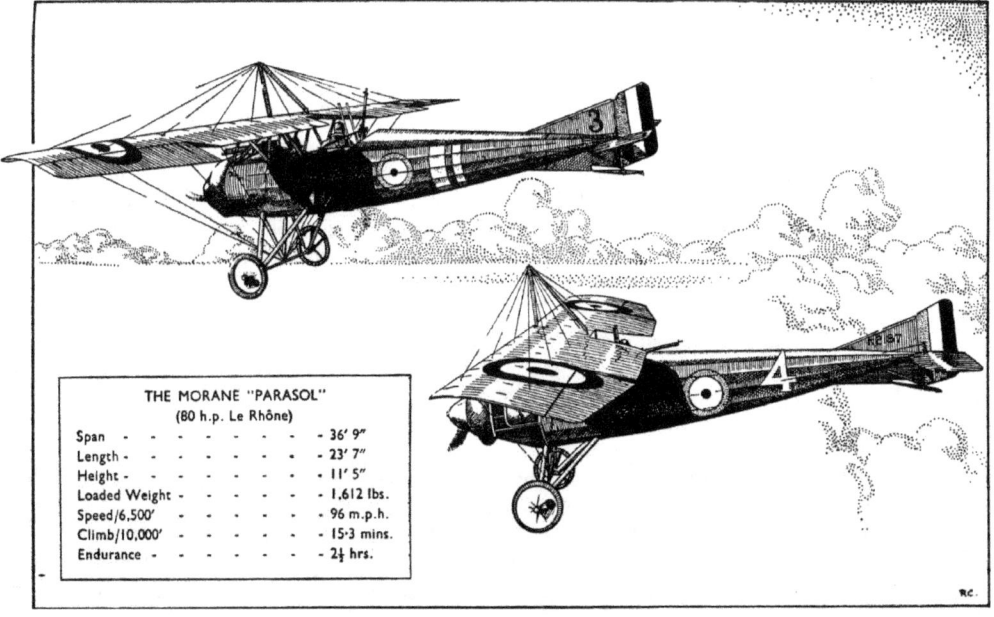

THE MORANE "PARASOL"
(80 h.p. Le Rhône)

Span	36' 9"
Length	23' 7"
Height	11' 5"
Loaded Weight	1,612 lbs.
Speed/6,500'	96 m.p.h.
Climb/10,000'	15·3 mins.
Endurance	2¼ hrs.

ZEPPELIN HAZARDS ~ SEARCHLIGHTS & LIGHTNING STORMS

Zeppelin im Gewittersturm, Blitze ~ Michael Diemer

Searchlights ~ Once over London, *Kapitanleutnant* Heinrich Mathy seeing his Zeppelin *L 31* spotted and coned by searchlight crews, skillfully escaped using parachute flares. He ordered his crew to drop all of their illumination flares out into the night sky to temporarily blind the British searchlight crews, and then helped by a mask of mist over the northern defences, was able to guide Zeppelin *L 31* away to descend elsewhere, unleashing his bomb load over London ~ killing 22 civilians and injuring 74.

Lightning ~ Airships were always liable to be struck by lightning in a thunderstorm and yet the risk of fire was minimal, provided the airship was below pressure height, and not discharging hydrogen. Not like *L 10* valving off gas during a thunderstorm when she was struck and went down in flames. Another encounter with lightning occurred during the mission of *L 59* the *Afrika-Zeppelin*, flying from Jamboli in the mountains of southern Turkey, down over the Mediterranean, across the Sahara, and down to German East Africa with a cargo of 30 machine guns, ammunition and medical supplies for General von Lettow-Vorbeck's forces in German East Africa ~ the longest flight of the Great War. During the night *L 59's* crew spotted lights off the southern point of Crete. They also knew a northeast storm was on the way after their radio operator reported electrical disturbances with his reception. As *L 59* passed the toe of Crete toward midnight, sporadic lightning gave way to a severe thunderstorm. A message came down from the forward lookout on the Zeppelin rooftop gun platform that it had begun to rain heavily. Clouds chased above and below the Zeppelin in different directions as gusts buffeted the airship and hail lashed the windows of the gondola. It became impossible to keep *L 59* at the height of 1,000m ordered by *Kapitanleutnant* Ludwig Bockholt and the *Afrika-Zeppelin* was forced under 500m several times. Another message came down from the roof ~ *the ship is on fire!* Zeppelin *L 59* lit up by lightning became a thrilling sight with the whole of its upper surface glowing from the static charged bluish white tongues of St. Elmo's fire. It flashed and streamed from the machine gun sights and supporting struts. Crewmen servicing the engines or attempting to walk forward along the narrow corridor were liable to shocks if they grasped a handrail. Just as quick, the lightning lessened and *L 59* continued its journey south to Africa.

Heat Turbulance ~ The *Afrika-Zeppelin* flew over the North African coast on 22nd November 1917, and set a course up the River Nile. The next morning, hot air thermals rising off undulating dunes below caused *L 59* to rock up and down incessantly. Subsequent cooling reduced the buoyancy of *L 59's* gas so much that she nearly crashed. Some of her crew suffered headaches caused by the bright sunlight radiating off the desert sand and fatigue from the midday heat, and then freezing at night.

ZEPPELIN HAZARDS ~ WEATHER

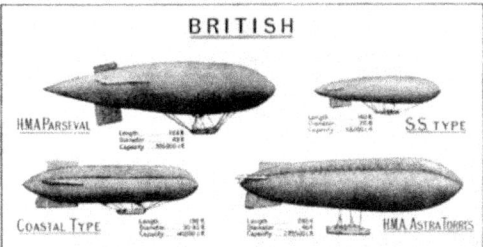

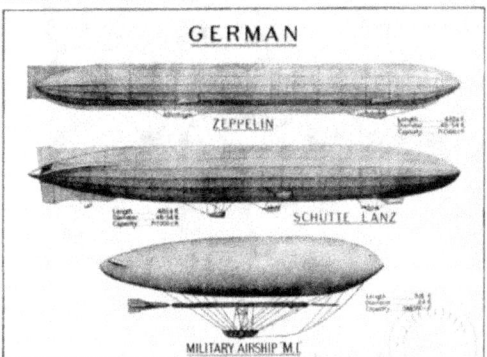

Iced Propellers ~ On the morning of 25th January 1915, Zeppelin *PL XIX* set off in freezing conditions to bomb the Russian Naval port of Libau in the Eastern Baltic. Ice began to form on the airships propeller blades which was soon flung thru the Zeppelin's outer envelope to cause a gradual leakage of hydrogen from its gasbags. Descending through dense cloud at 0845 hours, *PL XIX* bombed a factory on the outskirts of Libau, but was repeatedly hit by Russian ground fire. Passing over the Latvian coast, both engines failed one after the other. *PL XIX* buckled amidships and dropped into the sea seven miles offshore, where two Russian minesweepers captured her crew.

St. Elmo's Fire ~ During Zeppelin *L 11's* flight towards London in stormy conditions, its horrified crew saw the bluish white tongues of St Elmo's Fire a foot long burning on their machine gun sights, from the wire grommets on the caps of the forward lookouts on the airship's rooftop gun platform, and from commander *Kapitanleutnant* Treusch von Buttlar's fingers when he gamely thrust them out the control cabin window. Fortunately for these Germans, the Faraday cage formed by the duralumin girder framework protected *L 11's* hydrogen gas cells. But she it might have ignited, if *L 11* had been carried above the pressure height at which the expanding hydrogen in the gas cells would have been vented into the outer electrified air. Henceforth, all airships were ordered by Naval Airship HQ to attempt to circumvent all thunderstorms and if forced to go through them, to proceed as far under pressure height as the squalls would allow. The Zeppelin's gunners on the upper machine-gun platform were also ordered not to fire their guns if the airship was climbing and valving off hydrogen. While the hydrogen gas was stored securely in rubberized cotton cells, and the framework of their Zeppelins were made of duralumin to save weight, Zeppelin crews were ordered always, to wear felt boots over their flight shoes as a safeguard against striking sparks off any of the airships steel parts.

Frozen Compass ~ Over England on the night of 16th June 1917, *Kapitanleutnant* Franz Eichler, commanding Zeppelin *L 48* having just dropped his bombs on Harwich, turned for home unaware that his frozen compass was erroneously guiding them north. Eichler, believing he was over the English Channel on an easterly course, and flying away from peril decided to descend to 13,000 feet. In reality, *L 48* was flying north along the English coast as dawn was breaking. *L 48* was spotted by British ground observers, reported and intercepted over Suffolk by an outdated B.E.2c fighter, piloted by Lieutenant L.P. Watkins who fired a full magazine burst from his Lewis gun into *L 48* sending it to the earth below in flames. Two survivors from *L 48* suspected that their frozen compass was the cause.

ZEPPELIN HAZARDS ~ ANTI-AIRCRAFT ARTILLERY

PARIS ~ On 21st March 1915, the first raid was reported by a lady who lived near the Eiffel Tower.

I was awakened by fireman's bugles, and as we had been warned I had no doubt what the noise meant. For a long time nothing happened. The night was so clear and peaceful it seemed impossible there could be any danger. Suddenly there came reports from distant guns, and then series of vivid flashes from behind houses at no great distance, followed by a violent cannonade that made the windows rattle. Searchlights were playing in all directions, but at first nothing was visible except the ghostly outline of the Eiffel Tower. Then I noticed that several stars were obscured by what seemed to be a long grey cloud moving at a tremendous rate. It seemed more like a shadow than anything solid. What struck me most about it was its enormous length. When a searchlight fell on it, it was only a fraction of a second before it passed out of its field. I knew at once it was a Zeppelin. When I went back to the window the firing had increased in intensity, and the airship, which was far away behind the Eiffel Tower at what seemed a very great altitude, appeared to be replying to the anti-aircraft guns. From below the long grey shadow came a series of flashes ~ it must have been firing machine guns at the guns firing at it. Then, suddenly, the airship disappeared like a cloud, as suddenly and mysteriously as it had come.

French 75mm AA Gun ~ such as defended Paris and London from Zeppelins, mounted on a reinforced 1913 De Dion-Bouton.

THE MOBILE ANTI-ZEPPELIN SECTION

During 1915, the naval gunnery expert Admiral Sir Percy Scott was appointed to take command of the defences of London. He quickly imposed an almost complete blackout across southern England on any night suitable for raiding by enemy airships. Admiral Scott also planned a 104 gun anti-aircraft network complimented by 50 searchlights, and extra aircraft. At this time the most pressing need was to establish a mobile anti-aircraft unit, and equip it with a gun capable of firing a high-explosive shell, such as the French 75mm being used with great success in the defence of Paris. Within weeks, one truck mounted French 75mm AA gun became the focal point of a London's mobile anti-Zeppelin section created by Admiral Scott. He appointed Lieutenant Commander Alfred Rawlinson to command the fast response unit that included several truck-mounted searchlights and machine-guns. The gun was stationed at the Talbot Motor Works in Ladbroke Grove — then the Armoured Car HQ, and it was arranged that at the first warning of an impending raid received from shipping or coastal observers of an approaching enemy airship, Rawlinson and his mobile anti-Zeppelin section set off at high speed to manœuvre across the expected flight path of the Zeppelin and intercept it. About 1900, on the evening of Wednesday 13th October, the telephone rang in Rawlinson's private room. The Admiralty informed him ; *Some Zeppelins have just crossed the coast and are apparently making for London. They are expected to arrive here at about 2100 hours.* A period of intense activity followed and Rawlinson's section raced away toward their Moorgate Street gunnery site.

Everyone understood at once, the moment they saw us coming that an air raid was imminent. They did not, however, know where to go or what to do, though none of them had any doubt at all that the most pressing and most vital thing they had to do was TO GET OUT OF OUR WAY. I feel quite confident that no man who took that drive will ever forget any part of it and particularly Oxford Street, which presented an almost unbelievable spectacle. I had such an anxious job myself that I had no time to laugh, but I am sure I smiled all the way. After passing the Marble Arch the traffic in Oxford Street became much thicker. The noise of our sirens being as deafening as the glare of our headlights was dazzling, the omnibuses in every direction were seeking safety on the pavements. I kept my foot well down on the accelerator, and charged boldly without the least slackening of our speed. I also observed, out of the corner of my eye, several instances of people flattening themselves against the shop windows. Lieutenant Commander Alfred Rawlinson, Mobile Anti-Aircraft Section RNAS

Kapitanleutnant Joachim Breithaupt flying one of the new German naval airships, Zeppelin L 15 approached London from the direction of the West End, ascending to 11,000 feet. His two objectives were the Admiralty, and the Bank of England. The Admiralty is a conspicuous block of buildings in an exposed position at the apex of a rough triangle formed by the Mall and St. Jame's Park; seen from the air, with the great bend of the River Thames as a guide it could hardly be mistaken. However, the position of L 15 when Breithaupt dropped his first bombs was over National Gallery in Trafalgar Square, too late for an accurate hit on the Admiralty complex that he missed by over a third of a mile. Such inaccuracy may have been due to the simple nature of the bombsights used in early Zeppelins or if the Germans had neglected to adjust their bombsight to higher altitude, after increasing their height.

The most important objective of Breithaupt's mission had been missed, however the German bombs that L 15 released over Trafalgar Square were still spreading out and falling through the air. They descended with terrific impact striking Holborn, the Strand and London's theatre district killing 28, wounding 70, and just missing four famous theatres; the Lyceum, Strand, Aldwych and the Gaiety. The first struck the rear premises of the Lyceum Theatre, shattering the skylight and sending a shower of glass into the auditorium. The second blasted a huge hole in the middle of the road outside the Lyceum, bursting a gas main from which a roaring jet of gas burned furiously for hours. Another demolished the pavement opposite the entrance to the Strand, the next smashed into the road outside the Waldorf. When the dust cleared, a bus stood shattered and derelict. The driver, conductor, and a constable just alighting were all killed instantly. The body of a fourth victim was found, blown to bits.

A shock was now in store for Joachim Breithaupt. Rawlinson having just brought his gun to rest in Moorgate Street ~ took hurried aim at the rapidly approaching Zeppelin ~ and fired his first shot. It burst at a height of 7,200 feet, way too low to secure a direct hit, but it unnerved Breithaupt, taken entirely by surprise. Then the British had problems. Their French gun could not be elevated beyond 83° and at the critical moment when L 15 was coming towards his gun crew, brilliantly lit and a relatively easy target, Rawlinson heard ~ *Gun no longer bears, sir!* The Zeppelin had passed into the *dead circle* of the gun, from which it would not emerge for three long minutes. There was nothing the British could do, but stand by and watch the great Zeppelin fly over their quiet French 75mm AA gun.

THE MOBILE ANTI-ZEPPELIN SECTION

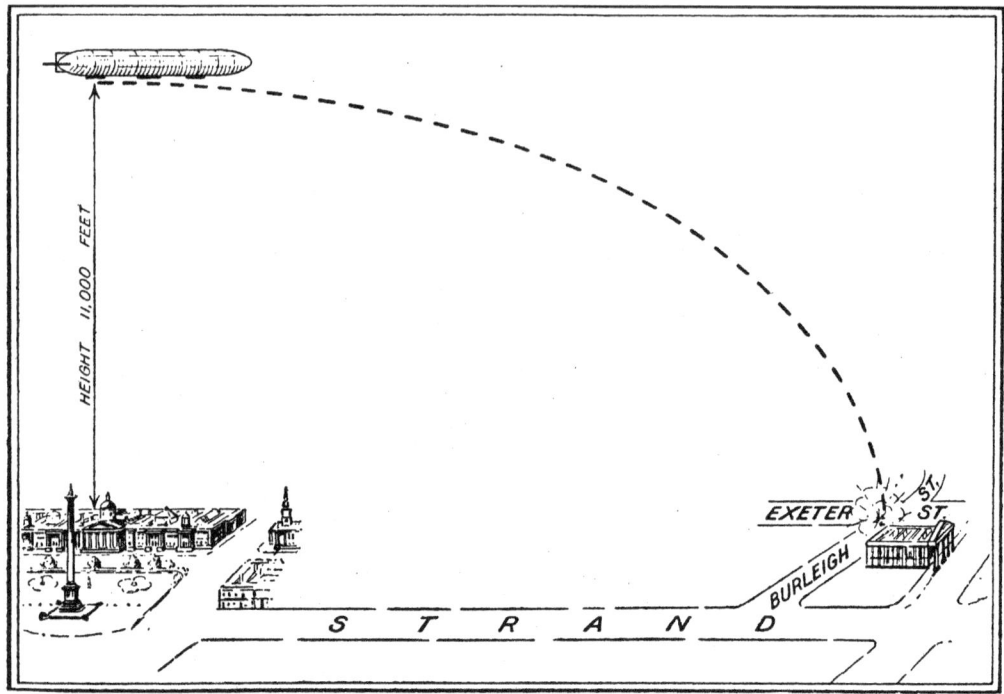

The position of Zeppelin L15 over Trafalgar Square, at the time when the Lyceum Theatre bombs were released. The Altimeter reading of Kapitanleutnant Breithaupt's Zeppelin L15 was 11,000 feet. Rawlinson, who fired twice at the enemy airship, estimated its speed at 50 mph. A bomb falling from 11,000 feet would take 26 seconds for its descent, during which it would travel forward 640 yards.

Breithaupt, realizing his danger and in a frantic effort to increase height, discharged his water ballast and dropped his remaining bombs that fell in Prince's Square (killing a horse). Before *L 15* disappeared from view in a cloud of mist and spray, Rawlinson's gunners fired their second shot. It burst high but perilously close. *As soon as I got the: Ready, sir! from the gun layer, I instantly gave them: Fire! When it finally burst high, it must have very considerably surprised the enemy who had been informed, with what had been great accuracy previously that there were no high-explosive shells in the London defences.*

At Sutton's Farm airfield, 2nd Lieut. John Slessor on anti-Zeppelin standby received a telephone call from Admiralty HQ warning of a Zeppelin approaching London. He warmed up his B.E.2c, took off into the night sky to begin his ascent over London, and later reported on the imperfect blackout.

The lights of the capital presented a wonderful spectacle. They did more. They illuminated quite effectively the great silver shape of Zeppelin L15. Long before I reached my patrol height I saw above me the impressively vast bulk of the airship. When I was still a good thousand feet below him, I saw streams of sparks as his engines opened up. The great bulk swung round and then ~ the most extraordinary sight ~ cocked its nose up at an incredible angle and climbed away from me.　　　　　　2nd Lt. John Slessor, No. 23 Squadron RFC

The crew of Zeppelin *L 15* first spotted the exhaust flames from Slessor's scout coming at them through the darkness. After seeing the British B.E.2c more clearly in a searchlight beam, *L 15* took immediate action to ascend, and successfully out climb John Slessor's scout. *Kapitanleutnant* Breithaupt after returning to Germany reported the unusually violent anti-aircraft fire and bursting shrapnel defensive fire (from Rawlinson's one effective gun). After this, a halt was called in the raids against London while a more formidable type of Super-Zeppelin was designed for use. Because of the effective use of a single artillery piece by Rawlinson's skilled gunners, for almost ten months the Germans avoided the capital and thereafter Zeppelin commanders were instructed to fly even higher to avoid the HE shells, which further diminished their accuracy. Worsening winter weather, and several serious storms also compelled Zeppelin HQ in Nordholz to repeatedly cancel Zeppelin raids.

SIR PERCY SCOTT ~ ON THE DEFENCE OF LONDON

On Wednesday, 8th September 1915, a Zeppelin came over London and dropped some bombs. Three days later, I received a letter from Mr. Balfour, who was then at the head of the Admiralty, asking me if I would take over the gunnery defence of London. Mr. Balfour suggested that the task would prove interesting, but at the same time he warned me that the means of defence at that time were very inadequate. I accepted the appointment. In selecting the ammunition to fire at Zeppelins a shell with a large bursting charge of a highly explosive nature was required so that it would damage a Zeppelin if it exploded near it; second, that all that went up in the air had to come down again, and that, in order to minimize the danger to the public from falling pieces, an explosive should be used in the shell which would break it up into small fragments.

General Galliene, who was in charge of the defence of Paris, had for the protection of his 49 square miles of city 215 guns. I had eight guns to defend our 700 square miles of the metropolitan area, no trained airmen, and no lighted-up aerodromes. Admiral Vaughan Lee, realizing the urgency of the matter, set to work. He undertook to get lighted-up aerodromes, trained men in night flying and soon had a bullet that would set a Zeppelin on fire.

The next thing was to get more guns. I knew that the Navy had some they could spare and which could be converted into anti-Zeppelin guns. Lord Kitchener very promptly gave me some; and with others that we picked up I found that in a very short time we had increased our number of guns to 118. But, unfortunately, mountings had to be made for these, which took a considerable time. The few guns we had for the defence of London were mounted permanently in positions probably as well known to the Germans as to ourselves. We had no efficient guns mounted on mobile carriages that could be moved about and brought into action where necessary.

The French, I knew, had some of their splendid 75 mm guns mounted on automobile carriages. I wanted the gun, so I ordered Commander Rawlinson, a very clever officer who spoke French like a Frenchman, to go over to Paris at once and either beg, borrow or steal a gun. I told him ~ You are to have it on the Horse Guards Parade, under Mr. Balfour's window, in less than a week! He was in a motor-car at the time. Looking at his watch, he said, I can catch the boat. I asked him if he did not want any clothes. He said, No. Please wire Folkestone to ship me and the car over to France. Thus he left, going at about fifty miles an hour down South Audley Street. That is the sort of officer that is wanted in war time! Twenty-four hours after leaving me he wired ~ Have a gun, two automobiles and ammunition. What he did is best described in his letter to me, as follows:

SIR ~ In obedience to your order that I should endeavour to obtain from the French Government a 75mm AA gun, mounted on an automobile, I proceeded to Paris. I first interviewed General Galliene, who in a most courteous and charming manner pointed out that, much as he would like to help London, he could not himself give me a gun, but he felt sure that General Joffre would give full consideration to anything that London wanted. General Joffre, without delay sent a telephone message the Minister of War in Paris telling him that I could have the gun complete with two automobiles and ammunition. The gun in my presence was tested and fired by a French crew, who also very kindly drove it to Boulogne and shipped it to London, where it arrived on the 21st. The whole transaction from the time of my leaving London to my return with French 75mm gun took four days. I attach photos of the gun & caisson. I have the honour to be, Sir, Your obedient servant ~ A. Rawlinson

Owing to the promptitude of Commander Rawlinson, we had this 75mm gun on the Horse Guards Parade, under Mr. Balfour's window. Although this was only one gun, its acquisition was valuable, as it showed what could be done, and how to do it. With the French gun as a guide we very soon mounted up eight of our own three-pounders on motor-lorries, which gave a start to the mobile section of our defence. There was an urgent need for mobile guns. I should have liked to copy the French auto-car mounting, which was a fine specimen of engineering, but our three-inch guns could not be adapted to it. We called in Mr. Whale, a very clever designer of Sir W.G. Armstrong, Whitworth, and Co., and a drawing was soon prepared. The design was arranged that the mountings be made of steel with a pair of wheels under it ~ being removable when the gun came into action.

I went over to France to see if the French would help me again. When I told General Galliene the number of guns we had, he laughed and expressed surprise that the Zeppelins did not come every day. He was a splendid officer and promptitude itself. Five minutes' conversation and it was decided that I should have 34 of the famous French 75 millimetre guns and 20,000 shells, with fuses complete. This brought our total up to 152 AA guns. They were rather a mixed lot ~ Mr. Asquith referred to them as rather a menagerie - but I went on the principle that any guns were better than no guns. The members of the Anti-Aircraft Corps, in fact, laid the foundations of the elaborate system of anti-aircraft defences which eventually taught the Germans that London was an unhealthy spot. The progress recently made in aviation and performances of the Zeppelins are sufficiently satisfactory to indicate that the time has arrived when the flying airship is a factor to be seriously reckoned with.

* * *

ZEPPELIN RAIDS AGAINST ENGLAND 1916

As the weather improved, the new commander of the German High Seas Fleet *Vizeadmiral* Reinhard Scheer who prepared to intensify the campaign against Britain with naval bombardments of the English coast and more Zeppelin raids, reconsidered his tactics. The Zeppelins would only be risked during the moonless half of each month. Bad weather or high winds had caused many raids to be cancelled or aborted and the Kaiser had said, *Royal palaces and historic edifices should not be bombed!* Naval Zeppelin Commander, Peter Strasser defined the tactics for future raids over southern England.
The airships will leave their bases by day and approach the English coast as darkness falls.
This way they can perform a scouting mission during their outward flight.
Night time raids offer the best chance of surprise ~ and concealment.
To maximize the advantages of night attack, raids will be generally confined to the period of the dark of the moon, running from roughly eight days before the new moon until eight days after it. During this time there will be no moonlight to help the enemy gunners when the raiders reach their goals.
This was later revised to ~ *Attack only with cloud cover, otherwise turn back.*
Strasser then asked Scheer for more airships to win the war.

From ~ Flotten Kapitan Peter Strasser
Commander Naval Airship Division, Nordholz.
To ~ Vizeadmiral Reinhard Scheer, High Seas Fleet.
Commander of the High Seas Fleet, 10th August 1916

The performance of the big airships has reinforced my conviction that England can be overcome by means of airships, as the country will be deprived of the means of existence through increasingly extensive destruction of cities, factory complexes, dockyards, harbour works with war and merchant ships lying in therein. The determining consideration for the establishment of 18 airships assigned to the Imperial Command was that the number was necessary for the scouting needs of the High Seas Fleet. This is still generally true.

On the other hand, the number of airships cannot be estimated high enough for a quick conquest of England. In the interest of a prompt and victorious ending of the war, attacks must be made with all the airships the building works can produce. I therefore respectfully request the Imperial command to affect an increase of the establishment of airships assigned to the Imperial Command from 18 to 22. This proposal naturally requires an early increment of four airship crews with four commanders and four executive officers. I am well aware of the prevailing personnel problems, but believe that the personnel must be made available if necessary through reduction in other areas, since airships offer a certain means of victoriously ending the war. Strasser

The first Naval Zeppelin raid of 1916 did not set course for the increasingly well-defended London. On 31st January 1916, nine naval Zeppelins rose into the air from Nordholz, and Tondern, for a raid along the west coast of England, with Liverpool as their main objective. It was the longest and most ambitious raid ever attempted up to that time. After a serious navigational misadventure, the embarrassed Zeppelin commanders instead headed for the Midlands; Britain's industrial heartland. The Midlands were unprotected by AA guns, without blackout precautions and were defended only by a solitary RFC scout sitting on an airfield near Birmingham. The German airships, free to fly through the night unopposed, dropped their bombs on targets of opportunity killing 72 and injuring 113. Apart from the casualties, Zeppelin scares drastically curtailed all nightly war production. On the receipt of air raid warnings; and there were many false alarms, work was suspended through vast industrial areas as factory managers blacked out their premises and often sent their workers home. The morale effects were out of all proportion to the material disrupting the nightly repose of millions of Britons and Londoners, many of who endured the nights in uncomfortable safety below in the Underground rail stations to protect their little ones. The unexpected dangers, the shattering noise of bombs, plus the blazing away of anti-aircraft guns all frayed the nerves, and British war production suffered from absenteeism. In this indirect way, the Zeppelins accomplished more than their bombs.

On the night of 31st March 1916, *Kapitanleutnant* Joachim Breithaupt returned to England in *L 15*. But, as his Zeppelin flew over Purfleet in Essex at 2145 hours, a British AA gun crew scored a direct hit on the German airship and their shell ripped through four of *L 15's* gas cells. As Breithaupt's stricken Zeppelin began to lose height, his crew frantically began tossing everything out to lose weight. *L 15* descending closer to earth, was spotted by Lieutenant Alfred de Bathe Brandon flying a B.E.2c of No. 19 Squadron RFC. Brandon flew above *L 15*, dropped incendiary and Ranken darts but was unsuccessful. A few hours later *L 15* came down in the sea off Margate at 0015 hours close to the Kentish Knock lightship where 16 of her crew were rescued by the armed-trawler *Olivine*.

OVER LONDON IN A ZEPPELIN

The weather chart on 25th April 1916, showed favourable conditions, and we had every reason to hope that the prospective raid on England would actually take place. Instructions had been given that our ship, LZ 97, was to be ready to leave the ground by 6.30 in the evening. We had landed at dawn after a longish cross-country flight, and refreshed ourselves for the new venture by a short sleep.

All that afternoon we were busy making the necessary preparations for our expedition. The men in charge of the filling operations hurried about their jobs, and the gas streamed hissing into the compartments of the envelope, while mechanics tested the revolutions given by the engines. Clouds of dust whirled high in the blast of air from the propellers. When the Commander made his appearance through the hangar doors the officer on duty reported that all was ready. At a signal given by a blast on a whistle, the landing party, which had been standing to for some time, seized the handrails of the gondolas. A command, a sharp tug forward, and the ship moved slowly out of its hangar. Gently guided by the ropes the ship glided out, its runners grating softly on their steel rails. A blast from the trumpet announced that the stern had emerged and that the ship was free of the hangar. A signaller, instructed by the Commander, gave directions to the crew by a flag, and shortly the ship was swinging gently in the wind on the broad landing ground 200 yards from its shed.

Hands off : ease up the guys!
The men at the ropes let go, and the landing party stepped back from the gondolas. For a moment the ship hung motionless before it soared majestically upwards. Hold on! Hand-rails and ropes were seized again, and the craft was pulled down. Then the Commander went on board, and after a hearty Good luck, hundreds of brawny arms pushed the gondolas upwards into the sky. We cast one brief glance round our home aerodrome, which lay flooded in the light of the setting sun, and then started the engines. Every man was filled with inexpressible joy. We were off to England !

A long journey lay before us, the first section being over conquered Belgian territory. After a short time Brussels had been passed, and darkness drew on apace. It was well into the night before we reached the coast, and for hours after we cruised over the English Channel, which could be seen dark green, almost black, beneath us. Night pressed down menacingly upon us, only stars glittered in the heavens, reflecting their light in the waves. But here and there beneath us were red flecks which we knew were not the reflected images of stars. They were lookout vessels and patrol boats, through whose funnels we could see, deep down into their glowing furnaces. Except for these there was no light at all. Thousands of feet above the oceans waters, passed our narrow vessel. The deep throb of its engines sounded into the still of the night, causing the gondolas and flying wires to quiver.

We continually checked our course and kept an unceasing watch into the night, but there was nothing to be seen until at last we could make out the coast of England. At that moment the moon came to our assistance. It rose above the dark green sea, a friend perhaps, but not one to be relied on, for as clearly as it showed to us our enemies below, it betrayed to them the presence of our ship in the sky. Again we made a brief comparison with the map, although we had already recognized the coastal contour as being near Blackwater, the very point which we had hoped to make. The reckoning of our course had been entirely free from error.

There is much uncertainty in these flights over the sea, crossing the Channel without meteorological reports of the air above England, particularly over the west of England, and consequently it was impossible to form an accurate judgment of the weather conditions, seeing that they depend mostly upon barometrical depressions advancing from the west. We had to rely solely upon observations at the coast itself, and therefore had to take into account the possibility of a strong wind suddenly springing up which might blow the ship out of its course, and when one is above the sea at night, there is no means whatever of discovering one's whereabouts.

Over England at last! Our hands are drawn to the bomb-release lever like iron to a magnet, but the time has not yet come. London is our objective, and there still remains a good two hours' flight before we arrive at our journey's end. We lean out of the gondola portholes once more, and pick out land marks and locate them on the map as well as we can from that height in the bright moonlight. Below us everything is as still as death, and the country is perfectly darkened. Not a gun is fired, not a searchlight directed at us. The English naturally do not want to give away prematurely the positions of their defence batteries and the towns which they protect.

Far, far away we discern a light and soon after a second. They lie on our course. A short calculation follows. We must be right over London. Impenetrable shadows envelop the gigantic city, only pierced here and there by minute pinpricks of light. Yet even so the various districts and the main streets can be unmistakably recognized in the moonlight. On emerging from the interior, where I had been testing the bomb-release mechanism, I am amazed at the clearness with which the ground can be seen. We know that the eyes below must also be watching us, but the silence remains unbroken. Did they really hope that we should not find their London?

OVER LONDON IN A ZEPPELIN

At high speed we steer for the city, the Commander standing ready on the bombing platform. The electric lamps which he has now switched on glow with a dull vari-coloured light. His hand is on the buttons and levers. Let go! he cries. The first bomb has fallen on London. We lean over the side. What a cursed long time it takes between release and impact while the bomb travels those thousands of feet. We fear that it has proved a dud until the explosion reassures us. Already we have frightened them ~ away goes the second, an incendiary bomb. It blazes up underneath and sets fire to something, thereby giving us a point from which to calculate our drift and ground speed. While one of us releases the bombs and another observes results, I make rapid calculations at the navigation table. Now the second incendiary hit is also visible. Its flames scarcely have leapt convulsively upwards in a shower of red sparks before we hear the shattering report of an explosion, so loud that it is plainly audible above the roar of the propellers. At the same time on come the searchlights, reaching after us like gigantic spider's legs : right, left, and all around. In a moment the bright body of the ship lies in the beams.

Hard aport! The steersman spins his wheel, and in a moment the great air ship obeys its helm. We are out of the dazzling rays and once more in the depths of night. But it is no longer pitch dark. The beams of searchlights fill the sky with a vivid light. They have lost us ~ strike wildly past us, catch us again, go over us ; one remains still, the others hunt around, crossing it or searching along it for the objective, while we steer in quite a different direction This mad frolic continues for hours on end. We lose all idea of the passage of time as we fly on, every half-minute releasing another bomb. Every explosion is observed, and its position pinpricked on the map.

It is difficult to understand how we manage to survive the storm of shell and shrapnel, for, according to the chronometer, we have spent a good hour under that furious fire. When London lies far behind us, we can still recognize it distinctly. The searchlights are still stabbing the darkness. More than 60 of them looking for the bird that had already flown. Silence closes in around us, and everything beneath seems stricken with death. Now we have to struggle against the freshly-risen wind, but the ship luckily is undamaged, and every engine intact. We therefore, grapple with the storm, as we have just done with our enemies. The last few hours and the events with which they had been filled are still fresh in our memories.

The English coast lies behind us, receding farther and farther into the distance, and the foam on the crests of the waves beneath shimmers in the moonlight as though it is phosphorescent. A vague twilight envelops us. It is dark inside the gondola, with the exception of the very faint spots of light from the pointers of the instruments. Many-coloured stars still dance before our eyes, the result of the dazzling searchlights. We are over the sea.

The man at the elevating wheel rubs his eyes, blinks, quickly slides open the shutter of his lamp, and flashes its rays on his instruments. The gondola is lit up, as the light gleams on the aluminium. Then hell is let loose ! They have long lain in wait for us down below there, and now the little dot of a gondola light has betrayed us. In a moment the searchlights of the warships in the Thames estuary have caught us and hold us fast. Again a withering blast of fire is directed against us ~ Put out that light! The Commander reaches over the steersman's shoulder and switches it off. But the ship, once caught, cannot get away from the searchlights, Shell after shell shrieks up at us, among them incendiary shells. They burst dangerously near. After ten minutes the light grew fainter and the firing dies away. Again we travel through the gloom and silence hour after hour.

The sky turns from indigo to grey, as dawn creeps up from the horizon. Many kilometres lie still before us, and the eastern horizon is red before we cross the Belgian coast at Ostend. Darkness still envelops the earth beneath, but up above already shines the light of day.

Keep a sharp a look-out for aeroplanes! The Commander orders. Whenever a German airship was reported from England, aeroplanes ascended from the aerodromes on the coast, and flew out over the Channel to lie in wait and intercept its passage to the Belgian coast. They knew well enough the route we took on our return.

Between Bruges and Ghent two hostile areoplanes are reported from the top platform. I take up my position with a machine gun on the starboard side of the forward gondola. I watch their approach, but they are flying too high. I cannot bring my gun to bear unless we get above them. Aloft, on the platform, the machine gun is chattering. A stream of flaming bullets flickers past us too short. The Commander orders a climb : they can't cope with us at that. They are, of course, faster than we are, but we can beat them at climbing. The distance between us increases, and they are left behind. We climb higher and higher. The gas blows off madly amid the rattle of the machine guns. Close to the Dutch frontier, on looking at the altimeter we find that we have broken the airship height record. The rest of our journey home is accomplished without further mishap, and we land at our own aerodrome after a flight of nearly twelve hours' duration. Our gallant craft has left its first war flight behind it, our bombs lie in the City of London.

Leutnant **Martin Lampel**

NIGHT FLYING AND NIGHT FIGHTERS

During 1916, the Home Defence Squadron was rapidly expanded to a Wing. The most effective aircraft against the Zeppelin raiders was the Royal Aircraft Factory B.E.2c night fighter, whose 90-hp engine turned a large four-bladed wooden-airscrew built up from seven laminations of mahogany, which were securely bound at the tips with fabric. From early 1916, B.E.2c night fighters were armed with four rack-mounted high-explosive bombs, a box containing 24 explosive darts or eight Le Prieur air-to-air rockets, after French airmen had downed Zeppelin *LC-77* in April, proving them in action. The most effective weapon against Zeppelins was the Lewis machine gun armed with five drums of incendiary, tracer, and explosive bullets, such as James Buckingham's invention: Buckingham's Mk VII flat-nosed cupronickel-jacketed bullet designed to punch large holes in the outer fabric of German observation balloons and Zeppelins. The Buckingham bullet had a unique explosive base that contained eight grains of yellow phosphorous. Fired out of a machine gun, the phosphorous ignited and if these bullets pierced a Zeppelins skin through to the gas cells of highly flammable hydrogen, all manner of fiery hell would erupt for the Germans. Supplies of the new incendiary ammunition were express delivered from factory to RFC airfields, for specific use against Zeppelins. The British also fitted illuminated gun sights to their night fighters to enhance the paired Lewis guns that proved far superior to the Vickers machine gun, as the exposed muzzle flash from the Vickers had temporarily night-blinded pilots who used it on night missions.

A pilot of a night-fighter Home Defence Squadron took off on night mission sitting in a basket weave seat wearing protective leathers, and secured with a wide safety strap around his midriff, perhaps with a little grease smeared on their face to combat the cold. There was no cockpit heating, no radio, no oxygen equipment, and no parachute. Small lamps were mounted in the instrument panel and the increment readings were daubed with radium. Identification lamps were fitted onto each wingtip and below them on brackets, landing flares were attached to the leading edge of the lower wings. At the end was a metal clamp, holding a magnesium compound flare. An igniter was fitted to the base of the flare that led wires to a battery and two brass buttons in the pilot's cockpit. These two brass buttons enabled the night pilot to either ignite his landing flares singly or together. A pilot's landing flares were lit from an altitude of 150 to 200 feet descending and burned for one short minute. The pilots were almost always the most experienced available. Aircraft were often wrecked on landing after night missions. Sometimes their aircraft caught fire, sometimes worse. There was no training in night flying, because it was looked upon as so dangerous that it had been listed as being reserved for war purposes only. Even for that, the Admiralty at one time considered abandoning it.

On the night of 2nd September 1916, an imposing air raid of 16 Zeppelins attacked London. The largest airship attack ever made on the British Isles. The 16 huge enemy airships carried 38 tons of bombs. The result was negligible. Four civilians killed and 11 injured. The Germans also suffered the loss of one airship and the surviving crews of their raiding fleet carried back to Germany the harrowing memory of Zeppelin *SL 11* set ablaze by a British scout. British counter measures; 50 searchlights, AA guns and night patrols had previously shown little success, but all this changed on the night of 3rd September 1916 when Zeppelin *SL 11* was *clawed down in flames* as Winston Churchill had predicted would happen by a British scout firing incendiary bullets. Lieutenant William Leefe Robinson flying a B.E.2c at 11,550 feet, at a range of 500 feet had concentrated the entire contents of his final Lewis drum into a single point. A dull pink glow was ignited deep inside of the airship. In seconds the entire tail assembly was roaring and spitting fire as the blazing Zeppelin slowly descended nose-first in a shallow dive earthwards that illuminated the night sky and ground below for 50 miles on all points of the compass. At 0230 hours the blazing hulk expanded and exploded with a flash that bathed the whole of London in its brilliance as the baked Huns at last vanished from the other Zeppelin airmen's horrified gaze. Lt. Robinson won the Victoria Cross for his determination in destroying Zeppelin *SL 11* that burned on the ground for two hours. His combat report to his commanding officer of No. 39 Home Defence Squadron RFC, describes his action.

I climbed to 10,000 feet in 53 minutes, and I counted what I thought were ten sets of flares ~ there were a few clouds below me, but on the whole it was a beautifully clear night. At about 0105 hours I noticed a red glow in North-East London. Taking it to be an outbreak of fire I went in that direction. At 0115 hours the searchlights picked up a Zeppelin over northeast London. I sacrificed height (I was at 12,900 feet) for speed and made nose down in the direction of the Zeppelin. I saw shells bursting and night tracer shells flying around it. When I drew closer I noticed that the anti-aircraft aim was too high or too low; also a good many some 800 feet behind ~ a few tracers went right over. I could hear the bursts when I was about 3,000 feet from the Zeppelin.

ZEPPELIN STRAFERS

French aerotechnicians of the famed 3rd Escadrille, with their Stork emblem (centre) and the formidable le Prieur rockets on the left. Yves le Prieur was the first Frenchman to earn a Black Belt in Judo in Japan. He also invented SCUBA and created the first dive club

I flew about 800 feet below it from bow to stern and distributed one drum from the Lewis gun along it. It seemed to have no effect. I therefore moved to one side and gave it another drum distributed along its side ~ without apparent effect. I then got behind it ~ by this time I was very close ~ 50 feet or less and concentrated one drum on one part, underneath rear. I was then at a height of 11,500 feet when attacking the Zeppelin. I hardly finished the drum before I saw the part fired at glow. In a few seconds the whole rear part was blazing. When the third drum was fired there were no searchlights on the Zeppelin and no anti-aircraft guns were firing. I quickly got out of the way of the falling blazing Zeppelin and being very excited, fired off a few Red Vérey lights and dropped a parachute flare. There was no retaliatory fire from the airship's gondolas ~ she might have been the Flying Dutchman for all the signs of life I saw. Lieutenant W.L. Robinson, No. 39 HD Squadron, RFC

German Zeppelin raids on Britain reached their peak in 1916 when they flew 22 missions, dropped 3,500 bombs, and killed 577 British civilians. As British air defences became more effective, the German raids declined. Every raiding Zeppelin carried incendiary bombs, but it was the high explosive bombs that caused the greater damage.

The defence of England, after responding to the Zeppelin threat absorbed over 17,000 officers and men who might have been deployed at the front, and the RFC had to deploy 110 aircraft at home to calm civilian fears. Because London drew the Zeppelins as a magnet, the British were able to concentrate their defensives, (and the Germans made the same mistake during the Battle of Britain). If the Germans had hit targets in the Midlands and the North, there was little to stop them and the moral effect would have been worse. Even so, the Zeppelins caused terrible psychological disturbance to the civilian morale. On some days after an airship raid, absenteeism in factories ran as high as 20%, causing huge losses to war production : some factories forecasted a drop of 16% off their expected yearly output for 1916.

First hand accounts from pilots combat reports show the effect of their aircraft in aerial defence. By September of 1916, the reign of the Zeppelin was almost over. Second Lieutenant Frederick Sowrey had been in the air for almost two hours on the night of 23rd September, wearing a flashlight on a strap around his neck to illuminate his dashboard, when he first sighted *Kapitainleutnant* Wolfgang Petersen's *L 32* hovering over Tilbury. As Petersen shook his airship free of the British searchlights it vanished from view. Again the searchlights spotted and coned the Zeppelin in their beams, its six revolving propellers glinting as Petersen turned his airship trying to shake off the lights. Eventually Sowrey got within firing range of *L 32* which began to rise ~ still in the apex of a trio of searchlights.

ZEPPELIN STRAFERS

During the night of September 23rd at 0045 hours I noticed an enemy airship in the east. I manoeuvred into a position underneath. The airship was well lighted by searchlights but there was not a sign of any gunfire. I could distinctly see the propellers revolving and the airship was manoeuvring to avoid the searchlight beams. I fired at it. The first two drums of ammunition had apparently no effect but the third one caused the envelope to catch on fire in several places, in the centre and front. All firing was traversing fire along the envelope. The drums were loaded with a mix of Brock and Pomeroy incendiary and tracer ammunition. I watched the burning airship strike the ground and then proceeded to find my flares. I landed at Sutton's Farm at 0140 hours. After seeing the Zeppelin had caught on fire, I fired the red Vérey's light.   2nd Lieutenant Frederick Sowrey

Sowrey focused the incendiary bullets from his third Lewis drum at the centre of Zeppelin *L 32* riddling its petrol tanks, igniting a deep rosy glow within the heart of the airship which made the Zeppelin glimmer like a huge red Chinese lantern. The blazing wreck of *L 32* struck the ground near Billericay, where Admiral Hall's men were first on the scene, and despite the heat they sorted through the charred wreckage of the control gondola. They were well rewarded with a usable copy of the German Navy Cipher Book that Petersen (against regulations) had on board *L 32*. With this invaluable item they quietly left the scene. For all of this, 2nd Lieutenant Frederick Sowrey was awarded the DSO.

On the cold misty night of October 1st 1916, *Kapitanleutnant* Heinrich Mathy commanding *L 31*, (the first super Zeppelin armed with ten defensive machine guns and carrying five tons of bombs) crossed the Norfolk coast and made for London. Over the city, the beams of a dozen searchlights pierced the sky, crossing and re-crossing on a point high overhead in a pyramid of coned light. Local guns spoke out sharply and the air was filled with shrapnel. Seeing this from afar, 2nd Lt. W.J. Tempest of No.39 HD Squadron, flew his B.E.2c swiftly to a height of 14,000 feet, and soared straight for the apex of the searchlights. Lt. Tempest described the fiery end of Zeppelin *L 31* over London as follows ;

On October 1st at 2200 hours I left the ground in a B.E.2c to patrol between Joyce Green and Hainault. Approximately at 2340 hours I first sighted a Zeppelin. My armament consisted solely of a Lewis gun and several drums of explosive bullets, as I had the rocket assemblage owing to its cumbersome nature removed, (which had greatly diminished the climbing ability of my aeroplane; for this I was severely hauled over the coals). I immediately made for her and fired one drum, which took effect at once and set her on fire at 12,500 feet. I then proceeded to North Weald to land and wrecked the machine on the aerodrome without hurting myself at 0210 hours. In fact I had dared to destroy a Zeppelin without the Le Prieur rockets.   2nd Lieutenant W.J. Tempest

Night-fighter squadrons based around London tried using new anti-balloon rockets designed by *Capitaine* Yves P.G. Le Prieur of the French Navy. To aim the rockets, pilots used a foresight ring fitted a convenient distance from the hindsight and set low enough to obtain an all round view of an airship, the ring size being the diameter of a Zeppelin at 200 yards. Pilots then aimed at the leading front half of the Zeppelin to allow for the combined speeds of both craft. When the pilot pressed a trigger switch in the cockpit all eight rockets were fired. Most pilots were unimpressed ~ as one made very clear ;

The trouble with those rockets was that they only had a range of about 50 yards ~ and if you were to destroy a Zeppelin you had to get so damned close to ensure hitting it you might as well not have had rockets at all and simply rammed the thing because you couldn't have gotten out of the way. They were terribly unreliable too. They were inaccurate in flight and would go all over the bloody show when you pressed the trigger.

The next raid did not come until 27th November 1916, when ten Zeppelins avoiding London revisited the Midlands and northern England. British defences had been improved and were waiting. Second Lieutenant I.V. Pyott shot apart Zeppelin *L 34* that fell in pieces to the sea off Middlesbrough. To the south, three pilots spotted *L 21* flying toward the coast and heading home to Germany. They attacked and sent the Zeppelin spiraling towards the sea as a blazing hulk. These successes against the Zeppelins in the autumn of 1916 gave the people of Britain a real sense of relief. They knew that the aerial threat had been overcome, and that whenever an attacking B.E.2c armed with incendiary and explosive ammunition got within firing range of a Zeppelin, it was finished. It was not even necessary to use a first-rate fighter. The dated B.E.2c sufficed, as the night-fighter that shot down nine Zeppelins.

Minor raids by the odd Zeppelin persisted until the middle of 1918 because of their hindrance to British war production. On 5th August 1918, Peter Strasser himself commanded the last sizable Zeppelin raid of the war ~ crossing the North Sea in daylight. A patrolling De Havilland DH4 piloted by Major Egbert Cadbury (of the famous chocolate family) spotted and intercepted the Zeppelins. After his gunner Robert Leckie shot *L 70* down in flames eight miles off the coast of Lowestoft, Strasser's Zeppelins all retired, dropping their entire 50-ton bomb load into the sea as they flew away.

THE B.E.2c NIGHT-FIGHTER & HANDLEY PAGE 0/400 BOMBER

THE B.E.2c (Top)
(90 h.p. R.A.F. 1a)
Span - 40' 9"
Loaded Weight - 2,100 lbs.
Speed/6,500' - 82 m.p.h.
Climb/10,000' - 53 mins.
and
THE B.E.2c (Bottom)
(90 h.p. R.A.F. 1a)
Span - 37'
Loaded Weight - 2,142 lbs.
Speed/6,500' - 72 m.p.h.
Climb/10,000' - 45 mins.
Endurance - 3¼ hrs.

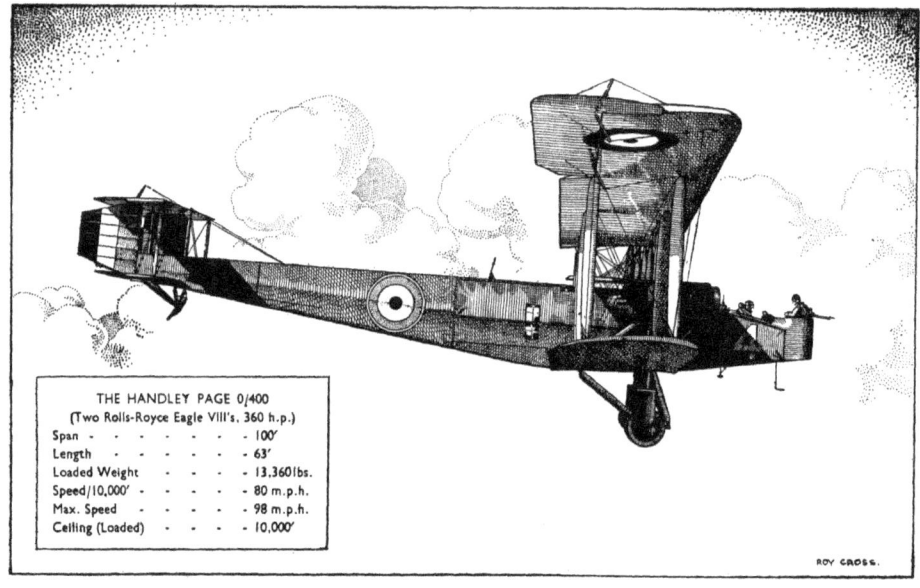

THE HANDLEY PAGE 0/400
(Two Rolls-Royce Eagle VIII's, 360 h.p.)
Span - 100'
Length - 63'
Loaded Weight - 13,360 lbs.
Speed/10,000' - 80 m.p.h.
Max. Speed - 98 m.p.h.
Ceiling (Loaded) - 10,000'

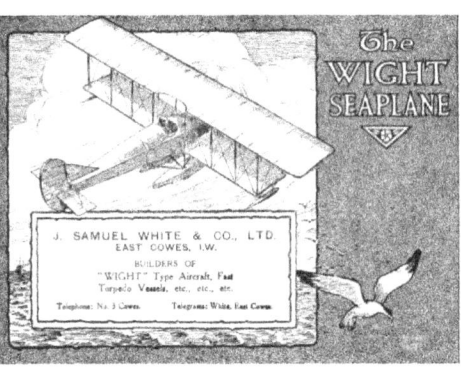

HOLLENDER'S FIRST AND MOST DANGEROUS RAID ON ENGLAND

At noon on 27th November 1916, three officers were sitting in the Casino at Nordholz. One of them, Kapitanleutnant Max Dietrich of the L 34, was celebrating his birthday, and his chair had been decorated with boughs of fir trees. The weather chart for that morning informed us that the storm which for days had been blowing over the North Sea, had at last died down. Kaptain Straffer, the airship's pilot, having inquired of his commanding officer whether there would be a raid on England that day, had been informed that it depended on the meteorological report which was expected from Bruges. We had almost given up hope when the adjutant rushed into the mess shouting ~ Gentlemen, orders to attack the industrial district of the English Midlands; splendid prospects ~ the first ship must be in the air by one o'clock at the latest! After that, nobody thought of eating ; there was rushing about, excitement, and the ringing of telephones. Oberleutnant Zee Kurt Frankenberg commanding L 21 suggested ~ Leave all the birthday things as they are. We'll have our celebrations tomorrow! At 12.45, L 21 left its shed and set out for England. Fifteen minutes later I started on L 22 followed by L 34. I remember the glowing faces of my men, for it was our first raid. Our dearest wish was about to be fulfilled.

In a short tune we reached Heligoland. There was not a cloud in the sky, and the eye was met by a magnificent spectacle. On all sides were the grey shapes of our gigantic birds of prey, flying together over the water. On the left were the airships from Ahlhorn and Hage ; in the middle our ships from Nordholz ; and on the right were those from Tondern. I counted ten in all, each setting its course for Flamborough Head. Darkness soon set in, and they gradually disappeared from sight, only L 21 remained visible for a long time, ahead of us. Great clouds of vapour floated at a height of 2000m in the west ; we climbed above them into the clear starlit sky. Suddenly appearing from out of the cloudbanks, L 36 bore down upon us only a few hundred yards away, then disappeared again like a ghost. Next, to our great annoyance, we were enveloped in thick mist, and the thermometer registered 16 degrees of frost. We strained our eyes to pierce the obscurity ahead, when in clear weather, the enemy's country should be visible.

10.15 in the evening. The slender sickle of the moon had disappeared beneath the horizon in a halo of orange light, and before me 15 nautical miles away, the English coast at Flamborough Head could be seen outlined. In the south a bright shaft of flame appeared, steadily increasing until it lit up the whole sky : where a Zeppelin had dropped its bombs, and the ship itself could be seen glistening in the beams of innumerable searchlights. The eyes of my men shone with pleasure, for they knew that we would very shortly be over the enemy's territory.

The next few hours slipped by like minutes. We reached our objective amid the dazzling rays of searchlights, the banging of guns, and the bursting shells which glowed as though red hot. Above it all we heard the explosion of our bombs while the airship shuddered throughout its entire length. It was like some wild phantasmagoria. Suddenly to the north of us, where one of our number had been caught by the searchlights, there appeared a crimson ball of fire, which rapidly increased in size. A minute later we recognized the glowing skeleton of an airship falling in flames. We wondered who it might be. (Max Dietrich's L 34 ignited by Lt. Ian Pyott, 36th HD Sqn. RFC)

By 1.30 A.M. the raid was over. As we made for home darkness again enveloped us. Far behind many patches of flame on the ground bore testimony to the success of our raid. While I was contemplating the scene the pilot suddenly exclaimed that the ship was rapidly losing height, and that he was unable to prevent it. This was hardly to be wondered at, for we discovered that the compartments inside the envelope had been riddled with shrapnel and splinters of shells, and the gas was pouring out. It was a moment that demanded a cool head. Owing to the darkness no part of the ship was visible. As quickly as possible I brought her down below 750m in order to ease the internal pressure and thus to lose less gas. Apart from this consideration the engines developed more power at that height. Everything superfluous was thrown overboard, and disappeared into the depths below. The machine guns, our ballast water, even fuel as much as we could spare, were sacrificed in this way. In spite of all I had but little hope of bringing the craft safely back to its base, and my spirits gradually sank as I thought of lying some where beneath the waters of the North Sea.

The men were still so enthusiastic over the raid they had just carried out that they were quite unperturbed, and did not fully appreciate the danger. I therefore quietly ordered the officer of the watch to get into touch with the Admiral of the Fleet by wireless, to report on our raid and to add ; The airship has been severely damaged by artillery fire and is urgently in need of assistance. Five minutes later we received an answer to the effect that torpedo boats and cruisers were being dispatched, and in the grey light of early dawn we met the Second Torpedo-Boat Flotilla, scouring the sea in search of us 60 nautical miles north-west of Borkum. A rising west wind helped us along, and the gas expanding as it grew warmer increased our lift. This turned the scale in our favour, for we reached the neighbouring aerodrome at Hage with the last litres of fuel in our tanks. Everything movable had been thrown overboard, and the engines could not have kept going for another half hour. There was no return to Nordholz on that day, for both our other ships had fallen victims to the terrible fire of the enemy. The birthday had become a day of death.   Kapitanleutnant Heinrich Hollender L 22

THE GERMAN VIEW ~ AIR RAIDS ON ENGLAND ~ FROM 1915-1918

Naturally the raids on England brought more recognition to the officers concerned than any other work. For the nation was made very familiar with them in our newspapers. On the night of January 19th 1915, the first raid by Zeppelins was made on the coast of England, and although carried out by only two ships, the L 3 and L 4 it was the cause of dismay and terror throughout the whole of England ~ and great rejoicing in Germany. This raid proved to us that the hitherto unapproachable island was accessible to German arms, and that here too the war could be carried into the enemy's country. On the other hand, it gave rise to the most exaggerated hopes, for many were led to believe that the naval airship would be able to achieve the impossible, to reduce the whole of England to ruins in a very short time, and thereby to decide the war. We ourselves never cherished exaggerated hopes as to the effects of our attacks, although, in spite of all English attempts to conceal the results, it was always obvious to us that the effects were great. Apart from the fact that we, of the Airship Service could personally observe the results of our raids, reliable reports on our work were continually coming through.

If the raids were really as ineffectual as England made out, it is difficult to understand why she considered it necessary to adopt the elaborate system of aerial defence which was built up during the war. The defences in operation throughout the country, and especially in the neighbourhood of the East and South Coasts, almost defy description with a great quantity of materiel in the shape of guns, munitions, aeroplanes, and searchlights which otherwise could have been used on the Western Front. Thus the air raids, in addition to their destructiveness, and doubtless their very great moral effect on the population, relieved the pressure at the front. The first sign of demoralization was the flight of all the well-to-do people from the East Coast, the depopulation of coastal places, Furthermore, the destruction caused was great, as each individual airship could carry up to three tons of bombs.

At first the raids encountered so little resistance, owing to the weak aerial defences that individual Zeppelins delivered attacks from a height of under 1000m. This condition, however, soon altered, for the strength of AA defences rapidly increased from month to month. In fact, from the summer of 1916 onwards, we suffered heavy losses, although the raids were carried out from a height of 3000m. Even then, in spite of a lavish expenditure of ammunition, and the use of powerful searchlights, AA guns never scored except with lucky shots. Meanwhile, however, the aeroplane pilots had learnt to fly by night, and it was an easy matter for them to spot an airship once caught by the searchlight beams ; almost impossible to escape, and by using incendiary ammunition, to shoot it down in flames. It was believed in England that by this means a stop had been put to Zeppelin attacks, as there were no raids from the end of November 1916 until March 1917. This hope was soon disappointed. An entirely new type of airship had been designed during this period, a type which could attain a height of 18,500 feet, from where the most powerful searchlights were utterly useless. Consequently both anti-aircraft guns and aeroplanes had to work in darkness, and the British defensive measures broke down.

Soon, however, other difficulties were encountered ~ the reasons for the diminishing frequency of the raids. A long flight in the intense cold sometimes as many as 40 degrees of frost were registered with a lack of oxygen, and the rarefied atmosphere, told upon the vitality of the airships' crews to such an extent that finally, many men were unable to endure the height and had to give up flying. Particularly when high flying was first practiced, and before the invention of the oxygen apparatus, it frequently happened that many members of the crew were rendered incapable by weakness or some other symptom of airsickness. On the L. 44, for example, while making a raid on Harwich in May 1917, so many were affected by air-sickness, that the ship drifted over the town completely out of control, and without a single engine running. It was not until they were well out to the middle of the North Sea that two of the engines were restarted, and the ship was able to return to its base.

The unfavourable meteorological conditions which prevailed at high altitudes, almost invariably a strong wind blowing from the west or the north, and the airship, which developed only one quarter of its engine power at that height, could scarcely make headway against it. Frequently therefore, even when favourable weather obtained at a lower level, our ships could not reach their objective. During 1918, when the airships were fitted with more powerful engines designed to develop their greatest power at high altitudes, this gave the Zeppelins a considerably increased radius of action, and enabled them to carry out flights of much longer duration. A raid then lasted from 20 to 30 hours, and when our giant aeroplanes in Flanders reached London and S.E. England, the airship's new sphere of action became the industrial regions of the Midlands and the North that were beyond the range of aeroplanes. The number of airships participating in each attack varied according to the number which were airworthy ; the biggest raid was successfully carried out by thirteen airships. In the Baltic our naval airships accomplished useful work through observation patrols for the Navy, and by raids into Russia. Naturally, however, since England was our most powerful antagonist on the sea, the greatest number of airship operations were carried out over the North Sea. Only the few who were fortunate enough to serve in the air as pilots or Zeppelin crews during the war, can realize how magnificent was the aerial arm created by Germany. Not one of our opponents ever achieved anything comparable to it.   Major Georg Neumann

GOTHA BOMBING RAIDS ON LONDON ~ AND PARIS

In 1917, the Gotha aircraft works in Thuringia were fabricating a new generation of aerial raiders. Their twin-engined biplane bomber design was capable of 125 kph with a range of 850km, and a maximum operational ceiling of 21,000 feet carrying six 200kg bombs. The Germans could now bomb England from their St Denis-Westrem and Gontrode airfields in Flanders. The first raid by the 23 Gothas of the *England Geschwader* (England Squadron), commanded by *Hauptman* Ernst Brandenburg, targeted Folkestone in Kent on 25th May 1917. They killed 96 civilians, injured 194, and destroyed much property. Next, a flight of 14 Gothas crossed the English Channel at an altitude of almost 15,000 feet on 13th June, and flew up the Thames in daylight, to appear over London at 1130. The Gothas flew in a loose diamond approach formation to optimize their collective firepower against any defending British aircraft. At 1140, Ernst Brandenburg in the lead bomber fired a white flare; the signal for his formation to scatter and attack individual targets. Within two minutes 72 bombs had crashed down within a radius of a mile of Liverpool Street Station, killing 162 people and injuring 432. The tragedy of the raid came when one 110-lb bomb detonated in the upper North Street School in Poplar. Exploding in a crowded classroom, it killed 18 small children, and injured another 34.

After unleashing their destruction, the Gothas resumed their defensive diamond formation and droned away. One Bristol F.2b attacked, but caught in a ferocious crossfire from the German Gothas, was driven off with a wounded observer. The ease that these huge bombers had penetrated the heart of the British Empire in daylight, and the appalling killing of civilians and property damage that they had inflicted, led to a huge uproar over the state of London's defences, especially as no Gothas were even damaged. It seemed to the British, that the Gothas could come over whenever they pleased and bomb London at their leisure ~ and that nothing was going to stop them. In numerous tactically similar raids, the Gotha *England Geschwader* took over from the Zeppelins that had by the middle of 1917 raided England 51 times. The Gotha bombers always came up the Thames Estuary towards their primary target of London; in either a diamond or box formation, at a height of 10,000 to 16,000 feet, which until the advent of the De Havilland DH. 4, was too high for most defending aircraft to reach.

The main problem of the Gotha bomber was their poor flying stability from near empty fuel tanks that caused 37 landing accidents; a larger number than the 24 Gotha bombers lost in action. The Gotha III and IV both had a plywood-firing tunnel through the centre of the fuselage, for their dorsal gunner to train his weapon down through the floor, and under the Gotha's tail to destroy any Allied machines coming at them from the lower blind spot favoured by stalking pilots. The development of the Gotha IV was greatly assisted by the capture of the British Handley Page 0/400 bomber, *The Bloody Pulverizer* in early 1917. The resultant Gotha IV was able to carry 500kg of bombs, and the Gotha Va had a biplane tail assembly, shorter nose, and nose wheel to improve landings on night operations.

In the autumn of 1917, the Gotha bombers attacked by night, and their raids continued right through the summer of 1918, particularly in April when they raided Paris on almost every clear night. The commander of the Paris sector, General Gallieni set up an outer ring of listening posts 100km from the city with an inner ring of 15 fixed AAA batteries, each consisting of two 75mm field guns, four machine guns and some searchlights ~ placed on the most expected routes into the French capital. As soon as the alert was sounded, a complete blackout would be imposed across the city. A second ring of listening posts was soon established in a semicircle some 20km from the centre. The early warning system worked well, but after the AAA struggled to find the correct range at night, the 45 aircraft of the CRP failed to intercept during a raid on 28th May 1915, when none took to the air. Standing patrols were introduced, but it proved hard to communicate details of the enemy bearing; with no ground to air wireless, details could only be transmitted by laying cloth panels on the ground. The next couple of years saw several Zeppelin raids : nothing like those deployed against London.

From January to November 1918 however, the German aviators flew 485 Gotha sorties over Paris. The first was on the night of 30th January, when 30 Gotha aircraft approached Paris in small groups, Only 11 reached the target to drop 77 bombs from between 1,000 to 4,000m. The French scrambled 57 aircraft to meet them; six got close enough to fire their guns, but all missed. Even more weapons; 105mm guns, new sound locators and extra searchlights were added to the French AAA air defence. while false lights were planned to dupe the raiders at Villepinte to the north.

DEATH FROM THE SUMMER SKY ~ HOW THE NEW RAIDER WORKS

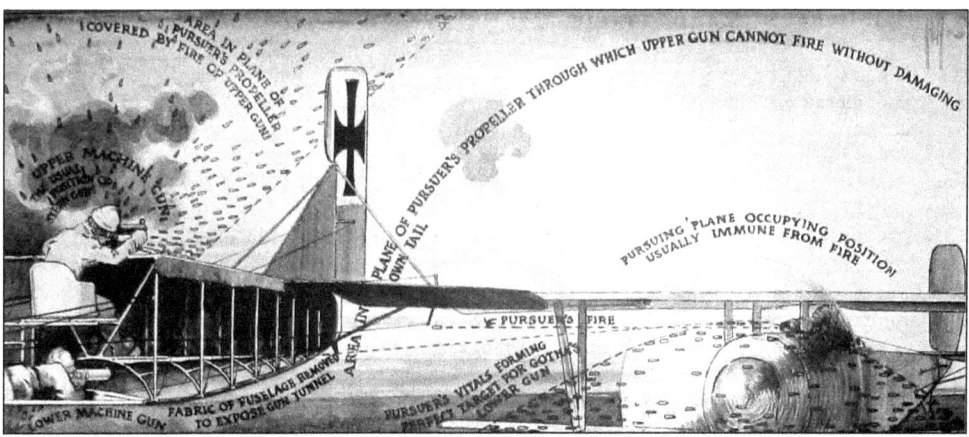

How the Gotha's gun turret removes the blind spot behind the tail in which a pursuing aeroplane can shelter.

The new raider, the Gotha giant biplane, diagrammed here, is typically Teutonic in its lack of finesse. Its simple strategy consists in massed formation with its fellows, and the delivery of an overwhelming machine-gun fire upon attacking planes. Once this formation is broken, it becomes a big, blundering target at the mercy of swift and handy fighting planes. It is also typically Teutonic in its imitativeness, a British giant plane which unfortunately fell into enemy hands a little while ago having provided many of the finer points of the Gotha. This fact is emphasised lest any should be tempted to see in the Gotha a fresh manifestation of the ingenuity of the Hun.   *The Graphic* ~18th August 1917

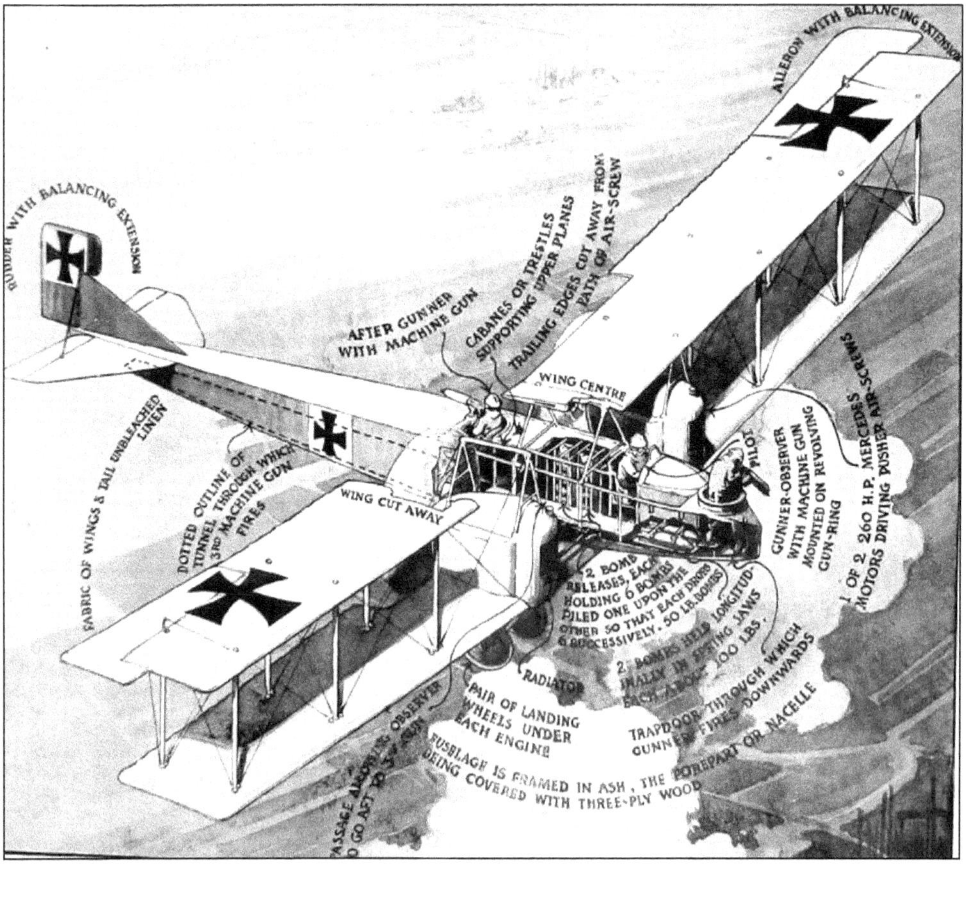

BRITISH HANDLEY PAGE 0/400 & GERMAN GOTHA G.V BOMBERS

Handley Page 0/400

Gotha G.V

INNOVATIONS IN THE AIR DEFENCE OF PARIS

Les nuits de Gothas en 1918 : Paris Défence Contre Avions (DCA)
The situation room at Paris AA GHQ 1918 ~ *L'Illustration*

From a total of 485 Gotha bomber sorties between January and November 1918, only 35 targeted the centre of Paris delivering about 12 tonnes of bombs. While the French claimed that this was due to the intensity of their barrage (120,000 rounds of mostly 75mm AAA were fired) many Gotha commanders aimed to drop their bomb loads on the industrialized northern suburbs like La Villette. Although central Paris escaped relatively unscathed, many factories suffered very significant damage, as did the railway junction at Creil. Attacks from Gothas killed 266 and wounded 603 in Paris, while 11 Gothas were shot down (all by AAA fire) on the northern approaches to Paris, via the Marne valley.

Once, during August of 1918, Gotha bombers destroyed a huge transport base near Calais and the disaster was of such severity that it was difficult to maintain transport on the Western Front, and by extension, impossible to send further supplies to minor theatres; a striking testimony to the value of air raids on *well-chosen* targets. It is perplexing ; why after this, the Germans made no sustained bombing attacks on Calais, Boulogne or Dieppe where disembarkations would have been hampered.

The targets that Zeppelin and Gotha commanders apparently sought, were often of no real military consequence, while they missed or neglected vital objectives within their reach. The Royal Arsenal, at Woolwich on their way to London, during three years was targeted once. *Kapitainleutnant* Mathy had boasted ~ *I will bomb London three times in succession or perish in the attempt*. Yet, on 13th October 1915, when Zeppelin *L 13* bombed Woolwich, they achieved no effective damage. One of *L13's* incendiary bombs fell outside the Main Gate of the Royal Arsenal where it burnt out. Until the spring of 1915, all British HE shells were filled there. Guns were made, re-lined, and repaired there. Both the British Army and the Royal Navy were dependent upon the Arsenal of Woolwich for an essential part of their munitions, and why the Germans did not concentrate on its destruction remains a mystery. Two other factories of vital importance were also well within reach of the German raiders; the Small Arms Factory at Enfield Lock, as well as the Royal Gunpowder Factory at Waltham Abbey (the main source of cordite, Tetryl and gun-cotton). Both were raided on occasions ~ but never hit. The Gothas by mid-1918 began to suffer from the attentions of fast climbing British S.E.5a and Sopwith Camel fighters. By the end of the war, there had been 52 Gotha raids over Britain, in which they had dropped 2,772 bombs that killed 857, injured 2,058, and destroyed swathes of property.

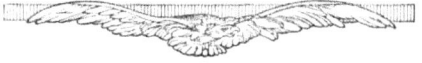

Lieutenant William Leefe Robinson VC ~ downed Zeppelin SL 11   *Lieutenant Wulstan Tempest D.S.O. ~ downed Zeppelin L 31*

Oberleutnant zur zee Werner Petersen and the crew of L 16 ~ shot down by 1ˢᵗ Lieutenant Frederick Sowery RFC, 23ʳᵈ September 1916

ANTI-AIRCRAFT ARTILLERY

Anti-aircraft artillery or AAA was practically unheard of at the beginning of the Great War, and perhaps because of this, the chosen materiel of anti-aircraft gunnery was both changed and affected from field modifications of fighting from positional defence, to a new role of more open manœuvre. Thus the priority of portability changed from a near neglected factor to one of paramount importance. For anti-aircraft service, the weapon had to be able to cover all elevations, with a traverse of 360°. Instead of firing at a stationary target, an enemy aircraft flew at a velocity of about 50 meters a second or one-sixth the average velocity of the actual projectile itself in the case of the French 75mm field gun. In October 1914, the British recognized the need for anti-aircraft defence of the British Isles and developed the QF 1-pounder *pom-pom*, (a 37mm version of the Maxim used in the South African War) that only fired contact-fused ammunition. Meanwhile, the German *Lichtspucker* (light-spitter) was redeployed in the AAA role. This 37mm five-barreled revolving cannon had been designed to fire five flares at low velocity. The projectiles that were known to Allied aviators as *Flaming Onions* from its vivid flight sequence of five bright flare shells would have been hazardous to their fabric aircraft, and to optimize their chance of a strike, German gunners always rapid fired all five rounds at once.

Early Warning. Anti-aircraft gun sites, and the tactically more flexible truck-mounted AAA guns, could be command coordinated from observation posts or through centres of anti-aircraft defense. Observation posts were supplied with field telephones and signaling alternatives for early warnings. An early warning gun was linked to the O.P. by telephone and ready to fire 2 rounds in the direction of approaching enemy aircraft. For directional early warning, it helped if the gun-base or the vehicle, on which the gun was mounted, were always aligned at the cardinal points of the compass.

Control of Fire. Fire discipline was essential. Most success came from withholding fire until the target was within 1,500 yards range, with reliance placed on a particularly effective first shot or burst. After the first burst, avoiding action taken by the enemy aircraft made further fire ineffective until the aircraft made another strafing or bombing run. It was essential for AA fire to be cut up into bursts of less than 20 seconds for the controller at the O.P. to hear the approach of any other low-flying aircraft. On the command, *Aircraft Action!* The AA gunners opened fire. Open sight firing used against aircraft attacking gun positions often spoiled their aim, compelled them to fly higher, and to change direction.

When it came to a choice of AAA projectiles, HE was preferred over shrapnel. However a Thermite Shell was the gunner's round of choice for use against Zeppelins. While ordinary bullets passed through the outer envelope of an airship without causing significant damage, Thermite shot a flaming shower of incendiary shrapnel. For night fire, searchlights spaced at 3,500 yards apart picked up targets, held them, and often enabled the guns to shoot down the enemy. The best technique for nocturnal AAA crews was for them to open fire only during those times when searchlight beams were exposed. The British further experimented, and fitted tracer to their shells for night firing.

Siting of Guns ~ With enemy attacks on motor transport invariably made on stretches of straight open road between defiles to cover their flight approach, light anti-aircraft guns needed to be sited to protect vehicles moving along such roads. AA guns were best sited away from the actual road to be clear of dust and any battle débris along the road, and to avoid being shot up during the engagement. On the other hand, not so far off that the target became a crosser instead a head-on. Gun pits were avoided as they often gave away the position on aerial photographs. For this same reason, frequent changes of gun position were sensible. Natural camouflage of foliage was preferred, with the crews often being reminded to remain inactive — except at targets within the requisite range of 1,500 yards. With no accurate means of measuring aerial range or the height of their target, gunners tried their best to estimate their fuse settings. At least, when AA guns were protecting their own spotting balloons, the altitude could be accurately referenced from the length of the cable, mooring the guarded balloon.

One of the most misguided anti-aircraft gun deployments seen on the Western Front was during August 1914, when a Belgian gun was mounted on the roof of the Antwerp cathedral for use against any approaching Zeppelins. The added elevation was not worth the risk of turning the historic building into a legitimate military target. At a desperate time for the people of Belgium, it only gave cause to boastful German claims of Allied duplicity in misuse of cultural and religious buildings. Warned by the fate of the surrender of Rheims' after a 177-shell bombardment, that damaged their own cathedral, the Belgians removed the gun before any damage was done to Antwerp's cathedral. If the elevated Antwerp gun had been seen, the Germans could so have explained shelling the cathedral.

* * *

GERMAN METHOD OF DIRECTING ANTI-AIRCRAFT GUNS

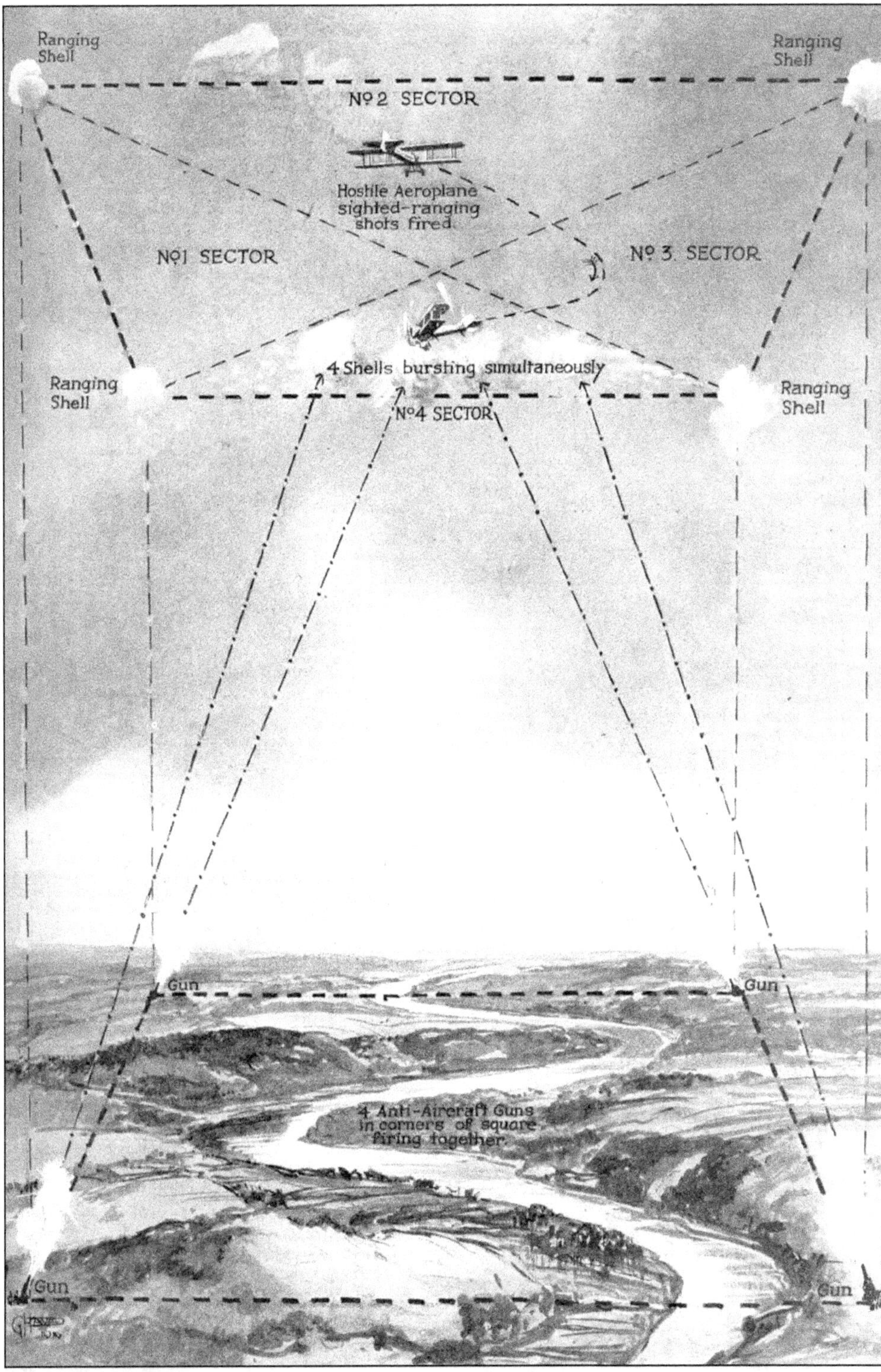

THE GEOMETRY OF SKY-PARALLELOGRAMS IN THE AIR

BRISTOL F.2b FIGHTER & De HAVILLAND DH.4 DAY BOMBER

THE BRISTOL F.2b FIGHTER
(250 h.p. Rolls-Royce "Falcon III")
Span - - - - - - - 39' 3"
Length - - - - - - 25' 9"
Loaded Weight - - - - 2,800 lbs.
Max. Speed - - - - - 125 m.p.h.
Climb/10,000' - - - - 11·5 mins.
One Vickers two Lewis guns

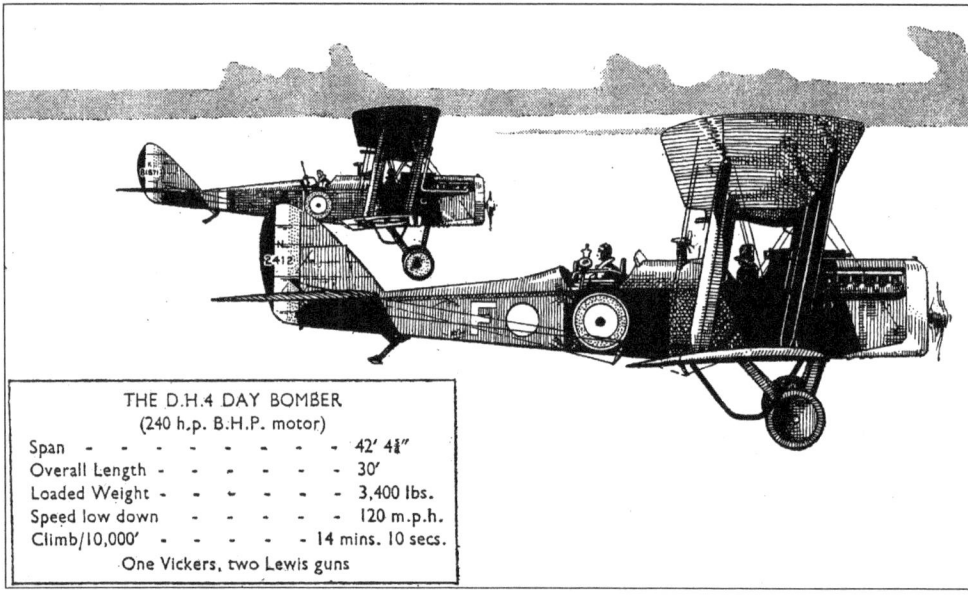

THE D.H.4 DAY BOMBER
(240 h.p. B.H.P. motor)
Span - - - - - - - 42' 4⅜"
Overall Length - - - - 30'
Loaded Weight - - - - 3,400 lbs.
Speed low down - - - - 120 m.p.h.
Climb/10,000' - - - - 14 mins. 10 secs.
One Vickers, two Lewis guns

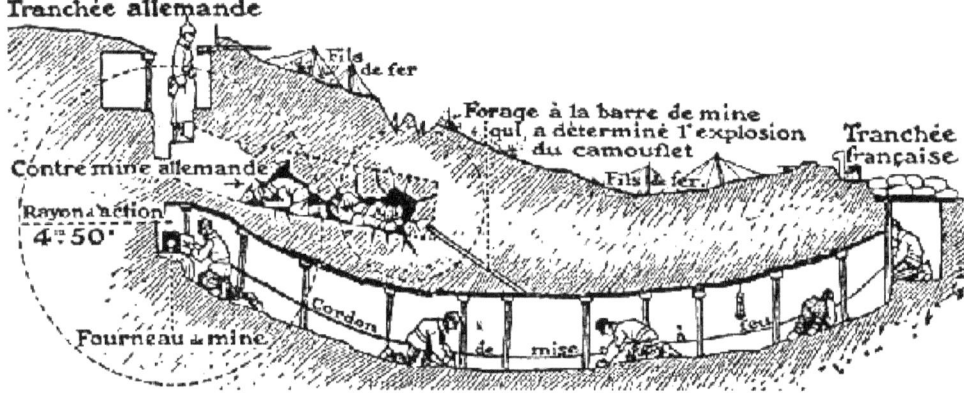

CHAPTER II

WINNING THE GREAT WAR

The role of the flight leader was to navigate and search the sky ahead, while his fellow pilots kept station and searched the surrounding sky. British formation flying was based on a flight of five or six aircraft flying in an array, with the flight leader in front as two others flew on either side to form a V. To the rear and above, were two other aircraft and behind them the sub-leader flew solo. In combat pilots operated in pairs. The leader to attack, while the wingman protected. When sighting the enemy, the leader manœuvred his flight into a good attacking position, preferably high and up sun. He needed to avoid silhouetting his formation against any high cloud that would make them easier to see, and to avoid flying over any under cast that might expose them to a higher enemy. Surprise was far more difficult to achieve with a formation than with single aircraft, as a formation was easier to see from a distance and jockeying for position took longer ~ all giving the enemy more time to discover their danger. Flight leaders knew the importance of the initial attack in formation, but once battle was joined, their pilots flew free to fight as they pleased until the recall signal was given. Most victories against enemy aircraft were achieved in the first pass, with the subsequent dogfight consisting of plenty of shooting, but far fewer hits. It was said, *he who has height, controls the battle,* referring to a better chance of surprise, more speed from a diving attack and a greater escape velocity.

Edward Mannock ~ When the war began in 1914, he was working as an inspector for a British telephone company in Constantinople. The Ottoman Turks, aligning themselves with Imperial Germany rounded up Mannock and interned him in a pungent prison. In April of 1916, the Turks repatriated Mannock, suffering from ill health and poor eyesight assessing him useless as a man. When Mannock arrived home in England, emaciated and ill from malaria, he was again declared unfit for military service, but his hatred for the enemy made him haunt recruiting offices until he was accepted for active service. In August 1916, he joined the RFC and proved a gifted pilot, flight-leader, and teacher of tactics who planned his operations so well that his formations were never taken by surprise. To avoid unnecessary risks to the lives of his airmen, Mannock always attempted to manœuvre for the best spot to launch an attack on the enemy. Again promoted, when he first took his new command of No. 85 Squadron RFC up to fight, he directed three to fly with him as decoys at 16,000 feet. Two other flights were to follow at different altitudes above. Mannock seeing ten enemy scouts approaching, dived with his decoys, and the Germans gave chase. As soon as he signalled to his flight of five next above, they darted at the enemy taking them by surprise. His third flight dived down to 2,000 feet to clean up the strays. The Germans lost five aircraft. Mannock claimed two victories and importantly, brought his entire flight home. Mannock's tactics were to flush, stalk, and kill his opponents, by patrolling offensively and exploiting the qualities of speed and surprise to the full. He gave the benefit of his experience in field lectures ~ then more practical tuition in the air, directing each pilot in turn to lead the flight. Then after each mission, he gathered his flight together for a helpful analysis of their tactics pointing out any errors. Once, after he had darted after a spinning enemy firing short bursts with little hope of hitting, someone asked ~ *Why?* Edward Mannock replied; *To increase his wind up!* A more nervous target was likely to make an error. He did. Mannock had him.

My man gave me an easy mark. I was only ten yards away from him ~ on top so I couldn't miss. A beautifully coloured insect he was ~ red, blue, green and yellow. I let him have 60 rounds, so there wasn't much left of him. I saw him going spinning and slipping down from 14,000 feet. Rough luck, but it's war, and they're Huns. To fight is not enough ~ you must kill.

When Lieutenant Ira Jones felt a bit down, unable to shoot any enemy scouts down after some good opportunities he talked with Mannock, who saw the problem and offered some practical advice ;

I've had a long talk with Mick about it. He thinks I am allowing too much for deflection. That is I'm aiming too much in front of the enemy. He has advised me to do a slight traverse, to sight about 5 yards in front of the engine, then to fire and while firing, to bring the sight back as far as the pilot, and then to push it forward again. If I do this, he says, I can't help but hit the machine somewhere. I must say it sounds reasonable enough. This is what he says he did at first. With experience, he says, I'll get accustomed to making the correct allowance for the enemy's speed and direction instinctively. Lieutenant Ira Jones

FLIGHT LEADER EDWARD MANNOCK

During patrols he not only mystified and surprised the enemy, but also the formations that he led. Once over the lines, he would commence flying in a never-ending series of zigzags, never straight for more than a few seconds. Was it not by flying straight for long periods that formation leaders were caught napping? As he tilted his machine from side to side, he scanned the sky above and below with the eye of an eagle. Suddenly his machine would rock violently, a signal that he was about to attack ~ but where was the enemy? His companions could not see them although he was pointing in their direction. Another signal and his machine would dive to the attack, the red streamer attached to his rudder fluttering faster than the heartbeats of those who followed it with taut nerves. A quick half-roll, and there beneath him would be the enemy formation flying serenely along: the enemy leader no doubt with his eyes glued to the west ~ the result a complete surprise attack.

Mannock would take the leader if possible in order to give his pilots coming down behind him a better chance of an easy shot at someone before the enemy formation split up, and the dogfight began. Having commenced the fight with the tactical advantage of height in his favour, Mannock would adopt dive-and-zoom tactics in order to retain the initiative. Woe betide any pilot who lost the initiative and got himself into such a mess that his comrades had to forgo their tactical advantage in order to extricate him from his perilous position! Whilst the fight was in progress, Mannock, in spite of being in the thick of it, would be summing up the situation the whole time. When the enemy had been demoralized and defeated, Mannock would give the signal to re-form, and would not leave the battle area until he was satisfied that all the machines had seen his signal.   Lieutenant Ira Jones

Mannock's Rules ~ Mannock always offered his new pilots practical advice; *Know your opponent. Never try to dogfight a Triplane in a turning battle, as it will get behind you every time. Dive and zoom is the only tactic that is safe. If the initial attack fails, it is best to climb rapidly and select another moment to attack.* But on his final sortie, he broke the old rule; *do not follow the Hun down too low.* A hail of heavy German AA fire from the ground pierced his machines fuel tank. Orange flames blazed off the right side of his fuselage. His left wing swung over, and his Scout plunged to the earth. Edward Mannock had won the Military Cross twice, was one of the few three-time winners of the DSO, and was posthumously awarded the Victoria Cross. Highly respected as a flight leader, and one of the great tacticians of his era, he wrote a set of practical rules for air fighting to be passed on to new pilots. His tactics were to be adapted according to the situation at hand, and were based on the premise: *The enemy must be surprised and attacked at a disadvantage, if possible, so the initiative is with the patrol. Combat must continue until the enemy has admitted his inferiority, by being shot down or running away.* Major Mannock's 15 rules were :

1: *Pilots must dive to attack with zest, and must hold their fire until they get within 100 yards of their target.*
2: *Achieve surprise by approaching from the East ~ from the German side of the front.*
3: *Utilize the sun's glare and clouds to achieve surprise.*
4: *Pilots must keep physically fit by exercise and the moderate use of stimulants.*
5: *Pilots must sight their guns and practice as much as possible as targets are normally fleeting.*
6: *Pilots must practice spotting machines in the air and recognizing them at long range,*
   and every aeroplane is to be treated as an enemy until it is certain it is not.
7: *Pilots must learn where the enemy's blind spots are.*
8: *Scouts must be attacked from above and two-seaters from beneath their tails.*
9: *Pilots must practice quick turns, as this maneuver is more used than in any other in a fight.*
10: *Pilots must practice judging distances in the air, as these are very deceptive.*
11: *Decoys must be guarded against ~ a single enemy is often a decoy ~ therefore,*
    the air above should be searched before attacking.
12: *If the day is sunny, machines should be turned with as little bank as possible,*
    otherwise the sun glistening on the wings will give away their presence at a long range.
13: *Pilots must keep turning in a dogfight, and never fly straight except when firing.*
14: *Pilots must never, under any circumstances, dive away from an enemy, as he gives his opponent*
    a non-deflection shot ~ bullets are faster than aeroplanes.
15: *Pilots must keep their eye on their watches during patrols, and on the direction and strength of the wind.*

FLIGHT LEADER JAMES McCUDDEN

PANCAKING ~ CONSISTS OF STALLING THE MACHINE JUST ABOVE THE SURFACE OF THE GROUND, AND DROPPING THE REMAINING YARD WITH AS LITTLE FORWARD SPEED AS POSSIBLE. A PANCAKE LANDING WAS MADE ON ROUGH GROUND, STANDING CROPS, WATER... WHERE AN ORDINARY LANDING WOULD RESULT IN A SOMERSAULT. THE ILLUSTRATION WARNS PILOTS OF THE ERROR OF STALLING TOO HIGH ABOVE THE GROUND OR THE OPPOSITE ERROR OF LANDING ~ BELOW THE GROUND ~ WITH THE NOSE CRASHED. (R.A.F. 09.04.1918)

Major James McCudden ~ This 23 year-old English airman was such an agile pilot that he could change his machines direction as quick as a snipe. After surviving three duels with Max Immelmann, on 27th December 1916, McCudden in his DH.2 met von Richthofen for a prolonged aerial waltz. None of the 300 bullets fired by the German hit. McCudden pushed his joystick forward sending his bird into a steep, sideways dive ~ praying that he could reach a cloudbank, before von Richthofen's aim improved. With his engine screaming at full throttle, he remembered his flying instructor's words: *Watch out for the Hun in the Sun, my son, and never let the bastards get on your back!* McCudden made it safely into the clouds, and never again forgot the golden rule. McCudden personally maintained his S.E.5a, *the mount of aces* in top condition. Twice he shot down four enemy aircraft on one day, one two-seater he shot apart at 400 yards, confident of his accuracy and his ability to judge his targets range. Most pilots he said underestimated the range at which they opened fire, especially novices, who fired bursts from twice the distance they thought they were from their target. By February of 1917, he had received the Military Medal, been granted a commission, and awarded the Military Cross. Returning to England in April of 1918, McCudden was promoted to Captain and awarded the Victoria Cross. On leave, he spent a lot of time with his friend, Edward Mannock competing for the affections of the glamorous West End dancer Teddie O'Neill, who McCudden took aloft on a joyride, at least once. Major James McCudden who was promoted to command No. 56 Squadron, and shot down 57 enemy aircraft in five months, was a keen observer of any patterns of German aerial behaviour.

The weather still continued very clear, cold and frosty, and every day I was up, waiting about over our lines for Hun two-seaters to come across after I had done my daily patrol. If patience and perseverance would meet their just reward I certainly would have got many more Huns than I did, for I was up at every opportunity studying the two-seater's habits, characteristics, and different methods of working. In fact, this branch of work alone, just studying the habits, and psychology of the enemy aero crews constitutes an education of great interest. My system was always to attack the Hun at his disadvantage if possible, and if I were attacked at my disadvantage, I usually broke off the engagement for in my opinion the Hun in the air must be beaten at his own game, which is cunning. I think the correct way to wage war is to down as many as possible of the enemy at the least risk, expense, and casualties to one's own side. It was easier to approach a Hun from a good position than to do the actual shooting ~ and having dirty goggles makes all the difference between getting or not getting a Hun.

CAPTAIN ALBERT BALL

Captain Albert Ball ~ When a man of 19, like many of the other early aces Albert Ball began his tactical repertoire ambushing careless or insensibly defended two-seaters. As he accumulated more flying experience, his confidence flourished, and he became fearless. The odds made no difference to Ball as he made solo attacks on entire formations of Roland two-seaters. With No. 11 Squadron RFC he refined his tactics, sensing safety in numbers : enemy numbers. The larger the formation of German aircraft, the safer he felt. His method was to dart straight into the midst of the enemy machines feeling they would be unable to shoot at him for fear of hitting their own fliers, while he couldn't miss with targets all around him. To spot any Germans on his tail, Ball was the first pilot to attach a mirror to his aircraft, fitting his on the upper wing. As a result of his fearless tactics, he often came home with his aircraft shot full of holes and crash-landed many times ~ always without a scratch to himself.

In three months of flying over the Somme from July to September 1916, Ball vanquished 30 of the Kaiser's aircraft to surpass both Boelcke and Guynemer and earn the admiration of the British Empire. The Nieuport Scout of Albert Ball was armed with a single Lewis gun fitted to the upper wing of his aircraft that he controlled with a cockpit cable adapted to the curved track of his Foster gun-mount, so that his Lewis could be pulled down for reloading or swung into position for firing upwards. This configuration suited Ball, who hunted alone on the edge of clouds, and relied on his excellent eyesight to pick up lone enemy two-seaters at long range. Once he spotted a victim, he would set off in pursuit using all the cloud cover he could. Stalk his opponent unseen, come up on his prey from dead astern, slightly underneath the enemy fuselage ~ the gunner's blind spot. Then at close range, pull his wing mounted Lewis gun back to almost vertical and fire a long upward burst from as near as 10 yards. Often he almost rammed the enemy, swinging his aircraft from side to side raking his opponents belly, fuel tanks, engine, observer and pilot with full magazine bursts of English lead from his Lewis. Albert Ball who flew in all sorts of weather, preferred to attack, then zip away.

I only scrap because it is my duty, but I do not think anything bad about the Hun. He is just a good chap with very little guts. Nothing makes me feel more rotten than to see them go down, but you can see it is either they or I, so I must do my best to make it a case of them. If fighting on the Boche's side (as you mostly are), never use your last drum, unless forced to do so. Keep it to help you on your way back. A Hun can always tell when you are out of ammunition and he at once closes with you, and if they are in formation, you stand no chance. Keep this last drum and when you want to get back, manoeuvre for a chance to break away, and if they follow you, as they mostly do, keep turning on them and firing a few rounds at long range. When this is done they nearly always turn and run. This gives you a chance to get home.
Captain Albert Ball
No. 60 Squadron RFC

Back at the aerodrome, Ball lived alone in an old wooden shack, which he called his *Dear Old Hut*, way out beyond his squadron's most remote hangar. Between sorties, it was there that he enjoyed the tranquillity of planting vegetable seeds, which his family frequently sent to him. Later on in the evening he enjoyed listening to the latest records on his windup gramophone. His favourite after dinner entertainment was to light a few red-magnesium flares outside of his hut and parade around in his pyjamas fiddling on a violin. By the end of September, Albert Ball recognised he had pilot fatigue. *I like this job, but nerves do not last long and you soon want a rest.* Ball requested his squadron commander to send him home to Nottingham for a while to recuperate. On his return to England, Albert Ball was only 20 years old when promoted to Captain, invested with the Military Cross, and personally awarded two Distinguished Service Orders at Buckingham Palace, by King George V.

Toward the end of February 1917, Albert Ball received a message informing him that he would be assigned to No. 56 Squadron as a flight commander and prepared for his voyage across the Channel, and back to France. His new squadron were being re-equipped with the latest S.E.5a powered by a 150-hp Hispano-Suiza V8 that provided a top speed of 138 mph and a ceiling of 18,000 feet. It carried a single synchronised .303 Vickers machine gun capable of firing through the propeller by means of the new Costantinesco synchronizer, plus an upper wing mounted Lewis gun on a Foster mounting fired over the propeller, as on his Nieuport. This arrangement enabled a pilot to fire at an enemy from below, as well as providing two forward firing guns. The great advantage of the S.E.5a was its superior performance at high altitudes. It proved to be one of the finest aeroplanes of the Great War, strong, manoeuvrable and stable, and when Albert Ball's personal S.E.5a came out of its packing crate it had been extensively modified. Its windscreen had been lowered to improve performance, a larger fuel tank was optioned, and a second Lewis gun was fitted to fire down thru the floor of his cockpit.

CAPTAIN ALBERT BALL

When Albert Ball boarded a troopship, for his voyage across the English Channel and return to France he would have thought about his new command, and becoming involved in two distinct styles of air fighting. As before, there would be solitary sorties, plus the flight fighting tied to his promotion. Other airmen knew him as the quickest scout pilot in France, but this style of aerial fighting would not be applicable when leading his flight. His flight formations of five S.E.5a's were going to be difficult to hide against a visual background of earth, cloud or the sun, far more so than a solitary scout. Therefore the element of surprise, the essence of a successful attack would become far more difficult. Captain Ball would have to wait for his stragglers and novices who were still learning how to hold their position in a steady flight formation. Whenever he required his flight to attack from a diving turn or a wide curve of pursuit, he would have to throttle back his own scout, for his flankers to have enough forward thrust in hand to keep abreast. His five-flight team attacks would take far more time than the previous flat-out stooping dives of his previous lone scout forays. More time was going to be used up, meaning less surprise, and a flight leader always had remember to protect his men in the air and be morally responsible for getting them home. Often, exceptional scout pilots did not always make good leaders and some superb individuals such as Georges Guynemer, Billy Bishop and Albert Ball, simply did not possess the temperament to lead, because they had acquired their fame and flying expertise by fighting alone, and therefore never fully developed the qualities of leadership in the air.
The S.E.5a has turned out a dud... It's a great shame, for everybody expects such a lot from them.
It is a rotten machine.   Captain Albert Ball

After his voyage back to France, and his return to the sky over the Western Front, on the evening of 7th May 1917, Ball pursued Lothar von Richthofen, forcing him to the ground. Then, flying off solo fired some flares that attracted the attention of another flight commander, Captain Crowe who flew towards him. In fading light, and flying through the same cloud, when they emerged Crowe saw Ball's red-nosed S.E.5a chasing an opponent in an easterly direction, into the face of an enormous low cloudbank over Fresnoy. Crowe saw neither Ball nor the enemy he pursued into the cloud again, and eventually flew home. Sometime after, *Leutnant* Franz Haller on the ground witnessed Albert Ball's undamaged S.E.5a emerge from the clouds in an inverted position with a dead prop *leaving a cloud of black smoke caused by oil leaking into the cylinders* ~ the engine had to be inverted for this to happen. The Hispano engine was known to flood its inlet manifold with fuel when upside down and cut out. Haller hurried to the crash site, recording that the crashed aircraft had suffered no battle damage, and when he searched Ball's clothing for ID, no bullet wounds were found on Ball's corpse. Soon after, a German pilot dropped a message ; *We have buried Capt. Albert Ball with full military honours, near Lille.*

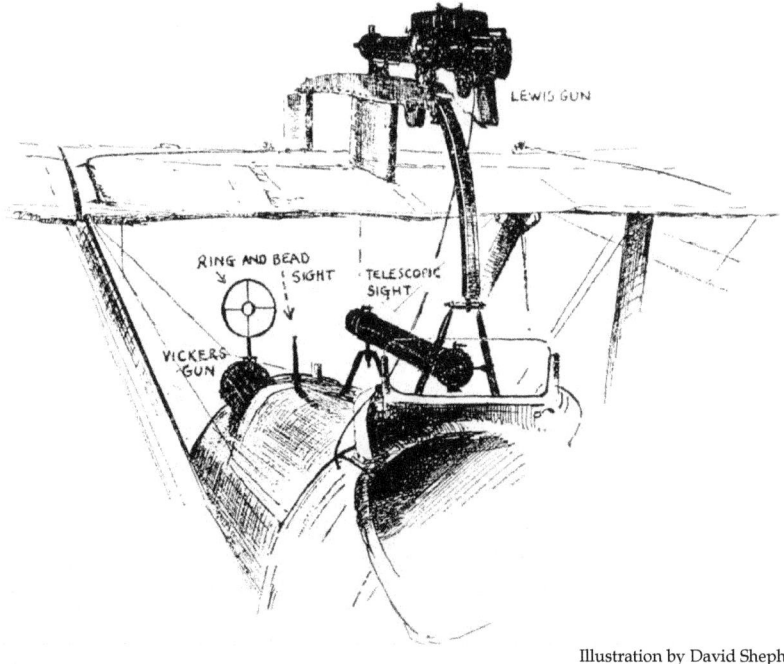

Illustration by David Shepherd
FULL CIRCLE 1964

ATTACKING TWO-SEATERS

The tactics for attacking enemy two-seaters during the Great War depended on whether they were alone. If so, pilots were instructed that the ideal point of attack was from below and to the rear, to shield them from the enemy observer's view. When flying in pairs, the wingman would approach from a flank and open fire at long range to distract the observer (top), whilst the leader (bottom) came up from behind and opened fire at 50 yards range. A successful surprise attack carried out in this manner enabled a fighter pilot to open fire at a decisive range before his enemy was even aware of his presence. The aim of the attacker was to get as close as possible and not to open fire until he was certain of a kill with his first burst of fire. A hostile two-seater when attacked from behind and below; if it survived would almost invariably turn with a view of bringing the observers gun to bear on the attacker. This manœuvre could be effectively countered by turning at first in the opposite direction and then, taking advantage of superior speed and handiness, to turn after the enemy and again come up under his tail. When attacking a two-seater from the rear, the aft gunner usually became a casualty first. When this occurred, the attacker closed in to try and kill the pilot or to complete the destruction of the enemy aircraft.

From: *Applied Flying 1922*

ATTACKING TWO-SEATERS

APPROACHING AN ENEMY TWO-SEATER

DEFENSIVE MANOUVRE OF A TWO-SEATER
From: *Applied Flying 1922*

When attacking alone, the best tactic was to dive on an opponent out of the sun. Although the target would be expected to bank and dive as soon as they were aware of being attacked, many aerial attacks were still decided in the first few seconds, *Beware of the Hun in the Sun* was always a valuable warning. The most skillful pilots flew on the very edge of clouds to render themselves less visible, while still being able to see below. Cloud cover enabled a pilot to swoop upon an opponent and then return to the convenience of near invisibility, but this did involve some risk of being ambushed by others taking advantage of clouds. To attack from the maximum altitude was also helpful, as actions tended to descend as they progressed, and often came down to almost ground level.

FLIGHT LEADER AIRCRAFT SIGNALS

Signaling in the air between the flight leader and the remainder of his formation consisted of hand signals or coloured signal cartridges and movements of the entire aeroplane. Hand signals were of little use in combat as they were easily missed. Coloured signal cartridges often used were: one red signal cartridge fired by the leader ~ *Close on me, prepare to attack!* ~ or in action ~ *Break off and reform.* Fired by any other pilot, red meant ~ *Help!* A green fired by the leader meant ~ *Forced to abandon flight, owing to engine failure or other cause* or a white ~ *Abandon flight, return to the aerodrome independently, owing to weather conditions.* Rocking the entire aircraft gently from side to side indicated that the leader was about to start a turn, and rocking fore and aft showed a dive or climb was imminent. Allied squadron leaders were easily recognizable in the air by streamers tied to their inter-plane struts while flight leaders had a streamer on their tail. Allied flight leaders tried individual markings such as Albert Ball's red-nosed prop spinner, but these markings found more favour among the Germans.

From: *Applied Flying 1922*

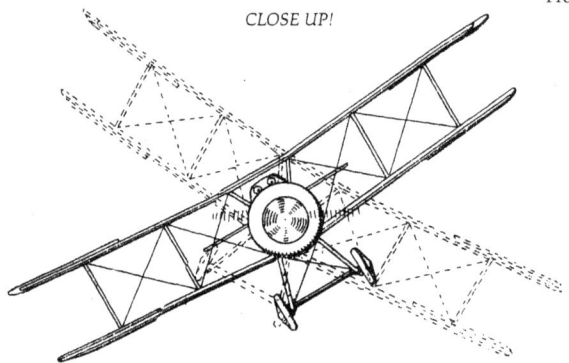

CLOSE UP!

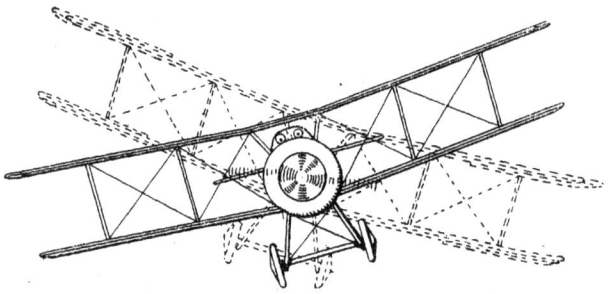

CHANGE DIRECTION!
Pilot banks twice in the required direction without turning. (Used by leader to warn his formation that he is about to turn).

ENEMY IS IN SIGHT!
Rocking the aeroplane laterally several times ~ could be made by any pilot as a warning to all.

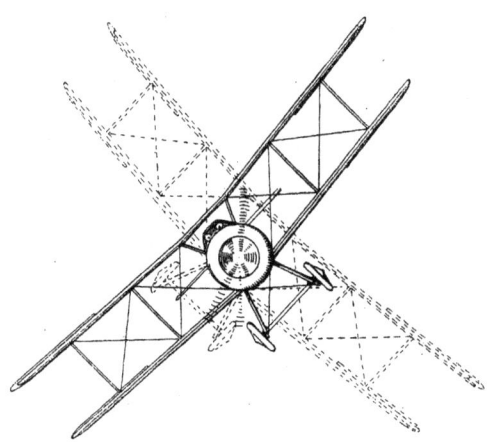

FLIGHT LEADER AIRCRAFT SIGNALS

OPEN OUT! *The flight leader rocks his aeroplane violently fore and aft. Used also as a preliminary signal to the order for ~ About turn!*

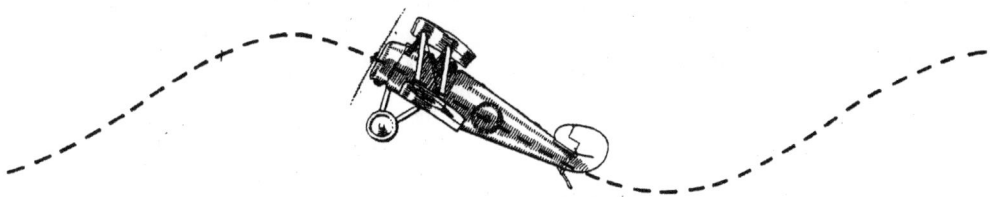

ABOUT TURN! *When the flight leader pulls the nose of his aeroplane up gently and then resumes level flight. (Used to warn his formation that he is about to half roll: the open out signal is to be given first, and then ~ About turn! From: Applied Flying 1922)*

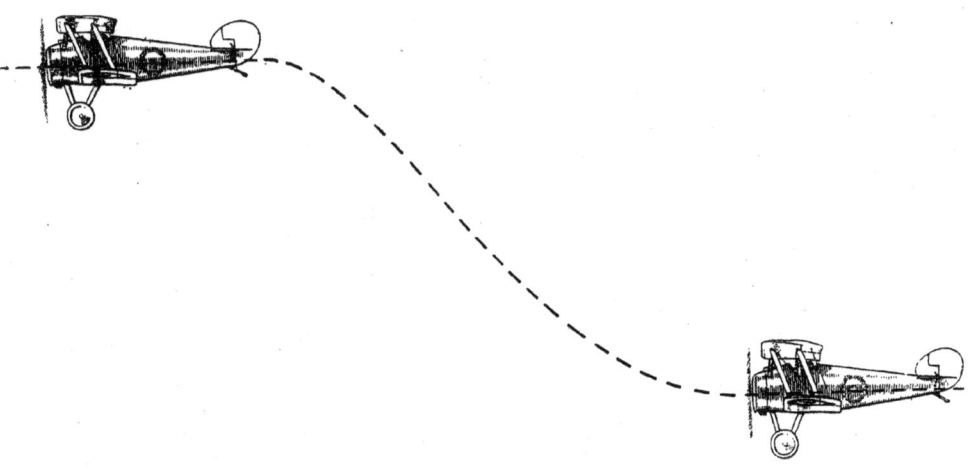

PREPARE TO DIVE! *The leader lowers the nose of his aircraft slowly, then flattens out again, and carries on at the same level.*

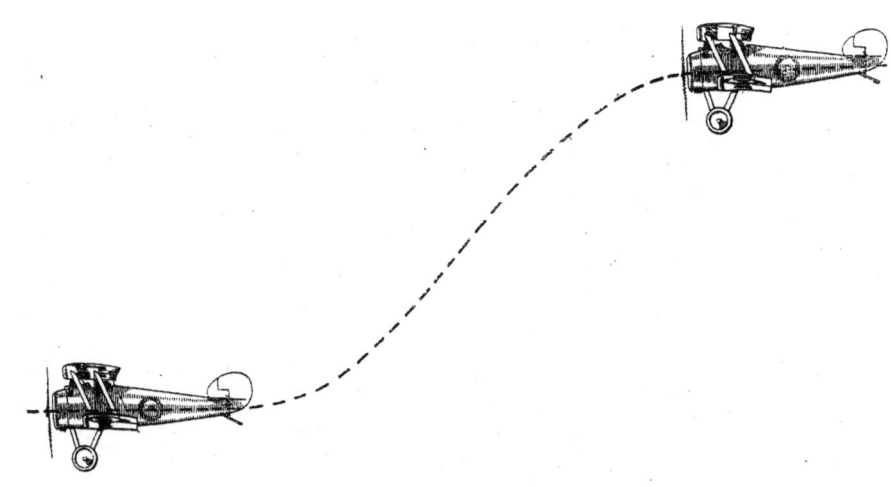

BLOODY APRIL ~ 1917

The winter of 1916-17 was so harsh that few sorties were flown. Before the Allies were ready to launch their next offensive, the Germans withdrew to their prepared defence in the Hindenburg Line during the last week of February and early April. This line helped German troops to hold out against the more copious Allies until their preparations for the Kaiser's decisive 1918 offensive were ready. Soon after, many mutinies broke out in the French Army that made the Allied situation precarious. Luckily the Germans were unaware of these mutinies and the delicate French frontline. In a whirl of aviation progress, the Germans introduced their twin-gunned Albatros D. II fighter. In the hands of pilots such as Werner Voss and Manfred von Richthofen the superb more streamlined Albatros was the best craft of its time, being a capable climber due to an up-rated engine, with twice the firepower of most Allied fighters. The D. II and the faster climbing D. III were so lethal, that during a particularly bad week of *Bloody April* in 1917, the RFC sustained casualties of near 30%.

The plywood construction of the Albatros was crucial to its success. The British and the French used primarily fuselages of longerons crossed with wooden and wire struts and ties covered with linen. When they used plywood, it was to cover flat or slightly curved areas. In crashes, a wrecked truss body became a crumpled mass of kindling wood, while the Albatros often held together well after crashes even when riddled with holes. On the Albatros, the plywood skin moulded and united as an integral part of the structure, added strength and stiffness to the airframe and minimized wind resistance. The 160-hp Mercedes engine thus drove the Albatros 20% faster than had been practicable with earlier forms of fuselage construction.   Richard Bickers

Although the German Albatros D.II scouts continued to dominate the sky over the Western Front well into the spring of 1917, from May the Allies began introducing new aircraft. These new machines included, the superb two-seater Bristol F.2a Fighter, the S.E.5a, the RNAS Sopwith Triplane, and the French SPAD S.13, which all gradually turned the tide of the air war in favour of the Allies for the remainder of the war, with only a few local upsets. Apart from the SPAD S.13, these aircraft all had only a single fixed machine gun on the fuselage, though the Bristol F.2a Fighter and the S.E.5a had an extra upper Lewis gun (flexible in the two-seater Bristol F.2a and wing-mounted in the S.E.5a). These were the precursors of future scouts, such as the Sopwith Camel which could carry twin Vickers machine guns once the new synchronizing gear which could be fitted to any type of engine had been invented and perfected by George Constantinescu. He further developed his single-gun synchronizer, with the assistance of Major C. Colley of the Royal Artillery into an improved twin-gun version of the synchronizer known as the CC gear. No. 55 Squadron RAF were first unit to be equipped with the twin-gun version of CC gear on their De Havilland DH.4's, (the best day bomber of the war that could also carry two 230-lb or four 112-lb bombs). No. 55 Squadron arrived in France in March of 1917.

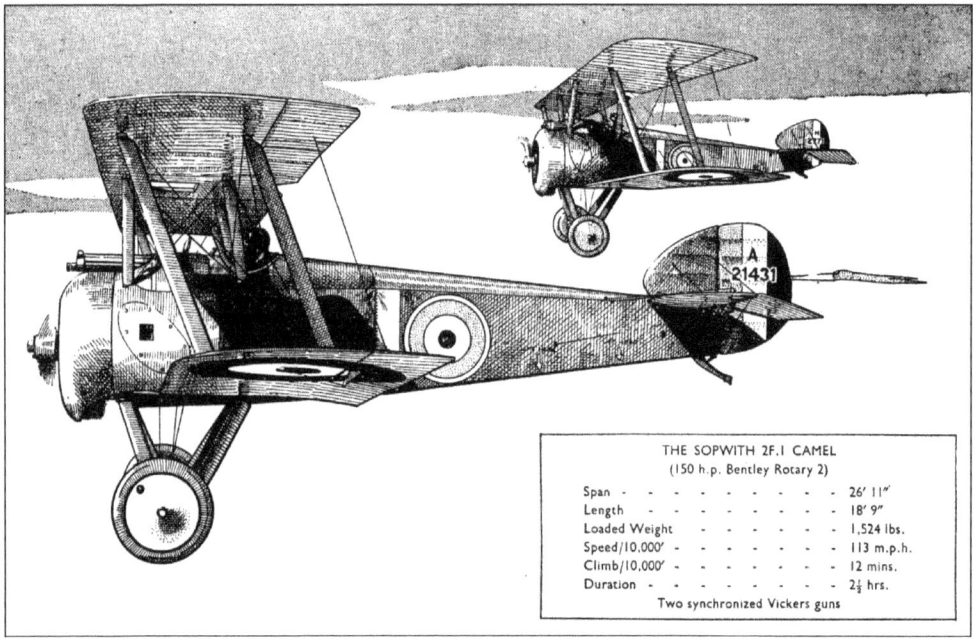

THE SOPWITH 2F.I CAMEL
(150 h.p. Bentley Rotary 2)

Span	26' 11"
Length	18' 9"
Loaded Weight	1,524 lbs.
Speed/10,000'	113 m.p.h.
Climb/10,000'	12 mins.
Duration	2¾ hrs.

Two synchronized Vickers guns

CAPTAIN BILLY BISHOP

William Bishop joined the 8th Canadian Mounted Rifles at the beginning of the Great War, and full of enthusiasm transferred to the Royal Flying Corps to qualify as a pilot in 1916. Assigned the tasks of offensive reconnaissance patrols from St Omer to the coast of the North Sea, as well as flying escort missions for Allied bomber aircraft or forward artillery observation sorties, Bishop recalled;

It was no child's play to circle above a German battery observing for half an hour or more, with the machine tossing about in the air, tortured by exploding shells and black shrapnel puffballs coming nearer and nearer to you like the ever-extended finger-tips of some giant hand of death. But it was just part of the never-ceasing war. Then, there is the fighting patrol work that goes on all hours. The patrol is not on our side of the line. It is far over the German lines to keep the enemy machines from coming too close, even to their own front trenches.

Flaming onions ~ Forbidden to fly below 1,000 feet, Bishop was directed to destroy an enemy observation balloon, however ~ as soon as he began his approach the Germans began pulling it down. Bishop dived and began firing from 550 feet and kept firing. The Germans returned fire from the ground with, *terrifying balls of fire shot from some kind of rocket gun*. Not quite. They were a form of incendiary and illuminating 37mm flare shells that seen from the air came blazing up in clusters of incandescent globules, to point out Allied aircraft to AA batteries, and sometimes set them on fire. Flying through the lethal arc of flaming onion cannon fire, Bishop's engine cut when he was down to 50 feet. Kicking into life again at 15 feet, Bishop flew off at full throttle over the heads of startled machine-gunners, through their sporadic fire, imaging his chances of survival better near the ground, (forgetting how Major Mick Mannock had fallen). Gifted with fine eyesight, mastery of his machine, and being a crack shot, Bishop downed 25 opponents in 12 days to receive the DFC. To celebrate, Bishop's mechanic painted the nose of his Nieuport XVII vivid blue. The Germans began to notice Bishop, calling him *Hell's Handmaiden,* and one particular *Jasta* even put a bounty on his head.

The great thing is never to let the enemy's machine get behind you ~ or on your tail. Once he reaches there it is very hard to get him off, as every turn and every move you make, he makes with you. By the same token it is exactly the position into which you wish to get, and once there you must constantly strive for a shot as well as look out for attacks from other machines that may be near. It is well if you are against odds never to stay long after one machine. If you concentrate on him for more than a fraction of a second, some other Hun has a chance to get a steady shot at you, without taking any risks himself.

I had found the one thing I loved above all others. To me it was not a business or a profession, but a wonderful game. I was flying over a layer of white clouds. When I saw a two-seater, I dived at him from out of the sun, firing as I came. My tactics gave the observer a direct shot at me from his swivel gun, and he was firing very well too, his bullets passing quite close for a moment or two. Then, they began to fly well beyond my wing tips, and on seeing this I knew that his nerve was shaken. I could now see my own bullets hitting the right part of the Hun machine and felt confident that the battle would soon be over. I pulled my machine out of its dive just in time to pass about five feet over the enemy. I could see the observer evidently had been hit and had stopped firing. Just after I passed I looked back over my shoulder and saw it burst into flames. A second later it fell a burning mass, leaving a long trail of smoke behind as it disappeared through the clouds. I thought for a moment of the fate of the wounded observer and the hooded pilot, into whose faces I had just been looking, but it was fair hunting and I flew away with great contentment in my heart. Captain William Bishop
No. 60 Squadron RFC

The Head-On Attack ~ called for very accurate timing. Two fighters, 200 yards apart, and darting at each other head-on with a combined speed of 180 mph were a mere two seconds away from collision. Few pilots had the determination to push their attack any closer. Georges Guynemer and Albert Ball, as the Allied masters of the head-on attack, fearlessly flew on knowing they would gain a free shot as their opponent broke away. Lieutenant Cecil Lewis RFC, (who later co-founded the BBC and would receive an Oscar during the 1938 Academy Awards ceremony for his part in the screen adaptation of George Bernard Shaw's play *Pygmalion*) described an aerial head-on aerial attack as…

Two machines approach head-on at breakneck speed, firing at each other, tracers whistling through each other's planes, each slipping sideways on his rudder to trick the other's gunfire. Who will hold the longest? Two hundred yards, 100, 50, and then, neither hit, with one accord they fling their machines sideways, bank and circle, each striving to bring his gun on the other's tail, each glaring through goggle eyes, calculating, straining, wheeling, bent only on death or dying.

BRITISH DIVE BOMBING FROM 1914-1918

The Louis Strange Bombsight ~ In 1914, Captain Louis Strange flying his B.E.2c toward the railway station of Courtai, was keen to test his bombsight, *consisting of a couple of nails and a few lengths of wire.* Beneath a 3,000-foot cloud base, through poor visibility and shot at by AA fire, Strange pointed the nose of his aircraft back into the safety of the clouds. After finding his way by compass he descended over Courtrai to rooftop level. A lone German sentry who began firing at his machine was seen and dispatched with a well-aimed hand grenade. Strange flew on, dropping his three 20-lb bombs onto an idle troop train. His bombsight had worked. His aircraft might have taken a peppering of 40 bullet holes, but Strange had incapacitated 75 troops and disrupted German rail traffic for at least three days. The bombsight of Louis Strange was widely adopted until a more scientific instrument was available.

The first reported dive bombing ~ British pilots flying over the Western Front at times attempted diving tactics on bombing raids and the earliest reported instance of this was that of Lieutenant Duncan Grinnel-Milne, No. 16 Squadron based near Lys in France during the winter of 1915. His Squadron's B.E.2c biplanes carried two 112-lb bombs under the wings and were trying out a new bombsight. Grinnel-Milne took off on his first bombing mission on 27th November 1915, toward a target defended by AA guns. When he arrived over the target area, he found a throng of aircraft jostling for position and being heavily fired on. Grinnel-Milne put his B.E.2c into a dive. With his engine on half throttle, he lined up his rail junction target ~ and released his bombs. He missed, hitting some nearby vehicles instead. Grinnel-Milne's first attack on the railway station had showed the advantage of dive-bombing over level bombing ~ a target could always be hit with more accuracy. He returned that afternoon, when the sky over the target was less crowded, and pressed in closer...
His first bomb hit a building. He then circled the target before diving to unleash his second.
I watched the bomb go down, diminishing to a pinpoint, then suddenly expanding again as it struck a building in the goods yard. A flash ~ bricks and dust and lots of slow spreading smoke... I flew home humming a tune.

Another example is that of Lt. William Brown of No. 84 Squadron RFC in France during March 1918. No. 84 Squadron were flying offensive patrols with S.E.5a fighters from Flez, near St Quentin and conducting experiments to find the most precise method of using aircraft for light bombing. Brown being the smallest pilot in his squadron, volunteered. No bombsight was used and the target was a 100-foot circle. Brown used wooden bombs, and found that in a diving attack he could hit the circle every time. On 14th March 1918, he was assigned a real target of four barges located on a canal near Bernot. For the mission described as ~ *Special Mission (Low Bombing Attack)* in No. 84 Squadron's combat reports, Brown's S.E.5a was fitted with four Cooper 20-lb HE bombs. He took off, climbed to 5,000 feet, and soon sighted the four barges near Bernot. Brown made two dives and missed twice. On his third attempt ~ *I dived straight for the barge,* his bombs hitting it amidships from a release height of 500 feet. Within a week his squadron was refitted, and the RFC added dive-bombing to its repertoire.

In response to continued daylight Gotha raids over London, the RNAS and RFC were combined to form the Royal Air Force on 1st April 1918 ~ and continued to conduct dive-bombing trials. The S.E.5a was the aircraft selected and armour plating was fitted to protect three test pilots from ground blasts along with an airspeed indicator to assist their targeting. Section Six carried out the air trials, under Captain C.E. Fairburn at Orfordness on the Suffolk shore, where a yellow flag planted in the shingle was the target (near the Orfordness Lighthouse that the Germans had tried to machine gun). Once in position over the target, the test pilots dived from 1,500 feet at 110 to 130 mph with their engines throttled back, and released their bombs from 800 to 1,000 feet. After a few preliminary dives to get the feel of it, the pilots looked along the left side of the engine cowling while they dived at the target, releasing their bombs the instant the target passed out of sight under the S.E.'s cowling. This method worked well and the tests resulted in 26% of the bombs falling within ten yards of the target. No sight of any kind was used and it was found that the pilot's aim improved quickly with practice. The three test pilots all concluded that it was far easier to drop the bombs in a vertical dive, that the standard machine gun sights indicated the target adequately if the aircraft was directed straight at the target, and that the lower the release, the more accurate the bombing.

The Strategic Air Offensive ~ Four squadrons of Handley Page long-range bombers formed the Independent Air Force in June of 1918, and their main objectives were the German chemical industry at Mannheim and Frankfurt; the iron and steel works in the Saar Basin; the machine shops of Westphalia along with the magneto works at Stuttgart; the submarine bases at Hamburg, Cuxhaven, Bremerhaven and Wilhelmshaven; and the accumulator factories within Berlin itself. Two squadrons of Sopwith Camels were incorporated for their protection, and the British air fleet came into being.

BOMB DROPPING APPARATUS OF GERMAN AVIATORS

HOW THE ROLAND BOMB-DROPPING APPARATUS IS WORKED

THE ROLAND BOMB-DROPPING APPARATUS CONSISTS OF THREE PARTS ~ THE BOMB TUBES, THE TUBE COVERS, AND THE PEDAL BOARD. THE BOMB TUBES ARE ARRANGED IN A PLATE LET INTO THE FLOOR OF THE FUSELAGE, A LITTLE BEHIND THE PILOT'S SEAT. THE PEDAL BOARD IS OPERATED BY THE FOOT OF THE OBSERVER. THE EXPLOSIVE CHARGE IS PLACED IN A STREAMLINE CASING ~ IN THE NECK OF THIS CASING IS A CIRCUMFERENTIAL GROOVE, AND WHEN THE BOMB HAS BEEN PULLED UP INTO THE TUBE IT IS ARRANGED THAT A FORK ENGAGES THIS GROOVE IN THE NECK, ALLOWING THE BOMB TO DROP. IN ORDER TO OBVIATE MISTAKES, THE PEDALS AND THE TUBE COVERS ARE PAINTED IN CORRESPONDING COLOURS.

BARON MANFRED VON RICHTHOFEN

Following Baron Manfred von Richthofen's transfer from the Imperial 1st Uhlan Regiment to the Imperial Air Service, he developed into a superb tactician who scorned aerobatics. After achieving 25 victories and for repeated heroism in the air, Kaiser Wilhelm II personally awarded him the coveted *Pour le Mérite*. Then appointed *Staffelfuhrer* and promoted to *Rittmeister* (Riding Master as he remained an officer of his Uhlan regiment), he expanded his formation by consolidating four *Staffeln* into *Jasta II* to create his *Flying Circus* who living in tents, could be moved overnight to other sectors to fight for aerial supremacy wherever required. Baron von Richthofen's formation was redesignated *Jasta I* during July 1917, and he saluted his devoted command, *for diminishing the number of our enemies.*

Inspired by Jean Navarre, who flew an all-red Nieuport XVII over Verdun, the *Rittmeister* had his own Albatros fighter painted blood red to taunt his opponents and become more visible as the leader. If his airmen could instantly spot his distinct machine, they could always follow his lead. Baron von Richthofen achieving most of his 83 kills in his red Albatros became known as the Red Baron, and the personal colour of von Richthofen became the colour of *Jasta I*. All highlighted in red; a tail, wing or stripe around the fuselage with a multi-hued display of other decorative colours with the acquired advantage of making it easier for ground and air observers to confirm their victories. Other German aviators were quick to adopt this practice and decorated their mounts reminiscent of knights of the air. Yellow became the colour of *Jasta 2*. Black and white checkerboards for *Jasta 3* and *Jasta 6* chose black and white stripes. Billy Bishop scoffed at them, calling such opponents, *the harlequins of the air*. From August, von Richthofen flew a Fokker Dr.1 Triplane, an aircraft that the Germans were inspired to create after examining two triplane fighters designed by Herbert Smith for the Sopwith Company. The first two Sopwith *Tripes* flown to France, alighted by some mischance behind the German lines. In this way, German engineer Reinhold Platz was provided with a new design to examine and reconfigure, but Reinhold Platz's Fokker Dr.1 (Oberursel 110-hp rotary) was at least 15 kph slower than the Sopwith Triplane (Clerget 110-hp), and would shed its wings if pushed too hard ~ Richthofen wrote;

Flying my Triplane for the first time, I attacked with four of my gentlemen a very boldly flown artillery-reconnaissance aircraft. I approached and fired 20 shots from a distance of 50 metres, where my adversary fell out of control and crashed this side near Zonnebeke. It climbed like a monkey and manœuvred like the devil.

Richthofen's preferred tactic was to direct *Jasta I* to attack whenever targets appeared and then fly above the melee to pounce on unwary stragglers. He preferred to kill with one straight pass from high out of the sun and when he did strike, Allied pilots rarely stood a chance. *Everything beneath me is lost*, he wrote and his boast was not empty. Whenever he found himself outnumbered, his tactic was to hang back over the German lines, as he put it ~ *let the customer come to the shop*. If German aircraft were forced down, it was over their territory with survivors rescued and their fighters salvaged, but Allied pilots were out of the war. Another German ace, Ernst Udet who fought from 1914 to 1918 to become an advocate of the parachute or *umbrella* as he called it, following the two occasions when it saved his life after he leapt out of tattered aircraft ~ described how it felt to fly with Baron von Richthofen.

Flying over the tops of the trees were several Sopwith Camels. These British single-seaters had the task of protecting the Roman Road, which was one of the main arteries of their communications system. I had no time for further observation for Richthofen set the nose of his red Fokker towards the ground and dived with the rest of us following close on his tail. The Sopwiths scattered like chickens from a hawk, but one of them was too late, the one in the Rittmeister's sights. It happened so quickly that one could hardly call it an aerial fight. For a moment I thought that the Rittmeister would ram him, so short was the space that separated them. I estimated it at ten metres at the most. Suddenly the nose of the Sopwith tilted downwards and a cloud of white smoke shot from the exhaust. The ill-fated machine crashed into a field close to the road and burst into flames. Richthofen, instead of changing direction as we expected, continued to dive until he was close above the Roman Road. Tearing along at a height of about ten metres from the ground, he peppered the marching troops with his two guns. We followed close behind him and copied his example. The troops below us seemed to have been lamed with horror and apart from the few men, who took cover in the ditch by the roadside, hardly anyone returned our fire. On reaching the end of the road the Rittmeister turned and again fired at the column. We could now observe the effect of our first assault; bolting horses and stranded guns blocked the road, bringing the column to a complete standstill.

Baron Manfred von Richthofen surrounded by four of his fellow aces of JASTA 11 airmen at Roucourt, France in 1918
Left to Right; *Vizefeldwebel* Sebastian Fesntner, *Leutnant* Karl-Emil Schaeffer, *Oberleutnant* Manfred von Richthofen, *Leutnant* Lothar von Richthofen, and *Leutnant* Kurt Wolff. The dog is Moritz ~ Richthofen's Danish hound.

Baron von Richthofen landing his Fokker Triplane

BARON MANFRED VON RICHTHOFEN AND JASTA 11

Baron Manfred von Richthofen in cockpit & Baron Lothar von Richthofen seated ~ September 1917

April 29th 1917. An all-red Fokker D.III with a mechanic's ladder against the cowling as backdrop for a group of Jasta 11 pilots at Roucourt. The Red Baron poses on a bicycle. L to R : Wolfgang Pluschow, Manfred von Richthofen with his brother Lothar von Richthofen just behind. An aircraft in the distant right appears to have crash-landed, with its swept-back tail in the air and lower wing resting on the ground. Far right : Otto Brauneck, and Baron Albrecht von Richthofen visiting his sons.

WERNER VOSS

Signed photograph of Werner Voss in his new Fokker Triplane

Leutnant Werner Voss with his Fokker F.I triplane, 103/17.

Voss decorating his new Albatros with a heart and swastika

THE DEATH OF BARON VON RICHTHOFEN

Near the French village of Villers-Bretonneux on 21st April 1918, von Richthofen leading his Fokker triplanes against a squadron of Canadian pilots became involved in a fight with two Sopwith Camels over the Australian lines. When Lieutenant Wilfred May's guns jammed, he abandoned the fight and headed for his Bertangles airfield at low-level. Richthofen quick to spot an easy kill chased him fearlessly close to the ground. May zigzagged and von Richthofen fired at every opportunity. The Red Baron now found himself tailed by Captain Arthur Brown and eyewitnesses on the ground saw Brown break off after firing a long burst at the Fokker. Brown thought he saw his tracer hit.

I was in the perfect position, above and behind. It was just a matter of straight shooting. I dived until the red snout of my camel pointed at his tail. My thumbs pressed the triggers. Bullets ripped into his elevator and tail planes. The flaming tracers showed me where they hit: a little short. Very gently I pulled on the stick. The nose of the Camel rose ever so slightly, easy now, easy! The stream of bullets tore along the body of the all-red tripe. Its occupant turned and looked back. I had a flash of his eyes behind the goggles. Then he crumpled, sagged in the cockpit. My bullets poured out beyond him; my thumbs eased on the triggers: the triplane staggered, wobbled, stalled, flung on its nose and then went down ~ von Richthofen was dead.   Captain Arthur Brown

The variance with this account, is that it does not concur with other reports detailing that von Richthofen flew on for more than a mile after Brown's machine gun fired, before turning sharply. It then passed over the Australian 53rd Battery position, where von Richthofen's Fokker triplane was also shot at from a Vickers machine gun operated by Australian sergeant Cedric Popkin, plus Lewis AA gunners Snowy Evans and Robert Buie. In 1957 Robert Buie wrote a letter to the *Central Coast Express*,

I am herewith enclosing the facts and truth of the shooting down of the Red Baron. I, myself, am the man responsible for his destruction. I was manning one gun and Snowy Evans the other. As the planes neared us, Evans opened fire, but the plane came on. I could not fire at the same time as I did not have clearance, but as soon as our plane was out of the line of fire, I started firing directly at the German pilot. Fragments came off the plane and it lessened speed. It came down a few hundred yards away. When the plane was reached, von Richthofen was dead. There were quite a few who tried to claim von Richthofen's downing. All the evidence was sent to British Army headquarters in France and a month later, while I was still in the line a dispatch came from General Rawlinson to the 53rd Battery and to me, giving me the credit for shooting down the German ace. I have the proof in my possession and I cannot see why the controversy goes on.

REPORT ON THE DEATH OF CAPTAIN BARON VON RICHTHOFEN: 1100 hrs April 21st 1918.
The following report is based on the evidence of eyewitnesses, written down after the events.
Capt. Baron von Richthofen was flying a single-seater triplane painted red and reported to be of a new pattern. When first engaged he was pursuing one of our own machines, reported to be a Sopwith Camel, in a W.N.W. direction. Both machines were flying very low, being not more than 150 feet up. They were coming swiftly towards the AA guns of the 53rd Battery. Richthofen was firing into the plane before him but it was difficult for the Lewis gunners to shoot owing to the British plane being directly in the line of fire. They accordingly waited their time until the British plane had passed. Richthofen's plane was not more than 100 yards from each when they opened fire. Almost immediately the plane turned N.E. being still under fire from the Lewis guns. It was now staggering as though out of control. The plane veered to the North and crashed on the plateau near the brickworks. The aviator was dead. The plane was badly smashed; it was a triplane painted dull red and was armed with two air-cooled machine guns. After the machine crashed, a troupe of German pilots flew over and circled above the spot until driven off by AA guns. After very careful consideration and the weighing up of all the evidence, it was proved beyond doubt that the Lewis gunners who brought down the plane were No. 598 Gunner W.J. Evans and No. 3801 Gunner Robert Buie of the 53rd Battery, 5th Australian Divisional Artillery.

Richthofen's blood-red Fokker triplane had side-slipped, hit the earth in a beet field, bounced breaking its undercarriage, slewing around as it did beside the Bray-Corbie road. After this bumpy, but perhaps controlled landing, von Richthofen, the top-scoring ace of the Great War was dead. Australian soldiers found him still strapped to his seat, his flying goggles broken with blood trickling from his mouth. The Red Baron had died from a single bullet that according to the trajectory of his wounds, had been fired from below, had pierced his side through to the spine, then deflected forward out of his chest into the folds of his uniform. When von Richthofen was shot down, he was wearing an appropriated British SIDCOT flying suit, designed by Australian pilot Sidney Cotton that comprised of a fur-lined suit with an outer skin of waterproof fabric and an insulating layer of airproof silk.
Richthofen was a brave soldier, a clean fighter and an aristocrat. May he rest in peace.   ~ The Aeroplane

POSITION OF THE STICK RUNNING THE ENGINE ON THE GROUND

DO NOT FORGET TO HOLD THE STICK WELL BACK AND SO PREVENT THE TAIL FROM RISING. IF THE STICK WERE HELD FORWARD THE LIFT OF THE ELEVATORS AND THE PULL OF THE PROPELLER WOULD COMBINE TO OVERTURN THE MACHINE.

GETTING OFF ~ THE LAST LOOK AROUND

ALWAYS LOOK BEHIND ON EITHER SIDE AND IN FRONT BEFORE OPENING OUT YOUR ENGINE TO TAKE OFF. THERE MAY BE ANOTHER MACHINE ABOUT TO LAND IN YOUR WAY. (De Havilland D.H.9)

R.E.8 WITH ENGINE FAILURE AFTER TAKEOFF AND TURNING BACK

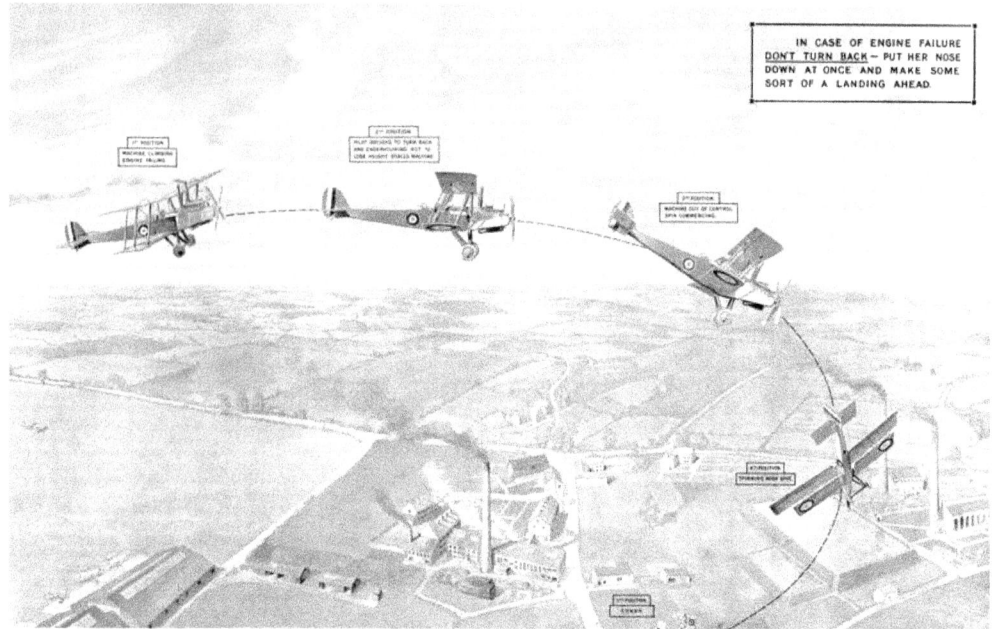

IN CASE OF ENGINE FAILURE <u>DO NOT TURN BACK</u> ~ PUT HER NOSE DOWN AT ONCE AND MAKE LANDING AHEAD

R.E.8 TRYING TO AVOID COLLISION WITH A TURNING BRISTOL

R.E.8 BIPLANE TRYING TO AVOID COLLISION AFTER TURNING INTO THE PATH OF A BRISTOL FIGHTER
DO NOT TURN SUDDENLY WITHOUT FIRST LOOKING AROUND
YOU MAY FOUL ANOTHER MACHINE TAKING OFF AFTER YOU

DE HAVILLAND DH.4 MAKING A BAD APPROACH AND LANDING

IN THE CENTRAL POSITIONS THE INEXPERIENCED BOMBER PILOT MAY MOMENTARILY LOSE HIS HEAD.
UNDER THESE CIRCUMSTANCES HE WILL DO WELL TO GET AWAY AGAIN AND TO HAVE ANOTHER TRY.

AVRO LANDING ACROSS WIND SHOWING CONSEQUENT RESULT

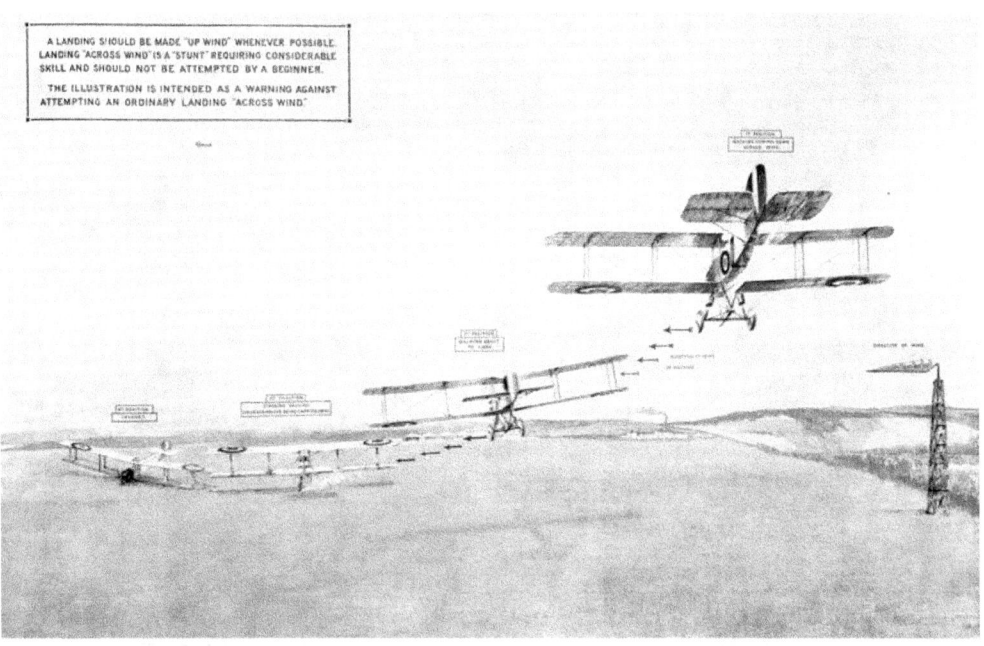

A LANDING SHOULD BE MADE "UP WIND" WHENEVER POSSIBLE.
LANDING "ACROSS WIND" IS A STUNT REQUIRING SKILL AND SHOULD NOT BE ATTEMPTED BY A BEGINNER.
THE ILLUSTRATION IS INTENDED AS A WARNING AGAINST ATTEMPTING AN ORDINARY LANDING "ACROSS WIND".

OUTMANOEUVERED ~ COMING UNDER ENEMY AIRCRAFT TAIL

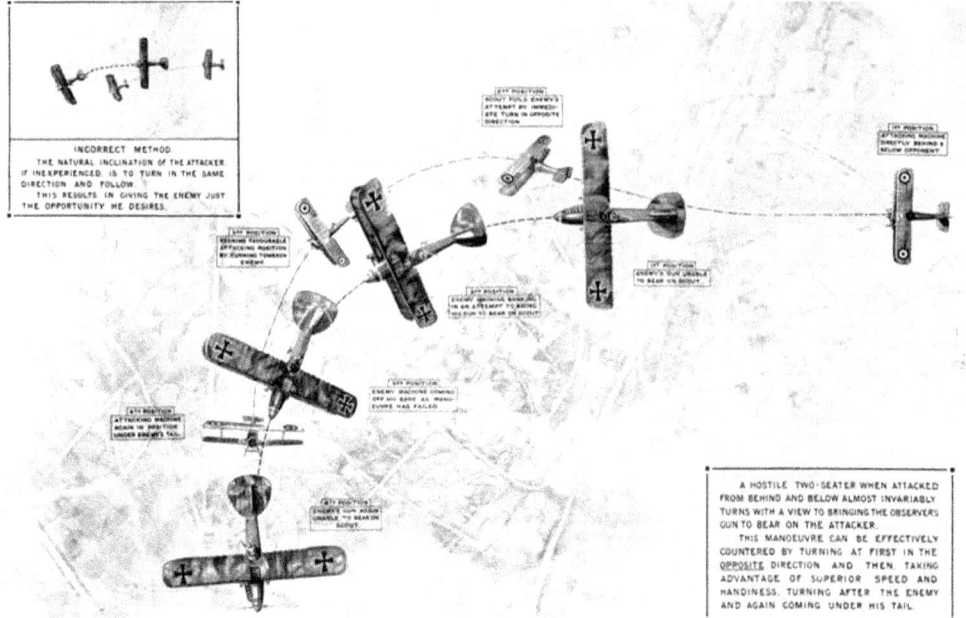

INCORRECT METHOD: THE NATURAL INCLINATION OF THE ATTACKER, IF INEXPERIENCED, IS TO TURN IN THE SAME DIRECTION AND FOLLOW. THIS RESULTS IN GIVING THE ENEMY JUST THE OPPORTUNITY HE DESIRES.
A HOSTILE TWO-SEATER WHENEVER ATTACKED FROM BEHIND AND BELOW ALMOST INVARIABLY TURNS WITH A VIEW TO BRINGING THE OBSERVERS GUN TO BEAR ON THEIR ATTACKER. THIS MANOEUVRE CAN BE EFFECTIVELY COUNTERED BY TURNING AT FIRST IN THE <u>OPPOSITE</u> DIRECTION AND THEN, TAKING ADVANTAGE OF SUPERIOR SPEED AND HANDINESS, TURNING AFTER THE ENEMY AND AGAIN COMING UNDER HIS TAIL.

LOW FLYING ~ AVOIDING BARRAGE

A SOPWITH SALAMANDER FLYING LOW OVER THE ENEMY LINES.
DON'T FLY THROUGH OUR OWN OR THE ENEMY'S BARRAGE UNLESS ABSOLUTELY NECESSARY
EVEN THOUGH IT MAY NECESSITATE FLYING LOW FOR A LONGER TIME OVER HOSTILE COUNTRY
CENTRE : RAF CRAFT RETURNING FROM A LOW LEVEL PATROL IN COMPARATIVE SAFETY BY AVOIDING BARRAGE
RIGHT: RAF CRAFT UNECCESSARILY RETURNING THRU BARRAGE THEREBY RUNNING A GREAT RISK OF BEING HIT

GOOD AND BAD LOOPING

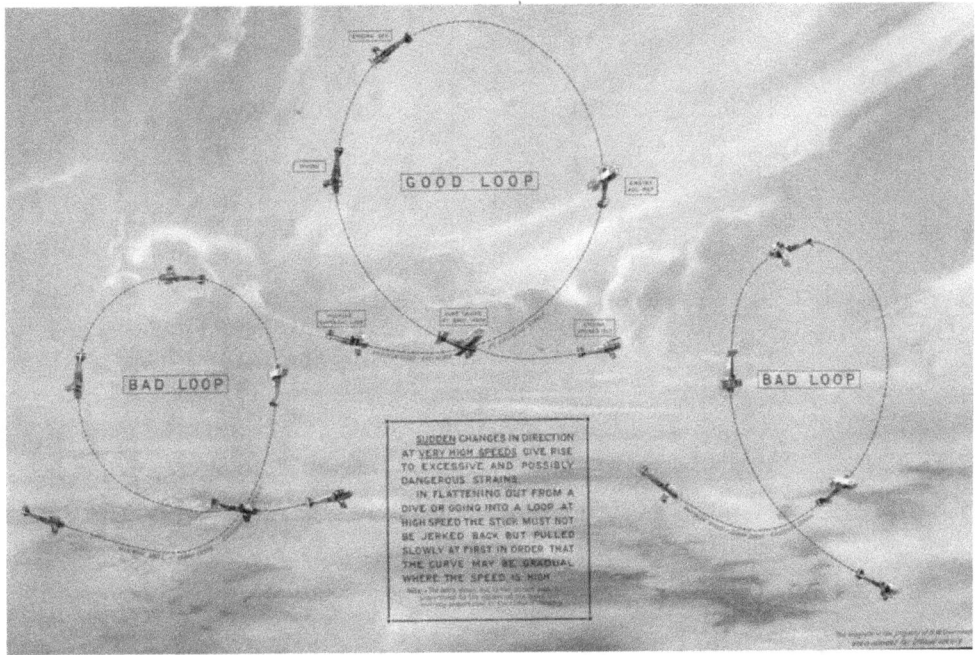

SUDDEN CHANGES IN DIRECTION AT HIGH SPEEDS GIVE RISE TO EXCESSIVE AND POSSIBLY DANGEROUS STRAINS.
IN FLATTENING OUT FROM A DIVE OR GOING INTO A LOOP AT HIGH SPEED THE STICK MUST NOT BE JERKED BACK
BUT PULLED BACK SLOWLY AT FIRST IN ORDER THAT THE CURVE MAY BE GRADUAL WHERE THE SPEED IS HIGH

RFC R.E.8 BIPLANE LANDING DOWNWIND AND OVERSHOOTING

REMEMBER ~ UNTIL YOU ACTUALLY TOUCH THE GROUND ~ YOU ARE CONCERNED WITH AIR SPEED ONLY.
THE CORRESPONDING GROUND SPEED WILL BE GREATER THAN THE AIR SPEED IN A DOWN WIND LANDING.
(IT WILL BE EQUAL TO THE AIR SPEED AT THE MOMENT OF LANDING PLUS THE WIND VELOCITY).
AS A DOWN WIND LANDING CANNOT BE MADE AT A LOW SPEED,
A LONG RUN IS REQUIRED TO BRING THE MACHINE TO A STANDSTILL

Wright Flying Boat

Sopwith 1 ½ Strutter on HMAS Australia

Friedrichshafen FF-49

CHAPTER IV

NAVAL AVIATION 1914-1918

An airplane-carrying vessel is indispensable. These vessels will be constructed on a plan very different from what is currently used. First of all the deck will be cleared of all obstacles. It will be flat, as wide as possible without jeopardizing the nautical lines of the hull, and it will look like a landing field. Clément Ader

The way of taking aircraft to the seas, was first proposed by the French aviator Clément Ader writing in *L'Aviation Militaire* in 1907, suggesting that a flush deck aircraft carrier could store its aircraft below decks, be able to raise them to the flight deck on an elevator, then steam into the wind to launch and recover them. Adler also foresaw carrier planes with folding wings, and recommended that the fuselage behind the pilot's position be left open for a quick exit, *in case of a fall into the water*. Naval aviation during the Great War, witnessed the provision for aircraft taking off from converted carriers, and seaplanes being mounted on capital warships to scout for the enemy. It was considered a very risky affair for a pilot side slipping onto a forward deck in a seaway. Eddies from the mast, funnels, and superstructure made it hazardous to land on a carrier's deck, and until the end of the Great War deck landing was not a naval operation. Flying off was more successful with single and two-seaters flown from the revolving gun turrets of warships ~ catapult was also successfully used.

Early on the first morning of the Great War, German seaplanes flew out over the waters of the Heligoland Bight with orders to, *reconnoiter seaward to the limit of their capabilities*. On the North Sea, where six days out of ten the sea is fairly turbulent, and one day out of five is much worse, the condition of the sea often prevented German seaplane patrols. However the Germans soon realized that the British had begun a blockade, closing the English Channel to their High Seas Fleet by basing the Grand Fleet of the Royal Navy at Scapa Flow in the Orkneys to guard the northern approach to the Channel. In August of 1914, Britain had the greatest navy the world had ever seen and their aerial requirement was long-range reconnaissance. The simplest way of looking over the horizon up until this time had been through the use a balloon tethered to a warship. If the lookout up in the crow's nest of a cruiser could spot a hull down enemy vessel a dozen miles away, experiments with tethered balloons let out to a height of 3,000 feet had seen the observer's horizon retreat to 50 miles.

Britain's naval airmen flying armed seaplanes from coastal stations routinely flew over the North Sea and English Channel on anti-submarine patrols. An aerial observer could often sight an enemy submarine or U-boat from five miles or more depending on atmospheric conditions, the amount of daylight, and the condition of the surface of the sea. Even if an aerial observer did not spot an enemy U-boat, once a lookout spotted the aircraft, an astute U-boat captain would order a dive anyway, assuming they had, and a 1914 U-boat could submerge in under two minutes. To prevent Allied patrol aircraft from targeting their submarines by mistake, British submariners adopted the tactic of marking their foredecks with distinctive white markings whose design was altered every two weeks.

The main problem in the development of the torpedo-armed seaplane was the weight of the torpedo : half a ton or more. But as soon as a British designed, Short Brothers seaplane air launched a 14-inch Whitehead torpedo in 1914, the Royal Naval Air Service ordered several Short 184 seaplanes fitted with the Renault 240-hp engine, and subsequent Short Folder Seaplane versions with folding wings could lessened in length from 56 feet to a mere 12 feet. Because of their historically happy relations with British industry, the Royal Navy often procured much better machines and better engines than the RFC. The Sopwith Pup (80-hp Le Rhone), and Sopwith Triplane (130-hp Clerget), were very good fighting aircraft. Both of these aircraft were a delight to fly and very popular with the pilots. The Pup had a speed of 106 mph and a ceiling of 17,000 feet, while the slightly faster Triplane could reach 20,000 feet. Five RNAS squadrons flew it, plus the all Canadian No. 10 Naval Squadron, known as the *Black Flight* commanded by the ace Raymond Collishaw. The Sopwith Triplanes of the *Black Flight* were recognizable by their black fins and cowlings and evocative names of *Black Death, Black George, Black Sheep, Black Roger,* and the *Black Prince*. The *Black Flight* claimed 87 German aircraft in three months of flying their Sopwith Triplanes. Collishaw flying *Black Maria* shot down 34 of them.

RECONAISANCE FLIGHT OVER THE BATTLE OF JUTLAND

If vessels of the German High Seas fleet steamed out of their moorings of Wilhelmshaven, they had to be found and tracked ; thus the Royal Navy's paramount. need of reconnaissance reports of the foe. The answer again, was for an aircraft carrier to carry seaplanes, with a speed of at least 21 knots to maintain station with the Grand Fleet of Admiral Jellicoe, and 24 knots for the Battle-Cruiser Fleet of Admiral Beatty. The conversion of three cross-channel steamers the, *Empress, Engadine,* and *Riviera,* into seaplane carriers was completed in September 1914, with retrofitted protective shelters for three seaplanes and the installation of handling booms to winch the Short seaplanes off their carriers, and onto the sea to takeoff and fulfill their reconnaissance role for the Royal Navy. Early on morning of 31st May 1916, elements of the German High Seas Fleet left Wilhelmshaven for the Norwegian coast.

Preliminary to the Battle of Jutland in the North Sea, the Admiralty sent a message to Admiral Jellicoe and to Vice-Admiral Beatty, ordering then to concentrate their forces, ready for eventualities eastward of the *Long Forties,* about 60 miles east of the Scottish Coast. As the warships of the Royal Navy steamed toward the prowling German High Seas Fleet, just one British aircraft would participate. After receiving a report from the light-cruiser HMS *Galatea* ~ *Enemy sighted,* confirming the presence of the German High Seas Fleet, Admiral Beatty on HMS *Lion* ordered the seaplane-carrier HMS *Engadine* steaming with his Battle-Cruiser Fleet in advance of the Grand Fleet ~ *Investigate with aircraft N.N.E.* where enemy warships had been reported. On the aft deck of the *Engadine* a Short seaplane was ready to go. Flight-Lieutenant F.J. Rutland clad in his flying gear sitting in his cockpit, after listening to his instructions waved the crew away, and warmed up his engine. The chain was hooked on to hoist him off and away onto the surface of the North Sea. Rutland wrote this report;

Report of Reconnaissance Flight   H.M.S. ENGADINE
                                          31st May 1916

Sir, ~ I have the honour to make the following report: - At 2:40 pm (GMT), in accordance with signal and your orders, Seaplane No. 8359 was got out and proceeded to scout for enemy ships. I was hoisted out at 3:07 pm and was off the water at 3:08 pm (Times were taken on board). The last information that I received from the ship was that the enemy was sighted in a N.N.E. direction, steering north. I steered N.10 and after about ten minutes sighted the enemy. Clouds were at 1,000 to 1,200 feet. This necessitated flying very low. On sighting the enemy it was very hard to tell what they were, and so I had to close within a mile and a half at a height of 1,000 feet. They then opened fire on me with anti-aircraft and other guns, my height enabling them to use their anti-torpedo armament. When sighted they were steering a northerly course. I flew through several of the columns of smoke caused through bursting shrapnel. When the Observer had counted and got the disposition of the enemy and was making his W/T report, I steered to about three miles, keeping the enemy well in sight. While the Observer was sending one message, the enemy turned 16 points. I drew his attention to this, and he forthwith transmitted it. The enemy then ceased firing at me. I kept on a bearing on the bows, about three miles distant of the enemy, and as the weather cleared a little I observed the disposition of our fleet, and judged by the course of our Battle Cruisers that our W/T had got through. At 3:45 pm a petrol pipe leading to the left carburetor broke, and my engine revolutions dropped from 1,000 to 800, so I was forced to descend. On landing I made good the defect with rubber tube, and reported to the ship that I could go on again. I was told to come alongside and be hoisted in. I was hoisted in at 4:00 pm. The visibility was about four miles varying to one, and this reduced the advantage of the seaplane's height. Also, the Seaplane having to remain close to the enemy increased the chances of them jamming the wireless. The messages, as sent were received in H.M.S. Engadine, …but it was not known if the messages had been received until our Fleet was sighted and their course observed. I could not keep both our Fleet and the Enemy's Fleet in sight, through low-flying clouds. The speed at which things took place prevented any receiving, the Observer being busy coding and sending all the time. The enemy commenced to jam latterly. The enemy's anti-aircraft firing was fairly good; the shock of exploding shrapnel could be felt, the explosions taking place about 200 feet away on one side, in front and astern.   Flight-Lieutenant F.J. Rutland

The German naval commander, Admiral Reinhard von Scheer received reports from all three patrolling Zeppelins that there were no large enemy naval forces on his left flank at 1630 hours as the two fleets began to fight, enabling him to enter the indecisive battle with increased confidence. During the night, the German High Seas Fleet withdrew to a rendezvous off Hornsriff, a shoal off the southern border of Denmark. Early in the morning of 1st June, Zeppelin *L 11* located the Grand Fleet, reported their position, and through worsening visibility of fog and mist, continued to shadow the British Fleet, assisting the German warships of the High Seas Fleet to return to Wilhelmshaven.

* * *

FLYING OVER AN ENEMY U-BOAT

NAVAL AIRCRAFT LIGHTER THAN AIR

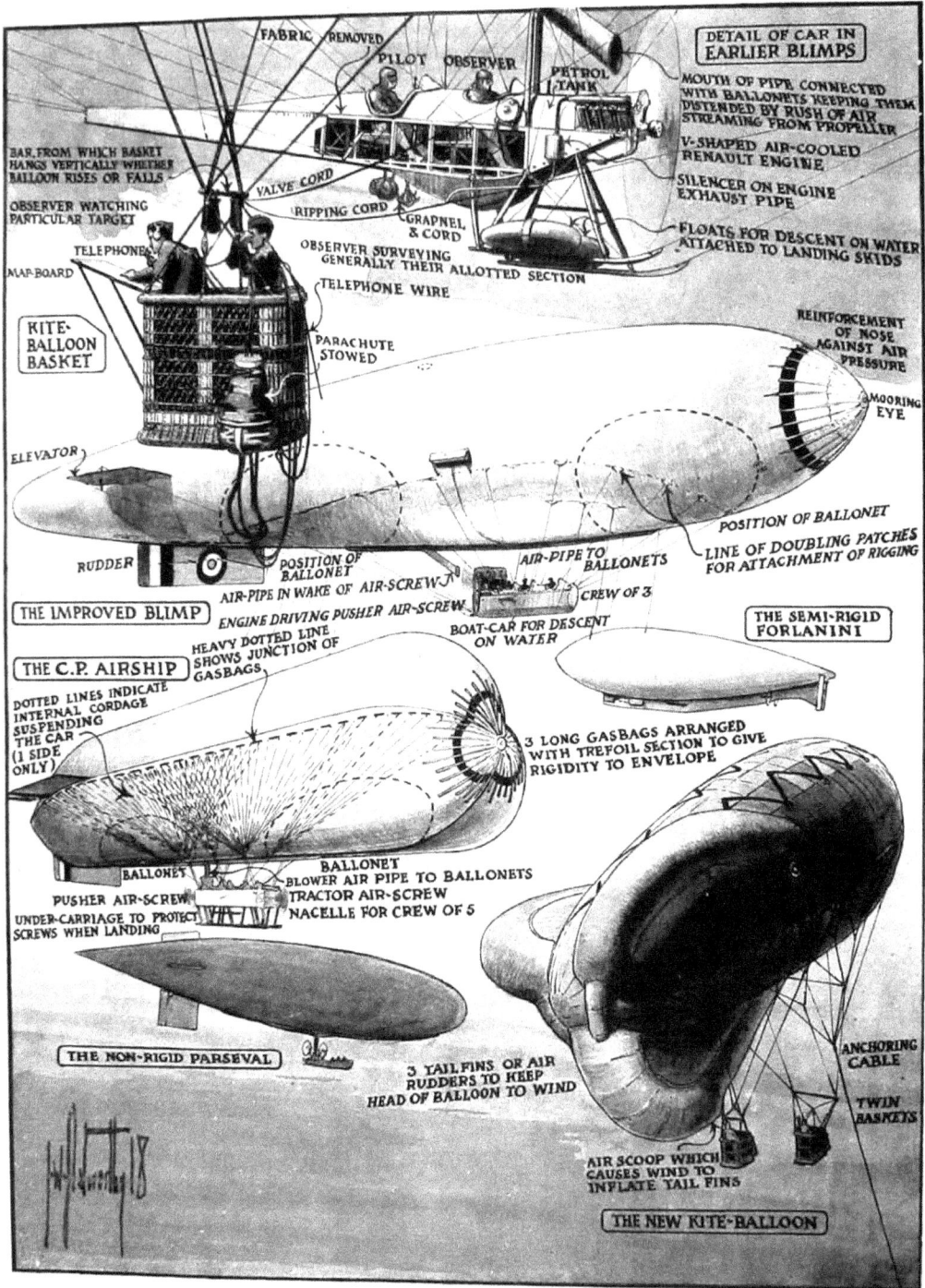

In England, all war airships are attached to the navy. Even kite-balloons have only recently been adopted by the R.F.C., kite-balloon sections having been lent for a long time by the R.N.A.S. to the R.F.C. The French Caquot balloon, however has lately been improved by the latter. The blimps, officially styled S.S. (Submarine Scouts), which have proved invaluable assets to the Allies in the U-boat war, were originally provided with discarded B.E. cars to speed up their output. The C.P. (Coastal Patrol), a larger ship, has been developed on distinctive lines from the French Astra-Torres type. Both have proved fast for their size.

NAVAL AVIATION

Hunting submarines ~ To acquire the initiative on anti-submarine patrols, RNAS aviators took every advantage of any quirk of weather or nature to surprise enemy submarines. Clouds, sea mist, and the glare of the sun, if skillfully used might help an aerial hunter approach his victim unseen. Idiosyncrasies of seagulls in different waters came to be known and were used. Flocks of sea birds seen from the air dipping swiftly to prey on a shoal of fish might have given the impression that they were flying in the wake of a moving periscope. However, it came to be known from the reports of submarine officers of the Royal Navy that gulls of the North Sea seldom bothered to investigate a moving periscope. However, other seabirds particularly those in the waters off the west coast of Ireland would follow in a periscope's wake for hours.

The Düsseldorf Raid ~ The Royal Naval Air Service, based at Dunkerque in early September of 1914 received orders to, *Deny the use of territory within 100 miles of the city to German airships and airplanes*. Restricted at first by bad weather, on 24th September, four machines flew from Antwerp to bomb the Zeppelin sheds at Düsseldorf and Cologne, however lost in a thick mist, one frustrated pilot wrote; *Looking for the sheds at Düsseldorf was like going into a dark room to look for a black cat*. Two weeks later, RNAS No 1 Squadron Commander Spenser Grey again failed to find the Düsseldorf Zeppelin sheds so dropped his two 20-lb bombs on the main Cologne railway station. Flight Lieutenant Reginald Marix flying a single-seat Sopwith Tabloid spotted the Düsseldorf hangar. He descended to 600 feet for his bombing run, let go his bombs hitting the main shed. The destruction was complete. The huge shed blazed with flames and thick black smoke belched from its roof. Within 30 seconds the roof fell in. The flames rose to a height of 500 feet revealing that an inflated Zeppelin had indeed been inside. After carrying out the first notable air raid of the war, Marix's shellfire damaged Sopwith was able to fly far enough away to land within 20 miles of Antwerp and return to the city with the aid of a bicycle borrowed from a peasant. Two days later Antwerp was evacuated and fell to the Germans

Coastal airships ~ In the early weeks of the Great War, the six airships of the RNAS were prepared and took off to patrol the waters over the English Channel to look for German submarine. If they happened to sight a U-boat, they would report its location to a Royal Navy direction finding wireless station that would direct the nearest corvette or destroyer to the area. More airships were needed, and Admiral Sir John Fisher wrote to Admiral Jellicoe of his idea for the naval airship as a submarine destroyer, and that he had ordered 50 ~ *even though everyone thinks I'm mad!* Admiral Fisher knew that airships at sea would not be targeted by enemy fighters or AA gunners, they could operate all day on submarine patrol, and carry enough bombs to tackle any identified U-boats.

The Submarine Scout or S.S. airship gondola was an adapted fuselage with the engine and propeller from the body of B.E.2c slung beneath the bag of a Willows airship and lovingly called a *battle bag*. Each S.S. airship carried a bomb load of 112-lbs, a pilot, a co-pilot, and a wireless operator. Able to stay aloft for eight hours, even if they seldom attacked a submarine, they could always call up attackers who could, and the Kaiser's submariners were noticeably wary of them, knowing that even if submerged, their U-boat might still seen by a sharp-eyed midshipman trained to look for the tell-tale signs of an oil trail or gulls flying in the wake of their periscope. Naval airships were soon operating from 12 coastal stations. The tactical value of the naval airships was in their ability to detect a submarine previous to its attack or to locate the submarine after a torpedo attack with the resultant destruction of the U-boat by depth charges from either the airship or surface escort. The U-boats of the Great War like the whale, could not remain under water indefinitely, so forcing them down to run submerged - and keeping them there, would exhaust their batteries, and eventually become a surface craft at the mercy of the smallest warship. Because of the S.S. airships limited range, the British Admiralty, introduced their largest coastal airship, the North Sea Class or N.S. airship that was 262 ft long, had two 250-hp engines, carried up to five Lewis guns and six 230-lb bombs, and could stay aloft on patrol for three days. Royal Navy airship NS 11 on one sortie, flew 4,000 miles with her crew of ten split into two watches, all comfortably enclosed within their 35 feet long gondola

The French Navy developed the Nieuport Astra-Torres AT2 airship for their anti-submarine service and at 120m in length and with a diameter of 18m was powered by two Sunbeam 300-hp V12s to give a cruising speed of 56kph, and a range of 1,500km. The enclosed gondola that accommodated a crew of 12, mounted one 75mm cannon plus twin machine guns, and carried a useful bomb load of either four 45kg or two 90kg bombs. (After the war, the Japanese purchased one in 1923 for trials).

OVER THE DARDANELLES AND IN THE AEGEAN

In the spring of 1915, the British and French kicked off their campaign to bombard the entrance forts of the Dardanelles Straits to clear the way for a disposable squadron of obsolete warships to steam up to Constantinople and knock the Ottoman Turkish Empire out of the war with the intent to reopen the supply route to Russia through the Black Sea and so sustain their struggling land forces. The Royal Naval Air Service were assigned to carry out air support missions during the campaign that would involve bombing raids against objectives on the Peninsula, enemy lines of communications and even bombing Constantinople. In this campaign, torpedoes were successfully used in action for the first time by seaplanes with three Turkish ships destroyed. But throughout, air operations in practical terms were limited in available resources of aircraft, spares, supplies, and in subsequent effectiveness.

Fleet Assistance, Reconnaissance, Spotting for Ship's Guns ~ First, airmen were required to correct the gun accuracy of Allied warships preparing to bombard the forts at the entrance of the Dardanelles covering the minefields in the Straits. Allied warships were re-equipped with improved streamlined naval balloons, designed by Captain Albert Caquot that could be flown in stronger winds and had proved more stable at sea. The aerial observers sent aloft in ship tethered balloons would be able to communicate with their ship by telephone and received detailed training on how to correct naval gunnery against shore targets on the enemy held Gallipoli Peninsula. As soon as the SS *Manica* was refitted with a hydrogen gas generator, compressor, winching gear and equipped with a supply of streamlined naval balloons, her officers set course to join the Eastern Mediterranean Squadron for the Dardanelles campaign. The need for carrying out a strong aerial offensive (especially by bombing) offered exceptional advantages, and had the necessary air power been built up from the beginning and sustained, the Anglo-French warships could have forced the Straits and taken Constantinople.

When the proposed Allied naval thrust through the Dardanelles failed to break through to Constantinople, primarily because of an unobserved line of naval mines lurking in Eren Keui Bay on 18th March 1915, when the warships *Bouvet, Inflexible* and *Irresistible* turning near the Turkish shore, hit Turkish mines, they were all holed with the *Bouvet* and *Irresistible* lost sunk. Instead of trying again, it seemed to some to have become necessary for a combined operation of landings at Cape Helles, Anzac Cove, and Suvla Bay. Aerial reconnaissance assumed even more importance and balloon ships, seaplane carriers, and an airship were all deployed to search for enemy mines and vessels, to photograph, machine gun and bomb Turkish trench, troop, gun positions and other missions.

The balloon observers above HMS *Manica* successfully reported the fall of shot for the Squadron's shore bombardment during the Allied landings on the Gallipoli Peninsula in April, and up to late May the successful use of tethered balloons from the aft deck of HMS *Manica* steaming close to the Gallipoli shore and overlooking enemy positions from a distance of 3,000 yards were of great use, but when the Turks brought long-range guns into position, *Manica* was obliged to keep 12,000 yards offshore and the balloonist's services were rendered slight for their purpose. Thereafter, two seaplanes from HMS *Ark Royal* observed for the ships guns. But, this was often frustrated when German wireless operators ashore interrupted the frequency of the British observers ~ sending rude messages to the Royal Navy.

Aerodromes of the Greek Isles ~ In the Greek Isles, on the islands of Imbros, Mytilene, Mundros, Rabbit Island, Long Island in the Gulf of Smyrna and Thasos off the Macedonian coast, flights of land based aircraft were established for further operations against the Turks. Commander Charles Samson RNAS and the airmen of No. 3 Wing of were detached from the seaplane carrier HMS *Ark Royal* to survey the island of Tenedos for another airfield. The sparse grass and sun baked rocky surfaces of Tenedos were turned into a fine aerodrome after his men acquired a steamroller and the help of some Turkish POWs. Samson's airmen lived in spacious bell tents and rooms created from aircraft packing crates. Their health benefited from a good supply of Greek figs, peaches, grapes, and raisins. Those ashore also traded various items for bread, eggs, and the occasional chicken or goat from the pleasant Greek islanders to supplement their rations of corned beef and hard biscuits. After setting themselves up, the island-based pilots flew missions to establish wireless communications with British warships, to correct the accuracy of their gunnery against the often very well concealed soldiers of the Ottoman Turkish Empire. Commander Samson, who once dropped bombs on a group of Turks gathering around propaganda leaflets he had just dropped, stipulated in his Standing Orders for No. 3 Wing that if his pilots were unable to fulfill their flights planned purpose, then another useful outcome must be found. Away from Gallipoli, there were missions flown to destroy the Berlin-Constantinople railway bridge over the River Maritza, and to bomb the Ferejik Junction on the same line, in Bulgaria.

OVER THE DARDANELLES AND IN THE AEGEAN

First Aerial Torpedo Attack ~ In May 1915, two Short 184 seaplanes were embarked wings folded and secured for sea in their hangar within the seaplane carrier, HMS *Ben-My-Chree* for operations off the Gallipoli Peninsula. On 12th August in the Gulf of Xeros in the darkness of early morning, Flight Commander C.H.K. Edmond's seaplane was pushed from the hangar, her wings spread and lowered noiselessly into the water. With enough fuel for 45 minutes of flying and knowing that the weight of the 14-inch torpedo would limit his altitude to 800 feet, he decided to fly without his observer. He started his engine and took off toward the Turkish Isthmus of Bulair. As he flew across the narrow neck of Bulair and over the Gallipoli Peninsula, Edmonds came under fire from the Turkish lines. Reaching the northern end of the Dardanelles at dawn, he descended low to attack a Turkish supply vessel seen off Injeh Burnu. Spotting the Turkish steamer, he glided down to 15 feet above sea level and released his torpedo, scoring a hit amidships. As he banked away, he saw the Turkish ship settling down in the water by the stern. A week later, Edmonds and Flight Commander G.B. Dacre took off without their observers, again to save weight, and headed for the Ak Bashi Liman, where three Turkish supply steamers had been reported. Edmonds lined up the middle vessel. Released and watched his torpedo hit, and circled his sinking victim. Dacre developed engine failure and having to taxi on the surface lined up his target. As he closed in, the Turks on board fought back with their old rifles. After he launched the weight of his torpedo at the enemy steamer that exploded and sank in seconds, Dacre's seaplane became airborne and he returned flying at 200 feet. (The full potential of this form of attack would not be realized until World War Two with the surprise attacks on Tarranto and Pearl Harbour). In June Erskine Childers, the secretive author of *The Riddle of the Sands* received a reserve commission in the Intelligence arm of the Royal Navy and served onboard HMS *Ben-My-Chree* during the Gallipoli campaign as an instructor in coastal navigation, and flew as a navigator-observer.

First submarine victory by aircraft ~ Over the North Sea or English Channel a submerged U-boat might have been practically invisible, but in the Bosphorus, underwater submarines were quite visible as were naval mines to at least a depth of 18 feet. On 15th September 1916, two Austrian Lohner seaplanes piloted by *Leutnant* Zeiezny and *Leutnant* Konjovic flying ten miles off Cattaro spotted the French submarine *Foucault* on the surface, and attacked. Their bombs damaged the submarines batteries and electric motors. *Foucault* began leaking and out of trim slipped below the surface. The Austrians, able to see the French submersible under the waves, circled overhead and waited. 30 minutes later, the French blew all ballast tanks and the *Foucualt* rose to the surface. *Capitaine* Léon Dévin opened the conning tower hatch and clambered out – to see the Austrian airmen diving at him. Four bombs burst close, and the diesels of the *Foucault* refused to start. Dévin ordered his men to abandon ship. Seeing this, the Austrian airmen ditched their remaining bombs a safe distance away, landed on the water and taxied toward the French submariners in the water. The Austrians letting the French crew cling to their seaplane floats, reported in, and waited for a rescue boat to save all 27 crew.

Hunting the battle-cruiser Goeben ~ When the *Goeben* steaming with the *Breslau* were observed and reported emerging from the Dardanelles on 20th January 1918, the British pilots of the RNAS finally had their historic chance to prove that armoured enemy warships were vulnerable to aerial attack.

We were wakened in the early hours of the morning by the shouts of ~ The Goeben's out! The Goeben's out! They had slipped out well before dawn, slipping through our minefields quite successfully. The dawn patrol had reported the Straits clear not knowing that the Goeben and Breslau were actually behind them out in the Aegean. Then they turned back, and that was when we came into the picture. There I was, about 25 miles away from where the Goeben was. The bombers were ready and got under way. I got a Camel to escort the bombers against attack from any enemy fighters from the mainland. We were hitting both ships with quite small bombs from a low height. In truth the bombs were not very heavy and they couldn't have done the armoured ships much harm but the crews lost their nerve and instead of proceeding in a straight line for the mouth of the Dardanelles they started to zigzag – and in a minefield that is fatal. We attacked with machine guns pelting everything we could at them. It couldn't do much harm unless you could shoot somebody through a slit. Anything you could do to rattle them ~ raking the decks with tracer. The Breslau hit five mines and began to sink most convincingly. So we concentrated on the Goeben and she hit at least two, and began to settle a bit lower in the water. By the time they reached the mouth of the Dardanelles the Goeben was going slowly and the Breslau had sunk. It wasn't our bombs that had sunk her but we'd bombed her on to the minefield so we claimed it as a RNAS victory. The Goeben was going slower. The only thing they could do, and they did it brilliantly, was to beach her as fast as they could at Nagara Point ~ a very good place to beach her. When they'd got her patched, their luck was in. By going full speed astern; they slid off and got away in the night. Captain Graham Donald RNAS

NAVAL AVIATION

The seaplanes and flying boats of the Royal Naval Air Service routinely flew anti-submarine patrols from bases on the British coast and from Dunkerque. Because of the their effectiveness against U-boats they developed in size into the largest British flying boat of the war, the Felixstowe F.2a which was powered by twin Rolls Royce Eagle VIII engines, and had a huge wingspan of almost 100 feet. The Felixstowe F.2a was heavily armed for action against U-boats with four machine guns and some were fitted with the experimental Davis recoilless cannon, a long tube that fired a shell at the target and a bag of sand out the other end with minimal detonation stress to the aircraft. Seaplane aircrew flying out of Great Yarmoth knew the trick of how to set their aircraft down in heavy fog. They carried a long stick, which was fitted in a socket vertically outside the hull and coupled to the control column. As the stick end dragged in the water, it pulled the column back and their craft settled on the water.

The year 1917 saw the start of unrestricted U-boat attacks; firing torpedoes submerged instead of using their deck guns for surface attacks. This had the unwanted consequence for the Germans of bringing the USA into the Great War on 6th April 1917. As the position became perilous, British air patrols had to be increased to hunt down the U-boats. Fortunately, wireless direction stations of the Royal Navy could identify the positions of the German U-boats by getting a fix on their transmissions. After these were plotted on the map, it showed that U-boats prowling close to the British Isles often passed the North Hinder lightship *on the surface* to conserve their electrical batteries, which powered their engines when fully submerged. Using this information, the British adopted *The Spider's Web* patrol pattern to prey on the U-boats. *The Spider's Web* was 60 miles in diameter centered on the lightship with eight arms each 38 nautical miles in length. Four F.2a aircraft could search the whole area in five hours, and in April the Felixstowe airmen sighted eight U-boats in one month, and bombed three. On 20th May 1917, two 500-lb bombs hit *UC-36* cruising inattentively on the surface. Soon after, the crew of another F.2a spotted a periscope on the surface travelling towards a lone cargo ship coming up the Channel. They dropped four 100-lb bombs on it, and advised nearby Portland what had happened. They sent out an armed yacht, and between them, *U-39* was destroyed.

As part of their U-boat offensive, the Germans tried to protect their U-boats from the attentions of Allied flying boats, and soon an air war developed over the English Channel, the North Sea, and other coastal areas. As 1917 went on, German resistance to Allied operations stiffened with an increase in German seaplane numbers that resulted in an entire close flying flight of six French F.B.A. seaplanes being shot down on 26th May 1917. The Germans naval aviators responsible flew Hansa-Brandenburg W.12 seaplanes from Zeebrugge. The Hansa-Brandenburg airmen employed a tactic derived from conserving aviation fuel. By alighting on a calm sea like pond ducks, they simply waited for Allied aircraft to appear. Other tactics recorded by one of their own were;

Opposing the enemy seaplane units in our area was comparatively easy, as our Hansa-Brandenburg aircraft were superior to the British flying boats in terms of speed, maneuverability, and armament. On the other hand, the drag of our twin floats put us at a serious disadvantage if we were ambushed by British or French land based fighters. Our best defence was to fly in tight formation, close to the water. Our two-seat reconnaissance aircraft were each equipped with a ring mounted machine gun for the observer, so by using this formation the combined firepower of five aircraft could be brought to bear on an opponent. Generally the Staffel flew in a wedge-shaped formation with the Staffel-Kapitan (Flight-Leader) in the leading aircraft. During any attack by fighters coming in from above and behind, the aircraft under attack went even lower so that a Sloping-V was formed and all machine guns had an unobstructed field of fire. It was a lesson learned through hard experience that made land based enemy fighters respect our formations, gliding along close to the surface, which gave a slight advantage against the classic attack from above and behind. Perhaps this explains why enemy forces inflicted only light casualties on our naval observation aircraft. Leutnant Fritz Stormer

On 1st September 1917, the leading German naval ace *Oberleutnant zur See* Friedrich Christiansen took command of the Zeebrugge formation to lead them on reconnaissance and bombing missions against England. After bombing Dover and Ramsgate he was awarded the Iron Cross and became the only seaplane pilot to be awarded the *Pour le Mérite* for having shot down 17 Allied aircraft, including three Felixstowe F.2a aircraft and an airship. On 6th July 1918, Christiansen flew up the mouth of the Thames to surprise and damage British submarine HMS *C25*, killing its captain and five submariners. After ditching his damaged craft in the North Sea and clinging to it thru a stormy night for nine hours, freezing, exhausted, and not quite expired ~ *U-10* surfaced nearby to rescue and revive Christiansen.

* * *

THE AMPHIBIOUS FLYING-BOATS OF THE R.A.F.

AT HOME IN TWO ELEMENTS ~ AIR AND WATER : THE FLYING BOATS OF THE ROYAL AIR FORCE

The flying-boats, whose whole body, or fuselage, floats on water, must be distinguished from the seaplane, which is merely an aeroplane with pontoon-floats and a tail-float, instead of a wheeled under-carriage and a tail-skid. In "All the World's Aircraft 1918", Mr. C.C. Grey defines a flying-boat as "A hydroplane with which wings, empennage, and aircrew have been combined. Originated in America by Mr. Glen Curtiss. The crew occupies a regular cabin in the hull. Wing-tip floats are fitted to prevent submerging of the wing-tips when rolling ; but the actual floatation is given by the hull, which fulfills the functions of a fuselage and floats combined." The drawing shows flying-boats launched, taking off, and in flight.

THE WOLF & THE WOLFCHEN

Wolfchen was the seaplane which accompanied Kapitän Karl-August Nerger on his cruise on the Wolf. This seaplane was a Friedrichshafen design, was fitted with a 150 H.P. engine, and was provided with a wireless installation and bomb-dropping apparatus. From November 1916 until February 1918, the Wolf and Wolfchen were at sea, and the seaplane contributed a good deal towards the success of the voyage. In 56 flights it displayed the Iron Cross to all the oceans of the world ~ to the Pacific, the Atlantic, the South Seas, and the Indian Ocean.

The repairs which had to be effected during this long voyage were very numerous, but, with the aid of the spare parts that were carried on board, and, to a certain extent, by making use of material taken from captured steamers, the machine was always kept in a serviceable condition, and ready to take the air whenever it was required. Very frequently it had to be completely taken to pieces, and as often re-assembled again. These operations were exceedingly difficult owing to lack of space, but could not be avoided lest the ship should be recognized as an auxiliary cruiser by the aeroplane on its deck. Also the machine had to be protected as much as possible from any damage which might be caused by exposure, particularly in the tropics.
Here are a few extracts from the Wolfchen's logbook :

PACIFIC OCEAN

May 24 1917, during reconnaissance flight, New Zealand was sighted 60 nautical miles to the west.

June 2 1917. Orders. Hold up the merchant steamer which has been sighted north of Raoul Island and bring her to S.M.S. Wolf. The ship had suddenly appeared while S.M.S. Wolf was lying close up to the island, engaged on repairs to her engines and trimming her bunkers. Wolfchen started at 3.30 PM and flew north. On reaching the steamer we spiraled down to within 100m of the deck and dropped the following message in English :
~ Steer south to German cruiser and do not use wireless. If not obeying orders, you will be shelled by bombs.
The second time we flew over the steamer, we dropped a bomb twenty yards from her bows. She changed course at once and steered for S.M.S. Wolf, escorted by the Wolfchen. After our threat, she did not dare use her wireless. It was the New Zealand ship Wairuna (3900 tons) bound from Auckland to San Francisco.
The ship and her cargo were worth many hundred thousand pounds.

June 16, 1917. Orders. Hold up the schooner which has been sighted in the west, and bring her to S.M.S. Wolf. Wolfchen started at 3 PM and flew west. We spiraled down from a height of 500 feet to within 250 feet of the ship. Our first two attempts to drop a message on the deck failed. Owing to the drift of the vessel, both fell some distance to the leeward. At our third approach a bomb was dropped from a height of 250 feet, close by the bows. The ship at once hauled down her top-gaff-sail, and displayed the American flag. Wolfchen ordered the vessel to steer S.E., and intimated that she would be bombed if she did not follow. At once she turned in the given direction, and the machine, circling overhead, led her to S.M.S. Wolf. It was the American four-masted schooner Winslow (567tons), with a cargo of coal, provisions, petrol, and wood from San Francisco.
Unfortunately the petrol was useless for flying purposes.

June 17. Owing to bad weather, Wolfchen was again dismantled and stored away on the afternoon.

INDIAN OCEAN

September 25 1917. Orders. Investigate a patch of smoke which has been sighted, and report on vessel, course, and distance. It was the Japanese boat, Hitachi Maru (6700 tons). The Wolfchen received a further order to support S.M.S. Wolf in holding up the vessel, and to bomb her if she committed any hostile act.

At the first shot from S.M.S. Wolf the steamer turned hard to starboard, apparently with the intention of escaping. Thereupon a bomb was dropped 30 or 40 yards ahead of her bows, and almost at the same time S.M.S. Wolf opened fire, so that the seaplane was led to assume that the steamer was resisting. We therefore flew up again and dropped another bomb from a height of 700 feet, which fell into the water close to the port side. The force of the explosion blew two men overboard. At that the steamer hove to. S.M.S. Wolf ceased fire and the Wolfchen ceased bomb-dropping, but flew over and round the ship until the prize crew had gone on board.

We landed close by the vessel and discovered that the nuts on the propeller bolts had worked loose, and that in consequence of the lash on the propeller shaft the engines could no longer throw true against the cranks. Having communicated by signals with the ship, we were towed in our aeroplane back to S.M.S. Wolf by motorboat.

Oberleutnant zur See **Friedrich Moll**

HMS FURIOUS ~ THE FIRST AIRCRAFT CARRIER

The Short Brothers next developed their Type 320 Seaplane (powered by a 320-hp Sunbeam engine) to deliver the more potent 18-inch torpedo. In June 1917, enemy seaplanes attacked a Short 320 that was reported by its carrier pigeon to be down in the North Sea. An attempt to rescue them by fast motorboats failed, with the result that this aircraft was captured and fell into German hands. Another time a U-boat came up alongside Short 320 floating with engine trouble north of Dunkerque. The rescue boat was just in time to see the aircrew taken prisoner, as the U-boat dived beneath the waves. In early 1917, the Admiralty found the answer to the unavailability of seaplanes for operations over the North Sea whenever the weather turned bad. It was decided that the battle-cruiser HMS *Furious* under construction would be reconfigured by removing her forward 18-inch guns and replacing them with a 228 feet takeoff deck forward of the bridge, but remaining a cruiser from her bridge aft to her stern, Hugh Popham described *Furious* as a thing half done, *like some latter day chimera or a species caught mid-evolution, she had the hindquarters of the past and the forequarters of the future.* Takeoffs from her forward flight deck became routine, but any landing was a nightmare. Squadron Commander Ernest Dunning flying a Sopwith Pup, carried out the first landing on the deck of a moving carrier. The captain said to him ~ *You may as well take a revolver and blow your brains out!* Dunning took off and then tried to land on *Furious* steaming at 26 knots, having to overtake, fly down the side of the carrier, dart around the central superstructure and land on the foredeck. With a steady 31 knots of wind blowing over the deck of HMS *Furious*, Dunning touched down on 2nd August 1917, slowed, the handling party grabbed Dunning's flimsy Sopwith Pup to stop it from being blown away, and the Royal Navy celebrated the event in great style. Four days later, in a gustier wind Dunning took off to repeat the feat in a tidier manner. But, as he took off, the Sopwith stalled, went over the side of the *Furious* into the sea, and poor Dunning drowned.

Funnel exhaust gases and turbulent wind eddies from the original superstructure made landing trials so hazardous, that *Furious* was redesigned in late 1917. The remaining aft guns were removed, the original centerline superstructure cut away, and the flight deck made continuous from bow to stern. A large lift gave access to hangars and aircraft maintenance workshops below. *Furious* had a series of longitudinal and traverse wires, raised by wooden pegs, and an emergency barrier of ropes suspended in an array from a folding gantry to protect her aircraft if they missed the arrestor wires. Still thwarted by some turbulence from *Furious's* residual superstructure, from 13 landings during April 1918, only three were free from damage or pilot injury. While HMS *Furious*, steamed out to sea the pilots of her eight Sopwith Camel 2F-1 fighters were instructed to complete their flights by ditching or by flying ashore ~ but not to land on HMS *Furious*.

On 19th July 1918, some 80 miles off Schleswig on the enemy coastline, Captain C.E.C. Robinson commanding HMS *Furious* changed course into the wind and ordered, *Full-steam ahead!* He was preparing to launch from the flight deck of HMS *Furious* her flight of eight Sopwith Camel fighters, each carrying two 50-lb. bombs. Their mission was to destroy the Zeppelin sheds at Tondern at the mouth of the Elbe. With their Camel's engines revving at full speed, one after another the pilots flew at full throttle down the flight deck off the bow into the wind, still not quite up to their full flying speed, to swoop down like swallows trying reach their required flying speeds. Just before they touched the water they lifted their noses and climbed away, only missing the crest of the waves by a few feet. Catching the Germans by surprise, Captain Jackson dived his Sopwith Camel toward the northernmost shed dropping two bombs. One direct hit on the middle and the other slightly outside of the shed. Captain William Dickson dropped his bombs and two more hits. The whole shed exploded and the flames soared up 1,000-feet. Next, Captain Bernard Smart dived on the second shed, releasing his bombs at 1,000-feet to hit the centre. The sheds containing Zeppelins *L54* and *L60* were destroyed in the subsequent blaze. Even though the Camel airmen had vanquished the Zeppelins, they had to evade the area and ditch. A British Destroyer rescued three pilots, four others landed safely ~ but one short of fuel over Denmark disappeared. Further experience with *Furious* resulted in her eventual evolution as a through flight deck carrier with an off-centre superstructure.

The Royal Navy's second aircraft carrier was originally laid down as the Italian passenger-liner *Conte Rosso* in 1914. The Admiralty purchased the basic structure in 1916 to be redesigned as a second flush deck carrier, HMS *Argus*. Early in the design stage, the funnels were deleted to reduce turbulence over the flight deck with all exhaust fumes ducted aft and vented from the stern. During September 1918, HMS *Argus* joined the Grand Fleet carrying the 20 Sopwith Cuckoo torpedo bombers of No. 185 Squadron. By October 1918, the carriers HMS *Argus, Furious,* and *Campania* with the converted cruisers *Courageous* and *Glorious* were able to launch 100 Cuckoos over the North Sea.

METHOD OF SIGHTING THE SOPWITH CUCKOO'S TORPEDO

THE METHOD OF SIGHTING THE SOPWITH CUCKOO'S 18" TORPEDO (DULY ALLOWING FOR A SHIP'S SPEED) IS APPARENT FROM THE ILLUSTRATION SHOWING THE TORPEDO AIMING SIGHT AS A SMALL RING NEAR THE PILOT'S EYE (FITTED BOTH PORT AND STARBOARD) THAT WAS USED IN CONJUNCTION WITH A TRAVERSE ROW OF BEADS LATERALLY DISPLACED CORRESPONDING TO THE SPEED OF A TARGET SHIP STEAMING AT 5,10,15 KNOTS OR MORE.

LANDING SHORT TYPE 184 FLOAT SEAPLANE DOWN WIND

IN LANDING A FLOAT MACHINE, CARE SHOULD BE TAKEN IN GETTING INTO THE WIND.
LANDING DOWN WIND IS A PRACTICE THAT SHOULD BE AVOIDED AT ALL COSTS.
IN ADDITION TO THE DANGER OF STALLING OWING TO MISJUDGING THE STALLING SPEED THE MACHINE ON STRIKING THE SURFACE WILL TURN ROUND INTO THE WIND, DIP A WING TIP AND PROBABLY TURN OVER.
EVEN SHOULD THIS NOT OCCUR THE WING WILL BE DAMAGED PLACING THE MACHINE OUT OF COMMISSION.

THREE TORPEDO BOMBERS CLOSING ON AN ENEMY SHIP

TACTICS.

1. AIRCRAFT FLYING AHEAD OF TORPEDO MACHINES PUT UP SMOKE SCREEN
2. TORPEDO ATTACKING MACHINES CLOSE THE TARGET AND DROP TORPEDOES
3. AFTER RELEASE TAKE ADVANTAGE OF SMOKE AND WIND TO GET AWAY.

RIGHT AND WRONG WAYS TO DROP A TORPEDO

Top: CORRECT DROP ~ DISTANCE THE MACHINE MOVES AFTER RELEASE OF TORPEDO ~ PATH OF TORPEDO IN AIR

Bottom: INCORRECT DROP ~ DOWN INCLINATION & INCORRECT DROP ~ UP INCLINATION

THE CAMPAIGN IN PALESTINE 1918

Air operations played a significant part in the British campaigns in Palestine and Mesopotamia, and also played a part supporting Colonel T.E. Lawrence in his legendary *Lawrence of Arabia* persona. Lawrence's aim of expelling the Ottoman Empire's soldiers from the Sherifian Arab tribal lands, together with his decent treatment of his mounted Arab force, his linguistic ability, imagination, perception and extraordinary personal energy all helped him maintain the coherence of his fascinating command. Lawrence's approach to destroying the Turkish lines of communication in his own words, *used the smallest force in the quickest time at the farthest place.* The loyal Arabs under his inspiration, advanced on the Hejaz railway from Aquaba in the south and towards the Dead Sea from the west.

No. 1 Squadron of the Australian Flying Corps (AFC) after attracting many ex-troopers from the Australian Light Horse arrived in Egypt during April 1916 to support the RFC and the Egypt Seaplane Squadron. Meanwhile, the British Army and the Desert Mounted Corps had been fighting in Palestine to prevent the Turks from cutting the Suez Canal, the British Empire's lifeline from the Mediterranean to the Red Sea and everywhere else, *East of Suez*. If the Turks had interrupted the flow of expected merchant and troop ships from South Africa, Australia, New Zealand or India steaming through the canal with vital troops and supplies, the effect on the British war effort would have been severe.

However all went well, and by September of 1918, the Turkish defences in Palestine began to collapse. General Edmund Allenby renewed his offensive on a decisive scale to remove Turkey from the war. His plan was to burst the Turkish frontline with a ferocious infantry attack on the coastal flank, and then his Desert Mounted Corps under General Harry Chauvel were to gallop through the breach and continue their thrusting advance towards Nazareth and the upper Jordan. Then to swing around to cut off the northern line of retreat and envelope the entire Turkish force 40 miles to the south. Colonel Lawrence's Arab army, after cutting the Turkish lines of communication, would attack the rail centre of Deraa farther east and complete the encirclement of the Turkish forces in Palestine. To assist the Arabs and to divert enemy suspicion, to the east bombing raids would be flown and the plan of action for September 19th for the 40th Wing of the Royal Air Force in Palestine was as follows : S.E.5a scouts of No. 111 and 145 Squadrons, to patrol over Jenin Aerodrome to prevent any attempted enemy air action from that quarter, and to attack with bombs and machine gun all targets in the area. D.H.9 bombers of No. 144 Squadron; to bomb El Afule railway station, and the Turkish HQ at Nablus. Bristol F2b Fighters and the Handley Page bomber No. 1 AFC; to reconnoiter and bomb.

The battle opened on the morning of 19th September and quickly developed according to plan. At 0430, British artillery began shelling the Turkish front, which shattered by 0730 hours, enabled the cavalry to move north along the coast. With no Turkish reserves to check them, the airmen flew low level bombing raids to destroy the enemy telegraph, telephone, and wireless communication centers. Captain Ross-Smith and his crew set off in their Handley-Page bomber carrying 16 x 112-lb bombs to smash the Ottoman HQ and telephone exchange at El Afule, and fly on to wreck the railway junction. Flights of D.H.9's demolished the telephone exchange at Nablus and because of this wreckage the Turks east of Nablus remained in complete ignorance of Allenby's attack for at least two to three days. Two S.E.5a fighters armed with bombs circled over Jenin aerodrome all day. If they spotted any movement, they bombed and strafed. Each pair of aircraft was relieved every two hours, and before departing each pair machine-gunned the German hangars. Hours later the mounted advance guard were already riding well ahead of the local Turkish retreat heading northwards from Nazareth to Jordan when Allied pilots spotted additional retreating Turks.

The first great news of achievement came from Flight-Lieutenant Cameron, accompanied by two other Bristol F.2b Fighters when they reported 600 vehicles, 5,000 infantry along with 2,000 cavalry and horse artillery retiring at a gallop and in disorder about Et Tire. The flight of three Australian aircraft dropped their 24 bombs on the fleeing Turkish masses hard to miss and fired over 2,000 machine gun rounds into disorganized groups of men and transport. At 1230 a second flight of three Bristol scouts repeated the attack on the Turks. Defiles along which the Turkish armies might retreat through had been noted in advance and were kept under regular observation by Bristol scouts flown by No.1 Squadron specifically equipped with long-range wireless to transmit such reports. It was not long before this wise preliminary bore results. The withdrawing Turkish 7th Army with their northern roads of retreat now blocked by British cavalry knew that their only possibility of escape lay eastward across the Jordan to the shelter of the Turkish 4th Army.

THE TURKISH 7th ARMY TRAPPED IN WADI-EL-FAR'A

On 20th September 1918, two Bristol F.2b Fighters having searched east of the Jordan and flying homeward, spotted far away and moving along the old Nablus-Beisan Roman highway excavated along one side of the deeply eroded watercourse of Wadi-el Far'a : a group of about 200 horse-drawn wagons, 80 horse artillery and 50 trucks. The Bristols flew across and dropped their bombs to block the road with a direct hit on a heavily laden truck. A blockage was rapidly created at the head of the column, and for the rest of the day with diminishing resistance, the beginning of a massacre was apparent. In the late afternoon a second pair of patrolling Bristol F.2b Fighters attacked the trapped line of Turkish transport along the roadside of the Wadi. Again targeting the leading Turkish troops and their vehicles, low-flying Bristol airmen of No. 1 Squadron AFC with a number of British flights attacked. The congested mass of 7,000 Turkish soldiers with their vehicles, guns, horses, and camels trapped in the valley were subjected to terrible bombing and deadly machine gun fire on the head of their column as it wound its way through the narrow mountain pass. The Bristol two-seaters swooped down on the Turks, dropping eight bombs that scored three direct hits on transport that further blocked the road. The effect of the others was equally destructive. Panic caused as much loss as bombs. Frightened by the explosions and bursts of machine gun fire from close overhead, horses bolted over the precipice on one side of the road while on the other, men ran in panic to shelter in the rocky hillside. Trucks abandoned by their drivers while still in gear and running, collided with bolting horses and went over the precipice. The Bristols of No. 1 Squadron AFC made six raids that day, dropping three tons of bombs and firing 24,000 machine gun rounds into the struggling soldiers of the Turkish 7th Army trapped on the heights of Wadi-el-Far'a.

The following day, the pass was blocked, both behind and ahead by overturned trucks, and horse wagons. Six or more aircraft attacked every 30 minutes with No. 1 Squadron AFC making four trips a day in their Bristol two-seaters to bomb and machine gun like conventional fighters as their armed observers strafed Turkish ground targets with their twin Lewis guns that turned retreat into rout, slaying the Turkish 7th Army as they attempted to escape north onto the highlands of Samaria from Nablus. Nine tons of bombs were dropped from the lowest of heights and 56,000 rounds from machine guns were sprayed into the panic-stricken Turks. Reports gave a terrible picture of maddened Turkish soldiers trapped among the rocks and precipices, trucks crashing into ravines or into tangled transport in front, horses stampeding. The retreating Turks were forced to abandon their transport and 87 field guns in the ravines, and found little shelter from attacking aircraft. Men were seen deserting vehicles in a wild scramble to seek cover. Maddened horses dragged many over the edge of the precipice. Others cut terrified horses from the wagons and rode in panic down the road.

In what many troops called the Battle of Armageddon, remnants of the broken Turkish 7th Army, mere handfuls of men in a tragic state of morale, still tried to pull back to the north on their own. The aircraft had so far outstripped their own ground troops, that it took the advance guard of General Harry Chauvel's Desert Mounted Corp three days more to reach the scene of carnage. When British cavalry passed along this road they were appalled by the sight as they rode forward to find abandoned or wrecked : four cars, 55 trucks, 75 carts, 87 guns, scores of wheeled wagons, water carts and field-kitchens. Meanwhile detachments of the Australian Light Horse pushed on from El Afule, northward for Nazareth and eastward to block the road of retreat to the Turks between Nablus and the Jordan to assure the capture of the survivors of the destroyed Turkish 7th Army. Some wandered across country and were then seen marching back toward Wadi-el-Far'a under a white flag. Other ragged parties of Turkish fugitives crossed the Jordan and wandered toward Deraa where they were met by British cavalry and the Arabs. They had no food or rations, having abandoned all stores and all further desire to fight. Other retreating Turkish units committed a number of atrocities against some hostile Arab settlements, and in return Arab irregulars took no prisoners. An entire Turkish brigade along with a detachment of Germans was massacred near the village of Tafas on 27th September 1918. The following day, fast riding Arabs caught another formation of retreating Turks in the open, and wiped them out in another merciless massacre. Very few of those Turkish soldiers escaped death or capture. General Edmund Allenby was now free to devote his attentions to the advance on Damascus.

The Turkish 7th Army ceased to exist and it must be noted that this was entirely the result of attack from the air.
Lieutenant-Colonel Richard Williams

THE CAMPAIGN IN PALESTINE 1918

Meanwhile, Colonel Lawrence's Arabs left Azrak to pitch camp at El Umtaiye and Um es Surah five miles east of the railway between Mafrak and Deraa. From their new raiding centers near to the Turkish railway, Lawrence's force on 17th September captured Tel Arar station north of Deraa, blew up the railway bridge over the neighboring wadi, and systematically destroyed segments along ten miles of railway between Deraa and Ghazale. The raiders continued the destruction of enemy railway in three directions from Mezerib, including nine miles of demolition on the hilly westward section of the Turk's main Nablus railway toward Tel-esh-Shebab : their supply line for their entire force west of the Jordan. Lawrence then organized the demolition of an important viaduct north of Nasib and destroyed several more miles of railway with *Tulip Bombs* ~ bombs planted under rails and sleepers and fired in sequence. Colonel T.E. Lawrence called this *tulip planting* and the term stuck. Further demolition continued around Mafrak during 19th September, the day of the big attack.

On 22nd September 1918, a Handley-Page 0/400 bomber arrived near Deraa to support the Arabs. When it landed at Deraa, it had flown from England with orders to proceed to the Sherifian HQ with stores of petrol, oil, bombs, ammunition, food and other supplies for the two Bristol F.2b Fighters attached to the Sherifian Arabs forces operating with Lawrence of Arabia. The Arabs fighting alongside Lawrence welcomed the giant Handley-Page and received it with the wildest enthusiasm. They sang, danced around it, and fired volleys from their pistols and other odd weapons into the air in an ecstasy of delight amusing Captain Ross Smith, a former sergeant of the 3rd Light Horse Regt., and now the leading ace in the theatre with 11 victories (who would become the first to fly between Cairo and Calcutta and be knighted for his post-war Britain to Australia flight in a Vickers Vimy). Captain Ross Smith made weekly bombing raids on the Turkish strong points of Hassana, Maghara, and Rohd Salem to eliminate or diminish the Turk's water supplies. Another frequent Handley Page bombing target was Junction Station where the railway from Jaffa to Jerusalem joined the line running south to Beersheba, along which travelled all Turkish reinforcements and munitions. Enemy railway stations and aerodromes were attacked day and by night whenever a full moon permitted, with flights of three Martinsyde Elephants being escorted by three Bristols, until the Turks evacuated.

In Palestine, British and Australian squadrons distinguished themselves by their daring and expertise flying a variety of aircraft on reconnaissance, photography, bombing, and fighter missions beyond the frontline over desert and mountains. An aircraft could travel as far in a couple of hours, as a horse could go in a week patrolling, harassing Turkish ground troops, delivering dispatches to isolated units and regularly supporting Allied troops who not only stopped the Turks from cutting the Suez Canal, but pushed them back. In quick succession, Allied forces fought their way over 360 miles of desert in 38 days destroying three Turkish armies in the process of taking the key cities in Lebanon and Syria. Ahead of the cavalry, Australian airmen flew over Damascus for the first time on 27th September. The railway station was filled with hundreds of rail rolling stock. On the roads southward from both Deraa and the Jordan parties of troops, and transport were in full flight. In the afternoon of 28th September, four Bristol aircraft raided the Damascus aerodrome and left it burning.

Next morning, a subsequent patrol of Bristol two-seaters reported the aerodrome no longer in use, and that the Turks were evacuating Damascus in haste. On the morning of 1st October, the 3rd Australian Light Horse entered Damascus and learnt that the remnant of the Turkish armies in Palestine and Syria had fled northward in a disorganized mob without transport or much equipment. Scouting for the cavalry and Rolls Royce armoured cars, Australian airmen flew over Homs, Beirut, Tripoli and Aleppo as their pursuit drove the Turks ever northward. Towards the end of October, they commanded the air as far north as Alexandretta and met with little opposition. On 26th October, British armoured cars supporting the Hejaz Arabs occupied Aleppo and British cavalry were N.W. of the town riding towards Alexandretta. An armistice on 31st October ended the war with Turkey. The victory would not have been so complete without the contribution of the airmen. They prepared it, consummated it and the worst scenes of destruction were their work. General Allenby credited them :

The victory gained in Palestine and Syria has been one of the greatest in the war, and undoubtedly hastened the collapse that followed in other theatres. You gained for us absolute supremacy in the air, thereby enabling my cavalry, artillery, and infantry to carry out their work on the ground practically unmolested by hostile aircraft. This undoubtedly was a factor of paramount importance in the success of our arms here. I wish you all bon voyage, and trust that the peace now attained will mean for you all future happiness and prosperity. Thank you.

* * *

R. Lufbery ~ Sketched at the Lafayette Escadrille field near Longpont as the aviator was getting into his Union suit preparatory to flying in a Chemin-des-Dance engagement.

Yanks with French Type of Anti-Aircraft

AVIONS ENNEMIS MITRAILLANT UNE PROGRESSION D'INFANTERIE.
MOORSLEDE, 29 SEPTEMBRE 1918.

THE END OF THE GREAT WAR

STRAFED! ~ HOW FIGHTER AIRCRAFT FLYING IN V-ECHELON CAN CONCENTRATE ON A TARGET
TRAINING POSTER: FROM AN ORIGINAL DRAWING BY G.H. DAVIS, MAY 1918 (RAAF MUSEUM)

As Kaiser Wilhelm II's German ground forces retreated, tottering from sustained Allied assaults on the Western Front, his fighter pilots now outnumbered four to one, would not concede the skies to their Allied opponents. Even the highly respected Fokker D. VII could not save the situation for them, and daily his airmen faced a worsening situation. In the last weeks of the war *Jasta 1* was renamed *Richthofen Jagdgeschwader* in honour of the fallen Richthofen, and *Oberleutnant* Hermann Goring with 20 victories (and holder of the *Pour le Mérite*) was appointed their new commander. With the war going against them, they fought savagely and regardless of cost. Goring's experiences and tactical thinking during 1918, would no doubt influence his concept of fighter operations in the next conflict, when he led the *Luftwaffe* ~ and even though Herman Goring ended the war with 22 confirmed kills he was not remembered well ~ he was the only member of *Jasta 1* never to be invited to their reunions.

Every day confirmed the growing importance of aviation in war. When it seemed that all other means of bringing Germany to her knees had ceased to progress, Allied air power continued to expand and improve. The ferocity of the war at this time can be appreciated by looking at the aces. Billy Bishop shot down 25 German aircraft during his last 12 days at the front. René Fonck twice vanquished six in one day, once destroying three in a 10 second burst.

Colonel William Barker's Sopwith Camel became the most successful fighter in the history of the RAF as he used it to shoot down 46 aircraft and balloons from September 1917 to September 1918. Barker trialed modifications to his Camel to improve its performance. When his Clerget's cooling efficiency was poor in hot weather, he cut more cooling slots into the cowling, and the poor upward visibility of the Camel led him to cut away progressively larger portions of centre-section fabric. *Billy Barker was such a gifted pilot that he needed not to get on the tail of his prey, but could riddle an enemy across his path, leading it with his machine guns as a hunter aims ahead of a duck on the wing or fire incredible deflection shots with his Camel in a roll, perforating enemy aircraft from the nose, back to kill the enemy airman in his cockpit.*

Louis Strange, who at the beginning of the Great War, had flown from Dover in his Henri Farman out over the English Channel with his improvised Lewis gun mount, had survived. After numerous promotions, Lieutenant-Colonel Louis Strange CO of 80th Wing RAF was woken up with the message; *Hostilities will cease from 11am today. No machines to cross east of the balloon lines.* Austria and Germany surrendered on 31st October and on the 11th hour of 11th November 1918 an Armistice came into effect. Those who struggled through celebrated a remarkable victory, and their national survival. As the war finally came to an end, the few Allied pilots still required to fly over the wonderfully silent Western Front must have pondered their future ~ one felt that :

The Heavens are devoid of Huns, there is nothing so unutterably boring as parading about in the sky in undisputed possession of that vast empty space. It is not only boring, but intensely chilly.

BRIGADIER GENERAL WILLIAM ~ BILLY ~ MITCHELL

William Lendrum Mitchell, born in 1879 in Nice, France, enlisted in the Army as a private in 1898, and as a young officer in 1908 observed Orville Wright's flying demonstration at Fort Meyer. Virginia. He then took flight lessons at the Curtiss Aviation School at Newport News, Virginia. After touring the battlefields of the Russo-Japanese War, Mitchell concluded that war with Japan was inevitable. Rising through the ranks, he ended the Great War as Chief of Air Service, IIIrd Army, and yet Mitchell did not share in the common belief that the Great War, had indeed been the war to end all wars : *If a nation ambitious for universal conquest gets off to a flying start in a war of the future, it may be able to control the whole world more easily than a nation has controlled a continent in the past.* Brigadier-General Billy Mitchell ~ as an influential supporter of air power after the war, asked the U.S. Army and Navy to allow him to demonstrate the use of air power by sinking captured German warships. *Sea craft of all kinds, up to and including the most modern battleships, can be easily destroyed by bombs dropped from aircraft, and further, the most effective means of destruction are bombs. Given a sufficient number of bombing planes ; in short an adequate air force ~ aircraft constitute a positive defense of our country against hostile invasion.*

An MB-2 strikes its target with an experimental phosphorous bomb blast ~ the obsolete battleship USS Alabama. On 27th September 1921, Mitchell's Air Brigade MB-2's bombed and sank the obsolete Alabama in Chesapeake Bay.

~ Entr'acte ~

Adolf Hitler age 35, on his release from Landesberg Prison on 20th December 1924. Hitler had been convicted of treason for his role in an attempted coup in 1923 called the Beer Hall Putsch. This photograph was taken shortly after he finished dictating Mein Kampf to deputy Rudolf Hess. Eight years later, Hitler would be sworn in as Chancellor of Germany, in 1933.

Riette Kahn at the wheel of an ambulance donated by the American movie community to the Spanish Consulate in Los Angeles. The Hollywood Caravan to Spain toured the U.S. to raise funds to help the defenders of Spanish democracy in the Spanish Civil War. Pictured here on 18th September 1937, it is outside Grauman's Chinese Theatre on Hollywood Boulevard.

CHAPTER V

THE SPANISH CIVIL WAR

The Spanish Civil War ~ On 18th July 1936, after a prelude of political unrest, violence, and murder, a treacherous fascist uprising broke out in Spain and spread across the Iberian Peninsula degenerating into one of the bloodiest civil wars in history. The first aerial combat occurred three days later when *Teniente* Miguel Garcia flying a Spanish Nieuport-Delange NiD52 fighter, shot down another Spanish aircraft of the same type, starting nine years of hostility in the skies over Europe. The Spanish Civil War was essentially a conflict between the extreme ideologies of General Franco's fascist Nationalists who reinforced with weaponry and soldiers from Germany and Italy, opposed the communist and anarchist Republicans supported by France and Soviet Russia. The Spanish Civil War soon became a testing ground for the weaponry of three dictators and purported *new* German theories of mechanized war (derived from victorious Allied tactics of 1918) were rebadged as *Blitzkrieg* or Lightning War. However, whereas Hitler and Mussolini were delighted to freely test their latest weapons in Spain, communist Stalin insisted on very high prices, and prompt payment in gold for everything purchased.

Operation Magic Fire ~ The success of the fascist insurrection was seen to depend on whether the tough, seasoned *regulares* of the Army of Africa stationed in Spanish Morocco could be transferred swiftly to Spanish soil. Franco appealed to Hitler for military aid, and received it in the form of the volunteer Condor Legion of the *Luftwaffe*, (Herman Goring commanding, after having been treated in Sweden as a drug addict and being permitted to refresh his flying skills with the Swedish Air Force). German involvement began with the airlift of Franco's troops and material from Tangier, in Spanish Morocco by 60 Junkers Ju-52 transports under the guise of a new airline, *Hisma AG*, across the Mediterranean from northern Morocco, across the turbulent air over the Straits of Gibraltar, to Seville in Spain to fight for Franco's Nationalists. Some German pilots made up to five return trips every day, with more than 40 rugged Moroccan *Africanista* troopers squeezed into the Ju 52s stripped down cabin designed for 17 passengers. By early October, 13,500 soldiers had been carried to Spain, along with 36 artillery pieces, 127 machine guns, and over 500 tons of ammunition. This airlift prevented the quick suppression of Franco's Nationalist revolt by the rightfully elected Spanish Popular Front parliament.

The first German Heinkel 51 biplane fighters arrived in Seville, the growing German base in Spain and were issued to Spanish pilots who achieved little. This soon compelled their German instructors to take over and fly operational ground support missions. These German fighter pilots routinely flew several sorties a day, but to reduce risks, they adopted the tactic of approaching in line, one close behind the other, so that each successive aircraft was screened by its predecessor. Sometimes the lead pilot would then loop around for another pass, with the trailing aircraft following on for a strafing pass. Alternatively, when ground-support fighters attacked a broad target such as a line of earthworks, pilots would approach from the rear flying side-by-side, watching for the signal from their leader… then all drop their bombs in unison, creating a row of explosions called the *Bomb Carpet*, which was first used by the Condor Legion when they attacked Republican positions near Oviedo.

The initial Italian contingent of the *Aviazione Legionaria* consisted of a *Squadriglia* of Fiat CR 32 biplane fighters flown by volunteer Italian pilots. The Republicans flew mostly French aircraft, and their pilots were a mixture of foreign volunteers drawn by the idealism of an anti-fascist crusade, and others who came for the adventure ~ it was a pity that they generally lacked training and effective organization. Against regimented Italian and German aviators, the Republicans were poor opponents until October of 1936, when the Soviets introduced their own pilots and their powerful Polikarov I-15 biplane fighter, which armed with four machine guns could fly 220 mph and was nicknamed *Chato* or Snub-nosed. Later the Soviets introduced improved Polikarov I-16 fighters to the skies over Spain. The Republicans called their new fighters the *Moscas* or Housefly, however their pilots were less dedicated than their opponents. Often a *Mosca* squadron was seen fleeing from a few Fiats. In early 1937 the main focus of the fighting shifted to the northern industrial areas of Vizcaya and the Asturias in the Basque region on Spain's northern coast. Soon the industrial north was in Franco's hands. As the war took on the form of retreat and gritty resistance, with an occasional counter-attack, 50,000 foreign volunteers arrived in Spain to fight in International Brigades, among them George Orwell who said: *People sleep peaceably in their beds at night only because rough men stand ready to do violence on their behalf.* Ernest Hemingway came to report, and later wrote, *For Whom the Bell Tolls*.

THE BOMBING OF GUERNICA

Increasingly, German bombers targeted Spanish harbours to prevent the landing of supplies, and important bridges to stop the movement of arms and troops. One such bridge was near the Basque town of Guernica; and so it was that the German commander of the Condor Legion, *Leutnant General* Wolfram von Richthofen, (a cousin of the late Red Baron) became interested in the development of a situation that began on 20th April, when two anarchist battalions, dissatisfied with their movement's representation in the government, went on strike. They then pulled out and left a 15-mile wide hole in the frontline east of Guernica ~ just the sort of occurrence a man like Richthofen would pounce on. The Spanish town of Guernica, like other Basque towns, made arms and there were two small arms factories located on its outskirts. One supplied ammunition to Republican groups and the other made the *Unceta* pistol. Although it was mainly a market centre for the fishermen and peasants of nearby villages, it was also the site of the Renteria Bridge over which a highway headed east towards Bilbao. On the Condor Legion's operations map, Richthofen drew a circle around this bridge.

At 1630 hours on 26th April 1937, two nuns posted as lookouts on the roof of their convent sighted a lone Heinkel III bomber of the Condor Legion's experimental squadron *K88* approaching up the Mundaca River valley from the north and the church bells signaled an alert. At 1640, the Heinkel piloted by Rudolf von Moreau, already famous for a record-breaking flight from Berlin to Tokyo and for dropping food into the besieged Alcaza, then flew off disappearing behind the hills west of Guernica. Von Moreau reappeared, approaching from the north, this time to come down to 4,000-feet as he had drawn no flak on his first pass. He flew on, over the bridge ~ then his bombardier released six bombs, which fell hundreds of yards further on. One sliced off the front of the Station Hotel another destroyed the back of the same building. The four other bombs fell in the station square. Von Moreau then flew off and returned leading three more Heinkel III bombers. Bringing them down to 3,000 feet over the sea he saw them on their way before pulling out himself. Flying in formation the three bombers dropped their load when they came to the bridge, a load that this time included incendiaries, but most of the bombs again fell much further on. Flights of Junkers 52 bombers followed the Heinkel III bombers. As the bombs dropped and Guernica blazed, the rising-smoke cloud forced later flights come down as low as 600-feet to differentiate the target town from the countryside. Together this force savaged the town for three hours, dropping over 50 tons of bombs and then machine gunned the streets, demolished the town centre, with at least 70% of the houses. As they dropped their bombs, they flew off and then returned, as one observer noted:

> *Just going back and forth, back and forth, machine-gunning. Sometimes they flew in pairs, sometimes in a long line, sometimes in close formation. It was as if they were practicing new moves.*

And they were… Next came the Heinkel 51 fighter-bombers from their aerodrome near Vitoria. After all of this destruction, amazingly the two arms factories were still not hit. The highway bridge to Bilbao survived – it was even used as a shelter. When the *Luftwaffe's* Condor Legion swept down on Guernica, it had been market day. Guernica had been full of people and more than 1,600 of them were killed in the first saturation bombing on European soil. It indicated that the next war would see formations of aircraft over the cities of Europe, showering them with explosives and perhaps even poison gas, as the Italians had recently used in Ethiopia.

On a cultural note, Pablo Picasso himself a Basque, who living in Paris at the time as he studied newspaper reports and perused the black and white photos of Guernica's destruction was deeply moved by the tragedy at a time when he was looking for a dramatic subject to fulfill his commission for the international exhibition in Paris. In six weeks he created his celebrated *Guernica*. From Paris his huge black and white painting toured Europe and the USA to raise money for the Republican cause. When Roland Penrose organized the exhibition of Pablo Picasso's *Guernica* at the Whitechapel Art Gallery in London in 1938 to protest the Spanish Civil War, the price of admission was a pair of boots in reasonable condition to be sent to Republican soldiers at the front. Working men left the gallery barefoot, having placed their boots beneath the picture as if a shrine. Franco hearing about this made it a criminal offence to own a postcard of the picture and denied his involvement, insisting that the Republicans had torched the town themselves. Later in 1940, when Paris was under Nazi occupation, a German officer visited Picasso's studio and looking for evidence of resistance activity, pointed to a photograph of *Guernica* on the wall asking ~ *Did you do this?* Picasso replied, *No! You did!* Later at the Nuremberg war trials, Herman Goring testified that the atrocity had been a deliberate experiment in aerial bombardment, *to test my young Luftwaffe at this opportunity in this or that technical aspect.*

THE BOMBING OF GUERNICA

The people of Europe and America were horrified to read an account of the bombing of Guernica, by the journalist George Steer published in *The Times* and *The New York Times* immediately after the attack: *At 2 a.m. today when I visited the town the whole of it was a horrible sight, flaming from end to end. The reflections of the flames could be seen in the clouds of smoke above the mountains from 10 miles away. Throughout the night houses were falling until the streets became long heap of red impenetrable debris… many people were forced to remain round the burning town lying on mattresses, or looking for lost relatives or children.* Guernica being only 35km from the port of Bilbao where the *Hamsterley* and two other British food ships were docked, the Basque authorities arranged to show the British crews the devastation. The master of the *Hamsterley*, Captain A.H. Still, described the ship's arrival at Bilbao and the trip to Guernica in the British *North Mail*: GUERNICA RUINS ~ *When we reached Bilbao there was an air raid and 70 people were killed. Later the Basque authorities placed motorcars at our disposal and showed the British ship captains the ruins caused by the bombing raids on Guernica. In one bombproof shelter I saw corpses of men and women lying huddled together. It was a terrible sight.*

Pablo Picasso's Guernica is so familiar, so large, so present. It's physically bigger than a movie screen. But what is the painting about? Is it an account of the Spanish town obliterated by Nazi warplanes ~ a piece of reportage? Is that why it's in black and white? This is why the painting has such an impact. Instead of a laboured literal commentary on German warplanes, Basque civilians and incendiary bombs, Picasso connects with our nightmares. He's saying here's where the world's horror comes from ~ the dark pit of our psyche. Simon Schama

German-made Stuka dive-bombers, part of the expeditionary Condor Legion in flight above Spain in January of 1938. The black-and-white X on the tail and Stuka's wings is the Saint Andrew's Cross the insignia of Franco's Nationalist Air Force. The Condor Legion was composed of volunteers from both the German and Italian Armies, and the German Air Force Luftwaffe.

Italy contributed 36 of these Savoia-Marchetti S.79 bombers to their effort to influence the outcome of the Spanish Civil War.

THE SPANISH CIVIL WAR 1936-1939

One of the most important lessons from the air fighting of the Great War was that the best formation for aerial combat was the open, abreast style, with a spacing of 50 or 60 yards between each scout so that pilots could keep station with each other; fly near their leader without the risk of collision, search the surrounding sky against the possibility of surprise attack, and turn inside each other to face an astern attack. Both British and German airmen under the constant and unforgiving hammer of war had learned this formula, which had been recorded in thousands of memoirs and memoranda and yet, this practical expertise was all lost with the ceasefire on the 11th November 1918.

New tactics for fighter aircraft ~ Although the low-wing Messerschmitt Bf 109 German fighter appeared a few weeks before the advent of the British Hawker Hurricane and some months before the Supermarine Spitfire, these three high-powered monoplane fighters marked the end of the reign of the biplane fighter. When these new Messerschmitt fighters arrived in Spain, they flew in a close wing-tip to wing-tip formation; totally unsuited to for combat from the lack of maneuvering space and absence of cross-over. The officers of the new *Luftwaffe* had failed to study the history of their Imperial forefathers and had overlooked the lessons of the past. As more Me 109 fighters arrived in Spain to fly with *Jagdgruppe* 88 led by the stylish Adolf Galland who flew in swimming trunks, and smoking a cigar in a 109 adorned with Mickey Mouse. His successor, Werner Mölders soon found that owing to the increased speeds, greater turning circles and restricted views from the 109's enclosed cockpit, (especially from behind) and so sensing a lone-pilot more vulnerable than ever, he formed his flights around the smallest fighting unit, the pair and refined the flying of a pair of fighters near line abreast 150m apart with the following advantages:

1: By searching inwards, each pilot could watch the blind spots of the other ~ behind and below.
2: If one aircraft was attacked from behind, the pilot could break outwards, his companion also breaking in the same direction: then, if the attacker followed through, he would find the second Messerschmitt on his tail.
3: If the lead Me 109 launched an attack, his number two dropped in behind to cover him, which enabled the leader to concentrate all his attention on his attack.

The perfect fighter formation, (as it continues to be) was based on a pair (or *Rotte*) of two fighters, and proved far superior. With more fighters, a double pair (or *Schwarm*) was used, doubling the fighting strength of the formation and the numbers of eyes available to look out. The *Rotte* of two fighters was exactly the same array as that devised by Oswald Boelcke, during the Kaiser's War, except that the spacing between fighters had been increased from the turning radius of Boelcke's Albatros of 55m, to the turning radius of Mölders' Messerschmitt 109 of almost 300m. In the summer of 1937, the Condor Legion received another 30 Me 109 fighters and thereafter, a 12 aircraft *Staffel* flew together as three *Schwarm*, either in line breast or in line astern. These formations were flown with elements at varying heights to give mutual cover and vision. Sometimes Werner Mölders led his *Staffel* from the flank, so that he could break away easily and get into a good attacking position, but usually he led from the centre. Mölders shot down most of his victims when he led his *Staffel* on fighter sweeps over Republican airfields, after luring enemy fighters up higher into the air so his Me 109 fighters had the tactical edge. Mölders also made great efforts to train his airmen and lift their morale, often ensuring that a novice had an easy shot at his first opponent. In 1938 an improved version, the Me 109B was dispatched to the Condor Legion to patrol the Spanish skies. The new Messerschmitt carried radio and could receive and transmit in clear and distinct speech. Werner Mölders kept his team completely in the picture ~ a substantial improvement over previous methods, where a leader signaled his orders by wing movements or firing coloured flares. In Spain, air fighting became articulate, which improved teamwork in the air, and enabled closer control from the ground. Werner Mölders on his return to Germany, detailed his experiences of the Spanish Civil War, in a lengthy report to *Luftwaffe* HQ, which German fighter tactics were thereafter based on.

The situation altered with the advent of fast monoplane bombers, such as the German Dornier 17 and their Heinkel III, the Italian Savoia-Marchetti 79, and the Soviet Tupolev SB. To intercept these innovative aircraft, fighter pilots required a very early warning or a large degree of luck. Fascist pilot, Angel Salas flying an Italian Fiat CR 32 fighter at high altitude was able to dive on and destroy a Soviet Tupolev bomber and *Tenente* Mantelli successfully repeated this tactic, shooting down another Tupolev SB. Some indication of the problem of interception is apparent from the minimal Italian loss of only 8% of their SM 79 bombers, as opposed to 24% of their fighters. As defensive ground tactics, the Republican side used of dummy aircraft, and regularly switched their aircraft to different airfields.

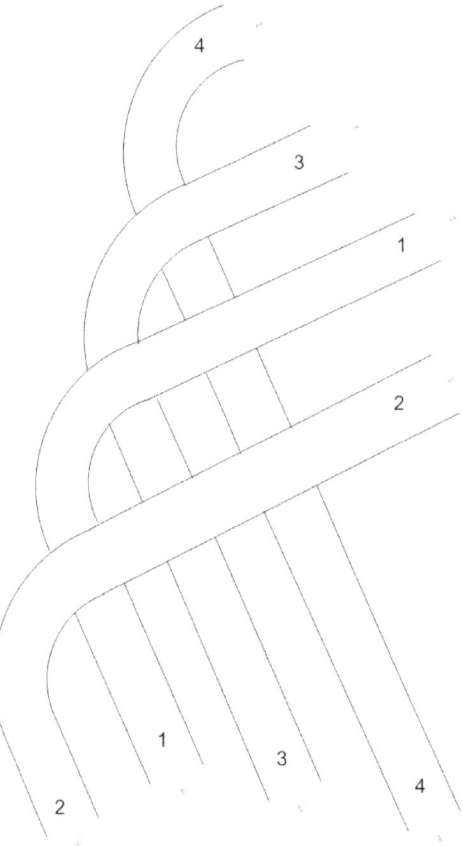

Werner Mölders ~ With other airmen in Spain, Mölders developed the formation known as the finger-four. This improved the all-round field of vision and combat flexibility of a flight Schwarm), enhanced mutual protection, and encouraged pilot initiative. In the finger-four, the aircraft assumed positions corresponding to the fingertips of an outstretched hand. The fighters flew in two elements (Rotten) of two aircraft each; two Rotten (four aircraft) made up a Schwarm (swarm) (Below ~ Werner Mölders in Spain)

THE SPANISH CIVIL WAR 1936-1939

The Stuka ~ The first flight of three Junkers 87A1 *Anton* or Stuka dive-bombers arrived in Spain in January 1938, to evaluate dive-bombing under operational conditions. The first Stuka flight, (known as *Kette Jolanthe*, after a caricature of a swine popular in Germany at that time), flying against targets with AA defences, proved able to score direct hits with 50% of their bombs within 75m of a target. Although only one *kette* of three Stukas was used in Spain initially, their crews were rotated so that a flow of battle experienced veterans returned to Hitler's Germany to pass on their tactical knowledge. They also, all received promotions up to their next rank. The Stuka was reliable, rugged, and sensitive to the controls, rock steady in a dive, and its wide all-round vision pleased those who flew it after the cramped cockpits of earlier aircraft. The secret of this two-seater was that its dive brakes allowed it to dive steady at a speed of about 500 kph. For accuracy, it was imperative that the pilot headed directly into the wind when plummeting in an 80° dive. Accuracy then depended on maintaining a constant dive angle and to assist, a protractor was etched into the plexiglass on each side of the Stuka cockpit canopy so that the pilot could read off his required angle during the dive. Stuka pilots in action, peeled off in rapid succession, plunged from the sky to release their bombs, pulled-out and climbed to circle in line astern. Then repeated the tactical approach on subsequent strafing attacks with their machine guns. In less than half an hour, a flight of Stukas could climb and dive perhaps ten times, and appropriately the Spaniards called them, *La Cadena* or *The Chain*.

Teruel ~ The Condor Legion called on their flight of Stuka dive-bombers to act as flying artillery to assist Nationalist forces engaged in the last stage of a harsh winter fight to relieve a surrounded Nationalist enclave at Teruel, the capital of the eastern province of the same name. On 17th February, He 51s and Stukas pounded their opponent's defensive perimeter, blasting an opening that the Nationalist troops were quick to exploit. By 20th February Franco's troops entered Teruel. The Republicans were still obstinately holding a few strongly fortified positions, and for their debut, the Stukas were sent in like birds of prey, swooping down towards their objectives with sirens wailing ~ to frighten the defending communists below, and pound them into submission. The following day, the Republican escape route was severed, and Teruel surrendered. The capture of Teruel was the first step on the Nationalists thrust out across the Northern provinces, to the Mediterranean coast on the Ebro front, through the Aragon and Catalonian offensives to envelope Madrid from the north and east. During the fascist advance across Aragon in the Spring of 1938, the Condor Legion targeted the towns of Albocacer, Ares del Maestre, Benasal and Villar de Canes ~ and then photographed them from the air and on the ground, to measure bomb patterns and the destruction caused, to assess the accuracy of Stuka bombing with 500kg bombs. The *Luftwaffe's* Condor Legion meticulously reported the effects of its new weapons and tactics. The fascist squadrons learnt to dart in, and strafe enemy trenches as soon as the artillery ceased to keep the Republican's heads down, while the Nationalist infantry charged the last few hundred metres. This became the norm. Republican commanders could not withdraw their troops and tanks, because of their exaggerated propaganda claims, the battles of Brunette, Belchite, Teruel and the Ebro were all tactical replication and Republican morale suffered. By late 1938 the Soviets were withdrawing. In February of 1939 the Republicans began evacuating Madrid. In March, Franco's fascists fought their way into the ancient capital, and the war was over.

The air war over Spain was fought in direct support of surface forces, and at first differed little from the encounters of the Great War. The many lessons of the Great War were still relevant: it was essential to keep a sharp lookout and seek the best position ~ the idea was still to get in close and then to deliver a fatal burst with the first few rounds. As aircraft speeds increased, surprise became even more achievable than in the Great War, due to the shorter time available between an aircraft becoming visible and reaching its firing range. A single fighter approaching head-on was virtually invisible at two miles. At Great War speeds, this gap would take over half a minute to close; but by 1939 this time had diminished to ten seconds. At this war's end the aviators of the Condor Legion had claimed over 300 Republican aircraft destroyed in action, and their combined experience enabled the *Luftwaffe* to revise their tactics, particularly after witnessing their devastating bombardments along the front lines, and the importance of close coordination between aircraft and ground forces. The Stuka again proved itself in action during the Catalonian campaign, and in the accuracy of its attacks against shipping ; an aspect that seemed to have passed unappreciated by the Royal Navy who were active around the Spanish coast at the time ~ the ship killing capability of the Stuka still came as an acute surprise to the Royal Navy. German Me-109 pilots in Spain reporting their 90-minute flight range to be insufficient, asked Willi Messerschmitt to design disposable fuel drop-tanks that could be jettisoned before a fight to extend their ranges. This was not done ; except for the Ju-87 R being equipped with them in 1940.

THE SPANISH CIVIL WAR 1936–1939

Leutnant General Wolfram von Richthofen found the Heinkel 51 biplane a very useful low-level bomber, especially when armed with clusters of small incendiaries known as *flambos*. These, napalm like *flambos* were a simple can filled with a mix of high-octane aviation fuel and motor oil fixed to a fragmentation bomb. *Leutnant* Adolf Galland was the first to use *flambos* in action to rout Republican militia from their trenches. Next, his nine-plane formation flew over in single file machine-gunning communists as they fled the torrent of blazing fuel flowing through their positions. Richthofen observed in Spain, that the secret to effective air-ground cooperation was for the close liaison between the local air commander and the officers commanding the ground forces in that sector of the frontline. Richthofen visited forward outposts on suitable hills, to ask the Spanish officers which enemy strong points needed to be targeted. Then, using a landline to advanced airfields, he ordered his *Staffels* to hit those targets, personally observing the outcome through his field glasses. As the pilots approached their target, the formation leader looked for smoke rising from the ground to determine the wind direction and align the attack run accordingly. After a pounding, the tanks began their advance, supported by a few loose flights of Heinkel 51 biplanes to pounce on any enemy war equipment that survived previous attention. If any further pockets of enemy resistance were reported to Richthofen, more fighter-bombers were ordered to the scene to quell the opposition, while overhead fighters continued flying protective patrols ~ and so on, until the day was won.

During the summer of 1938, when it became obvious to the Soviets that a Nationalist victory was inevitable, (and the Republicans were unable to purchase further aid), Stalin directed his remaining Russian aircrews and advisors to pull out, and to tail off the amount of war equipment being supplied to the Republicans. Soviet pilots serving in Spain, observed the Germans in action adopted a *Zveno* four aircraft formation of two *Pary* or pairs of their own, but as their machines were without radios, the Soviet spacing between aircraft needed to be closer for their visual signaling techniques that were unchanged since 1917. In 1939 it was recommended that the entire Soviet fighter force adopt the new Spanish *Zveno* formation of two *Pary*. However as most Soviet airmen who had flown over Spain had been killed in Stalin's purges, their recommendations perished with them, and were not carried out.

During October 1938, the volunteer International Brigades were disbanded, and some Italian troops were also withdrawn. Interestingly, Italy was the country that learnt the least from the Spanish Civil War. The Italians still believed in the desirability of retaining biplane fighters, together with the conviction that their lightly armed, high speed bombers could out-run defending fighters, which was proven disastrously wrong on several occasions. They overlooked the context that their bombers had operated in an atmosphere of air superiority and ignored the fact that their fighter, the Fiat CR 32 had done no more than hold its own. The Italians felt that the maneuverability of their CR 32 would keep their pilots out of harm and that their armament of only two machine guns was adequate, overlooking the fact that the new fighters being fashioned by other countries placed greater emphasis on climb, speed and rate of turn ; plus having self-sealing fuel tanks and armour protection, and that twin guns were now being superseded by batteries of up to eight or a combination of cannon and machine guns. After Barcelona surrendered, Franco's fascists rounded up their anarchist and republican adversaries and shot thousands, after having already executed over 200,000 prisoners of war during the conflict.

* * *

It is one of my duties to put myself under the German skin, if I can, and to report to you how they feel. Would the German nation willingly go to war for General Franco in Spain, if France had intervened on the side of the Republican government? The answer that Hitler gave himself was that the French would not, and he was consequently convinced that no democratic French government would be strong enough to lead the French nation to war for the Czechs. If we handle him right, my belief is that he will become gradually more pacific… But if we treat him as a pariah or mad dog we shall turn him finally, and irrevocably into one.

Sir Neville Meyrick Henderson
British Ambassador to Germany

So, by his own admission, if Adolf Hitler had been opposed earlier, he would have backed down. But for three years, Germany and Italy were free to send airmen, troops, and war material in their subjugation of a sovereign European nation. The entire lesson of the Spanish Civil War and WWII is that if the ambitions of Nazi Germany had been countered *before* Hitler had time to rearm, future horrors against humanity that could hardly be exaggerated, might not have occurred.

LUFTWAFFE ACES

Werner Mölders ~ 115 kills

Adolf Galland ~ 104 kills

Walter Nowotny ~ 258 kills

Eric Hartman ~ 352 kills

Eric Hartman and Ursula Paetsch ~ married in December 1944

Hans-Ulrich Rudel teaching his preferred attack method on a model T-34 Soviet tank

Heinz Schnaufer ~ 121 kills

Otto Kittel ~ 267 kills

ALLIED AVIATORS OF THE EUROPEAN THEATRE

Douglas Bader ~ 20 kills   Adolph Sailor Malan ~ 27 kills   John W.C. Moffat ~ disabled the Bismarck

James Howard USAAF ~ 12 kills   Francis Gabreski USAAF ~ 28 kills

Katya Budanova ~ 11 kills   Lillia Litvyak ~ 12 kills & 4 shared kills over 66 combat missions. First female fighter pilot to shoot down an enemy plane ~ first female fighter ace, and holder of the record for the greatest number of victories by a female fighter pilot.

CHAPTER VI

BLITZKREIG AND THE BLITZ

German ambitions for the geopolitical domination of Europe led to their demand for the ethnic German populated Sudetenland region of western Czechoslovakia, and their plan to invade Poland. Luftwaffe *Leutnant* Bruno Dilley would lead the first air strike of the coming war, and as part of their preparation, Dilley and the airmen of his flight boarded a pre-war Berlin-Konigsberg Express to cross the Dirschau rail bridges, and study their assigned target. From the train they could clearly see detonator wires strung along the riverbank of the Vistula. These two bridges were key points on the rail link, and the assaulting German Army intended to use them on their invasion route. If attacked however, Polish Army engineers guarding the bridges had orders to detonate the demolition charges rigged to the spans of the Dirschau bridges to destroy the structure, and blunt the German *Blitzkrieg*.

On the dawn of the attack at 0425 hours on 1st September 1939, in dreadful weather with curtains of fog and mist, *Leutnant* Dilley led his flight of three Ju 87B Stukas into the air. Flying at treetop level, they were on their way to sever the detonation wires leading from the Dirschau railroad station to the pair of steel bridges spanning the Vistula. Two minutes from Dirschau, Dilley spotted the silver gray of the Vistula River beneath his wings and banked turning downstream toward the bridges. His two wingmen followed, skimming along the river at low level all the way. Seeing the Dirschau bridges appear through the mist, Bruno Dilley transmitted, *Target ahead!* Easing their Stukas to the left, at 0435 the three pilots, one after another pressed their bomb-release buttons before the bridge, blasting the area where the wires were strung, then soared over the bridges in a long climbing turn to the left. Whilst the dive-bombers bombed the detonation wires to secure the bridges, German armour thrust into Poland. Polish engineers partially rewired to detonate and destroy one bridge, but German troops captured the remaining bridge as planned, and continued their invasion of Poland.

The *Luftwaffe* pushed on to deliver a series of heavy air strikes against Polish airfields that involved 2,130 bombers and 1,215 fighters that succeeded only in destroying a few non-operational aircraft. The Polish Air Force had already wisely dispersed their combat aircraft away to small, camouflaged fields to avoid destruction on the ground. Safe for now, but they lacked any early-warning communications, they could not be coordinated as a fighter screen intercept the prowling *Luftwaffe*, and could achieve little in repelling their opponents. Many Polish aircraft were destroyed or put out of action in the first week, yet some always fought on. During the Polish campaign, Stuka flights cooperated with the German ground forces, as *Luftwaffe* liaison officers advancing with the German Army relayed their needs to airborne Stukas in the role of flying artillery to hit assigned targets within 30 minutes of a request. The fitting of howling *Jericho-Trompette* sirens further enhanced the psychological effect of the swooping Stukas with their bird-of prey appearance that mortified many inexperienced troops.

On the first day, a Polish cavalry formation was reported at Wielun, and destroyed by 30 Stukas, as was a Polish infantry regiment disembarking from a troop train at the railroad station of Piotrkow which was wiped out by a flight of Stukas. So the tactical momentum went against brave Poland. Once their early-warning posts and airfields were over-run, the Poles were forced to withdraw east across the Vistula in growing confusion. Two days after the initial assault on Poland, both Britain and France declared war on Germany then hesitated, taking no direct action against their new adversary. The way to Warsaw was open. When the forward German tanks of the 35th Panzer Regiment reached the outskirts of Warsaw, they met bold resistance from accurate Polish artillery fire within the city. After struggling to advance through the suburbs, the Germans withdrew with heavy casualties. By 25th September, the capital was surrounded, and to prevent further street-fighting casualties, the *Luftwaffe* began round-the-clock phosphorus incendiary bombing missions to set Warsaw ablaze. By evening, Warsaw was a wreckage of flaming rubble, and its surviving citizens were facing starvation and typhoid. The Germans thrust east to link up with the soldiers of the Soviet Union, who by secret agreement, annexed eastern Poland after German troops secured the western half of the country. After a brave fight, against overwhelming force, Poland surrendered on 30th September 1939 ~ but not before General Klakus asked all Polish airmen to escape into neutral Rumania, then to travel via Italy or Greece to France and offer their services to the British RAF to avenge the injustice against their homeland. The RAF welcomed 145 volunteer airmen from Poland into their No. 303 *Kosciuszko* Polish Squadron RAF (recognizable by their checkerboard tails), and they would claim 126 German aircraft.

* * *

WESTERN EUROPE 1940

The Royal Air Force after assessing the precarious situation in Europe, accurately predicted coming events in a memorandum issued by Fighter Command in November 1939 advising all airmen:

The possibility cannot be excluded however, that the enemy might over-run the Low Countries at a very early stage of the war and by using aerodromes or refueling grounds in Holland and Belgium, might be able to send over fighter escorts with his bombing raids on England. I want all concerned to be thinking ahead, and to carry out simple tactical experiments, so as to be prepared for various eventualities.

British aviation took a hard knock on 14th December 1939, when 12 Vickers Wellington bombers attacked German naval installations on the coast of the Heligoland Bight. Unknown to the British at that time, the Germans had the searching eye of an early form of radar, called *Freya*, which gave them a useful eight minutes warning time of approaching enemy aircraft. Intercepted by Me 109B fighters and being insufficiently armed, the British lost five from their original 12 bombers. Four days later, a formation of 24 Wellingtons was directed to bomb Wilhelmshaven in perfect weather, but were intercepted by defending flights of Me 109s and Me Bf 110 *Zerstorers* (Destroyers) and suffered. German fighters attacking from all angles shot down 12 and damaged three that returned to England only to crash-land near their base. The Germans lost two Messerschmitt fighters. These confrontations were appalling for the British, who had to accept that their unescorted bombers could not operate by day against German fighters. Accordingly, the RAF policy of long-range bombing became one of night operations for the rest of the war (with few exceptions). One Bf 110 pilot reported:

The Messerschmitt Zerstorer, the Bf 110 is easily capable of catching and destroying this English type quite easily, even at low speeds. This provides scope for multiple attacks from any quarter, including frontal beam. This attack can be very effective if the enemy machine is allowed to fly into the cone of fire. The Wellington is generally very prone to catching light and burns readily.   Hauptman Günter Reinecke

Norway ~ The Western Front settled down during the winter, partly from the effects of the appalling weather, and little occurred until the middle of spring. When Norway was attacked in April of 1940, the pattern of German *Blitz* warfare included two new elements. Airborne troops: parachutists and troops air-landed from Ju 52 aircraft and silent gliders. These troops successfully captured the Norwegian airfields. The *Luftwaffe* commander in Norway, *Leutnant-General* Hans-Jurgen Stumpff then requested new twin-engined Junkers 88 light-torpedo/bombers with a top speed of 300 mph a cruising radius of 750 miles for distant anti-ship missions. Stumpff also requested a squadron of the new Ju 87 R (*Reichweite* or range) Stukas that carried a pair of drop tanks on their under-wing hard points, that extended their radius of action from a 156 miles to 400 miles, enabling them to hunt warships. The accuracy of the Stuka against ships was fearsome. Four *Luftwaffe* pilots from I/St G 1 were awarded the Knights Cross for action against Allied shipping during the Norwegian campaign, Hozzel, Grenzel, Mobus and Schaefer. First damaged on 19th April was British cruiser HMS *Suffolk*.

During the Allied evacuation of Namsos on 3rd May, the Stukas sank two Royal Navy destroyers, plus the French *Bison*. The Polish liner *Chrobry* was then hit, set ablaze, and sank beneath the waves. On 30th the AA sloop HMS *Bittern* had her entire stern blown off by an accurate Stuka attack and sunk, repeatedly asserting the *Luftwaffe's* faith in their Stuka dive-bombers. After Norway, it is curious that the Germans overlooked the vision of fitting drop tanks to other aircraft types in future campaigns.

The *Blitzkrieg* attack through Western Europe was a repeat of the successful tactics used in Poland. After accurate attacks on enemy airfields and fortifications to weaken the frontline at selected points, German tank spearheads thrust thru piercing the Western Front, as dive-bombers pounded frontline troops and artillery, cut their supply lines, pounded their reserves and smashed any formation that looked like making a stand. On 10th May 1940, the Germans struck, and in quick succession ; Denmark was occupied as the German Army thrust thru neutral Belgium and Holland to reach their final goal: France. Even though French airmen flying Poetz 63 recon aircraft had spotted and reported the main panzer thrust coming thru the forested Ardennes on 11th May toward France ~ the intelligence officers of the French Ninth Army refused to believe their aviator's reports. Due to their decisive disregard, the French lost their only opportunity to bomb the panzers whilst they were vulnerable on the narrow Ardenne roads. The French *Armee de l'air* assisted by a few RAF Hurricane squadrons, lacking any effective early warning system, fought at a disadvantage as their aerodromes were gradually bombed into oblivion and continually strafed. Six fresh Hurricane squadrons were sent to France.

DUNKIRK 1940

As the German Army swept into France, French fighters were diverted to strafe invading columns of German armoured vehicles to delay them, with the flawed consequence that there were fewer fighters to oppose the *Luftwaffe*, handing the Germans air superiority. On 13th May, 700 German aircraft launched an attack on the French positions defending the Meuse River south of Sedan, the selected river crossing point for the main German panzer force. Over 200 Ju 87B Stukas precision bombed the French positions and by the end of the day the Germans were over the Meuse and surging around fleeing soldiers who had thrown away their rifles or abandoned their field guns whenever the Stukas appeared. These river crossings were the key to the German campaign in France, and essentially, German tactics during the Battle of France were a larger scale report of the invasion of Poland. The dual task of the *Luftwaffe* was to support the army, and to neutralize opposing air forces.

On 21st May 1940, General Heinz Guderian commanding the German 2nd Panzer Division and known to his men as *Schneller Heinz* or *Hurrying Heinz*, reached the English Channel, severed the Allied front and turned north to confront the British and French armies in Flanders now trapped with their backs to the sea. Inexplicably on 24th May, Hitler ordered General Gerd von Rundstedt's panzer force pursuing the BEF to halt 25km from the coast around Dunkirk. By the time the order was rescinded, a strong Allied rearguard had been formed to hold the German armour off the beaches. Hitler was wary of the boggy terrain around Dunkirk that might engulf his panzers, still required to tackle French forces to the south regrouping in an attempt to save Paris. Luftwaffe commander, Herman Goring, implored Hitler ~ *Leave the destruction of the enemy surrounded at Dunkirk to me.* Hitler approved. The German aircrews were fatigued, and their aircraft needed overhauling.
Wolfram von Richthofen, the commander of Stukas held a different view;

Unless the panzers get moving again, the British will have time to evacuate the pocket by sea, no one can seriously believe that we can alone stop them from the air.

Over Dunkirk, the RAF flew from their home bases in southern England with their newest fighter, the Spitfire that had been held in reserve until now. Air Vice Marshal Sir Keith Park, the commander of RAF fighters in southern England, ordered his 32 squadrons of fighter aircraft to take turns of flying watches of 30 minutes each over the beaches of Dunkirk. Both the Hurricane and the Spitfire having been built as defensive fighters (with relatively small fuel tanks) could only fight for a short time over Dunkirk before having to return to England to refuel. The evacuation of the trapped Allied force began on 27th May 1940 as a picture of massed troops and material on long sandy beaches while a gathering of ships and small boats waited offshore ~ all easy bombing targets. The Stukas starting their dives from a safe position behind the German lines, screamed down out of the cloudy skies to bomb and strafe the rescue vessels, all overloaded with troops as they headed out to sea. Ship after ship was struck and many sunk, often with Allied AA guns silent for fear of hitting RAF fighters. On 28th May, British prime minister, Winston Churchill at a meeting of the British War Cabinet promised; *Whatever happens at Dunkirk, we shall fight on.* That same day over Dunkirk, the *Luftwaffe* suffered losses that exceeded those of the previous 10 days. The RAF blunted the *Luftwaffe's* attacks over the beaches of Dunkirk, and the soft deep beach sand muffled the Luftwaffe's bombs, so British casualties were lighter than expected. By June 4th, the last day of the evacuation, 338,226 British and French soldiers had been evacuated to England.

Hitler considered Dunkirk a victory and ordered church bells rung all over Germany for three days. Wiser heads clearly saw Dunkirk the as decisive failure of the *Luftwaffe* with 300 dead aviators and 189 German aircraft lost, who could have tipped the scales of victory the other way during the coming Battle of Britain. Over the beaches of Dunkirk, the RAF had lost 99 fighters, with 75 pilots killed or taken prisoner. The final stage of the Battle of France began on 5th June, and within three weeks it was all over. The French called for an armistice and the Germans possessed France to the north of a line between Bordeaux and Lyons, the *Zone Occupée*. Vichy France existed to the southeast.

The single weaving section ~ used by the RAF in France, involved the separation of the squadron into two flights of six aircraft, each with one weaving aircraft. Six loose pairs and two aircraft on separate weaving courses to cover the rear of the squadron. All fighter manuals of this time emphasized the constant need to check behind every few seconds. Also they reminded pilots to make certain that an opponent had not got on ones tail, and of the foolishness of flying straight and level for more than a few seconds in hostile skies.

German *Luftwaffe* medium bomber Heinkel He-111 ~ pilots and aircrew below…

Crashed He-III during the Battle of Britain

Focke-Wulf Fw 189A2 rear gunner position ~ and pilot in forward cockpit

Fw 189A-2 *Uhu* reconnaissance plane at Bordeaux-Merignac in France   FW 189A-2 rear gunner mount

Focke-Wulf Fw 189 *Uhu* (German : Owl ~ called *The Flying Eye* by the Allies) was a twin-engine, twin-boom, three-seater tactical reconnaissance/army liaison aircraft designed with a twin-boom configuration to maximize the field of view in the cockpit pod.

A formation of German Dornier Do 17Z light bombers flying over France on 21st June 1940

Do-17 cockpit

Me-110 interior sketch ~ and in flight...

Me-110 cockpit November 1940

THE BATTLE OF BRITAIN 1940

In the summer of 1940 Britain stood alone, expecting their fascist opponents to invade at any time. *The British have lost the war, but they don't know it. One must give them time, and they will soon come round.* So Hitler said to his closest military adviser General Alfred Jodl. In Britain church bells hung silent, only to be rung if enemy landing craft or paratroopers were sighted. Aged veterans and those previously medically unfit for active service, were now welcomed into the ranks of the Home Guard, to patrol miles of barbed wire strung along the English coast. All would be won or lost in the air.

The *Luftwaffe* was ordered as a preliminary to Operation *Sea Lion* (the planned invasion of Britain) to first blockade the English Channel, and to destroy the Royal Air Force ~ and when this had been done to shield their invasion fleet from the Royal Navy. The German Navy, which had still not recovered from the losses it had suffered in Norway were worried, which caused Hitler to waver. He decided that the invasion of Britain should only be undertaken only as a last resort, a mopping up operation after the British had been broken by the submarine blockade and air attacks of the *Luftwaffe*. One of the more experienced German pilots, Major Adolf Galland, who had flown 367 sorties in action, was of the opinion ~ *The Royal Air Force will prove a most formidable opponent.*

The British possessed a radar-aided ground controlled interception system, which enabled Fighter Command to make efficient use of its numerically inferior force. The 53 coastal radar stations of the *Chain-home* network would detect the bearing and range of enemy aircraft up to 120 miles away. Behind them were the Observer Corps, with their network of observation posts equipped with visual and aural detection devices: they were the only means of tracking German raids once they had crossed the British coastline. Its watchers and plotters elaborated upon radar information identifying enemy aircraft types, which radar could not and provided information on the nature of the threat, with its height and course, along with warnings of solitary attackers heading for vulnerable points. Reports transmitted to Fighter Command HQ at Bentley Priory were then transmitted simultaneously to the Fighter Group and Sector Stations concerned. The Group Controller then allocated squadrons to intercept raids and the squadrons were scrambled by their Sector Controller and directed to meet the threat with a time lag of only six minutes between the first radar observation and the plot appearing on the map. Each Group was sub-divided into Sectors: each having an airfield after which the Sector was named, plus up to three subsidiary airfields, with a maximum number of four squadrons of day fighters. A device code-named *Pip-squeak* operated by one fighter in each squadron transmitted a signal for 14 seconds in every minute for ground stations to take bearings on, to enable the Sector Controller to keep track of his squadrons in the air. The RAF would operate over home territory, where pilots could fly more sorties, and often survive to fight another day if they were shot down; whereas all German pilots shot down, became prisoners-of war.

The opening phase ~ The Battle of Britain began on the rain swept morning of Wednesday 10th July, with the German plan to lure RAF fighters out over the English Channel in response to attacks on British ports and coastal shipping. The day began with a lone Messerschmitt 109E making a weather and shipping sortie, to locate any coastal convoys in the English Channel. A few 109's took advantage of low cloud to slip across the Channel to test their machine guns and cannons against the barrage balloons, and dive-bomb Dover harbour. The Stukas made their rendezvous with their fighter escorts at 10,000 feet to fly over the Channel to bomb coastal convoys and attack any RAF fighters sent up to intercept them. This reduced the reaction time of the RAF to a few minutes. The climax came on 25th July, when a swarm of 60 Stukas attacked a 21-ship convoy in the Strait of Dover. Only two ships escaped undamaged after five were sunk: the *Ajax, Coquetdale, Empire Crusader, Henry Moon*, and *Summity* for the loss of two Stukas. The following day, two Royal Navy destroyers, HMS *Brilliant* and HMS *Boreas* came out of Dover to attack E-boats, and both took direct hits from diving Stukas. After this, the Admiralty suspended all daylight convoys through the Strait of Dover. By the end of July, accurate German bombers had sent 40,000 tons of British shipping to the bottom of the English Channel, but the flaw with the German plan, was that by the time intercepting RAF fighters had arrived on the scene, their bombers had often already dropped their bombs and were returning home. Another upset for the Germans, was that the RAF fighter pilots had strict orders only to intercept, if German dive-bombers were present. These German fighter sweeps failed to bring the British to battle and the last air actions of this phase were fought between 9th and 10th August. These preliminary engagements over the English Channel cost the *Luftwaffe* 192 aircraft with another 77 damaged, whereas RAF Fighter Command lost 169 Spitfires and Hurricanes. German intelligence continued to badly underestimate British fighter production.

THE BATTLE OF BRITAIN ~ 1940

In Messerschmitt 109s the German tactics always followed the same pattern, a quick pass and away, sound tactics against Spitfires with their superior turning circle.   Squadron Leader Alan Deere
No. 403 Squadron RAF

The older Messerschmitt 109 E *Emil* variant lacked both an artificial horizon and a gyrocompass. The Spitfire offered its pilot a superior view from its bubble shaped canopy and could out-turn the Me109 with ease (as could the Hurricane at medium heights). If the Me 109 tried to turn away, the Spitfire could cut the corner with ease and gain a firing position. Still, as always to destroy a rival, it was generally essential to get in close. Only in the dive was the Me 109 superior and its half-rolling evasive tactic followed by a steep dive to the deck was often resorted to, knowing its direct-injection carburetor gave it the edge as the Spitfire's engine drained of fuel in such angles, and tended to cut out. However, by half-rolling the Spitfire, fuel from the carburetor could be thrown into the engine instead of out to restore normality. As always, pilot ability was paramount. Unless an aviator could fly his aircraft to its limit, the excellence of his aircraft would avail him little against a skilled opponent.

Deflection shooting ~ There was not much time for fighter pilots of 1940 to adjust their aim flying fighter aircraft with speeds approaching 400 mph. The new reflector sights were superior to the old ring and bead sights of the Great War. When their movable horizontal range bars touched the wingtips of target, a pilot knew he was in effective range. But, he still had to judge how much *lead* or deflection needed to hit his adversary. Inexperienced pilots hose-piped their machine guns from side skidding fighters or opened fire from absurd ranges. Many could fire a fatal burst at an enemy from dead astern, but when they needed to make a deflection shot at angles of more than a few degrees, they had to be content with a *probable* or *damaged*, failing to make a kill. Because the typical RAF fighter pilot was not a superb deflection shot, the Spitfire's eight guns were harmonized to 450 yards for a *broad area of lethal density* to give an average shot the greatest chance of destroying his opponent. But this wide fire effect grouping limited the finer marksmen of Fighter Command who found the area of lethal density, not particularly lethal. Squadron Leader A.G. *Sailor* Malan decided to realign the eight guns of his Spitfire to converge on a spot 200 yards ahead of his aircraft.

Sailor demonstrated their hitting power, when one moonlight midsummer night, he watched the searchlights near Southend holding enemy bombers in their beams for several minutes. Sailor climbed towards a Heinkel making towards the coast at 8,000 feet, well illuminated by the searchlights. Opening fire from 200 yards the South African kept his thumb firmly on the tit, saw his bullets strike home, and with his windscreen covered with oil, broke away. He watched the German spiral down and then climbed after another Heinkel, also coned by the searchlights. This time Sailor fired short two-second bursts as he closed; when the enemy turned he fired again with just the exact amount of deflection required, for the Heinkel began to burn and crashed in flames. Highly satisfied with his two kills, Sailor returned to base and telephoned his charming wife who was waiting for the arrival of their first child, whose godfather would be Winston Churchill.   Lt. Johnnie Johnson

Most Spitfire pilots followed suit. Squadron Leader A.G. *Sailor* Malan, the South African ace who would shoot down 32 enemy aircraft during the Battle of Britain, was considered the master of fighter tactics in the RAF after he issued *Ten of My Rules for Air Fighting* to his pilots of No. 74 Squadron RAF. *Sailor's Rules* became part of the official tactics of the RAF Fighter Command. Another superb tactical innovator was Wing Commander Douglas Bader, commanding the Duxford Wing who avidly read the books of the airmen of the Great War to modernize their tactical lore on the importance of height, surprise, teamwork, and accuracy. He adopted the tactics of the German *Schwarm,* and then reshaped their flight pattern to the *Finger Four* formation, with the positions of his British fighters similar to the fingertips of an outstretched hand. If the fighter pilots of his *Finger Four* flights were bounced, the two pairs of fighters split off in opposite directions. At a time when squadrons were being sent into battle in pairs, Bader kept pressing for bigger wings of three, four, or even five squadrons of fighters. Together Douglas Bader and *Sailor* Malan perfected British fighter tactics and the *Finger Four.* On the German side, bombers and fighters often flew in equal numbers, so that a wing of Me 109E fighters escorted a group of 25 Heinkel III bombers, or Stuka dive-bombers. Werner Molders led his Messerschmitt wing on forays and frequently bounced Spitfires as they climbed to tackle the bombers. Mölders positioned his fighter escort, 4,000 feet higher and about a mile behind the bombers. But increasingly, darting flights of Spitfires would exploit the gap of a few thousand feet between the bombers and Mölders escorting Me 109 fighters, to attack the bombers before Mölders wing arrived.

THE BATTLE OF BRITAIN ~ 1940

RAF Plotting Room   *RAF Fighter Command*

Rules are for the obedience of fools and the guidance of wise men. Wing Commander Douglas Bader

THE BATTLE OF BRITAIN ~ 1940

Forward machine gunner in the nose of a German Heinkel He III bomber, while en route to England in November 1940

Adolf Galland discusses fighter tactics,
(Werner Molders on his left)

THE BATTLE OF BRITAIN ~ 1940

For maximum effect, German fighters needed to range ahead of and to the sides of their bomber formations to break up British fighter squadrons before they could engage. However, Goring alarmed at mounting bomber losses, directed closer fighter escorts that denied the Germans their most effective role of free-chase fighters. Me 109 fighters also had a limiting short range. Over England, a Messerschmitt pilot had to continually look for a red *low fuel* light on the instrument panel. Once seen, he had to return to France. Messerschmitt fighters could have been fitted with long-range jettison fuel tanks (as carried by anti-shipping Stuka Ju 87R variants offering them an additional range of 250km) that had been requested during the Spanish Civil War to enable more than 15 minutes flying time over London. Now the bombers could go only where fighters could accompany them, and this was not far: North London for Me 109s based in the Pas de Calais, or Southampton for the Me 109s in Normandy. The *Luftwaffe* had to destroy RAF fighter defences in the south, then range further inland and extend their offensive so their bomber formations could fly unopposed over England; reduce the great cities, sap civilian morale, and ultimately clear the way for a German invasion.

The second phase ~ Operation *Adlertag* or Eagle Day on 13th August saw the focus of the German attack shift to the radar stations and forward fighter airfields of southern England when heavy attacks by 485 bombers were launched across the Channel. All coastal radar stations were vulnerable, but the *Luftwaffe* always aimed their bombs at the high transmitting and receiving towers; difficult targets to hit, when they could have disabled the few radar stations by aiming at the control centers situated in the open nearby. Imaginative German intelligence officers were of the opinion that the nerve centers of the stations were hidden in bombproof underground shelters and overlooked these obvious targets, which could have been easily destroyed. Although much damage had been caused, little effect was discernable to the *Luftwaffe* and Goring on 15th August ordered his aircrews to abandon their attacks on the radar stations. Had more effective raids been mounted against them, the British early warning system would have been rendered ineffective. Was Goring going to continue to let the magic eye see? Did the Germans know that radar stations could be easily jammed or spoofed? Had their scientific advisers informed Goring that the radar horizon could not follow the curve of the Earth? Had they suggested that if their aircraft gathered at low altitudes over France and crossed the Channel 100-feet above the waves, they should be undetectable by radar? Had the Germans done enough homework?

A bigger problem for the British was that since 8th August, the output from British factories was no longer keeping pace with losses. Between the 8th and 18th of August the RAF had downed 367 enemy aircraft for losses of 213 (of which 30 were destroyed on the ground) but fewer than 150 new Hurricanes and Spitfires were supplied in this period. Fighter Command was facing a very anxious situation where it might waste away in a few weeks. By the end of the second week of September 1940 there were no fresh squadrons to replace the tattered RAF fighter formations in the decisive Southeast.

Goring ordered that *Luftwaffe* operations be directed against the RAF to initiate decisive battles in the air to destroy Fighter Command. If the Spitfires and Hurricanes did not respond, they would be destroyed on the ground. German bomber *Gruppen* would be supported by three fighter *Gruppen*, one flew ahead on a *Freie Jagd* (Free Hunt) to clear the target area, while the others were assigned close escort and high escort roles. Another success came after a pair of Ju 88s made a daring raid on the Brize Norton airfield near Oxford. Waiting until the British fighters defending the airfield were replenishing fuel and ammunition, the Ju 88s swooped toward the runway, lowering their landing gear as if to land, in a ruse hoping to be mistaken for friends. The raiders then climbed sharply, switched to attack profile, and unloaded their bombs with devastating effect setting two hangers ablaze and destroying numerous fueled-up aircraft. A low-level raid of nine Dorniers trying a similar approach hit Kenley airfield. Flying over at tree height, the Dornier airmen carried out a formidable low-level bombing raid, cratering the runways of Kenley and coming around to successfully strafe its dispersed fighter squadron on the ground. Between 13th and 18th August 1940, 34 fighter airfields were attacked. The basic German tactics were working, but could have achieved more effective surprise if small compact formations of fighters and bombers had crossed the Channel at low-level to avoid being detected by radar, then remaining at tree-top height and changed direction every 20 to 30 miles to confuse the Observer Corps, and then attack Fighter Command airfields two or three times every day until further blows were unnecessary. Two years after the Battle of Britain, the Germans finally exploited this tactic ~ low-flying small formations of fighter-bombers to avoid radar detection to bomb and strafe the towns on Britain's south coast. Other times their radio discipline was so poor, that they were overheard by Observer Corp aural devices, and RAF fighter pilots were told where to find them.

THE BATTLE OF BRITAIN ~ 1940

The third phase ~ from 24th August, German attacks concentrated on the main centres of fighter production and the fighter airfields. The *Luftwaffe's* well-escorted bomber force continued to compel the British fighters to intercept and be engaged while their airfields were destroyed. Furthermore, the German bombers were getting through and destroying RAF airfields, with fewer losses. By September 1940, the *Luftwaffe's* tactics were beginning to work and the Royal Air Force was in peril. At the beginning of September, stocks of German bombers were increasing in number and although their numbers of single-engine fighters had decreased by three per cent, there were ample trained aircrews. The *Luftwaffe* could carry on, but could Fighter Command? Between 24th August to 6th September, 103 RAF fighter pilots were killed and 128 hospitalized, a rate of loss that could not be countered by replacements. Six vital airfields in southeast England had been badly damaged and the number of squadrons available to intercept the *Luftwaffe* had been reduced because of the growing necessity to fly constant patrols over RAF fighter airfields. Fighter Command was reeling ~ and on 7th September, Alert No. 1 was issued : *Invasion imminent and probable within 12 hours.* The moon and tide would be favorable from 8th September. The Germans chose September 24th as their day.

At the height of the aerial onslaught, Winston Churchill addressing the House of Commons declared; *Never in the field of human conflict was so much owed by so many to so few.* The comment of one fighter pilot was that the Prime Minister's accolade must surely refer to his Squadron's mess bills! Then, on the night of 24th August, one Heinkel III bomber, by mistake dropped some bombs across the southern outskirts of London. These were the first bombs to fall on central London since the German Gotha bombers had attacked the metropolis in 1917. Prime Minister Winston Churchill ordered four retaliatory raids over the next ten nights on Berlin, and on the first raid, 29 out of a force of 81 Bomber Command aircraft dispatched were able to bomb Berlin. Hitler was livid, and ordered the *Luftwaffe* to change their objective to massive reprisal raids against London. Field Marshal Sperrle, one of the *Luftwaffe's* air-fleet commanders opposed the plan, certain that two more weeks of attacks would break the RAF ~ and he was right. By not focusing on the continual destruction of British fighters and their airfields, the flawed directive of bombing London gave the defenders a valuable breather, and ensured the survival of RAF Fighter Command. The RAF had won, at the price of London's suffering. Had the German airfield attacks continued Dowding would have had to pull his fighters back to airfields north of London, beyond the range of the Me 109s and out of reach of the bombers. The third phase worked in favour of the defenders as the British fighters rose to defend their capital, and the German fighters failed to destroy them. As always, the Spitfires engaged and held off the Messerschmitts, while the Hurricanes went for the enemy bombers. Losses in this phase amounted to 262 RAF fighters, while the *Luftwaffe* lost 378 aircraft, with another 115 machines damaged.

The attacks on London and its suburbs continued from September 7th until October 5th. It was the crux of the Battle of Britain, for it gave Fighter Command the breathing space to repair its battered airfields and to restore its communications. From mid-September on, Fighter Command's strength, instead of ebbing away began to pick up and the fight could be continued until victory was achieved. The *Luftwaffe* had three fatal flaws at this time : their lack of sophisticated radar on the Channel coast; the short range of their Me 109 fighters; and the flawed leadership of Hitler and Goring.

The fourth phase ~ beginning on 7th September, vast daylight bomber raids flew against London, with formidable fighter escorts, but these streams of German aircraft heading for a known target, presented the defence with an ideal target (repeating the same tactical blunder of the Zeppelins of the Great War when they routinely headed for London) and the German fighters could only linger over London for a few minutes before fuel shortage forced them to turn back. Of British concern, the entire production of Bren light machine guns came from a single factory on the outskirts of London, and had this factory been destroyed, there would have been serious repercussions for the army, but no hits occurred. The battle reached its climax on 15th September, which Winston Churchill called the culminating date (now commemorated as the Battle of Britain Day). The Luftwaffe now realized that they had failed to gain the air superiority needed for the invasion of Britain, and on 10th October 1940, as autumn gales swept the English Channel, Adolf Hitler cancelled Operation *Sea Lion*. Fourth phase losses for Fighter Command were 380 : 435 for the *Luftwaffe*, with another 161 aircraft damaged. Dowding had always expected Hitler to succumb to the temptation to strike at London instead of the airfields and factories. Indeed he counted on it, predicting sometime after the success of Dunkirk, *The nearness of London to the German airfields will lose them the war.*

BEWARE OF THE HUN IN THE SUN

IN A SURPRISE ATTACK THE ENEMY MAY COME OUT OF THE SUN WHERE IT IS DIFFICULT TO SEE HIM. REMEMBER TO LOOK FOR THIS WHEN ABOUT TO ENGAGE ANOTHER AIRCRAFT THAT MAY BE A DECOY.

SHEER CARELESNESS

NEGLECT TO CHECK YOUR OXYGEN OR OTHER EQUIPMENT MAY BRING DISASTER ON YOURSELF AND YOUR FORMATION.

RAF FIGHTER COMMAND GROUPS AND SECTORS ~ 1941

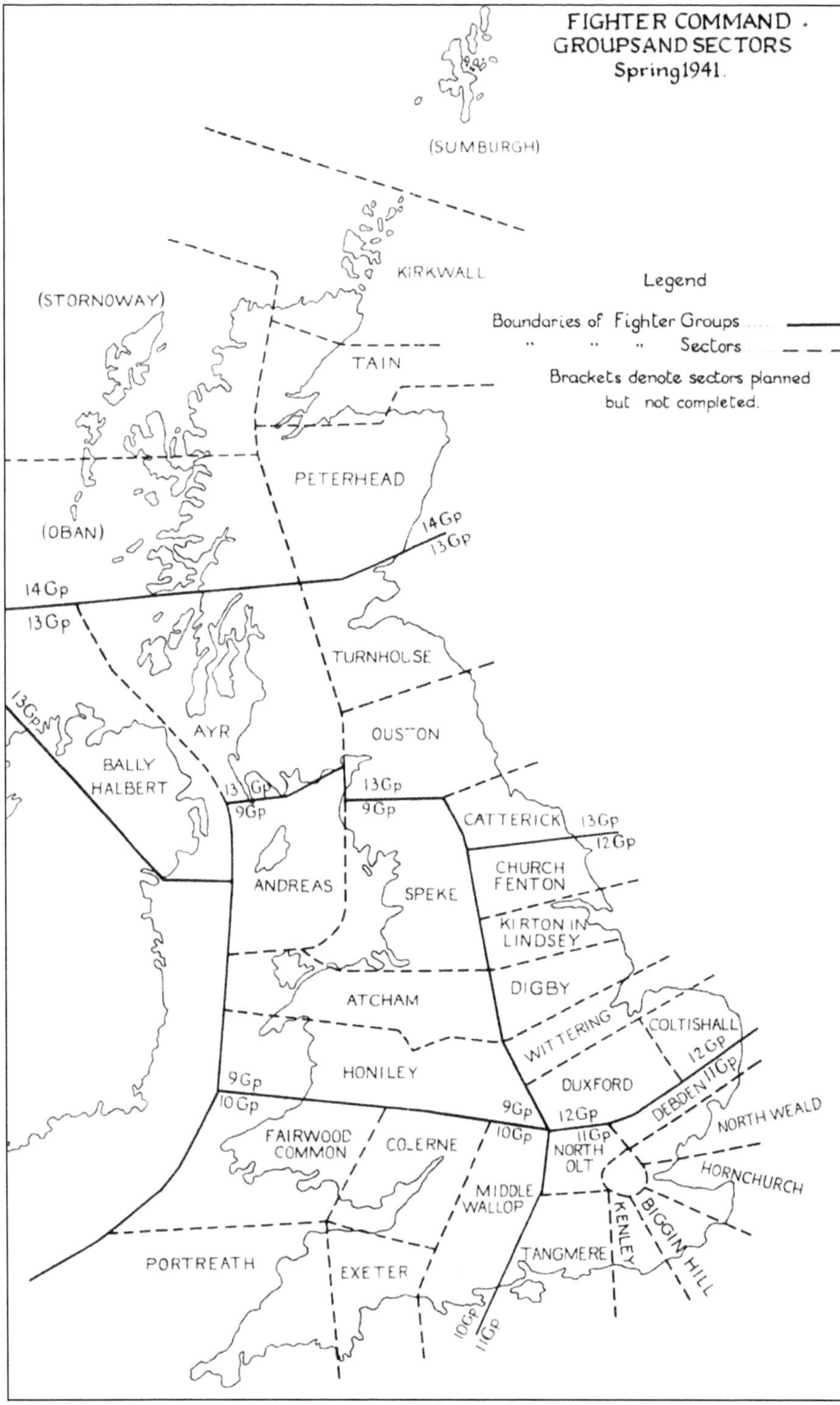

TEN of MY RULES for AIR FIGHTING

1. <u>Wait until you see the whites of his eyes.</u>
   Fire short bursts of 1 to 2 seconds and only when your sights are definitely 'ON'.

2. Whilst shooting think of nothing else, brace the whole of the body, have both hands on the stick, concentrate on your ring sight.

3. Always keep a sharp lookout. "Keep your finger out"!

4. Height gives <u>You</u> the initiative.

5. Always turn and face the attack.

6. Make your decisions promptly. It is better to act quickly even though your tactics are not the best.

7. Never fly straight and level for more than 30 seconds in the combat area.

8. When diving to attack always leave a proportion of your formation above to act as top guard.

9. INITIATIVE, AGGRESSION, AIR DISCIPLINE, and TEAM WORK are words that MEAN something in Air Fighting.

10. Go in quickly – Punch hard – Get out!

THE BATTLE OF BRITAIN ~ 1940-41

The fifth phase ~ of the Battle of Britain came in October, and was as distinct as the other phases were; as the Germans tried yet another plan, and failed. In this phase the *Luftwaffe* converted a number of fighter aircraft to fighter-bombers for intruder missions over southern England. Bombs were attached beneath Messerschmitt 109s and 110s to make fast hit-and-run raids at high altitude. With the final outcome of the Battle of Britain decided in favour of the RAF, the German medium bombers were reassigned to night attacks over London, and the Stukas were mostly dispersed to Italy and Germany in readiness for planned campaigns in southeastern Europe. On October 20th Goring ordered a halt to daylight bombing raids on London and planned a night bombing campaign to be primarily directed against London. *Luftwaffe* losses in this phase were 325 downed with another 163 damaged, while Fighter Command lost 265 ~ the Battle of Britain was Germany's first decisive defeat of WWII.

Long after the policy of crashing through with heavy bomber formations had been abandoned owing to shattering losses incurred, the battle went on. Large fighter formations were sent over, a proportion of the fighters being adapted to carry bombs, in order that the attacks might not be ignorable. This last phase was perhaps the most difficult to deal with tactically. Air Chief Marshal Sir Hugh Dowding, Fighter Command

Head-on attacks ~ RAF Squadrons No. 32 and No. 111 began experimenting with fearless head-on formation attacks. The attacking squadron spread out into loose sections line abreast and went straight at the enemy bombers. High closing speeds meant that the time available for the escorts to intervene was so short, that British frontal attacks were often carried out without interference. The German bombers, with large glazed noses provided no protection to their crews, and a determined head-on attack often broke up their formations ~ after which they could be picked off, one by one.

German bombing tactics ~ Once over the target, the Germans adopted one of two bombing methods, depending on whether or not there were defending fighters in the area. If there were not, and enemy AA fire was ineffective, the bombers attacked individually at low to medium level, and were often able to making several dummy runs before releasing their bombs : if there were no fighters, but AA fire was severe the bombers forced to higher altitudes would still bomb in the same unhurried way: as the Germans were ardent believers in individual bombing for each crew to drop their bombs as accurately as possible. If there were fighters about, German bombers flew close to provide each other with fierce covering fire, and released all bombs on the radio command of their mission leader.

The Blitz ~ was the closing stage of the Battle of Britain. From November 1940, German bombers forced by the growing strength of Fighter Command, switched to night raids against London. They then had to find their targets in the dark with navigational aids. The British countered by hiding the target with decoys, diverting the attackers with sophisticated electronic devices to jam or distort German electronic aids, and shooting down German bombers with AAA fire and night-fighters fitted with airborne-interception radar. Throughout the winter of 1940-1941, the German bombers raided targets in south central England. London was the main objective, and the campaign reached its climax in the period between 19th February and 13th May 1941 when their bombers adopted the *Krokodil* (*Crocodile*) tactic : a long stream of bombers dispatched along the same route at intervals of four minutes at altitudes between 3,000 to 6,000m. Inflicting considerable damage, bombers arriving later were able to bomb the fires started by earlier arrivals, or by a vanguard force as occurred in the devastating raid on Coventry on 14th November 1940, after it was marked by *Kampfgruppe 100* using the *X-Gerat* radio navigation aid. Late in 1940 the Germans abandoned the *Krocodil* tactic, in favour of concentrating all their bombers into a tight formation to saturate the defences by passing through a large number of aircraft in the shortest time possible. Where *Krokodil* attacks had lasted up to 10 hours, later attacks called for a short 20-minute attack of severe application. German night operations over England were most successful when they combined a generous use of incendiary bombs with a concentration of bombers over the target in the shortest possible time.

Nuisance Raids ~ replaced mass raids from 14th May. Although it was not clear at the time, one reason for the end of the *Blitz* was not the growing successes by the British night-fighter arm that had shot down 96 German bombers, but that Hitler was looking elsewhere. The few remaining Stukas in France carried out a few harassing raids over the English Channel and southern coastal towns, and by the New Year had withdrawn south to prepare for a Mediterranean campaign. The Battle of Britain had cost the RAF 414 pilots and more aircraft, but the *Luftwaffe* had lost 1,733 aircraft and their crews.

THE BATTLE OF THE ATLANTIC

Although the Battle of Britain had to be won first, the vital contribution of RAF Coastal Command during the Battle of the Atlantic was equally important to the Allied cause. In the North Atlantic, German U-boats and anti-shipping aircraft were attempting to sever Britain's lifeline to America of convoyed merchant shipping. British national survival was at stake. Winston Churchill after the war wrote ~ *The only thing that ever really frightened me during the war was the U-boat peril.*

The Enigma Code ~ For decades after World War II, part of the Allied victory at sea remained secret. The supposedly unsolvable German *Enigma* encryption machine that looked like a portable typewriter was capable of generating more than ten trillion code combinations. As often, the brilliant technical achievements of the Germans were not wisely used. They often unwittingly used formulaic phrases such as, *Heil Hitler* to begin or end messages, and knowing these obvious ten letters was the first step to cracking an encrypted message for the British code breakers in the grounds of Bletchley Park ; a country house in Buckinghamshire, codenamed Station X. To avoid alerting the German Navy that their *Enigma* codes had been broken, the indirect tactic was to avoid over targeting the U-boats, and hunt their supply vessels to reduce their capacity to intercept convoys. In February 1941 the captured German trawler *Krebs* off Norway had two *Enigma* machines on board and the naval settings for the previous month. This enabled the German Naval Enigma to be decoded, and Station X decrypts were all codenamed ULTRA. In May 1941, the German weather ship *München* was seized and her *Enigma* codebooks for June retrieved. But both times, German sailors partially destroyed the *Enigma* machines.

Operation Primrose ~ On 9th May 1941, Captain Baker-Cresswell on the destroyer HMS *Bulldog* spotted U-110 surfacing. He fired his forward guns and all Lewis guns from the bridge at the U-boat. *Kapitanleutnant* Fritz-Julius Lemp certain that his U-boat would be rammed and take its secrets to the bottom, failed to set scuttling charges and ordered; *Everybody out!* Baker-Cresswell who had intended to ram, recognized an opportunity for capture, and ordered Lieutenant David Balme to lead an armed boarding party to row across and seize the half-submerged U-boat. U-110 was stripped of everything portable, including the first captured working *Enigma* code machine. The bookshelves; signal books, code books an navigation manuals were all carefully transferred to HMS *Bulldog*. Everything was kept as dry as possible, as the codebooks and signal books were printed in ink that disappeared on contact with seawater. U-110 later sunk while being towed to Scapa Flow. All sailors who took part in the operation were sworn to secrecy, and the U-boat crew were kept secure for the remainder of the war.

Before the war, when *Lufthansa* had ordered a four-engined transport aircraft from Focke-Wulf, their designer Kurt Tank created the Condor ~ only losing a bet of a case of champagne by 11 days, that he would have the Condor flying within a year. The *Luftwaffe* seeing its potential, ordered a reconnaissance version, and from mid-1940, basing their long-range Focke-Wulf Condor aircraft in Bordeaux-Merignac in France, flew missions to scout and report any merchant vessels to patrolling German U-boats. Admiral Doenitz, commanding the U-boats then introduced *wolf pack* tactics, so German submarines at sea prowling for opportunist attacks could congregate to attack convoys. Once reported, a Condor continued to shadow the convoy of merchant ships and attack any straggler not keeping up, with its bombs and cannon. Initially, the Condors only faced light shipboard AA guns, and caused such havoc flying anti-shipping sorties that they were known as the *Scourge of the Atlantic*.

The first British counter-measure consisted of Hawker Hurricane fighters mounted on catapults on CAM ships (Catapult Aircraft Merchantman) in convoy. If a Condor or Ju 88 bomber approached the convoy, the fighter was fired off the CAM ships bow ramp with a huge rocket. On 3rd August 1941, after a Hurricane claimed the first Condor, the danger of long-range Condor patrols was diminished. Pilots of Hurricane fighters carried on CAM ships knew that they could not land on their launch-ship. Having shot down the enemy, the pilot had to consider whether he had enough fuel to reach the nearest land aerodrome. If not, as was often the case when flying over the Atlantic, he had to bale out or otherwise ditch into the sea as close as possible to a ship that could retrieve him from his rubber dinghy while his aircraft sank. His Hurricane, ditched at sea and lost, was an operational expense willingly accepted as an alternative to losing merchant ships. Pilots were rotated out of CAM ship assignments after two round-trip voyages to avoid the deterioration of their flying skills from lack of flying time at sea. Although 12 of the 35 CAM ships were sunk during 170 round-trip voyages, nine combat launches of Hurricanes claimed eight enemy aircraft.

CAM SHIP CATAPULT ARMED MERCHANTMAN WITH HAWKER HURRICANE IN THE LAUNCH POSITION
Hawker Sea Hurricane Mark I on the fo'c'sle catapult on board Catapult Armed Merchantman SS EMPIRE DARWIN. Note the flaps pre-selected in the take-off position. The catapult was angled to starboard over the ship's bows to prevent the rocket blast to the superstructure, and reduce the risk of the pilot being run over by the ship, should the aircraft ditch.

Hurricane fighter prepared for launch ~ when rockets would propel it along the rails to get it airborne.

Hawker Sea Hurricane Mark I lowered onto the training catapult for a training launch.
To the aft of the catapult are some of the firing rockets used to power the launch cradle.

The Hawker Sea Hurricane being catapulted from the CAM ship ~ note the long flame from the rocket assistors.

HUNTING THE BISMARCK

Early on the 21st May 1941, RAF reconnaissance aircraft reported seeing the German battleship *Bismarck* and heavy cruiser *Prinz Eugen* under the command of Admiral Gunther Lutjens moored in Korsfjord, Norway suggesting that they were preparing to make a foray out into the North Atlantic against Allied convoys. An RAF reconnaissance Spitfire confirmed their breakout on the 22nd of May.

The following day, the carrier HMS *Victorious* (Captain Henry Bovell) set course for Scapa Flow to join the Home Fleet, while HMS *Suffolk* and HMS *Norfolk* spotted the enemy in the Denmark Strait. On the morning of the 24th, the British Battle Cruiser Squadron cornered their fox, and then only one minute into the gunnery engagement, *Prinz Eugen's* 8-inch guns hit the British flagship HMS *Hood*. The *Hood* turned to bring her full broadside to bear against the Germans. Seconds later *Bismarck* fired a salvo that plunge pierced *Hood's* 3-inch armoured deck. HMS *Hood's* forward gun magazines detonated and within four minutes she was gone. HMS *Prince of Wales* turned hard to starboard to avoid the wreckage of the *Hood* to suffer seven hits that killed 14 sailors. With her guns jammed the *Prince of Wales* withdrew under the cover of her own smoke. The carrier *Victorious* with four escort cruisers separated to chase the *Bismarck*, and launch her torpedo attack aircraft, to try to at least score a hit to slow the mighty battleship. In high seas south of Iceland, Captain Bovell's *Victorious* launched nine Swordfish biplanes to target the *Bismarck*. Through a heavy rainsquall, Lt-Cdr. Eugene Esmonde led his flight at a steady airspeed of 85 knots. Three Swordfish equipped with Anti Surface Vessel (ASV) radar detected a vessel 16 miles ahead. Hoping they sighted the *Bismarck* through a gap in the clouds, they lost it seconds later. Their contact proved to be USS *Mordoc*. The *Bismarck* six miles to the south detected the Swordfish, and with it the Royal Navy lost their ace of surprise. Esmonde led his Swordfish attack, but the *Bismarck* evaded all eight of their torpedoes. Then Lieutenant Percy Gick flying Swordfish 5F came in low on the *Bismarck's* port bow. Launching his torpedo, it struck the *Bismarck's* armoured belt amidships, reducing her speed down from 28 to 24 knots.

Admiral Lutjen's fatal mistake ~ At 0300 in the morning of 25th May, Lutjens turned the *Bismarck* hard to starboard away from the pursuit to steer a wide looping course to cross stern of the British. After which, the radar operators on *Suffolk*, accustomed to losing contact for short periods as their captain zigzagged to avoid submerged U-boats, lost radar contact at 0500 hours and had to report: *Have lost contact with the enemy*. Lutjens tactic had worked and the *Bismarck* was making her escape. Lutjens then imprudently neglected to preserve radio silence and sent two unnecessary transmissions detailing the sinking of the *Hood* and a pointless report of *Bismarck's* damage. Predictably, listening posts in Britain received and plotted the bearings of the two German transmissions that soon revealed the *Bismarck's* position, plus a decryption signaling the *Bismarck* course setting for the Bay of Biscay. The only aircraft carrier now within striking distance was HMS *Ark Royal* (Commander R.S. Jessel) coming up from the south, but the battleships of the Royal Navy could not tackle the *Bismarck* if the Swordfish of the Fleet Air Arm could not hit and slow their target down. HMS *Ark Royal* carrying nine Swordfish of 818 Squadron (Lt-Cdr. T.P. Goode) set course to follow the *Bismarck*. At 1910 hours, on 26th May as *Ark Royal* heaved through waters so rough that the stern of her flight deck was rising and falling 60 feet, the Swordfish pilots took off into the surrounding storm to search for the *Bismarck*.

Swordfish Tactics ~ The optimum torpedo attack procedure called for a dawn or night approach from a height well above any enemy AA guns, then for a pilot to dive until he could see the target over the top of his upper wing, before leveling off at 25 to 50 feet above the water for a short run-in. (Alternatively, if two 500-lb bombs were carried, they needed to be released during a shallow dive). Originally designed as a torpedo-reconnaissance aircraft with a crew of three (or two carrying a bulky recon radio and long-range fuel tank in the spacious cockpit) the Swordfish had an armed operational range of 550 miles or reconnaissance range of 1,030 miles. At 1925 hours, with 70% cloud cover from 2,000 to 5,000 feet. Lt-Cdr. T.P. Goode used a coordinated approach tactic using sub-flights of three Swordfish flying in from different angles to make the *Bismarck's* 50 AA gunners divide their fire, and to make it even harder for the German skipper to evade their torpedoes. With no break in the cloud cover down to 2,000 feet, each sub-flight was directed to return to *Ark Royal* independently. At 2055, when Goode estimated that they were in a good approach position, he nosed down to dive...

Visibility was limited ~ *a matter of yards. I watched the altimeter go back. When we reached 2,000 feet I started to worry. At 1,500 feet I wondered whether to continue the dive. At 1,000 feet I felt sure something was wrong, but still we were completely enclosed by cloud. I held the formation in the dive, and at only 700 feet we broke cloud, just when I was running out of height.*

HUNTING THE BISMARCK

Sighting the *Bismarck* and flying into a storm of AAA fire, Flight Lieutenant John W.C. Moffat flying Swordfish 5C launched the decisive torpedo at 2105. It detonated against the stern of the *Bismarck* as it turned hard to port, tearing a huge hole in the hull, and decisively slowing the great battleship for the Home Fleet to catch. Moffat's Swordfish had shown its supremacy over the *Bismarck*. Other Swordfish airmen flew over the wave tops, ready to strike the desperate German target, but were confronted by concentrated AAA arcs of tracer and gunfire so deadly that any further attempt was suicidal. Ditching their torpedoes to turn away, they were still shot at by the *Bismarck's* gunners until seven miles out. Five Swordfish were hit by the *Bismarck's* AAA fire, and three were damaged. Flying over frigid waters and through the roughest weather, they returned to HMS *Ark Royal* at 2205. Now only six of *Ark Royal's* Swordfish remained serviceable ~ Swordfish 4C returned with 175 holes.

Two Swordfish remained aloft for five hours to shadow and report on the damage to the *Bismarck*. They transmitted that the enemy battleship had made two slow circles and was staggering off with two jammed rudders to the N.N.E. The *Bismarck's* two rudders that had been hard over in an evasive turn as Moffat's torpedo hit, were now stuck at 12° to port. When the last two Swordfish returned to the *Ark Royal* at 2345, they confirmed that the *Bismarck* had been hit on the port side, then on the starboard quarter, and later by a hit on the port quarter. When HMS *King George V* and HMS *Rodney* caught the *Bismarck* on the 27th May and opened fire from long range at 0900 Admiral Sir John Tovey bellowed ~ *Get the range! Get closer! I can't see enough hits!* And rightly so, an admiral during a fleet action who felt himself in possession of any advantage, either tactical, material or moral would strive to close the firing range to make the gunfire of his ships more decisive. The heavy guns of HMS *Rodney* and HMS *King George V* bombarded the slow and listing *Bismarck*, destroying the bridge and ship's command, but firing over 2,800 shells that scored at least 400 hits could not sink the *Bismarck*, and still its naval ensign flew. At 1020 HMS *Dorsetshire* was ordered to close in and fire three 21-inch torpedoes into the *Bismarck* to send the enemy battleship to a watery grave 600 miles west of Brest. The destroyer *Maori* rescued just 115 survivors from *Bismarck's* crew of 2,094. One obsolete Fairey Swordfish biplane had succeeded in slowing down and disabling the *Bismarck*, revealing the carrier as the decisive weapon, soon to replace the battleship as the ultimate warship. This would be confirmed six months later during the Japanese attack on Pearl Harbour.

* * *

Zigzagging ~ Merchant convoys generally steered a variable zigzag course, known only to the Commodore of the convoy, who conveyed changes to the merchant and escort warships by visual signals, and every change of course by convoy vessels upset the calculations of waiting and gathering enemy submarines. Although German U-boats during the Great War had operated individually and scattered over the ocean like a chequer board, the more innovative German submariners of 1942 began cruising coordinated patrol lines, where any variable zigzag alteration of course by a convoy away from a suspected or sighted U-boat, found it merely steaming into the patrol zone of a another U-boat.

Escort Carriers ~ With swathes of ocean that could not be patrolled by either shore-based aircraft or a convoy specific CAM ship aircraft, there were very few aircraft carriers available to fly support. Small escort carriers, just half the length and one third the displacement of larger fleet carriers and nicknamed *Woolworth* carriers by their Royal Navy crews were promptly designed and commissioned. Their role was to specifically to provide air cover in those unprotected areas of the North Atlantic to protect convoys against U-boats and to ward off enemy long-range Condor reconnaissance aircraft. The American made escort carriers supplied to the Royal Navy were somewhat different to those made by the British, having no ice-cream machines. They were considered luxuries on vessels that served grog and alcohol anyway. Aircraft were perhaps the best anti-submarine weapons of all and small escort carriers like HMS *Avenger* and *Biter* helped to close the *Mid-Atlantic Air Gap*, that precarious zone of ocean where land-based air cover could not reach.

The first aircraft to capture a German U-boat ~ was a Lockheed Hudson bomber of No. 269 Squadron flying missions from Iceland, whose crew sighted *U-570* breaking the surface on 27th August 1941. They zoomed overhead, and straddled it with depth charges as it started diving. As the sea spray from the underwater explosions subsided, the U-boat was spotted once more on the surface, with one sailor waving his Captain's white shirt as a flag of surrender. Relays of aircraft were then called in to keep watch over *U-570* until a destroyer arrived on the scene and took the captured submarine in tow.

*Operations Room at Coastal Command HQ at Eastbury Park in Middlesex.
A WAAF officer plotting the course of an aircraft, as two WAAF clerks update the position of a convoy.*

Heinkel 59 rescue seaplane

THE BATTLE OF THE ATLANTIC

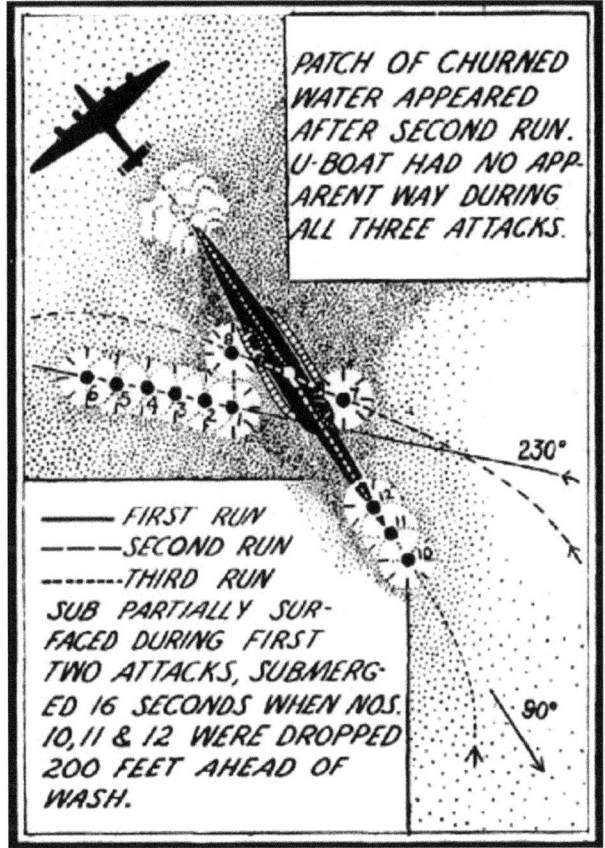

Combat report of anti-submarine action by aircraft. Tidewater Tillie is the B-24 Liberator bomber in which Lt. W.L. Sanford and his crew of the 2nd Anti-submarine Squadron have recently executed two attacks on enemy submarines, which resulted in one probably sunk and one known sunk. The first attack, illustrated by the above diagram, took place on 10th February, 800 miles west of St. Nazaire while the squadron was operating out of Great Britain. While patrolling at 300 feet at the base of a solid overcast, the left waist gunner sighted a U-boat on the surface off the port bow about four miles away. When first observed, the conning tower was clearly seen, but as the aircraft approached it disappeared and about forty feet of the stern was seen projecting out of the water at an angle of 20 degrees. As the aircraft attacked no churning was visible from the screws of the apparently motionless U-boat. Six Mark XI Torpex depth bombs spaced for 19 feet were released from 200 feet at 200 mph. The entire stick overshot ; the first bomb was observed to explode about 30 feet starboard of the submarine, as the tail-gunner Sgt. Yuschak fired 75 rounds at the exposed part of the sub's hull. As the pilot circled to port, the U-boat settled back on an even keel with the conning tower visible and both decks awash.

A second attack on the still motionless submarine was made with three more depth bombs. The tail-gunner fired another 75 rounds and saw the first depth bomb explode on the port side, while a second exploded to starboard. The U-boat appeared to lift slightly, lurching with the force of the explosion, and then remained motionless on the surface. While Lt. Sanford circled to make a third run the sea was seen to be churned just astern of the U-boat, and the conning tower settled beneath the surface without way, sixteen seconds before the last three depth bombs were released. The detonations occurred 200 feet ahead of the patch of disturbed water, but no plume resulted. Instead, a dome shaped bubble appeared followed by a large circular slick of brown fluid. Nothing further was seen and 30 minutes later the B-24 set course for base. When first sighted the U-boat apparently was attempting to dive at too steep an angle without sufficient way. This gave the pilot opportunity to maneuver for two additional attacks which resulted, according to Admiralty assessment, in ~ Probably Sunk.

THE BATTLE OF THE ATLANTIC

On 22nd March 1943, the Liberator *Tidewater Tillie* commanded by Captain William L. Sanford was flying on anti-submarine patrol in the Canary Islands area 600 miles south-west from their base at Port Lyautey on the French Moroccan coast. The B-24 bomber, well camouflaged with Mediterranean Blue on its upper surfaces and Cloud White underneath was patrolling at 1200 feet flying through scattered clouds when co-pilot 2nd Lt. Harlan Jackson sighted a broad wake about five miles off the starboard beam. Captain Sanford continued flying ahead into the next cloud. Made a right hand turn to follow the wake, and began losing altitude. Emerging from the cloud, a surfaced U-boat was spotted. Because of the bombers subtle camouflage and Sanford's tactical use cloud cover, the attack run was a surprise and the enemy lookouts did not see the Liberator until it was too late. Sandford continued his attack run, with the sun behind him toward the U-boat. With the Liberator at 200 feet and making 200 mph, bombardier Ralph Jones released his four bombs, as the B-24 passed over the U-boat.

As Sanford's Liberator flew over the U-boat, three German sailors were seen in the conning tower of what was identified as U-524 : one attempted to man the AA gun. Tail-gunner Edward Yuschak fired 75 rounds of .50 cal into them as *Tidewater Tillie's* bombs straddled the U-boat. Explosions enveloped the after portion of the U-boat, which continued on for a few seconds, then began to settle by the stern. With the entire bow section from the conning tower forward projecting out of the water at a 45⁰ angle, it took less than one minute for U-524 to sink. Several survivors were seen clinging to debris and a large oil slick developed… then a brown, paint like substance came up in the middle of the slick. Possibly rusty bilge oil discharged when the U-boat began to break up. Both of Captain Sandford's attacks displayed superb anti-submarine tactics and attested to value of aircraft on anti-submarine operations. The Battle of the Atlantic was a principal campaign through World War II, since everything that happened elsewhere, on land, at sea, and certainly during the Battle of Britain, ultimately depended on its outcome. The U-boats were defeated by mid-1943 and for the remainder of the war, anti-submarine aircraft continued to sink more U-boats than the surface warships destroyed.

After the British captured a copy of the German *Triton* codebook from U-559 in the Mediterranean, used to transmit messages to and from their U-boats at sea, from May 1943 they went over to the offensive and sunk almost 30% of the U-boat fleet. To preserve the secret of Station X's cryptanalysts, who were giving out frequent U-boat target directions and to encourage the misinterpretation that British attacks on U-boats were due to a large number of reconnaissance planes; all German vessels first had to be sighted by an aircraft, then an easily intercepted signal had to be transmitted from the aircraft to confirm their sighting: before any attempt to destroy an enemy vessel. Then came the added advantage of Microwave Radar on aircraft, which was introduced into the search for U-boats. It was so sensitive that it could detect the periscope of a submerged U-boat. It could even perceive a U-boat cruising on the surface through cloud, so that a looming anti-submarine aircraft could swoop on an unwary U-boat and begin dropping bombs or firing rockets capable of penetrating a U-boat hull. Part of the advantage of using microwave radar from RAF aircraft was that, although German U-boats had the means to detect shortwave radar, they had no device to detect microwave emissions.

Straddling U-347 ~ Because of the German submariners need for speed, either in traversing areas of sea under air patrols or to catch a reported convoy, U-boats began to spend more time on the surface; confident in the protection of their powerful AA armament. Patrolling anti-submarine seaplanes tried to drop their depth charges on both sides of an enemy U-boat in an attempt to assure at least one hit. The water-hammer effect of the explosions from both sides, especially if they struck near enough could shatter an enemy submarine hull. In one of the best examples of this ; flying a Catalina on an anti-submarine patrol over the northern waters close to the Arctic Circle on 17th July 1944, Flying Officer John Cruickshank sighted U-347 on the surface. In the face of fierce AA fire he maneuvered into position and flew in to release his depth charges. They failed to drop. As he climbed and turned to come in again the Catalina faced accurate AA fire from the U-boat and was repeatedly hit. The navigator/bomb aimer was killed, then the co-pilot and two other members of the crew were injured. Cruickshank himself was struck with two serious wounds in the lungs and ten penetrating wounds in his legs as his damaged Catalina filled with the fumes of exploding shells. He did not falter. He skillfully pressed home his attack and released the depth charges himself, perfectly straddling U-347 that sank beneath the waves. More than a thousand Allied merchant ships had been sunk by U-boats before the spring of 1944, when Royal Navy convoy escorts of destroyers and light escort-carriers had finally vanquished the German submarines ~ and won the vital struggle for control of the seas.

* * *

CHAPTER VI

THE EUROPEAN THEATRE

Following Italy's entry into World War II in June 1940, and their unprovoked attack on Greece, Britain moved to establish a forward base at Suda Bay on the north coast of Crete. The Commander of the British Mediterranean Fleet Admiral A.B.C. Cunningham assessing the changing situation directed the 21 Fairy Swordfish biplane torpedo-bombers on board the carrier HMS *Illustrious* to attack the enemy warships moored in the Italian naval base at Taranto as a precursory operation to important troopship and supply convoys from Gibraltar to Alexandria scheduled for November.

As a preliminary, the Royal Air Force carried out a detailed reconnaissance of Taranto harbour. When the report and photographs of this mission reached Captain D.W. Boyd, commanding HMS *Illustrious* they showed five battleships, 14 cruisers, and 27 destroyers moored in Taranto harbour. On the 10th November 1940, when the phase of the moon would suitably favour the venture, an RAF Short Sunderland flying boat crew flying from Malta, reported all six Italian battleships in port. Captain Boyd then signaled Admiral Cunningham that he was moving into position to attack. During the evening of 11th November, HMS *Illustrious,* 170 miles southeast of the Italian mainland steamed through the dark waters of the Ionian Sea at 20 knots, with a three-quarter moon directly astern, and increasing revs to full speed created a 30 knot wind over the foredeck, to get her Swordfish airborne.

Two waves of Fairy Swordfish aircraft were launched. The first wave of 12 aircraft, six armed with torpedoes, two with flares and bombs, plus four with bombs, were led by Lieutenant-Commander Kenneth Williamson and flew off into the night at 2040 hours. The second wave of nine Swordfish, five armed with torpedoes, two with bombs, and two with bombs and flares, led by Lt-Cdr. John Hale took off at 2130 hours from a position 275 km (170 miles) southeast of Taranto. The Swordfish were carrying 18-inch Mark XII torpedoes set run 300 yards before they would detonate either on contact or by magnetic proximity to a ships hull. Then at 2250 hours, below and off to starboard, British aircrews spotted the flashing light from the lighthouse on Cape Santa Maria di Lucia, on the heel of Italy. When Taranto harbour came into view, surprise had been lost as the enemy AA guns began to fire…

The lighting-up party ~ Two flare dropping Swordfish broke off and flew at 7,000 feet along the northeast shore of the Mar Grande to drop a line of brilliant blue-white light parachute flares. Firing off one every half mile to outline the warships below at anchor, to light the way for the torpedo aircraft. The flare-droppers also carried four bombs apiece to contribute to the general confusion.

The first wave ~ The main force of Swordfish banked out to the west to begin their long, gliding torpedo runs. Lieutenant Commander Williamson went in first. From 4,000 feet descending to 30 feet before flattening out almost mast height to release his torpedo at the anchored *Conte de Cavour* hitting it between the bridge and B turret. The 24,000-ton battleship then began to sink. Slowly submerging to rest on the muddy harbour floor. One Swordfish was lost when a sudden burst of Italian machine-gun bullets scythed into Williamson's Swordfish as he banked to starboard. His disabled plane plunged into the water near the floating dock, and they were captured. Next Lieutenant. N.M. Kemp flying 30 feet above the waves, darted at full throttle over the western breakwater, and flew straight at the massive battleship *Littorio*. At 1,000 yards Kemp dropped his torpedo and pulled away in a steep climb. His torpedo hit the *Littorio* on the starboard bow and ripped open a 50 by 30 feet hole in its hull. Lieutenant H.A.I. Swayne following dropped his torpedo 400 yards from the *Littorio,* hitting her on the port quarter punching a hole 5 by 20 feet just seconds after Kemp's success. The *Littorio* settled bow first to the muddy floor of the harbour. The first wave then flew south through the night sky and headed for HMS *Illustrious* waiting for them off the island of Cephalonia. Within 50 miles of HMS *Illustrious* they heard her homing beacon and at 0100 HMS *Illustrious* doing 23 knots turned into the wind and waited… then at 0112, a blip appeared on the radar screen : then many. At 0120 the first wave of Swordfish began touching down, with only one aircraft from the 12 raiders failing to return.

Searchlights ~ Although searchlights formed part of the defences of Taranto, the Italians neglected to use them on the night of 11th November. If they had, fewer of the slow flying Swordfish biplanes would have survived their mission. Without searchlights to hold their attackers in their revealing beams, the gunners were firing more or less blindly at aircraft they could hear, but only rarely target.

The second wave ~ south of Taranto, at 2120 Captain D.W. Boyd kept HMS *Illustrious* headed into the wind to fly off the nine Swordfish of the second wave led by Lieutenant Commander J.W. *Ginger* Hale. At 2300 hours and still far out to sea heading for Taranto at 8,000 feet, they began to see a blaze of lights over the enemy harbour. As Italian listening devices detected them, their shore batteries began firing well before this second wave even came within range. Near midnight Hale directed his two flare dropping planes along the eastern shore of the harbour. Their observers dropped 24 flares, and a brilliant magnesium glare to light up the entire harbour. Hale led, diving from 5,000 feet, jinking from side to side to confuse the Italian gunners, all of whom firing so continuously that their gun barrels glowed red. Hale leveled off, began his dash, and aimed for the *Littorio*. He dropped his torpedo at a range of 700 yards. It hit, and tore a third gaping hole in the *Littorio* 30 by 40 feet wide. Lieutenant C.S.C. Lee skimmed over the water after Hale, and went for the *Caio Duilio*. He dropped his torpedo and hit it 30 feet below the waterline, ripping a hole 36 by 24 feet wide. As the water poured in, her crew beached her to prevent her from sinking. However, she would never sail again. Lee made his escape skimming between the cruisers *Zara* and *Fiume*, who fired on each other as he zoomed through the space between them. Then the second waves flew off into the darkness and soon after the skies over Taranto were empty. With a few flashes, some Italian AA batteries were still firing at wisps of smoke and imagined engine noises. Although the Swordfish had dropped 11 torpedoes set to run at 30 feet to detonate on contact or magnetic proximity, to pass under the Italian anti-torpedo nets surrounding their warships that extended down 26 feet ; yet, even when launched properly, many British torpedoes failed to function. Only five torpedoes hit and detonated.

Taranto was an operation upon which perhaps the whole course of the war in the Mediterranean depended, and in a raid known in Italian history as *La Notte di Taranto (*Taranto Night) that lasted only 23 minutes, three of Italy's six capital ships were put out of action. In the inner harbour a heavy cruiser and a destroyer were also hit. This operation devastated the Italian Navy and removed the threat of Italian naval intervention from the Royal Navy's operations in the central Mediterranean as the balance of power shifted back in favour of the British. The day after this brilliant British attack on Taranto, the remainder of the Italian fleet steamed north to seek refuge in Naples. The British Fleet Air Arm raid then unintentionally inspired the Japanese plan for attacking Pearl Harbour. After the success of the Taranto raid, Japan recalled from London her assistant naval attaché, Commander Minoru Genda, and requested him to prepare a detailed report on Taranto and send it to Japanese naval intelligence with a full description of the British raid. Japanese Admiral Isoruku Yamamoto closely studied Genda's report of the raid on Taranto and when Genda arrived back in Japan he was ordered to carry out the planning for a surprise attack on the American Pacific Fleet at Pearl Harbour.

* * *

The Battle of Cape Matapan ~ that commenced on 27th March 1941, off the S.W. coast of the Greek Peloponnesian Peninsula was a prime example of the use of tactical air power. The Fleet Air Arm of the Royal Navy launched an air attack on the Italian fleet, to induce an Italian admiral to reduce the speed of his vessels, so they could be caught in a decisive naval action. The battle commenced when the 1st Battleship Squadron of the Royal Navy, commanded by Admiral Sir Andrew Cunningham then covering troop movements to Greece, received a report of an Italian fleet heading out into open seas. As the source of this information was an ULTRA decryption from Station X, a visual sighting by a directed reconnaissance aircraft was a prerequisite before any of the Italian vessels could be engaged. The Italian vessels were dutifully sighted by a Sunderland at 1220 on 27th March steaming east of Sicily on course for Crete. Cruising ahead, the Italian 3rd Cruiser Division of three Italian cruisers was followed by the 18th Cruiser Division of two light cruisers. Then came the battleship *Vittorio Veneto* with an escort of four destroyers. None of the Italian vessels were equipped with radar, but the British were. A Swordfish spotted the Italian Fleet again at 1930 in the evening ~ still on the same heading.

On the 28th, the carrier HMS *Formidable*, escorted by a flotilla of destroyers launched her aircraft to find the Italian vessels. At 1510 the *Vittorio Veneto* was sighted heading west. Minutes later a flight of three Albacore torpedo aircraft took off from HMS *Formidable* to target the Italian battleship, while their two escorting Fulmar fighters prepared to strafe her decks. Approaching low, and unseen… however once heard, the Italian AA gunners hit back with an intense barrage. One torpedo slammed into *Vittorio Veneto's* stern. Shaken, she slowed in the water and took on 4,000 tons of seawater. But her crew operating her flood control system expertly, got their ship moving within minutes. First up to 10 knots with her starboard screws, then the Italian flagship recovered half speed and escaped to Italy.

THE BATTLE OF CAPE MATAPAN

Lieutenant *Tiffy* Torrens-Spence finding a gap in a billowing enemy smokescreen, flew through to identify the Italian heavy cruiser *Pola*, and launched his torpedo from close range. Hit and holed, seawater flooded into the *Pola's* boiler room. She slowed to six knots and stopped dead in the water. Not sunk, but unlikely to escape. A.B.C. Cunningham pressed through the night with his heavy ships, reforming his ships to line-ahead for an expected night action. At 2220 using their radar at night, HMS *Valiant* reported a contact blip on her radar screen as a stationary vessel, 700 feet long, 5 miles off her port bow. It was the Italian 1st Cruiser Division coming to rescue the *Pola*, after responding to her request for a tow, and believing the British fleet to be far off, still had their guns trained fore and aft. The commander of the Italian *Pola* then mistook two closing British battleships for the *Fiume* and *Zara* and fired a recognition flare. Through their binoculars, the British could see illuminated Italian sailors walking the decks relaxed, enjoying a few more quiet moments with their after dinner smokes…

The British destroyer HMS *Greyhound* assisting the gunnery of Cunningham's battleships used her searchlight to dazzle the unsuspecting Italians and illuminate the night sky for 24 British 15-inch guns to fire a broadside of amour piercing projectiles into *Fiume* from 3,800 yards. Her aft turret took a direct hit and toppled overboard. A second broadside was fired into the *Fiume* 30 seconds later and she was gone in a huge orb of orange flame. HMS *Valiant* and *Barham* fired five 15-inch broadsides into the lead Italian cruiser *Zara* that listed heavily and ablaze from stem to stern, still had her turrets pointing fore and aft. After an engagement of less than five minutes, Cunningham's squadron of three battleships destroyed three heavy cruisers (*Fiume*, *Zara* & *Pola*) and two destroyers (*Alfieri* & *Carducci*) with over 2,400 officers and men before their opponents fired just one shot in response. All possible, because Lieutenant Torrens-Spence's torpedo had brought the *Pola* to heel. Of all the Italian warships, only the destroyer *Gioberti* escaped. As the sailors of the destroyer HMS *Jervis* rescued the *Pola's* damp captain he said ; *Either, that pilot was mad or he is the bravest man in the world.* After the Battle of Cape Matapan, what was left of the Italian fleet, never sought a naval action again. Taranto was a shock to the Italians, but Matapan was the strategic end of the Italian Navy, which only came out in force two years later and that was to surrender to Admiral A.B.C. Cunningham at Malta.

Hitler decided that, despite the continued defiance of Britain, his German forces would invade Soviet Russia sometime during 1941. But when his fascist partner Benito Mussolini sent his Italian forces across the Albanian border on 28th October 1940, launching an unannounced attack on Greece, events would transpose the Russian campaign forward with schedule slippage of fateful consequence. Hitler initially postponed his planned invasion of Soviet Russia for one month, for his forces to assist the Italians in Yugoslavia. The Germans reached Salonika on 9th April 1941, and within five days had broken through the Allied line in front of Mount Olympus. As in France and Poland, the Germans made aggressive *Blitzkrieg* thrusts with armoured columns under commanders who stayed well forward, exploiting every advantage with little regard for flank security. On 27th April 1941, the Swastika flew over the Acropolis of Athens, after the Germans had achieved in three weeks, what the Italians had failed to achieve in six months ~ however, the Germans had lost six very fateful weeks.

Operation Mercury ~ as an early summer brought bright sun and clear skies; perfect flying weather, Hitler issued his directive for the airborne invasion of the Mediterranean island of Crete for 25th April. German forces planned to take it by airborne assault reinforced by landings from the sea. To begin, an Italian tanker arriving at Piraeus before 17th May was made available to ensure adequate fuel supplies for the entire operation. Then, attack by dive-bomber and fighter units would eliminate any Allied aircraft on the ground, and strafe their AA defences preceding the operation; caring not to destroy the Cretan aerodromes for their own air-landing operations. The German XI *Fliegercorps* was responsible for ferrying the paratroopers to Crete using 530 Ju 52's and 70 DFS 230 light assault gliders, all together 8,100 *Fallschirmjager* would be dropped on Crete. On the morning of day one, the first wave of 7th *Flieger* Division, Group West would parachute land 1,680 men to seize the airfields of Malemé, 2,460 men at Cannae. Then in the afternoon, once the *Luftwaffe* Ju 52s aircraft had returned from Crete, they would refuel and return again to the island with the second wave. The 1,380 men of Group Centre would attempt to capture Retimé and 2,360 men of Group East would move on Heraklion. Then, three mountain regiments would be air landed on the first captured airfield to support the paratroopers, after which Stukas and Messerschmitt fighters would fly to land in Malemé and Canea. By capturing Crete, the Germans hoped to keep the RAF away from their Rumanian oilfields, and acquire new bases to further disrupt British shipping and fleet movements.

THE BATTLE FOR CRETE

German intelligence had under estimated the Allied strength on Crete at one undersized division of two brigades, plus an unknown number of troops recently evacuated from Greece. The aerial reconnaissance flights of the *Luftwaffe* had failed to reveal many camouflaged Dominion positions, within vineyards and olive groves, which had been checked from the air by RAF observers and further improved on. The Germans expected 5,000 opponents. But on Crete there were 40,000 under equipped and battle weary defenders waiting for them, with two tanks to protect each aerodrome. Some officers wanted to mine the aerodromes, but were not allowed, as it was hoped that British fighter aircraft should return as soon as possible. Therefore all three aerodromes had to be preserved. The defenders armament in AA and coastal defence guns was below what was required. On receiving from Egypt a shipment of 100 incomplete field guns captured from the Italians, one officer wrote;

Sufficient to say that, others came without their instruments, sights, some came without ammunition, and some of the ammunition without fuses. The gunners, men of infinite resource and energy set to work and made a sighting appliance out of wood and chewing gum. Another lot of gunners made out charts, which enabled them to shoot without sights or instruments. Nobody groused and everybody got on with the job.

Then there was the problem of sketchy communications, bad roads, and insufficient transport. The island garrison of the New Zealand Expeditionary Force under General Bernard Freyberg plus an assortment of Greek, British, and Australian troops were planning to stay put and establish a permanent refueling base for the Mediterranean Fleet at Suda Bay. The three aerodromes at Malemé, Canea, and Heraklion were not developed, and anyway there were few aircraft available to use them, but the British plan was to construct advanced landing grounds, from which their bombers could attack targets in Italy and bomb Italian forces in Greece and North Africa. Winston Churchill cabled to General Sir Archibald Wavell in Egypt, and to General Freyberg on Crete, a recent ULTRA decryption of German intentions: *It seems clear from our information that a heavy airborne attack by German troops and bombers will soon be made on Crete. Let me know what forces you have on the island and what your plans are. It ought to be a fine opportunity for the killing of paratroops. The island must be stubbornly defended.*

Malemé ~ The Commander of the New Zealand 5th Brigade, Brigadier Edward Puttick was right to feel anxious about the area left under manned west of the Tavronitis River entering the sea close to the perimeter of the Malemé airfield, and it was unfortunate that his intention to place a Greek battalion there had not been put into effect before the attack came, as it had consequent results when the German attack eventually came. Puttick also wanted to keep all AA guns at Malemé concealed (Freyberg did not agree) and withhold their fire until the actual airborne invasion. Had Puttick's suggestion been adopted, each AA gun would have shot down more slow flying Ju 52 transports, killing hundreds. The static defence of Malemé airfield was assigned to 22nd Battalion, commanded by Lieutenant Colonel Andrew, with orders to cover the airfield and its approaches with small arms fire, withholding mortar fire until the enemy landing began. If a landing was made, support and reserve companies were to launch an immediate counter attack, and then resume their defensive positions.

General Kurt Student from his HQ in the Hotel Grande Bretagne in Athens directed the paratroops of Group West under the command of General Eugene Meindle to drop around the airfield of Malemé on the north coastline west of Canea, to silence any AA guns and defending troops; especially those on the overlooking Hill 107 near the airfield. Once Malamé had been secured, Ju-52 aircraft would fly in to land the 5th Mountain Division. On 20th May, the Germans began their early morning routine of strafing and bombing to cause the defenders to anticipate that this day would be the same (although ULTRA reports detailed otherwise) and when the expected lull came, the allied troops began having breakfast, but this time the aerial bombing and strafing attacks resumed with ferocity. Most of the AA guns were put out of action and the defenders were forced to seek shelter. Bombs were dropped at the approaches to the aerodrome to put the telephone lines out of order. At 0815, the first wave of the German airborne assault began, with continued intense bombing that was lifted in a few corridors for the 1st Battalion *Sturm* Regiment to descend by glider, each carrying 12 men, through a corridor of thick protective dust to swoop down near the aerodrome. Their mission: to silence any still active AA batteries. The blanket of dust and smoke partially hid the landing of some 40 DFS 230 gliders west of Malemé, as well as flights of five gliders at various points between Malemé and Suda. 15 minutes later, the tri-motored Ju-52 transports dropped their parachutists in waves of 200 at 15-minute intervals, along with their equipment marked by different coloured parachutes to denote specific weapons, such as their signal pistols, (which could be also used to fire explosive grenades over a useful distance), radios, loaded rifles, mortars, machine guns, and other supplies.

THE BATTLE FOR CRETE

Looking out to sea with the field glasses, I picked out hundreds of planes. There were the huge, slow moving troop carriers with the loads we were expecting. First we watched them circle counter-clockwise over Malemé aerodrome and then, when they were only a few hundred feet above the ground, as if by magic white specks mixed with other colours suddenly appeared beneath them as clouds of parachutists floated slowly down to earth.

General Bernard Freyberg VC

Parachutists dropping east of Malemé met with deadly opposition, and most were unable to recover their weapons containers; having to rely on their Luger pistol or MP 38 submachine gun, (neither weapon had a great range), four hand grenades and a knife they carried. Until the German paratroopers reached the weapons containers, they could be picked off at long range. Those descending west of the Tavronitis River or on the Aghya Plain did so with little opposition and gathered to form their fighting units. By mid-morning groups of German parachutists had linked up with others on the slopes of Hill 107. Lieutenant Colonel Andrew decided to lead 22nd Battalion supported by the two Matilda tanks under his command in a counter attack. The two tanks emerged from their camouflaged lair and lumbered down the road towards the Tavronitis Bridge 30 yards apart. However these tanks were worn out crocks. The turret of one Matilda suffered from its turret crank only enabling it to traverse clockwise ~ and then its ammunition did not fit. The other tank continued unscathed through German small arms fire. Reaching the dry riverbed, it passed under the bridge and pressed on northwards for another 200 yards until it bellied down stuck in rough ground, its turret jammed and useless. Its crew had to abandon it. The New Zealander's counter-attack stalled and failed. Throughout these actions, the dive-bombers were unable to silence any British artillery, which particularly well camouflage in order not to reveal their position, withheld their fire whenever any German aircraft were in sight. By evening, not one aerodrome had been captured by the Germans.

In the evening, the Germans gained their first glimmer of triumph. West of the red earth of Malemé surrounded by small green hills of olive groves, *Leutnant* Horst Trebes and the Regimental Surgeon, Dr. Heinrich Neumann, assembled the glider assault troops. Charging up Hill 107 firing pistols and throwing stick grenades in a close action, they were just able to dislodge the New Zealand defenders from the crest of Hill 107 and capture the key position overlooking Malemé airfield. Flinging themselves down onto the Cretan earth, fearing a counterattack, Dr. Nuemann remembered: *We were so short of ammunition that had they done so, we should have had to fight them off with stones and sheath knives.* After a favourable report of this was sent from Malemé to Athens, General Kurt Student decided to focus every resource of the *Luftwaffe* on securing Malamé airfield, and its surrounds.

The next morning on 21st May, a few Ju 52 aircraft were able to make landings on the beaches near Malemé and bring in urgently needed weapons and ammunition to the assault paratroops in that area. As a few antiquated Allied field guns still prevented any landing on the aerodrome proper, the Ju52 aircraft safely flew in behind enemy positions dominating the airfield to land the IIIrd Battalion of the 5th Mountain Division at 1600 to team up with airborne units to secure the western end of Crete. By 1700 they had captured Malemé. A captured British Matilda tank was used to clear the aerodrome of burnt out and damaged aircraft. As soon as a landing strip was cleared, 30 German aircraft landed and left without interruption, as others landed on the beaches to the west. The tactics of the soldiers of the 5th Mountain Division were to climb along paths that were not even real trails and over heights previously considered to be unscalable, and then to attack their opponents in the flank or from the rear at points where he expected them the least. Throughout the struggle for Crete they adhered to their motto ~ *Sweat saves Blood.* Soon the German forces began pushing the defenders back to Canea. On the fifth day after their landing, the mountain troops outflanked the Allied position east of Malemé and on the next day entered Canea, the capital of Crete, and occupied Suda Bay.

Canea ~ Simultaneous to the Malemé operation, the 3rd Regiment and the Field Engineer Battalion of German parachutists dropped to advance on Canea and Suda Bay. Astute preparations for the coordination of close support attacks were carried out, so that one glider company landing near Canea area had a flight of Stukas to support them for three minutes, bombing the AA positions and a group of houses which were the glider company objective. Messerschmitt 109 and Me 110 fighters also flew in to strafe the AA batteries and enemy ground troops for the glider troops. Elsewhere around Canea, it was a disaster for the initial forces of descending Germans, savaged by the defending Greek and New Zealand troops, with one battalion losing almost 400 of their 600 men. General Kurt Student said; *The defending infantry, particularly the New Zealanders put up a stiff fight, though taken by surprise.*

THE BATTLE FOR CRETE

Meanwhile, planners of the forthcoming German invasion of Russia, anxious of the slippage in their timetable caused by the continued need for the involvement of *Luftwaffe* elements over Crete, passed the command of the Crete operation to General Julius Ringel (CO of 5th Mountain Division) who was flown in on the evening of the 22nd May. On the 23rd the Germans further consolidated their hold on the Canea-Malemé coastal road strip and moved easterly across Crete.

Heraklion and Réthimnon ~ The defenders of this sector were those worst supplied, having no AAA. Then there were only five rounds per anti-tank rifle and 80 rounds for each 3-inch mortar. The Vickers machine gunners had only 16 belts of ammunition each, and everyone had rations for just ten days. However there was improvisation, so buying pigs, goats, chickens, and eggs from farms and villages supplemented their dire situation. The West Australians of the 2nd/11th Battalion hired milking goats and kept them in their lines. Used meat and vegetable tins were used as pannikins, old herring tins became dixies, and spoons were whittled from wood. Most soldiers' uniforms were in need of some sort of repair and few men received any further replacement items on Crete. The few improvised Greek battalions were even worse off, having about ten rounds per rifle and old French machine guns.

A second wave of Germans arrived in the afternoon to attack Réthimnon at 1615, and Heraklion at 1730. British defenders near Heraklion, who wisely reserved their AA fire earlier not to reveal their positions, opened fire and shot down 15 Ju-52 transports. Soldiers on the ground watched as the German transports exploded and their occupants tumbled out ; *Like plums from a burst bag. I saw one aircraft flying out to sea with six men trailing from it in the cords of their chutes.* The Germans failed to take Heraklion and Réthimnon airfields from their Allied defenders. With heavy casualties, some parachutists withdrew to a defensive position in an olive oil factory east of Réthimnon to at least dominate the Heraklion-Suda road and sever allied communications to prevent Allied reinforcements from moving westwards. But, the Australian 2nd/1st Battalion (Lieutenant Colonel Ian Campbell) attacked and forced the few German survivors to flee toward Heraklion. After regrouping his forces during the night the German commander at Heraklion found just enough extra troops to set out to capture the airfield early the following morning. At daybreak his German soldiers closed in on the British position. Not a shot was fired. Warships of the Royal Navy had evacuated most of the Heraklion garrison during the preceding night. The isolated group of Australians and Greeks defending Réthimnon had almost exhausted their supply of food and ammunition, when on 30th May the Australians heard engines in the distance… seeing two German *Panzer II* tanks approaching, their officers gave them the choice of surrendering or trying to escape.

In the early morning of 26th May, with resistance crumbling everywhere, General Freyberg decided that the only option was to retreat and evacuate. So, orders were distributed to his forces, to retreat to the southern coastal fishing village of Sfakia. After severe fighting, and repeated encounters with Allied rearguards and hidden snipers, the German forces reached the south coast of Crete on 1st June 1941. The 11-day campaign on Crete came to an end, with the evacuation of almost 15,000 Allied survivors over 31st May / 1st June and their exit voyage to Egypt. The invasion had been expensive for the *Luftwaffe* losing 271 Junkers 52s (a quarter of their transport aircraft) and 6,453 soldiers. The evacuation had cost the Royal Navy three cruisers and six destroyers sunk, with three battleships, one aircraft carrier, seven cruisers, and four destroyers damaged. The Royal Navy tactic of using destroyer screens gave them effective protection against attack by German torpedo-carrying aircraft. During the evacuation of the forces on Crete, whenever an air attack was expected, a screen of destroyers were arranged so attacking aircraft of the Germans had to pass close over the destroyers of the Royal Navy at a very low altitudes — just as they flattened out to drop their torpedoes, the enemy aircraft presented a perfect target at very short range to the rapid firing AA guns of the British destroyers.

The Germans had won their epic first airborne invasion in history, not so much by landing with daring in the face of the enemy, but from their imaginative ability to supply and support their forces on the ground. For the first time an air force had defeated a first-rate navy, showing how costly it would have been to keep Crete supplied if the Allies had won. Unless Britain could neutralize the *Luftwaffe*, the loss of Crete would still have been inevitable. Their Balkan adventure with the Italians had imposed upon the Germans a fatal six-week delay for their planned invasion of Soviet Russia after the loss of 271 extremely useful Ju 52 transport aircraft which would be desperately missed in Russia. It would not be until 29th May 1941 that Hitler finally set June 22nd 1941 as the date for the invasion of Russia to begin, codenamed Operation *Barbarossa*.

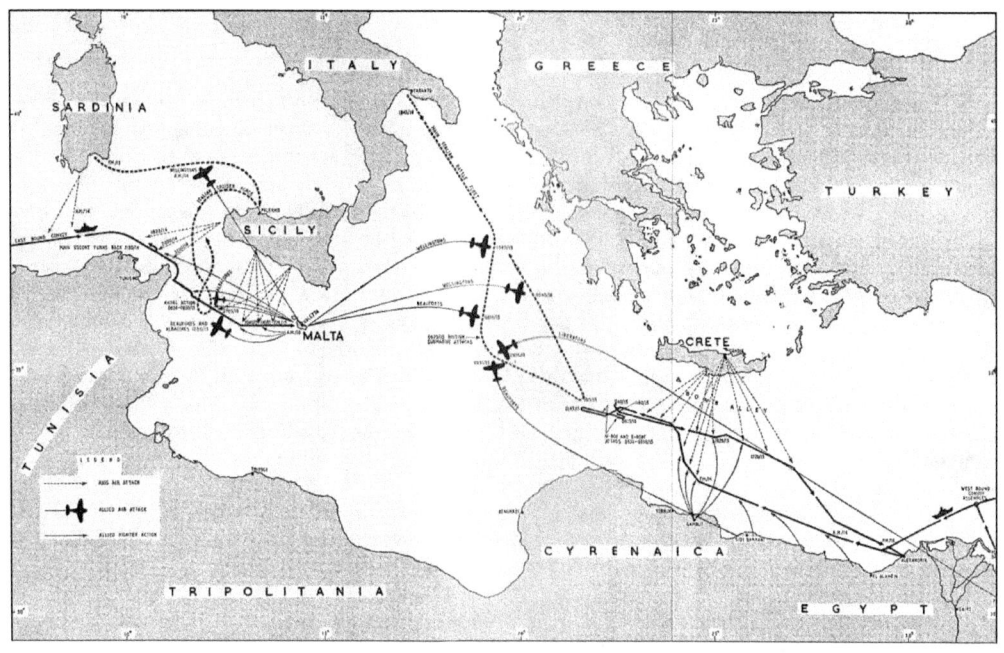

The Malta Convoys of 14-16th June 1942 (Operations Harpoon & Vigorous)

Radius of action of aircraft from Malta in relation to axis shipping routes, Summer and Autumn, 1941.

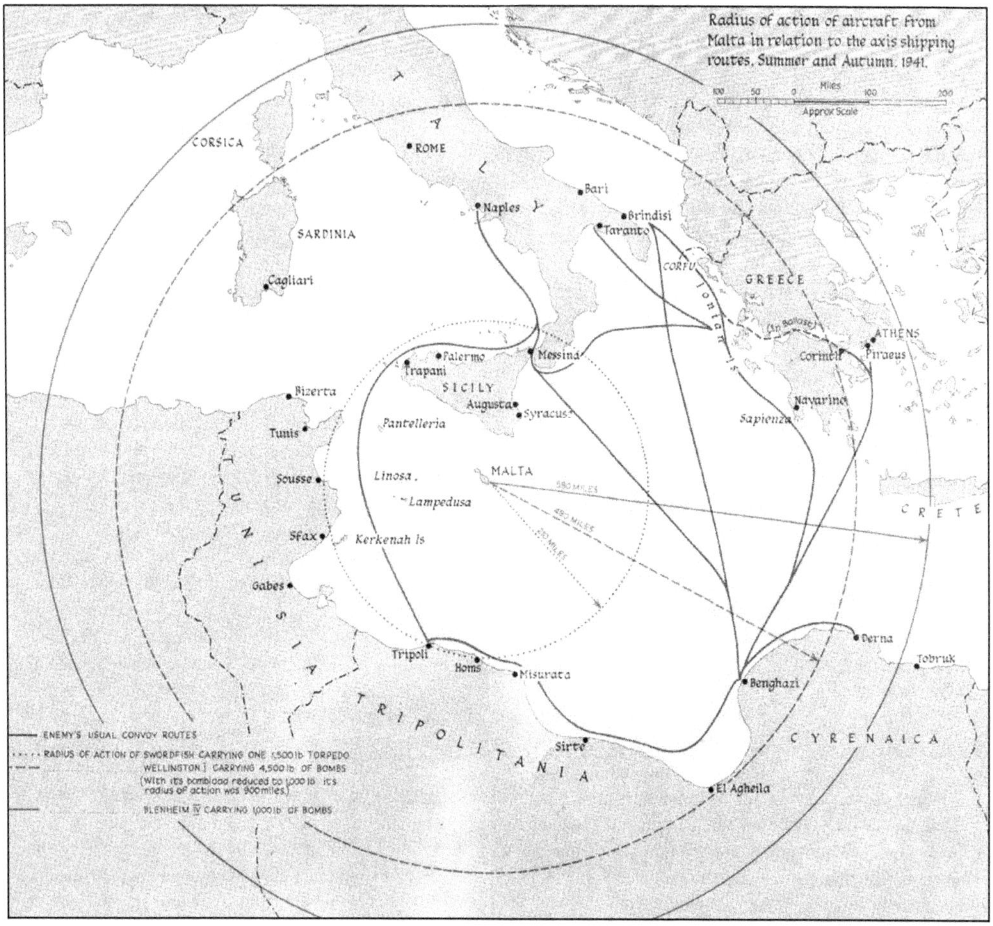

MALTA

Because of their geographical position, British forces based in Egypt had to be supplied by shipping convoys running the gauntlet of Axis naval and airpower from Gibraltar via the small Mediterranean island of Malta to Alexandria ~ or by the long way, around the cape of Good Hope. The strategic key to events in the Mediterranean thus became Malta off the southern extremity of Italy sitting astride the main Italian supply route to North Africa. As the British began to disrupt the flow of Italian vessels carrying military supplies to North Africa, it became inevitable that Malta would come under severe attack, especially as Italian shipping continued to suffer even more from Royal Air Force squadrons and Royal Navy warships based at Malta. The Italians and Germans decided to bomb and starve Malta into submission by attacking its ports, towns and cities, and to attack any Allied shipping attempting to supply the island. Preliminary attacks made by the Italian *Regia Aeronautica* saw their fighters fly V-shaped formations of five aircraft to locate their rivals and out-turn them until a firing position was gained. Italian fighters also escorted flights of Fiat Br-20 and Savoia-Marchetti 79 bombers raiding the port of Valletta. In June 1941, Air Vice Marshall Pughe Lloyd appointed to command RAF forces on Malta, and to maximize the sinking of Italian supply vessels wrote ;

The sinking of any of them, might have meant the loss to the enemy in the desert of at least ten tanks, two or three batteries of artillery, one hundred motor vehicles and sufficient spares for one hundred or more aeroplanes, food for a month for one thousand men, and ammunition for one hundred guns for a battle.

Allied air bases on Malta directed air searches for enemy shipping based on ULTRA information received from Bletchley Park, detailing precise locations of Axis convoys, so that the end of 1941 was sinking 60% of enemy shipping carrying supplies from Italy to North Africa. In North Africa, Axis aircraft were so short of fuel that they could only fly one sortie per day, and Rommel retreated to shorten his supply line. The Germans alarmed by mounting Italian shipping losses deployed *Luftwaffe* formations in Sicily to subjugate Malta. As Italian and German escort fighters approached, British Hurricanes and Spitfires scrambled from their airfields of Hal Fra, Luqa, and Ta'Qali to intercept. Malta at its nearest point is only 40 miles from Sicily, and with a warning time of less than 15 minutes, every second counted when scrambling. Whenever a raid was detected, British fighters took off fast, climbed hard, and headed south into the sun and away from an oncoming raid. When enough altitude had been reached, they turned, and came thrusting back over the island with the glaring Mediterranean sun behind them. American fighter pilot Captain Reade Tilley (seven kills) who flew Spitfires with the RAF over Malta, wrote an interesting paper in 1943 entitled ; *Hints on Hun Hunting* that contained the essentials for flying a fighter (particularly over Malta) from a personal perspective of a pilot who had learned the hard way in a tough finishing school. Reade Tilley's advice included :

Rapid Takeoff ~ When fighters are scrambled to intercept an approaching enemy, every minute wasted in getting off the ground and forming up means 3,000 feet of altitude you won't have when you need it most. Thus an elaborate cockpit check is out. It is sufficient to see that you are in fine pitch (propeller setting) and that the motor is running properly. As you roll down the runway take a quick look up for the man ahead of you. When you have sufficient indicated air speed, give him about six rings of deflection (of the gunsight) and you will be alongside in a flash. Don't jam open the throttle and follow along behind as it takes three times as long to catch up that way. If you are leading, circle the drome close in, throttled well back waggling your wing tips like hell for identification. The instant you are in formation, get the cockpit in fighting shape ~ trimmed for climb, oxygen right, check engine instruments ~ gun button to fire. Now you are ready for action.

Tactical Formation ~ The formation which is now used in Malta and which for nearly a year has achieved so much success is Line-Abreast with aircraft flying in fours line abreast at intervals of 150 to 250 yards. Where possible the three fours of the squadron keep together. They may fly sections line abreast stepped down into sun, a flat V or line astern stepped down. The vital advantages of the successful line abreast formation are obvious. There are no stragglers. With everyone searching inwards, he covers the blind spots of at least two other aircraft and everyone's tail is covered. When attacking bombers, everyone can help themselves to a target, and the return fire is split more effectively. It is easier to reform after attacks than any previous formation. Flying formation can be done at greater speeds then previously, and fuel consumption is more uniform.

When you attack ~ a series of three second bursts with new aim and angle of deflection each time is most effective. Don't cease attacking just because the enemy aircraft is beginning to smoke or a few pieces fall off; then is the time to skid out for a good look behind, before closing in to point blank range and really giving it to him. When actually firing at an enemy aircraft you are most vulnerable to attack. When you break away from an attack, always break with a violent skid just as though you were being fired at from behind, as maybe you are!

MALTA

The Sun ~ Always note the bearing of the sun before taking off. Then if you get into a scrap miles out to sea or over the desert and a cannon shell prangs your compass, you may be able to save yourself a lot of unnecessary walking or paddling. When flying low over water or desert, adjust your height so that you can easily see your shadow on the surface, and then watch the water for other shadows sneaking up behind yours, as they may harbour an unfriendly feeling towards you.

Clouds ~ are of most use to a fighter pilot who is in trouble. If you are shot up or the odds are impossible, it is great stuff to hide in. Layer clouds are most useful as you can pop in, or dive out below to take a look. Remember that it is not healthy to maintain a straight course when there are gaps in the cloud. If you are being pursued turn 90° in every cloud you pop into. If it seems in order, a quick 180 may put you in a position to offer some head-on discouragement to the pursuer. Never fly directly on top of layer cloud, as you stand out like a sore thumb to an unfriendly element, even those as far as ten miles away, if they are slightly above you. On days when there is very high layer cloud, fly halfway between it and the ground in order to spot fighters above you. Higher layer cloud is perfect for defensive fighter work, because you can see the enemy formations and distinguish between fighter and bomber long before they can see you.

Combat ~ When stalking the wily Hun, bear in mind that he seldom puts his eggs all in one basket but usually splits his aircraft into several groups, each group stacked up-sun at 3,000 to 5,000 feet. Keep an eye on the sun; you will be safer. The Germans are masters at using stooge decoys who would probably be as helpless as they look, if half the Luftwaffe was not keeping a jealous eye on them from the sun. Enemy aircraft do not fly alone: they fly in pairs or fours. If you can see just one, have a damn good look round for his pal before you go in to attack... and remember ~ look out behind!

Boxing ~ is a tactic the 109s are very keen on. The 109s come over the top and split into two groups, one either side of you. Suddenly one group will peel down to attack from the beam. You turn to meet the attack; the other group will come in and sit on your tail. If you are leading a section or squadron you can fox them easily by detailing half your force to watch one side and half the other.

Radio Discipline ~ The Squadron Leader is the only man who uses the RT for transmission when the squadron is in pursuit of the Hun. There is no need for you to say anything, just keep your mouth shut, and reflect on the ground controller's messages to the leader. So keep your eyes open and your mouth shut until you spot the enemy, then your moment has come. The procedure : make your voice purposely calm, slow and unexcited ~ Hello Red Leader: 109's at four o'clock above ~ or ~ Red Three to Red Leader: aircraft at nine o'clock our level. Sometimes the enemy aircraft are not seen until they are actually attacking. The message must be instantaneous and precise. If it is incoherent or garbled because you are excited, the man being attacked may get a cannon shell first. The proper procedure is ~ 109's attacking Red section ~ or if you see one man being fired at ~ Look out Red Four ~ or ~ Red Four break. Any one of these messages spoken clearly is perfect. Just be sure to designate the man being attacked correctly. The one sure way to lose friends and help the enemy is to give a panic message over the RT at the critical moment ~ Look out; there's a 109 on your tail! ~ said in a screech is usually sufficient to send every Spitfire within a radius of 50 miles into a series of wild manœuvres. If no call sign used, every pilot in every squadron responds automatically. Far better to say nothing at all and let one pilot be shot down than to break up several formations for the Huns to pick off at his leisure.

Despite Axis efforts, Malta never fell and became a bastion of British light naval and submarine forces ~ as well as maritime strike aircraft that targeted enemy convoys supplying the Afrika Korps.

Then, in the eastern Mediterranean on the night of 30th October 1942, came an event that would shorten the course of the war. After the British destroyer, HMS *Petard* had depth charged and forced U-559 to the surface; her German crew abandoned her after opening the seacocks. From HMS *Petard*, Lieutenant Anthony Fasson with Able-Seamen Grazier and Tom Brown dived into the sea, to swim across to the foundering U-boat. Fasson climbed onboard and climbed down U-559's conning tower to make his way to the submarine Captain's cabin. There he found an assortment of secret looking documents that he gathered up, (including the Weather Short Code Book and Short Signal Book) and passed up to Tom Brown in the conning tower. Fasson and Grazier then went back to recover more from within U-559, but were trapped when the wreck plummeted to the seafloor. Their bravery facilitated a decryption breakthrough that enabled Station X decryption analysts to decode all U-boat transmissions through the German four-wheel Enigma that had been unreadable during a previous ten-month intelligence blackout, and win the campaigns in the Mediterranean and the Atlantic.

OVER THE EASTERN FRONT

Clandestine photo-reconnaissance flights over Russia flown by Dornier 215 and Junkers 86P aircraft (modified for top secret high altitude flight) from 1934 to late 1940, returned with detailed information on the deployment of Russian air and ground forces. That these flights continued unopposed and unremarked, emphasized the impotence of the Soviet air defences, and implied that *Luftwaffe* strength in the air would contribute to a swift German victory. When they did invade Russia on 22nd June 1941, they achieved complete tactical surprise, due to the complete absence of any Soviet advance warning system. The German invasion started with a massive *Luftwaffe* raid on Soviet airfields by 637 bombers escorted and supported by 231 strafing fighters. Because the Soviet aircraft were lined up neatly on 60 airfields, on day one of *Operation Barbarossa*, the Soviets lost 1,811 aircraft on the ground and another 322 destroyed in the air for the nominal *Luftwaffe* loss of only two aircraft. Soviet aircraft proved inferior, their pilots were poorly trained, and their tactical repertoire limited.

Taran (ram attack) ~ became the forlorn tactic adopted by some Soviet aviators after they realized their mounts were too underpowered to take on the *Luftwaffe*. Deciding on their method of either directly ramming (*taranyy udar*) an enemy aircraft or to disable its controls, by using their propeller to slice off the tail controls of an enemy bomber, or try to cut off the wing or tail of an enemy aircraft, (both of which gave them a reasonable chance of survival) these brave Russian pilots knew that the demise of their obsolete fighter for an enemy bomber was very rational. On day two of the invasion, one Me 110, Ju 88 and He III fell from ramming attacks by three Russian fighter pilots flying obsolete Polikarpov I-16s and two of these three Soviet fighter pilots survived. More than 200 Soviet ramming attacks against German aircraft were reported with 17 Soviet pilots surviving two or more ramming attacks. Lieutenant Victor Talalikhin who sliced the tail off a Heinkel-III, lived to tell how it happened:

I managed to strafe the bombers left engine, and it turned away losing height. At that moment my ammunition ran out and it occurred to me that although I could still catch it, it would get away. There was only one thing for it ~ to ram. If I'm killed, I thought, that's only one, but there are four fascists in that He-III bomber. I crept up under its belly, to get at its tail with my propeller, but ten metres away, a burst of fire hit my machine and shot my right hand. Straight away I opened the throttle and flew right into it.

Lieutenant Victor Talalikhin baled out of his Po I 16 and was decorated, *Hero of the Soviet Union*. By the end of the first week of hostilities 4,000 Soviet aircraft had been destroyed. The Germans had lost 179. By July, the Germans were closing in on Minsk, forcing the Russians to ever more desperate measures, such as the committal of large bomber forces *without* fighter escort. From September onwards the Soviets were running out of men and machines to sacrifice. As the Russian winter arrived, rain, snow, and mud effectively stopped all movement on the ground, while poorly drained airstrips became inoperative. When December came, the German offensive bogged down and froze in the dreadful winter. General Hermann Hoth's 3rd Panzer Group halted by the Volga Canal, 20km short of Moscow (the result of those lost six weeks from the Balkan campaign). The severe cold weather stalled the panzers and began to ground the *Luftwaffe*. In worsening temperatures of 40 below zero, the serviceability rate of *Luftwaffe* aircraft diminished to 25%. Radios became inoperative, tyres became brittle and cracked, and oil became as thick as molasses. Legendary Stuka pilot Hans-Ulrich Rudel, *The Eagle of the Eastern Front* who sank three Soviet warships ; including the battleship *Marat*, also destroyed 70 landing craft, and 519 tanks during 2,530 combat missions recalled, *Engines no longer start, everything is frozen stiff, no hydraulic apparatus functions, to rely on any technical instrument is suicide.* Rudel had a 100,000-ruble bounty placed on his head by Joseph Stalin.

The Night Witches ~ In 1942 the Soviets formed the 588th Night Regiment of volunteer lady pilots, initially to fly aging Polikarpov Po-2 biplanes on night missions, specializing in nuisance tactics, (their mechanics and bomb loaders were also all women). Flying at high altitude toward the German lines these Russian ladies, cut their engines for a silent approach, glided towards the invaders and showered them with grenades and petrol bombs. One of these young ladies, Ukrainian teenager Natalya Meklin, flew 840 missions in less than three years to be decorated a *Hero of the Soviet Union*. Flying whatever aircraft they could, they became so efficient at flying harassment bombing missions that the Germans called them *Nachthexen* (Night Witches) and three further regiments were formed from the 1,000 brave women who were inspired to volunteer. The leading Russian female ace was Lieutenant Lillia Litvyak, a flying instructor before the war who joined the 437th Fighter Regiment to fly combat missions in YAK-1 fighter aircraft. Litvyak became the first woman in history to shoot down an enemy plane and achieved fame as the *White Lily of Stalingrad* with 12 victories. She preferred flying with her cheerful friend Katya Budanova, who before the war, had been a carpenter in Moscow.

OVER THE EASTERN FRONT

Walter *Nowi* Nowotny scored his first three kills on 19th July 1941, downing three Soviet I-153 fighters over Saaremaa : but the engine of his Me-109 *White 2* took a critical hit, causing him to lose altitude over Soviet held territory. Not wanting to crash land and be captured by the Soviets, Nowotny turned sharp and headed out toward the Baltic Sea to ditch his 109 in the Gulf of Riga. Climbing into his rubber survival dinghy with neither food nor water and no other option, he began using his hands to paddle southwards to the coast. During his second night adrift, with waves splashing into his dinghy Nowotny was almost run asunder by an unaware Soviet destroyer. Exhausted from his paddling, he fell asleep and woke on the third day to find the Baltic currents had carried him close to the shore. Paddling towards the sandy beaches of the Latvian coast he made it ashore and was rescued by two Latvians. As the focus of the northern front became the German siege of Leningrad, the superstitious Nowotny (the first ace to score 250 victories : 194 in 1943 alone) thereafter always wore the same trousers ~ his *Victory Pants* that he had worn during those three days in the Gulf of Riga, (the only exception being on his last sortie, when he was killed flying his Me 262). Walter Nowotny, recalling his second mission flying over Leningrad wrote;

It was a clear blue sky, and filled with Soviet fighters attempting to attack our bombers. I picked an I-18 and made a sharp turn, putting my Me-109 in a good position. A few bursts sent him burning to the ground. The remaining fighters tried to escape, but my Messerschmitt was faster. Flying above the docks on the mouth of the River Neva, I got the backboard plane in a finger-four formation into my gunsight. Two bursts of fire and the Rata blew up. Fuselage and wings tumbled down on fire. The Flak fired fiercely from below. I made a 180-degree turn and spotted four I-18s attacking our bombers from behind. Pulling up the nose of my plane, I made one of the Soviet fighters pass through my bullet tracers. The success stunned me. He immediately went into a steep dive, started spinning and left a thick black trail of smoke. This was my sixth victory today. Number seven didn't last long. I was just about to return home, as suddenly a Rata pulled up beneath me.
I pushed my stick forward, and seconds later the enemy went down in spirals.

His fellow airmen knew Major Eric Hartmann on the Eastern front during October 1942 as *Bubi*. But his Soviet opponents knew him as *The Black Devil*. Eric Hartmann's first rival on 5th November was a heavily armoured Sturmovik IL-2. Hartman knew that he had to shoot out this opponent's oil cooler from underneath. Killing at very close range ~ he flew through the debris of his victim, damaging his mount, and having to make a forced landing, Hartmann learned two things that day; get close to shoot… and break away fast. Hartmann would become the world's highest scoring fighter ace with 352 victories using his tactical précis of *See-Decide-Attack-Break*. Eric Hartmann explained;

I knew that if an enemy pilot started firing early, well outside the maximum effective range of his guns then he was an easy kill. But, if a pilot closed in and held his fire, and seemed to be watching the situation, then you knew that an experienced pilot was on you. Also I developed different tactics for various conditions, such as always turning into the guns of an approaching enemy, or rolling into a negative G dive forcing him to follow or break off, then rolling out and reducing air speed to allow him to over commit ~ and then take advantage of him.

Oberleutnant Otto Kittel, the fourth greatest ace with 267 aerial victories, preferred the Focke Wulf 190 considering it far more robust in combat, and that it had several advantages over the Me 109. The FW with its stronger engine was faster than the Me 109 below 6,000m, and the air-cooled FW coped better than the liquid-cooled 109 if shot up, and could survive more significant battle damage. Otto Kittel's approach was, *Take the safe route, and avoid ill-considered and wild offensive tactics.*

The contributions of Soviet ace Alexandr Pokryshkin, as an aerial tactician on the Eastern Front began the instant he veered away from the outdated Soviet fighter tactics prevailing thru 1941. Soviet pilots were expected to circle over their assigned patrol sectors at low speed, but Alexandr Pokryshkin urged an aggressive approach of ~ *speed, altitude, maneuver, and fire*. Encouraging his pilots to fly with confidence, verve, and less predictable tactics, Pokryshkin developed much advice to improve Soviet fighter performance. He always personally attacked the flight leader of German fighter escorts after discerning that by shooting him down, the demoralizing effect on the others often caused them to scramble home. In 1943, he developed the pendulum flight pattern for patrolling airspace using flights of different fighter types stacked in altitude. Pokryshkin's fighter pilots learnt to counter the altitude advantage of the Me 109 by flying at full throttle in the fighting zone ~ even if this used their fuel faster and reduced their range. The Soviet Air Force also abandoned the three-fighter section for the *pary* (pair) and *zveno* (four) as recommended during the Spanish Civil War.

THE NIGHT WITCHES

Natalya Meklin ~ 982 flights

Captain Marina Raskova, Soviet pilot and navigator
First woman navigator in the Soviet Air Force 1933
Commander, 125th Guards Bomber Aviation Regiment

As one of the first women to be awarded Hero of the Soviet Union, Marina Raskova used her personal connections with Stalin to convince his GHQ to form three combat regiments of women. Following a speech by Raskova on 8th September 1941 calling for women pilots to be allowed to fight, Stalin ordered the formation of the all female 221st Aviation Corps of women pilots, engineers, and all ground support staff.

THE NIGHT WITCHES

Planning a raid ~ Storm-navigators Tonia Rozova, Sonia Vodyanik, and Lida Golubova.

Squadron commanders of the Soviet 588th Night Bomber Regiment ~ Night Witches ~ plan a combat mission.

The women flew durable wood and canvas Polikarpov Po-2 biplanes ; a 1928 training aircraft. It was used in various missions during the war, including supplying partisans behind German lines. The women (two per plane) would fly at ground level until near their target, ascend, and then shut off their engines, gliding down over their target so as not to draw anti-aircraft fire. This created a rustling sound that reminded Wehrmacht soldiers of broomsticks, so they nicknamed them Nachthexen ~ Night Witches. The planes carried six 50kg bombs which could kill several men at a time in a tent. The unit's real usefulness was in spreading fear and disorientation among the enemy with their nuisance raids. Some 30 women died during missions. They carried no parachutes, and in general flew so low that one would have been useless anyway. At their best, the Night Witches had 40 two-person crews, flew over 23,000 sorties, and dropped 3,000 tons of bombs.

BOMBER COMMAND

The Allies decided on a joint day-night bombing offensive, to over stretch enemy German defences, in which the British would attack German cities by night to snap German morale as the Americans hit key factories by day. But the joint British-American attacks all too often, bombed the homes of civilians causing needless suffering, while factory targets often escaped effective damage. Indeed, Albert Speer said after the war; *It would have been better to limit RAF's bombing efforts against key German industrial targets such as the U-boat yards in Bremen* ~ which would have ended the war much earlier. The role of the British fighters in the American daylight raids was to fly as escorts for the initial and final legs of the bomber's route, and to fly fighter sweeps over German airfields in France, to divert enemy *Luftwaffe* fighters which were otherwise needed to tackle the American bomber stream.

When German pilots attacked incoming day bomber formations, they preferred quick hit-and-run strikes using cloud cover or the sun to gain a good position for a fast diving attack. Straight in, then flying down out of range before climbing back up to make repeated attacks on bomber formations. Another tactic was to dive behind the bombers, pull up hard and into them from below. Adolf Galland's tactic was to fly a shallow climb to overhaul the bombers and their escorting fighters. Galland's unexpected approach frequently took escorting fighters by surprise and denied them time to accelerate in pursuit before he reached the bombers from below. Such fighter attacks were timed to coincide with other German attacks at the flanks of the higher Spitfire squadrons to distract them.

The Pathfinders ~ Allied bombing accuracy improved from August 1942, after the Pathfinder Force was formed from the five squadrons of Bomber Command that had been most successful in locating targets. At low-level, they preceded the main bomber force over Germany to illuminate the target area with flares, and then mark the intended aiming point with incendiary bombs. At first the Pathfinders were only a minor success. The fires they started to mark the targets with often spread that distorted the aiming point, and the Germans also lit decoy fires to lead the bombers astray. The RAF answered with distinct target indication bombs that would explode with specific fireworks colour. The role of *Master Bomber* was flown by a highly experienced pilot who continued to circle the target throughout the raid, giving instructions by radio to approaching bombers as to which part of the target marked with coloured flares to aim at. This precarious task maintained accurate bombing even when smoke and flames obscured the target and markers. By the end of 1942, the Pathfinder Force had accurately marked targets on 75% of raids when the sky was clear or partly overcast, but if clouds shrouded the target, they acheived little success.

Deception ~ The *Luftwaffe* devised a simple, effective system of dividing their airspace into boxes, each of which was patrolled by night fighters having been warned of the approach of enemy bombers by their *Freya* ground radar interpreted by a ground controller. Powerful radio transmitters were constructed in Britain to counter German broadcast control and German speakers gave false orders to confuse the night fighters. During one raid against Kassel, the German controller warned his fighters, *Beware of another voice... Don't be led astray by the enemy.* On another occasion after much contradiction from an excellent imposter, a German controller became abusive to which the RAF officer retorted ; *The Englishman is now swearing!* The German countered: *It is not the Englishman who is swearing, it is me!* Another night, the Germans attempted to contest the deception by using a female controller, but the quality of the British RAF deception was such that they had a young German-speaking lady primed.

Project Window ~ As the struggle between the British bombers and German night fighters took place amid exploding AAA shells, searchlight beams, and the glare of burning cities, RAF Bomber Command attempted to jam enemy radar to reduce the growing effectiveness of German night fighters. After British scientists discovered that the German air defence *Freya* radar system could be saturated by dropping thin strips of metal foil, Churchill ordered ~ *Open the window!* ~ on the night of 24th July 1943, as a combined British and American bomber force approached the coast of Europe to bomb Hamburg, some bombers began to release drifting clouds of tin foil at the rate of one bundle a minute, (a tempo they would maintain until the bombers were on their way home). As the German *Freya* radarscopes began showing a blizzard of white dots, German air controllers sent night fighters on wild goose chases as the Allied bombers droned on unscathed and undetected toward Hamburg. The Allied bombers dispatched to bomb Hamburg, turned this charming German city into a firestorm of howling winds in excess of 150 mph that uprooted trees, and vacuously sucked building debris and people into the flames, killing over 50,000. Of the 800 bombers that flew to Hamburg, only 12 failed to return, demonstrating once again, that complex technology can be offset by simple counter measures.

BOMBER COMMAND

German night fighter tactics ~ After Hamburg, the German night fighters adopted *Wild Boar* tactics; ignoring their own flak, and hurtling into British bomber formations caught in the glare of their own incendiaries. The most skillful German airman of the night, Heinz Schnaufer, with 121 victories including 114 four engined RAF bombers, preferred to fly his Me 110G night fighter a little ahead of his prey, then throttle back to climb until his sight ringed the bomber. Few of the British bombers that Schnaufer targeted ever saw his blackened fighter before he opened fire. Those that did went into the twisting corkscrew tactic used by bomber pilots ; to both evade enemy night fighters and to enable more accurate fire from their rear gunner. A corkscrewing bomber was difficult to hit at night, twisting and turning, altering its speed, height, and course. Schnaufer taught night fighter pilots to stay with the bomber, and wait until it was changing direction at the top of a climb, when its wings would be level, to acquire a non-deflection shot. Pilots flying Ju 88 night fighters equipped with heavy caliber upward firing cannons used the same approach technique, but opened fire from about 100m below the bomber. Heinz Schnaufer who claimed that he never lost an Allied bomber in a corkscrew, destroyed seven RAF bombers in 20 minutes during the evening of 21st February 1945.

Under frequent daytime attack, American bombers fought their way through to hit their targets. Although the number of U.S. bombers shot down was relatively low, many of their bombers that did return were fit only for salvaging spare parts or the scrap heap. The developing firepower of new German fighters was demonstrated on the 17th August 1943 raid on the Schweinfurt ball bearing factory and the Messerschmitt factory at Regensburg. A large amount of damage was done, but the American 8th lost 59 of 363 bombers sent, with scores damaged. Worse came when Schweinfurt was targeted again on 14th October 1943. From 291 raiding American bombers, 60 were lost, with 138 badly damaged. An unsustainable loss rate of 20.6%. The Americans needed a long-range fighter to confront the Messerschmitt 109 and Focke Wulf 190 fighters. It came during early 1944 in the shape of the new American P51D Mustang fighter with sufficient range to fly to Berlin and back and the necessary speed to keep up with a Focke Wulf in a dive. To begin, the Mustang fighters remained close to the American bombers and worked in relays. Then 16 squadrons of Mustangs were assigned to patrol up and down the air route of the American bomber stream, stalking and eliminating enemy fighters.

The Focke Wulf 190 outclassed the Spitfire V in almost all aspects. Its acceleration, horizontal speed, rate of climb, and dive were superior and its superb rate of roll enabled it to alter direction faster. The FW 190 could also turn tighter than the Spitfire, by applying 10° of flap with a burst of acceleration to overcome its extra drag. The more potent Spitfire IX could match the FW 190 in speed and climb, to narrow the margin of its quicker acceleration and diving. The Focke Wulf still had the edge in its rate of roll. Later versions of the Spitfire had their wings clipped to improve their rate of roll, but this caused a loss of turning ability. *Sturmgruppes* of heavily armed FW 190s with armoured engines and cockpits attacked Allied bombers from astern, pressing to within 100m.

The loss rate to Bomber Command on moonlit night raids over western Germany averaged 4.6%. But some nights were catastrophic. Over Nuremburg on 30th March 1944, German night-fighter pilot *Leutnant* Fritz Lau identified a British bomber formation in the distance from their flaming hulks falling through the night sky. German aircraft comprised of components containing magnesium that blazed white as they fell ~ but RAF bombers burned dark red. Lau only saw two white fires that night, and would lose count of the flaming red wrecks of Lancaster bombers falling from the sky. That night, Germany lost 10 fighters, but from 795 Allied bombers, 94 were lost with 545 aircrew presumed dead.

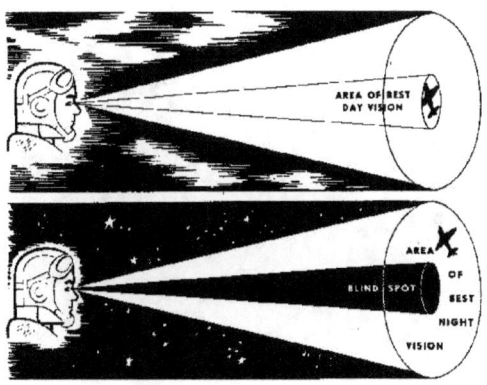

Aircrew vacancies again

RAF Avro Lancaster over Cheshire, England in October of 1941.

RAF bombing objectives ~ Naples in Italy being the farthest target.

German fighter pilots with B-24 model showing its defensive arcs of fire

B-24 over St. Malo, France

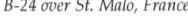

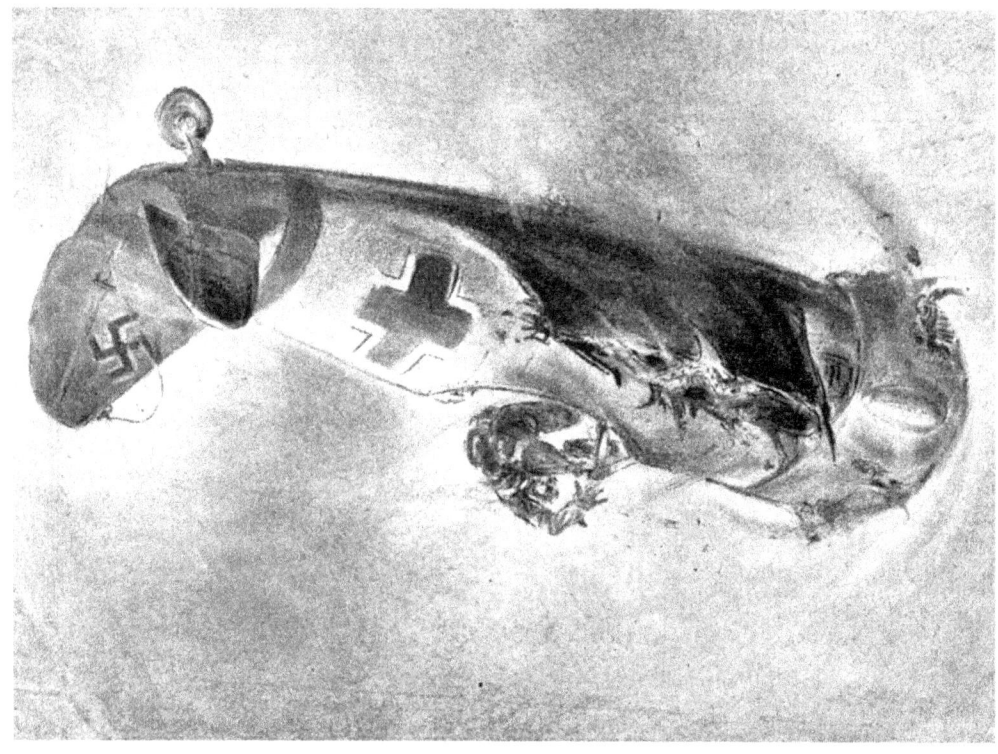

A series of illustrations ~Ich ramme from Berliner Illustrirten 1944 ~ Artist: Helmuth Ellgaard (Source: Familienarchiv Ellgaard)

B-17 flying back to its base ~ after a mid-air collision with a German fighter

THE MESSERSCMITT 163 KOMET ROCKET FIGHTER

Deep inside Germany, after aero-scientists discovered that swept-back wings added speed, they developed a rocket-powered interceptor based on the aerodynamics of a glider design of the 1930s. Then in 1941, German engineers equipped it with a futuristic rocket engine to appraise its potential as a lightning interceptor. On 2nd October 1941, Messerschmitt test pilot Heini Dittmar was sitting in the prototype *Komet*, being towed by a Me-110 to a height of 4,000m for an aerial launch. Igniting his rocket engine powered by the chemical reaction of two volatile fuels (theoretically rated at 3,750 pounds of thrust) Dittmar accelerated his *Komet* ever faster to become the first person to cusp the speed of sound. Experiencing the atmospheric compression of transonic buffeting he continued with a touch less than full throttle to record a speed of 1,004 kph. Later when Heini Dittmar test flew the improved Me 163B rocket interceptor over Peenemunde on 6th July 1944, he power dived to break the sound barrier setting a new speed record of 1,130 kph, and land with part of his tail fluttered away. People on the ground confirmed hearing the sonic boom of Dittmar's historic flight.

In the summer of 1944, the *Luftwaffe* declared their new wonder weapon operational. The first unit flying Messerschmitt 163B rocket interceptors, JG 400 was formed at Brandis, 16km east of Leipzig in Saxony. JG 400's role was to defend the Luena synthetic fuel factory that was often targeted in late 1944 by American bombers. The Messerschmitt 163B *Komet* at full throttle had a maximum level flight speed of 959 kph with a flame on operational flight duration ranging from just 12 to 22 minutes, depending on the *Komet* airman's application of throttle. The improved C series *Komet* incorporating a dual chamber Walter rocket engine with a cruise chamber beneath the main rocket, enabled the pilot to extend his flight by a few more minutes. Although the Me 163C achieved an altitude of 12,000m in less than two minutes, the unpressurized cockpit of the Messerschmitt *Komet* limited its operational flight ceiling to what its pilot could physically endure in those few minutes at high altitude, accessing oxygen from a mask without losing consciousness. *Luftwaffe* pilots hoping to fly the *Komet* trained in altitude chambers to harden them against the ordeal of flying through the thin air of the stratosphere without a pressure suit. Additionally, they endured special low-fibre diets prepared to minimize the effects of gas in their gastrointestinal tract that would expand rapidly during their dramatic ascents.

The tactics of *Komet* airmen flying these new interceptors were to wait on the ground until the drone of enemy bombers was heard approaching in the distance. Then alone or in pairs they ignited their rockets to streak through the sky at 1,000 kph in 70° zoom climbs to intercept the Allied bombers. A *Komet* pilot made his first pass, climbing through the incoming bomber stream at 9,000m to soar up to 12,000m, bank earthward and dive through the bomber stream once more, giving the pilot two or three openings to fire his heavy cannon shells before returning to the airfield. On August 24th 1944, *Feldwebel* Siegfried Schubert, the finest pilot of the *Komet* armed with two 30mm cannons mounted in the fuselage either side of him (containing 50 HE rounds per gun) downed two B-17s in five minutes. The record for a *Komet* pilot. By the close of 1944, Allied bomber pilots learnt to dread the vapor trail emission of lethal *Komets* in a climb ~ and the *Luftwaffe* only ever lost six *Komets* to enemy fire.

Clock Position ~ Allied bomber aircrews relied on the practice of calling out positions of attacking enemy fighters by referring to the imagined face of a clock, with the bomber being central. To define the approach altitude of an enemy fighter, Allied airmen added the supplementary descriptions: *low* (below the bomber), *level* (at the same altitude as the bomber), and *high* (above the bomber). Whenever ~ On your six ~ was heard, it referred to a rearward fighter, while ~ 12 o'clock high ~ described the preferred frontal approach of a Messerschmitt or Focke Wulf fighter, coming in from directly above (especially against the lightly armed nose of a B-17). *Komets* diving through a *12 o'clock express* were fleeting targets, closing so fast that the electric gun turrets of Allied bombers couldn't track them.

To begin, accidental explosions during landings were frequent. Experienced pilots overcame this by diving, then pulling up within the defence perimeter of their airfield protected by AA gunners on the ground, to circle bleeding off the *Komet's* excess speed. Preferring to be fuel empty landing, *Komet* airmen reverted their interceptors to gliders to swoop down and land on their airfields. The fuel mix of their rocket interceptors could dissolve human flesh in seconds, so they all wore flight suits, boots, and gloves all made from non-organic material, as any organic material would burst into flames on contact with the fuel. The *Komet* that flew at near sonic speeds proved a serious threat to Allied airmen, and the technological innovations of the *Komet* designers advanced the advent of the jet age.

THE MESSERSCHMITT 262 JET FIGHTER

In April 1944, the *Luftwaffe's* new Messerschmitt 262 *Schwalbe* (Swallow) twin-engine jet fighter came into service, armed with four 30mm cannon and ready for action. Sufficient had been constructed to form a squadron at Lechfeld to evaluate the tactics to make the most of the Me 262 jet fighters high speed and superior climbing performance. As *Schwalbe* pilots approached speeds of 950 to 1,000 kph (near the sound barrier) their jets began to pitch and shudder becoming almost uncontrollable. For the first jet pilots, this speed represented the upper limit of velocity, and they quickly learned to check their speeds in flight. Hans Mutke claimed to have broken the sound barrier on 9th April 1945 in his Me 262, reporting the now familiar transonic buffeting, followed by a restoration of normal control after particular speed was surpassed, then a resumption of atmospheric fluttering as he reduced his throttle. Mutke also reported engine flameout. Messerschmitt 262 pilots abandoned the *Schwarm* formation of four aircraft, as keeping formation required continual speed adjustments and their twin Jumo turbojet engines (rated at 1,980 pounds thrust each) tended to flame out at combat altitudes if they reset their throttles too fast. As abrupt shifts in speed became too risky, *Schwalbe* pilots adopted a wide three-plane formation that required less attention to their throttles. Soon the four 30mm cannons of the new *Schwalbes* were destroying highflying Allied bombers that had for so long droned over Germany. An attacking Me 262 was away so quick that Allied fighter escorts only had seconds to react. Pursuit was futile, because of their astonishing speed and climb.

The roller coaster attack ~ The command of the first *Schwalbe* Messerschmitt 262 wing, was given to the proven leader and ace Walter Nowotny with 258 victories. The airmen of *Kommando Nowotny* approached in flights of threes, spaced 150m apart and 2,000m higher than the American bombers. Using their superior speed, flights of three *Schwalbe* jets swooped in fast shallow dives until 1,500m astern and 500m below targets and then soared through the escorting American fighter screen with slim risk of interception. At the bottom of their dive, with closing speeds of 350m a second, way too fast for accurate shooting, they mushed off their excess speed with a sharp pull-up at the bottom of their dives. Then leveling out, they were perfectly placed for attacking: 1,000m astern, and overtaking the bombers at 150 kph. Then darting in, the *Schwalbe* pilots raked the underbellies and tails of the American bombers with 30mm cannon rounds ~ streaking away so fast that the American gunners in their bombers had little hope of keeping them in their sights. Then Me 262 pilots circled the bombers, to repeat their attack maneuvers until low on fuel.

Rat catching ~ No. 83 Group Commander, Harry Broadhurst knew that his piston-engined fighters were facing obsolescence, and that his RAF airmen could not tackle the Messerschmitt jets in the air. With few options, he ordered his pilots on frequent strafing attacks to hit them on the ground. But the Germans were waiting for them with AA detachments seemingly everywhere. American fighter pilots preferred *rat catching* ~ the term they applied to their tactic to counter Luftwaffe *Schwalbe* jet fighters by flying standing patrols near the approaches of known Me 262 airfields and to ambush them as they took off or landed. Their distinctive airfields were easy to spot being wide enough for three jets to take off together line abreast, as well as having three sets of parallel scorch marks made by the hot exhausts of their jets. This became the best interception tactic, as the endurance of the Me 262 was short and on final approach to touch down was extremely vulnerable, flying at slow speed and almost out of fuel. Many Me 262s were shot down, but Walter Nowotny countered with standing patrols of Me 109 fighters and some ferocious actions were fought. Adolf Galland witnessed the fall of Walter Nowotny in November 1944, when Nowotny attempted to land his damaged *Schwalbe*, but crashed, and perished in the resultant jet fuel explosion. *Rat catching* prevented the small numbers of *Schwalbe* jet fighters from becoming far more dangerous ~ as it was, Me-262 jet pilots claimed 509 Allied aircraft destroyed for a loss of less than 100 of their own.

Mastering the elements of fighter combat always took time, and American pilots of World War II had a saying, *Fly five, and stay alive*. Five missions gave a novice the sense of situational awareness to understand and evaluate all the relevant factors happening in an air-to-air engagement, and survive. Another saying of the time was, *Lose sight, lose the fight*. Sighting the enemy and keeping them in sight has always been critical to winning any air-to-air engagement. Successful fighter pilots often combined having very fine eyesight to spot enemy aircraft at long ranges together with the ability of being able to fly their aircraft to the edge of its performance possibility ; knowing their aircraft as thoroughly as possible to get the most out of it, along with a thorough understanding of tactics to maximize his aircrafts potential while negating the advantages of his opponent's aircraft over his own. Allied tactics continued to offset almost any new advantage the Germans introduced to the skies over Europe, and ensured their eventual success.

VICTORY IN EUROPE

Generalleutnant der Jagdflieger Adolf Galland, tangled with Herman Goring, and then approached Adolf Hitler to request more jet fighters to oppose the concentrated bombing raids over Germany. Having made a powerful enemy during the last weeks of the war in Europe, Galland's telephones were tapped, he was spied on, and his activities investigated. Goring became even more hostile towards him, accusing him of disloyalty and using unsound tactics. Goring relieved him of his command and prepared a trial in which the collapse of the *Luftwaffe* fighters would be directed at him. Galland made no defence and asked to be sent back to the front as a jet pilot. As the news of Galland's troubles spread thru the *Luftwaffe*, Goring called a meeting of the most highly decorated *Luftwaffe* fighter pilots. Galland's sacking was criticized, and Goring suspecting that Galland had fueled the unrest called them *mutinous dogs*, threatened to have some shot, and adjourned the meeting. The Nazi sympathizer, Gordon Gollob composed false evidence to use against Galland, and the official reason for Galland relocating to his retreat in the Harz Mountains was ill health.

Hitler, who respected Adolf Galland, had not heard of his strife, but when he did he soon ordered, *All this nonsense stops immediately!* And asked Galland to lead *Jagdverband 44* a unit of Messerschmitt 262 jet fighters. Some of Galland's friends, including ten holders of the *Knight's Cross* joined him. Eric Hartmann was asked, but would not leave his unit. Galland managed to gather together 20 jets. Exploiting the awesome speed of the twin-engined *Schwalbe* fighters, they avoided Allied fighters to concentrate against the bombers using their lethal racks of 50mm cannon slung under each wing, supplementary to their standard 30mm armament. Just one properly aimed 50mm round was enough to destroy a bomber. Adolf Galland ended the war as he had begun, leading a flight of fighters.

Bringing the aircraft onto the field and taking off became more and more difficult. Eventually it was a matter of luck. The magic word, jet, had brought us together. We wanted to be known as the last fighter pilots of the Luftwaffe. We could do nothing but fly and fight and do our duty as fighter pilots to the last.

On 28th April 1945, Adolf Galland claimed his 104th victory against a B-26 bomber, but making the mistake of a novice, he forgot to pay attention as he watched his victim crash. A short burst from the guns of a P-47 Thunderbolt flown by U.S. Lieutenant James Finnegan hit. Galland hurt in the knee, coaxed his crippled *Schwalbe* back to Munich, to find an airfield under attack from more Thunderbolts. With no choice, he landed with a flat nose wheel under heavy fire, abandoned his jet on the runway, and took shelter, ending his last mission.

In the closing days of World War II in Europe, as RAF and US bombers ceased operations for lack of worthwhile targets, the remnants of the once formidable *Luftwaffe* burned on the ground. *Generaloberst* Robert Ritter von Greim (a decorated tank buster of 1918) responding to an order from Hitler, flew from Munich into Berlin with the famed female ace (and his companion) Hanna Reitsch. Their *Fieseler Storch* aircraft was hit by AA fire over the Grünewald, then under attack by Soviet forces. After von Greim was wounded in the leg, Hanna Reitsch took over the controls of their aircraft, landing the *Storch* on an improvised airstrip in the Tiergarten near the Brandenburg Gate, to attempt to make their way through the ruined city to the *Fuhrerbunker* beneath. Hitler promoted von Greim to *Generalfeldmarschall*, assigning him to lead the *Luftwaffe* in its last days, replacing Herman Goring, whom Hitler dismissed for treason, after having contacted Western Allied forces without his consent to negotiate a ceasefire. On 28th April, Hitler directed von Greim to leave Berlin, and for Hanna Reitsch fly with him to Plon, to arrest Heinrich Himmler for treason. During a particularly intense Soviet bombardment, Adolf Hitler again refused to leave Berlin, giving Reitsch and von Greim a vial of poison. Hanna Reitsch hoped to fly out with the children of propaganda minister, Joseph Goebbels who were living as a family besieged in the underground *Fuhrerbunker*, but the fanatical fascist parents declined their children of their future. That night, flying the last plane out of Berlin, the couple took off from the rough Tiergarten airstrip. Seen by the advance guard of the Soviet 3rd Shock Army, they began shooting at their aircraft, thinking Hitler might be escaping. After the surrender of the Hitler's Third Reich on 8th May, during a Soviet-American prisoner exchange, von Greim was given over to the Soviets. Fearing torture and execution from Joseph Stalin's feared NKVD, von Greim committed suicide while in Soviet custody in occupied Salzburg on 24th May 1945. His last words before biting on a glass vial of potassium cyanide were ~ *I am head of the Luftwaffe, with no Luftwaffe.*

ALLIED ESCORT FIGHTER RANGES

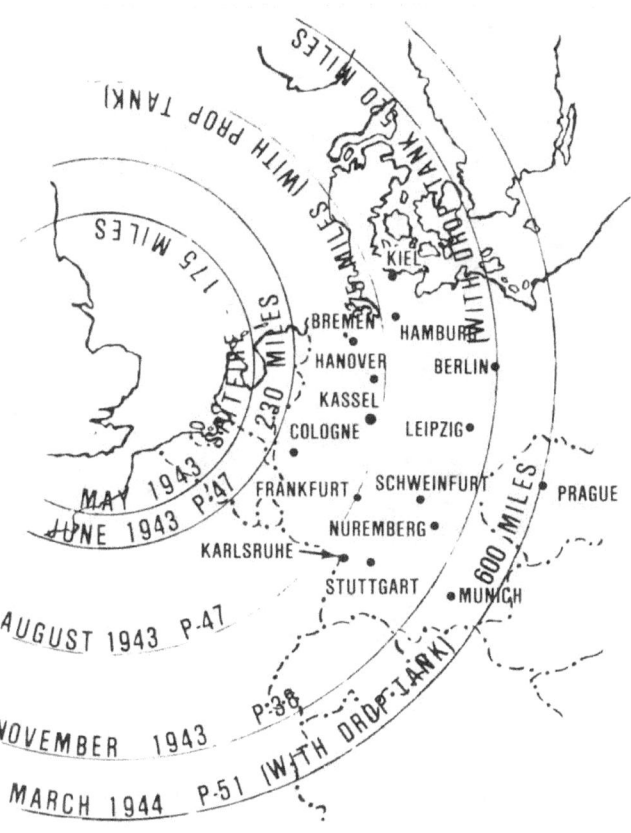

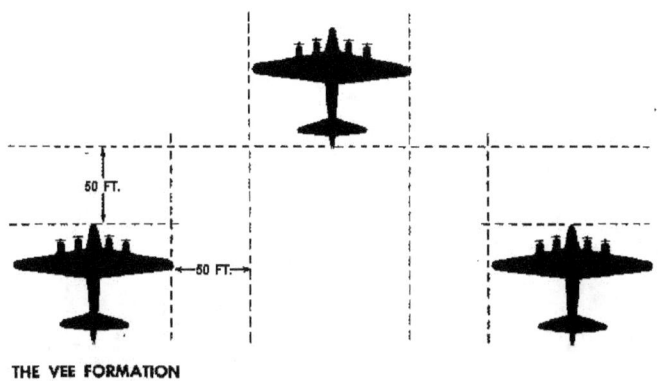

THE VEE FORMATION

From the B-17 Pilot Training Manual

Photograph of the 7th December 1941 attack on Pearl Harbour taken by a Japanese pilot during their first wave of bombing. This view looks east over Ford Island. The plume in the centre is a torpedo strike explosion against the USS Oklahoma.

Chart from ~ Reports of General MacArthur, prepared by his General Staff.
The chart shows the routes of the Japanese attack force to and from Pearl Harbor and the routes of the two American carriers.

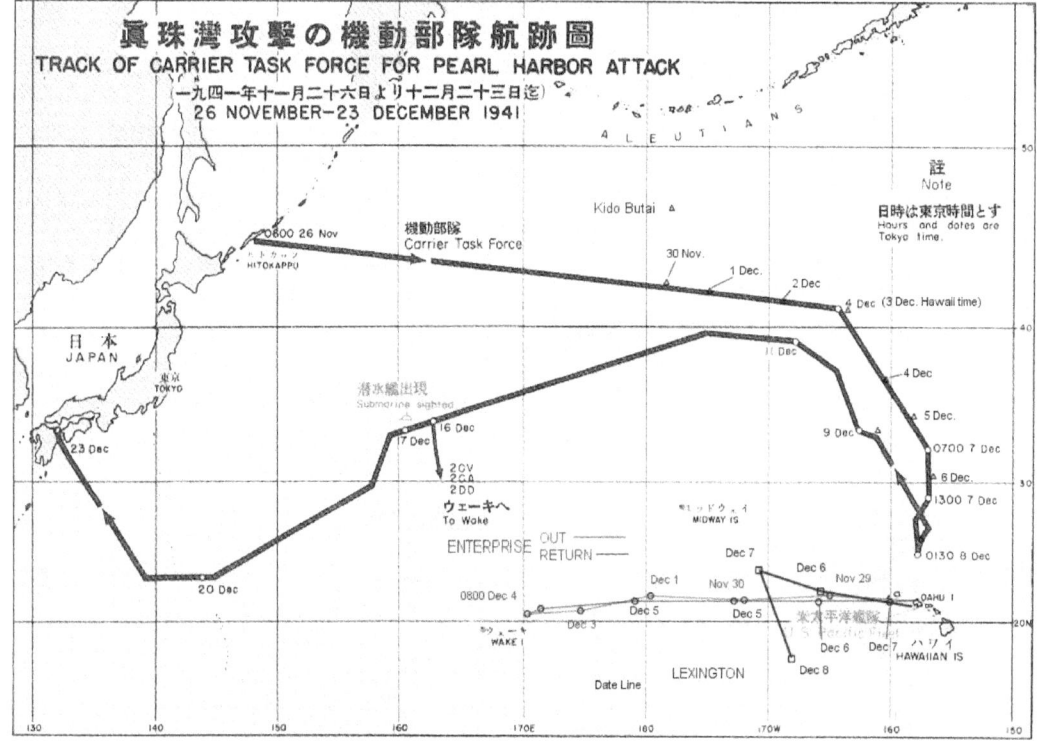

CHAPTER VIII

THE PACIFIC THEATRE

Asia for Japan by Winsor McCay

Following an 1895 dispute over Korea, Japan fought a successful naval war against China and from this triumph their creation of an advanced navy, with aspirations of greatness began. On 8th February 1904, under the cover of early morning darkness, destroyers of the Imperial Japanese Navy launched a surprise attack against the Russian Pacific Fleet moored in Port Arthur on the toe of Korea, sinking two Russian battleships plus an old cruiser. In February of 1905, during the decisive Battle of Tsu-Shima Strait, Japanese destroyers firing 74 torpedoes neutered Russian sea power in the Far East, securing their sealanes of supply to mainland Asia to conduct a rapacious war in China. During 1932 aggressive Japanese forces in Manchurian N.E. China, set up a puppet government and occupied Shanghai, and then refused to consider any proposals for a peaceful settlement. Beginning an internal campaign of public animosity towards foreign nations, even proposing war if necessary, the Japanese extended their Manchukuo boundaries by subjugating the province of Jehu in China following their expressed foreign policy, which sloganeered as the *Greater East Asia Co-Prosperity Sphere*, in practice saw the forceful acquisition of territory and the murder of 10 million Chinese civilians. America was outraged in 1937, after Japanese aircraft bombed and sank their gunboat, the USS *Panay* and three U.S. flagged merchant vessels on the Yangtze River, and without mercy, machine-gunned the survivors. The United States demanded formal apologies, and Japan responded by expressing their regret. Nevertheless, Japanese occupation forces in Manchuria, continued to harass and assault American citizens, destroy their property, and attack missionary hospitals to persuade all foreigners to evacuate.

Admiral James O. *Joe* Richardson, Commander in Chief of the American Pacific Fleet, objected to U.S. Navy warships being transferred to Pearl Harbour from San Diego, California as a deterrent to Japanese expansion, pointing out that it might have the opposite effect : that a forward naval defense was neither practical nor useful, and that the U.S. Pacific Fleet so deployed would be the first target, in the event of war with Japan, and that Pearl Harbour was vulnerable to both air and torpedo attacks. For ten years, the U.S. Navy held attacks on U.S. Army defences at Pearl Harbour, and always won. But President Roosevelt asserted ~ *relocating the fleet to Hawaii will exert a restraining influence on Japan.*

My view at least as early as October 1940, was that affairs had reached such a state that the United States would become involved in a war with Japan. If the Japanese attacked Thailand or the Dutch East Indies we would not enter the war, but the Japanese could not avoid making mistakes and that as their area of operations expanded ~ sooner or later they would make a mistake and we would enter the war. President F.D. Roosevelt

Admiral Richardson who continued to consider the U.S. Pacific Fleet at Pearl Harbour to be vulnerable to an attack by Japanese carrier-based aircraft, risked his career by making two trips to Washington to confront the President personally on key issues of basing the Fleet at Pearl Harbor. After *Joe* Richardson voiced his objections to Roosevelt at a White House luncheon on 8th October 1940 he was subsequently fired for going out of channels, and in time, replaced by Admiral H.E. Kimmel.

PRELUDE TO PEARL HARBOUR

November 1940 ~ U.S Assistant Naval Attaché in London, Lieutenant-Commander John Opie III went aboard the Royal Navy carrier HMS *Illustrious* when she departed Great Britain for Alexandria. During the coming months, Opie was on several RN vessels and the Australian cruiser HMAS *Sydney*. After the Battle of Taranto, Lieutenant-Commander John N. Opie travelled to the American Legation in Cairo to draft his report to the Office of Naval Intelligence (ONI) that included a copy of the report written by the CO of HMS *Illustrious*, Captain Denis Boyd. Under the heading *Lessons*, Opie wrote :
- AA fire is not effective.
- Low flying planes attacking ships limit shipboard gunnery for fear of hitting friendly ships.
- Strain on pilots was intense ; doubt that they could have made a second attack
- Some believe that ships should put to sea on moonlit nights, rather than try to defend in harbor.
- RN has given up on high level bombing, and prefers torpedo attack to dive-bombing.

November 14th 1940 ~ *The New York Times* ran a six-column article on page one of the Taranto raid. *The Washington Post* featured the Taranto raid as front-page news and *TIME* devoted three pages to it. After reading newspaper reports of the raid on Taranto, Secretary Knox of the U.S. Navy wrote to Secretary Stimson the War Department noting that Taranto was a warning of what could happen if Japan launched a surprise attack upon Pearl Harbour.

December 1940 ~ Berlin. Japanese Embassy Assistant Naval Attaché LCDR Takeshi Naito traveled south to Taranto to evaluate the damage inflicted and discuss the attack with Italian naval officers.

January 7th 1941 ~ Opie's documents sent from Cairo via diplomatic pouch arrived in Washington. (But it was more than a month before LCDR Herbert Eckberg sent his one-page summary from ONI to the new Pacific Fleet Commander Admiral *Kim* Kimmel who replaced Admiral J.O. *Joe* Richardson). Copies were also sent to War Plans, Ordnance, Aeronautic, and other bureaus - on 14th February 1941).

January 7th 1941 ~ On this same day, LCDR J.N. Opie who continued observing RN operations in the Mediterranean was on board HMS *Warspite* when the nearby carrier *Illustrious* was attacked by 30 *Stuka*s that achieved six hits from 500kg bombs. (Her armoured deck saving her from destruction). Opie wrote in a personal letter to his superior, Captain Alan G. Kirk, U.S. Naval Attaché in London ; *I am not trying to drum up my own trade but I honestly feel that I should fly to Hawaii and talk to the boys there on war experiences and how to train to meet the lessons learned.* Opie received no answer to this letter.

January 1941 ~ The U.S. embargoed all shipments of oil, scrap metal and other goods to Japan to slow Japanese expansion into Southeast Asia (Japan received 80% of its petroleum from the U.S.), making the acquisition of resources critical for Japan. The Dutch East Indies held rich deposits of oil, tin, and rubber. In 1941, the Netherlands had been overrun and Britain was fully engaged in her war with Germany. Only the U.S. Navy stood in the way of a *Southern Advance* advocated by the Japanese.

January 1941 ~ Ambassador Joseph C. Crew, the American representative in Japan, sent a secure cable to the State Department in Washington, reporting that Tokyo was awash with rumours that Japanese military forces were planning a surprise attack against Pearl Harbour in case of strife with the United States. (The State Department replied that these rumours were unfounded).

February 14th 1941 ~ A three-page letter was posted from Admiral Stark to Kimmel on Hawaii, regarding the dangers of aerial torpedo attacks against the U.S. Pacific Fleet moored in Pearl Harbour.

March 3rd 1941 ~ Rear-Admiral Aubrey W. Fitch made 35 copies of Opie's report and addressed them to the most senior officers in U.S naval aviation citing ~ *recent developments in the European war* ~ to raise an awareness of the success of torpedo attacks in RN operations against moored enemy ships.

March 11th 1941 ~ Japanese Embassy Washington. Ambassador Admiral Kichisaburo Nomura at a party asked Rear-Admiral Kelly Turner if he would like to converse with him at greater length. Admiral Nomura telephoned Turner the following morning and they arranged to meet the same day. In his report of events, Kelly Turner wrote : *Nomura stated ~ My mission is to prevent war between Japan and the United States, the best interests of our two countries are to maintain peace. I am exploring the ground as best I can in order to find a basis on which our two nations can agree. The radical elements of the Japanese Army are to blame for the war in China and other strong measures. The senior officers of the Japanese Navy have been, and still are in favour of peace with the United States. In my opinion, the presence of the U.S. Fleet in Hawaii forms a stabilizing influence for affairs in the Pacific.* Turner : *I believe he is fully sincere, that he will use his influence against further aggressive moves by the military forces of Japan.*

PRELUDE TO PEARL HARBOUR

March 27th 1941 ~ Takeo Yoshikawa the 1933 honour graduate of the Imperial Japanese Naval academy at Eta Jima who had retired early because of stomach trouble was recalled to duty in 1937, given four years of intensive English lessons, and sent to Hawaii on a top secret assignment. Arriving on the Japanese liner *Nitta Maru*, the sublime passenger Yoshikawa came ashore on Pier 8. He rented a second storey apartment overlooking Pearl Harbour and masqueraded as an obscure clerk in the Japanese consulate. He frequented the *Suncho ro ~ Spring Tide* teahouse on Alewa Heights which overlooked Pearl Harbour, where the owner kept a telescope on the balcony for curious clients. Takeo Yoshikawa took glass bottom boat rides across Kaneohe Bay on the windward side of Oahu's Koolau Mountains to confirm that it was too shallow for a fleet anchorage. He made observations on underwater obstructions, tides, and beach gradients while on swimming expeditions, then gathered more information by taking the U.S. Navy's own tugboat to listen to local gossip. He traveled to Maui and reported that the American fleet no longer used the old whaling port of Lahaina as an anchorage. As the airport at Honolulu conveniently offered tourist flights over Oahu, he invited a Geisha friend on a scenic flight that overlooked every part of Pearl Harbour. Takeo Yoshikawa's two most useful reports were that the fleet was in port every Sunday, and that departing air patrols never appeared to go north. The habit of the U.S. fleet being in port for shore leave on Saturday night meant that on Sunday morning, the American warships were predictable targets. In May, when Yoshikawa was requested to obtain photos of the anchorage plan of U.S. Navy ships in Pearl Harbour, he simply went to a souvenir shop and purchased an assortment of postcards with aerial views of Pearl Harbour showing the U.S. warships in their regular moorings, sent them to Tokyo via the diplomatic pouch and copies of those postcards were in the cockpits of every attack aircraft on 7th December 1941.

April 29th 1941 ~ LCDR Opie returned the U.S.A. to report to the CO of ONI, and deliver a lecture which was written up as an intelligence report and sent to the main offices of the Navy Department, and Kimmel on 2nd May 1941. Opie began his lecture with : *Three outstanding lessons : materiél differences are small, spirit is what counts, and training must go on in war. Radar is the greatest development in materiél. The Royal Navy's armoured-deck carriers have gained protection at the expense of reducing the complement of aircraft. AA gunfire is the bug-a-boo of the war. The only answer against planes, is planes.*

May 1941 ~ Members of Japanese military mission visiting Berlin, headed by Admiral Koki Abe traveled to Rome for a few days to talk with *Supermarina* officers, before continuing south to Taranto to see the Italian Fleet. The visiting Japanese showed incredible interest in the aerial torpedo attack against the Italian warships anchored at Taranto on the night of 11/12th November 1940.

July 20th 1941 ~ Washington. Japanese Ambassador Nomura arranged another meeting in the afternoon with Rear-Admiral Kelly Turner to convey an important item of Japanese Army news. Ambassador Nomura had earlier tried to visit Admiral Stark, but not finding him at home, he called on Rear-Admiral Turner's residence with the purpose of assessing U.S Navy response to his news : *Within the next few days Japan is expected to occupy French-Indo-China. This occupation has become essential. It is essential that Japan has uninterrupted access to necessary raw materials.*

July 1941 ~ U.S. Navy Lieutenant-Commander Albert K. Morehouse on board the HMS *Ark Royal* reported on the dive details of the British torpedo. The ONI received a copy of Morehouse's report on 22nd July, and Admiral Kimmel was sent a copy - however there is no evidence that he ever received it. The Morehouse Report should have ended the idea that shallow water provided protection against an aerial torpedo attack ~ but Kimmel still believed that torpedoes could not run within Pearl Harbour. This was because ; even though U.S. officers serving as neutral observers with the British Royal Navy were acquiring and forwarding up to date intelligence reports about successful aerial torpedo attacks being made in European naval operations ; officers in staff positions repeatedly failed to keep up with changing tactics and technology, and deliver the intelligence reports needed by the fleet commanders.

<u>Torpedo Dropping Depths.</u> *The war Plans office of the Eleventh Naval District made several enquires of units at N.A.S. San Diego regarding the depth to which an aircraft torpedo would dive on its initial drop. The security of fleet anchorages was in question. The information supplied by Torpedo Squadrons varied, but ten to twelve fathoms was believed to be the shallowest dive that would result from an aircraft torpedo launching.*

Records of the RN 18" Mk. III indicate that this torpedo may be dropped in water as shallow as <u>four fathoms.</u> This shallow dive is only possible when the torpedo is set in the low speed (27 knots) setting. In high-speed setting, the torpedo dives to approximately three fathoms beyond its depth setting. In low speed setting six feet deeper than the set depth is the average launching dive depth.   Prepared by : A.K. Morehouse, LCDR, USN.

PRELUDE TO PEARL HARBOUR

August 1941 ~ A report prepared by Rear Admiral Patrick Bellinger and Major-General Frederick Martin (responsible for Hawaiian air defence) was sent to Washington. It predicted that Japan would attack Pearl Harbour before a declaration of war using aircraft, and submarines, and that their carriers would cross the North Pacific in the empty Vacant Sea away from shipping lanes. The same week, the chief of naval operations in Washington predicted that Japan would attack on a Saturday or a Sunday.

September 1941 ~ Japanese torpedo technician Captain Fumio Aiko, solved the problem of making Type 91 *Thunder Fish* torpedoes run shallow by fitting wooden stabilizing fins. Genda succeeded in launching torpedoes in 36 feet of water, shallow enough for Pearl Harbour's average depth of 40 feet. Genda and Fuchida instructed pilots to release their torpedoes from a height of 40 feet at 150 knots.

When we received new torpedoes, we were told that these were the most sophisticated torpedoes in the world. Because of the breakwaters in Pearl Harbour, there was only about 300 yards in which to line up our attack runs. This meant we would be dropping at a range of only 200 yards. Pearl Harbour was also very shallow, so even if we dropped from the ideal altitude of 15' the torpedo would still sink to a depth of about 30' or more, and might get stuck in the mud. To prevent this, the metal fins of the torpedoes were fitted with breakaway wood fins designed to create enough drag to prevent the torpedo from sinking too deeply.   C.P.O. Mori Juzo IJN

December 1st 1941 ~ IJN Chief of Staff Osami Nagano advised Commander of the Combined Fleet, Admiral Isoroku Yamamoto; *Japan has decided to open hostilities against the United States, the United Kingdom, and the Netherlands early in December. Should it appear certain that Japanese-American negotiations will reach an amicable settlement prior to the commencement of hostilities, all elements of the Combined Fleet are to return to their bases in accordance with separate orders. The Kido Butai will proceed to the Hawaiian Area with utmost secrecy and, at the outbreak of the war, will launch a resolute surprise attack on, and deal a fatal blow to the enemy fleet in the Hawaiian Area. The initial air attack is scheduled at 0330 hours, X Day. Upon completion, the force is to return to Japan, re-equip, and re-deploy for Second Phase Operations : to crush hostile naval and air forces in Asia, the Pacific and Hawaii, seize the main U.S., British, and Dutch bases in East Asia and capture and secure the key areas of the southern regions. On the home leg, be alert for tracking and counterattacks by the Americans, and return to the base in the Marshall Islands, rather than the Home Islands*

December 6th 1941 ~ The final report of Takeo Yoshikawa advised: *All American battleships are in port, no balloons, and no torpedo-defence nets deployed around the battleships. Lexington left harbour yesterday. Enterprise is thought to be operating at sea with her planes on board. All carriers are at sea.* While continuing diplomatic negotiations with Washington, Japanese officers in Tokyo were planning military offensives against British Hong Kong, Malaya, Singapore, the Dutch East Indies, and the Philippines. At that time, the USN had two carriers in the Pacific. USS *Enterprise* was returning to Pearl Harbour, while USS *Lexington* was on her way to Midway. The *Saratoga* was at anchor at San Diego, while *Wasp, Yorktown, Ranger, Hornet,* and the escort carrier *Long Island* were attached to the Atlantic Fleet.

Admiral Isoroku Yamamoto ~ proudly bore the scars received when he lost the second and third fingers on his left hand as a naval cadet during Admiral Togo's victory over the Tsar's Russian Navy at the Strait of Tsu-Shima in 1905. Now as an admiral himself, he began in early 1941 to assemble some trusted officers to plan Operation Hawaii, which he also named Operation Z, after Admiral Togo's historic banner at the battle of Tsu-Shima. When the time came for action, Yamamoto directed the Japanese Fleet to steam far north of the shipping lanes, refuel at sea, and head south 500 miles from Honolulu at sunset. A full night of high speed steaming would put his battle fleet 200 miles from Pearl Harbour at sunrise in an area rarely visited by American patrol planes. The attack was set for 7th December 1941 Hawaiian time, to take advantage of a bright moon for a predawn fly off.

PEARL HARBOUR 1941

The first blow is half the battle ~ Japanese proverb

Admiral Isoruku Yamamoto, commander of the Imperial Japanese Navy directed that the coming undeclared attack on Pearl Harbour would be coordinated at sea, in preserved radio silence by using visual signal lamps. On 26th November 1941, six Japanese aircraft carriers; *Akagi, Kaga, Hiryu, Soryu, Shokaku* and *Zuikaku*, two battleships, and their warship escorts led by Vice-Admiral Chuichi Nagumo left the Kurile Islands for Hawaii. Included in Yamamoto's plan were both conventional and midget submarines. The larger Japanese submarines were to wait off Oahu to target any U.S. warships evading the air attack. The midget submarines were to penetrate the harbour and torpedo warships. On board the Japanese flagship *Nagato*, Vice-Admiral Ugaki noted in his war dairy, *Senso Roku* (Records of the Seaweeds of War) that attached to the deck of the mother submarine I-24 was her midget submarine that carried a crew of two, plus two torpedoes ~ each with a 300kg warhead.

A surprise attack on X-Day will be an entirely unexpected storm. How much damage they will be able to inflict is not the point. The firm determination not to return alive on the part of these young Lieutenants and Ensigns who smilingly embark on their small submarines cannot be praised too much. The spirit of kesshi-tai (do-or-die) has not changed at all. We can fully rely on them.

Japanese Submarine I-24 drawn by a captured Japanese prisoner from I-24 shows how the midget submarine was was carried to Pearl Harbour for the attack on 7th December 1941.   - US Weekly Intelligence, December 1944

The delays of the U.S. Intelligence Service (still unprepared for war) in communicating decoded Japanese war-warnings meant that they were unable to transmit important reports to Hawaii in time. Even the sighting of an unidentified midget submarine attempting to penetrate the entrance to Pearl on the morning of 7th December, and the USS *Ward* sinking it at 0637 by depth-charges, failed to raise a general alert. At 0702, radar operators on Oahu detected a group of aircraft 40 miles north coming in, but were reassured that it was an expected flight of B-17s due in from the west coast of America.

Early on the Sunday morning of 7th December 1941, the first Japanese air strike led by Commander Mitsuo Fuchida prepared to takeoff. His pilots were being briefed on the missions flare signals : one flare meant that the attackers had achieved surprise and that the torpedo aircraft would go in first; two flares meant that there was opposition and that the dive-bombers and fighters would lead the way. The Japanese aircraft launched from the heaving decks of six carriers at 0600 consisted of 104 Nakajima B5N2 *Kate* torpedo-bombers, 135 Aichi D3A1 *Val* dive-bombers, and as the Japanese attack formation approached with total surprise, Fuchida fired a flare. The torpedo-bombers approached from the east and west in four groups low over the naval base at 0800 before lining up on their targets. With no torpedo nets in the water to intercept enemy torpedoes, US Admiral William *Bill* Halsey said, *Before we're through with them, the Japanese language will be spoken only in hell.*

Undisturbed by AAA, most *Kates* made perfect torpedo runs in flights of two or three from 70-feet at 100 knots at their targets. Sinking four U.S. battleships, the *Arizona, Oklahoma, West Virginia,* and *California,* they damaged three others, and hit numerous other warships. Another 105 strike aircraft, supported by 78 Mitsubishi A6M Zero fighters attacked. Altogether the U.S. Pacific Fleet lost 18 warships and 187 aircraft in an attack that went on for two hours. When the Japanese aircraft returned to their carriers, they flew off to all points of the compass to confuse the Americans, but most turned south to mislead the Americans into believing the Japanese carriers lay in that direction; then headed north to where their carriers really were. The U.S. Pacific Fleet had been neutralized for six months, but the American aircraft carriers USS *Lexington* and *Enterprise* with their escorting cruisers and destroyers were not at Pearl. The shallow water was another factor favouring the U.S. Fleet that enabled salvage teams to raise and repair most partially sunken ships ~ the Japanese lost 29 aircraft.

JAPANESE AVIATORS OF PACIFIC THEATRE

Minoru Genda
Planner of Pearl Harbour attack

Mitsuo Fuchida
Leader first air wave attack on Pearl Harbour

Junichi Sasai ~ 27 kills
The Prince of Rabaul

Clean-up Trio: Hiroyoshi Nishizawa 87 kills

Saburo Sakai 64 kills

Toshio Ohta 34 kills

Takeo Taminizu ~ 32 kills

Kenji Yanagiya ~ 11 kills

Matsuo Hagiri ~ 13 kills

JAPANESE AVIATORS OF THE PACIFIC THEATRE

Hiroyoshi Nishizawa ~ 87 kills

Junuchi Sasai ~ 27 kills

Saburo Sakai wearing his life preserver

Junuchi Sasai with a crashed U.S. Curtiss P-40 in the Dutch East Indies.

AMERICAN AVIATORS OF THE PACIFIC THEATRE

Thomas McGuire ~ 38 kills

Lieutenant E.H. *Butch* O'Hare in his Grumman F4F Wildcat ~ with 5 kills

First Lt. John Bolt 1943 ~ 6 kills

First-Lt. James Morehead ~ 8 kills confirmed.

Commander Clarence Wade McCluskey

Major Richard *Dick* Bong ~ 40 kills

AMERICAN AVIATORS OF THE PACIFIC THEATRE

Lieutenant-Commander John S. Thach C.O. of VF-3 Squadron

Commander David S. McCampbell ~ 34 kills

Alex Vraciu ~ on the *Lexington*, showing his Philippine Sea tally 19th June 1944

Major Gregory *Pappy* Boyington ~ 28 kills
CO of USMC *Black Sheep Squadron* VMF-214

Dick Bong poses with *Marge*, his Lockheed P38J-15 *Lightning* with a photo of his fiancée, Marjorie-Ann is affixed to its nose.

Lieutenant-Commander James *Jimmy* Doolittle

THE JAPANESE ADVANCE ON AUSTRALIA

COMING? THEN HURRY! (AWM: ART04297)   KEEP THEM FLYING! (WALTER JARDINE RAAF PUBLICATIONS UNIT 1942)

For some weeks, the Japanese carriers were busy supporting landings to establish the perimeter of their new empire. On 20th January 1942, aircraft from the Japanese carriers *Akagi, Kaga, Shokaku,* and *Zuikaku* attacked Rabaul on the island of New Britain. They returned during the next two days and on the 23rd provided air support for the invasion of Rabaul and nearby Kavieng on New Ireland. Japanese bombers were then based at Rabaul to be within striking range of Port Moresby on the south coast of New Guinea's Huon Peninsula. If the Japanese captured Port Moresby, its position on the Coral Sea would make it an excellent staging point for their proposed invasion of Australia. But first, Japanese forces invaded N.E. New Guinea on 8th March 1942, to capture Lae and Salamaua ~ and these enemy movements gave the United States time to transfer USS *Wasp, Hornet,* and *Yorktown* to the Pacific.

Darwin ~ Four carriers of Vice-Admiral Nagumo's strike force, escorted by two battleships, five cruisers and some 20 smaller warships entered the Arafura Sea under the cover of darkness. Early on the morning of 19th February 1942, some 130 miles from Australia's northern coast, the Japanese carriers turned into the wind to launch 135 aircraft, led once more by Mitsuo Fuchida to bomb and strafe the port of Darwin on the north coast of Australia. Their mission was to neutralize Darwin as a base, as a prelude to the Japanese invasion of Timor. The Nakajima Kate attack bombers circled, and then made a run over the harbour at 14,000 feet pattern bombing the shipping moored in the harbour. Val dive-bombers bombed and strafed individual ships as Zero fighters swooped low to strafe. Several ships were sunk and damaged. Most alarming was MV *Neptuna* with its cargo of 200 depth charges exploding in the harbour. On 25th April 1942, 24 Japanese bombers approached from the direction of Bathurst Island, but well reported by a vigilant observer, Darwin had ample warning. The 8th Fighter Squadron flying Warhawks scrambled to destroy 11 Japanese bombers without loss, with First-Lieutenant James Morehead in *White 61* shooting down three enemy twin-engined bombers, but thoroughly shot up by escorting Zeros during the fight, had to crash land near Adelaide River.

O'Hare saves the Lexington ~ The pressure on Australia was relieved by the presence of the carriers USS *Yorktown, Enterprise* and *Lexington,* which diverted the main Japanese thrust away from Australia. The day after the first raid on Darwin, the Japanese attacked the USS *Lexington,* which had been assigned the task of penetrating enemy-held waters north of New Ireland. On 20th February and some 450 miles from Rabaul, a jagged V blip of unknown aircraft was detected by the USS *Lexington's* radar. An hour later, at 1645 they were 47 miles west and closing fast ~ a fighting patrol of six F4F Wildcats were launched toward the radar reference. Lieutenant Edward *Butch* O'Hare leading the second section of Lieutenant Commander Thach's VF-3 fighter squadron found himself between the *Lexington* and a flight of eight Japanese bombers at 1,500 feet nine miles out — heading straight for the carrier. O' Hare's first diving attack on the far right side of the Japanese V formation shot up a *Betty's* right engine and wing fuel tanks. He then peppered the opposite bomber with hits so concentrated that the nacelle of the *Betty* popped out of its mountings. O'Hare next shot the lead *Betty* that exploded into fragments, soon destroyed five bombers, and damaged a sixth. When Lt-Cdr. *Jimmy* Thach arrived on the scene with the rest of his flight, he saw three Japanese *Betty* bombers falling in flames at the same time. Lt. *Butch* O'Hare who had skillfully fired an average of 60 rounds to destroy each *Betty* bomber became the first American flying ace of WWII, and was awarded the Congressional Medal of Honour.

Top : Japanese G4M2 *Betty* bombers carrying out a low level attack. Middle : *Betty* bomber interior, cockpit, and crew.
Bottom : Imperial Japanese Navy Mitsubishi G4M2 *Betty* somewhere in the S.W Pacific ~ note frontal radar antenna.

THE DOOLITTLE RAID ~ APRIL 1942

Four months after their raid on Pearl Harbour, the Japanese people living on their home islands probably felt safe from retribution as their naval and ground forces were robustly establishing their new empire of the seas. However, in an atmosphere of anxiety that followed the raid, U.S. military leaders were continually thinking of some potent way to hit back at the remote Japanese homeland. Early in 1942, Captain Francis Low came up with the best idea to give the Japanese a nasty surprise, a carrier strike on Tokyo. The first problem was that no naval aircraft had sufficient range, so Captain Donald Duncan took the initiative to the air force. They suggested the B-25 Mitchell medium bomber, and the perfect man to lead the mission, Lieutenant Colonel James *Jimmy* Doolittle, who in 1922 became the first to fly coast-to-coast in less than a day. At once, Doolittle began training Mitchell bomber pilots for short takeoffs at Eglin Field, Alabama. Still unaware of their mission, Mitchell's men completed their training, flew to the naval air station of Alemeda in California, to see the dockyard cranes hoisting their 16 modified Mitchells onto the flight deck of the aircraft carrier USS *Hornet*. After Doolittle addressed the group and informed his men of the mission, every airman volunteered. Each Mitchell bomber would carry three 500-lb HE bombs, and one 500-lb incendiary cluster bomb.

Admiral William Halsey planned for USS *Hornet* to launch the strike force of 16 bombers 400 miles from Japan, but at 0738 on the morning of 18th April 1942, while the task force was 650 nautical miles from Japan, an enemy fishing boat, *Nitto Maru* sighted the *Hornets*' task force and radioed a warning to Japan. USS *Nashville* sank the boat by gunfire. The Japanese officer who had commanded the boat committed suicide rather than be captured, as five of his surviving crew were rescued by the *Nashville*. Now, a decision had to be made. To enable the *Hornet's* navy fighters elevation to the flight deck to protect her 14 support vessels from a possible Japanese attack, the bombers either had to takeoff or be pushed overboard. The greater distance created fuel issues, but as before, every airman volunteered to fly regardless, and to takeoff earlier than envisaged. At 0800, Admiral Halsey knowing that Japanese carrier aircraft might attack his group, ordered Doolittle to takeoff immediately. Halsey turned the *Hornet* into the wind, as the airmen rushed to their bombers. Their engines were started, and at 0824, with Doolittle at the controls, the lead Mitchell took off. One hour later the 16th bomber cleared the *Hornet's* flight deck, and Halsey quickly turned the carrier task force back to Pearl Harbor.

In a spectacle of low-level flying, the B-25s flew towards Japan in single file at near wave top level to avoid radar detection. Their surprise was perfect. It was noon Tokyo time, when Doolittle's lead Mitchell flew over the Japanese coastline, as groups of Japanese civilians were returning to work after an air raid exercise. Doolittle climbed to 12,000 feet and dropped his first bomb over the centre of Tokyo. The other bombers separated to set course for their assigned targets ; then attacked Tokyo, two Mitchells headed for Yokohama to strike the new carrier *Ryuho* (delaying her launch till November) while four individual B-25s attacked Yokosuka, Nagoya, Kobe, and Osaka. Tokyo was stunned.

The B-25 of Captain Edward York, which low on fuel set course for the closer landmass of Russia. Landing 25 miles north of Vladivostok, they were interned, but bribed smugglers to take them to Iran. The remaining 15 Mitchells flew southwest along the southern coast of Japan, over the East China Sea towards eastern China. Few of Doolittle's squadron theoretically had any hope of reaching far away China, if it were not for a chance tail wind as they came off their targets that increased their airspeed by 25 knots for seven hours. Due to their premature launch, the Mitchell commanders had to decide; whether to ditch offshore (two crewmen perished swimming to shore) or to fly on through the night and when their fuel ran out, to bail out in the dark over the mountains. Doolittle's crew chose to fly. After their 13-hour flight, they parachuted into China. Doolittle landed in a paddy on top of a pile of dung that happily saved his previously injured ankle from breaking. Of the eight American airmen captured by the Japanese, three were executed after a sham trial, one died of malnutrition and the surviving four were tortured, starved, and kept as prisoners until the end of the war. The Japanese having seen U.S. land-based bombers, began searching for Doolittle's supposed airfield of departure.

A jubilant President Roosevelt stirred Japanese confusion by suggesting they look in *Shangri-la*, the mythical Tibetan hideaway in James Hilton's novel *Lost Horizon*. The great consequence of the Doolittle Raid was that it ended the disagreement between Admiral Osami Nagano who proposed *Operation FS* to isolate Australia to bring it under Japanese control, and Admiral Yamamoto who believed that Japan would not be able to achieve its goals without first defeating the U.S. Pacific Fleet. The Doolittle raid proved that Yamamoto was right, and compelled the Japanese to expand their defensive perimeter, causing them to pause and revise to their pattern of conquest, all of which prompted the Japanese to commence their Midway plan of operations two months later. Before this, the ruling Japanese régime and their GHQ continued planning to isolate, and to subjugate Australia.

Lieutenant-Colonel James *Jimmy* Doolittle performs a full-throttle B-25 *Mitchell* bomber takeoff from the bow of USS *Hornet*

OPERATION FS ~ THE JAPANESE PLAN TO ISOLATE AUSTRALIA

Captain Sadatoshi Tomioka as CO of Plans Division ~ Japanese Naval General Staff confirmed : *What I worried about most was Australia. I knew that America's military strength would expand enormously by 1943, but as long as that colossal military strength was pinned down on the Hawaiian Islands and in the United States, I believed that Japan could hold its conquered Pacific territories. That American air power would be ineffective without forward air bases from which aircraft could operate. However, with Australia as a gigantic forward base, the Americans would be able to deploy their massive military strength within striking distance of Japan's southern defensive perimeter and use the islands of the South-West Pacific as stepping-stones to recover the Philippines and then move against Japan's home islands. To remove this danger to Japan's southern defensive perimeter, I argued that it was not sufficient to isolate Australia from the United States by blockade; Japan must also ~ invade and occupy ~ key areas of the Australian mainland. I won powerful support when the Chief of the Navy General Staff, Admiral Osami Nagano, accepted this argument.*

IJN submarines deploy to isolate Australia ~ November 1941. The Imperial Japanese Navy deployed four vessels of the 6th Submarine Squadron south to contain and sink as many ships as possible with torpedoes outside of Darwin's harbour. The IJN 6th Submarine Squadron under the command of Commander Keiyu Endo consisting of : *I-121, I-122, I-123, I-124* departed from Hainan Island, China on 1st December 1941, and received the coded signal *Climb Mount Niitaka* on 2nd December 1941, notifying her that hostilities would commence on 8th December. On that day Lt-Cmdr. Kishigami Koichi commanding *I-124* ordered his crew to lay 39 mines off Manila Bay. On 10th December at 1000, the British ship SS *Hareldawins* was sunk by a mine from *I-124* ~ as were U.S. vessels *Corregidor* and *Panamanian Delight* on 17th December. On 10th January, *I-124* cruising with *I-123*, continued south according to their strategic plan: four IJN submarines would by the combined use of their weapons; and the mere potential of their presence, alarm the shipping in Darwin's harbour and deter other vessels from approaching to paralyze the port of Australia ~ and commence the isolation of Australia. On the 19th the Japanese Commander of the Southern Area reported: *I-124 approaching Darwin Harbour*. His signal was intercepted by codebreakers who warned: *Japanese submarine operating near Darwin*. The next day, Koichi attacked the Australian corvette HMAS *Deloraine* by launching a torpedo at her, but alerted to the approaching torpedo, the *Deloraine* steamed full speed at the sub and depth-charged her at very close range. *I-124* was sunk in 48m of water with the loss of 80 Japanese submariners.

10th January 1942. Japanese Army General Staff and Japan's PM Hideki Tojo knowing Australia posed a threat to Japan, resolved at an Imperial Japanese Liaison Conference to : *Proceed with Southern Operations, while blockading supply from Britain and the United States and a strengthening the pressure on Australia ultimately with the aim to force Australia to be free from the shackles of Britain and the United States.*

21st January 1942. Chief of Naval General Staff, Admiral Nagano wanted to invade Australia. Japanese PM General Hideki Tojo addressed the *Diet* ~ and called on Australia to surrender to Japan. In Australia, Tojo's demand for surrender was ignored. After the fall of Singapore on 5th February 1942, Tojo repeated his demand on Australia to surrender.

4th March 1942. At a meeting of the Japanese Imperial War Council, the Japanese Navy requested troops for the invasion of Australia ~ but the generals refused, seeing no need to commit troops and logistical resources to the conquest of Australia after the easy capture of Rabaul and the bombing of Darwin suggest that Australia had little with which to defend itself from invasion ~ Japanese Army resources are already over-extended by rapid territorial conquests and need time to consolidate gains.

7th March 1942. Army Vice-Chief of Staff, Lt-General Moritake Tanabe announced details of the : *Fundamental Outline ~ expanding existing war achievements with supplementary operations such as the invasions of Hawaii, Ceylon, and Australia. That the Army and Navy are studying.* Navy HQ agreed to the Army plan to sever Australia's lifeline to the United States and then to press for Australia's surrender. The Japanese Army did not rule out support for an invasion, if Australia did not surrender.

15th March 1942. With Emperor Hirohito's approval, Japanese GHQ formally resolved to extend Japan's southern defensive perimeter from Port Moresby in Papua to Fiji and Samoa to isolate Australia from the United States. This plan was given the Japanese code-reference *Operation FS* and placed under the overall direction of Vice-Admiral Shigeyoshi Inouye at Rabaul. PM Hideki Tojo, agreed that Australia must be reduced. Generals Asaka and Terauchi supported Tojo, but emphasized that unless Fiji, the New Hebrides and New Caledonia were soon captured, their plan had little chance of success and all Japanese gains in the S.W. Pacific would become increasingly precarious. *In order to lessen the losses of naval operations against the Australian mandates, a simultaneous attack by the landing of a massive expeditionary force at several points on the Australian mainland should be undertaken.*

Japanese GHQ plan of attack to isolate and invade Australia ~ provided by Chinese Intelligence in 1942.

28th March 1942. Tojo repeats his demand in the *Diet* for a third time for Australia to surrender. Three days later, three Japanese midget submarines penetrated Sydney harbour and torpedoed the Royal Australian Navy depot ship, HMAS *Kuttabul* killing 19 Australia and two British sailors.

October 1942. U.S. Office of Strategic Services (OSS) reported that Japan was planning an invasion of Australia in mid-1942 based upon information passed to an OSS agent by a member of the neutral Spanish diplomatic staff in Tokyo. The OSS report quoted a Japanese Imperial War Council meeting :

The costly lesson learned by the Japanese during their attack at Midway has resulted in the opinion, apparently held by Admiral Suetsugu that surface operations even with carrier air support cannot successfully be carried out within 750 miles of a strong enemy base of shore-based aircraft. He believes that the Japanese possessions in the Western Pacific can only be made impregnable by concentrating on the strengthening of their air defence and that only Australia remains as an obstacle to the creation of a perfect defensive chain. However, he appears to have felt that Australia has already been reinforced to such an extent that operations designed only to isolate it would cause such heavy losses that its subsequent invasion would be rendered impossible.

Forces on Japanese Mandated Islands and New Guinea to defend the left wing against the Americans. Right wing to be defended by mining the Malacca Straits and by naval forces defending the Indian Ocean approaches. Part of the forces based on Java, to stage a diversional attack landing of Darwin and engaging the American and Australian forces around Katherine and Birdum. The main force, proceeding via the Sunda Straits, to land at Freemantle and Perth ; to then occupy the country west of Esperance-Sandstone, and then to advance eastward.

THE BATTLE OF THE CORAL SEA

Once U.S. and Australian code breakers of Y Intercept Service in *Raleigh*; an ivy clad mansion in Melbourne were able to crack 15% of the Japanese JN-25 code, Admiral Chester Nimitz knew the Japanese intended to attack Port Moresby during the first week in May. After Port Moresby, Japan planned to then establish a seaplane base in the Louisiade Archipelago 240km southeast off the tail of New Guinea to sever the sea lanes across the north of Australia linking the Indian and Pacific basins. The Japanese intended to base other squadrons there to attack the Australian ports and airfields in Queensland at Horn Island, Cooktown, and Townsville enabling them to isolate Australia without actually conquering its continental territory. Nimitz knowing, *how to be at the right spot at the right time*, stripped Pearl Harbour's defences bare, to attack the Japanese as they entered the Coral Sea off the N.E. coast of Australia in the first action where aircraft carriers would engage each other. The first report came on 3rd May from Australian coast-watcher D.G. Kennedy on Santa Isabel Island after he sighted Japanese vessels heading for Tulagi. Onboard the USS *Yorktown*, Rear-Admiral Frank Fletcher; a tattooed crusty old salt called *Whisky Jack* by his friends, turned his carrier group N.E. and increasing speed up to 27 knots, ordered air strikes against the Japanese in Tulagi Harbour. Over Tulagi, U.S. pilots saw their windows and gun-sights fog over as they swooped from higher cooler atmosphere into lower warmer air. Most of their bombs missed, hitting only the destroyer *Kizazuka*, and four landing barges. Now the Japanese knew ~ that at least one U.S. carrier was close by.

On 7th May, aircraft from *Lexington* and *Yorktown* discovered the Japanese carrier *Shoho* north of Misima Island, where *Shoho* assisted her attackers by turning into the wind to launch her aircraft. The straight steaming *Shoho* made an easy target for the American pilots who sank it. Lt-Cdr David Dixon radioed: *Dixon to carrier, scratch one flattop*. The Japanese fleet turned away, now vulnerable to land based bombers without their fighter support, now at the bottom in the hulk of the *Shoho*. Then came the embarrassing episode of three American B-17 land-based bombers searching for the Japanese invasion fleet that targeted the cruiser HMAS *Australia*. After more masterful maneuvering by HMAS *Australia's* Admiral John Crace, who had just avoided the bombs of 44 attacking Japanese aircraft; *Fortunately their bombing, in comparison with that of the Japanese a few minutes earlier was disgraceful!*

On 8th May 1942, both fleets found each other simultaneously 320km apart, and both launched their torpedo-bombers and dive-bombers to attack the carriers and screening ships below. Except for the tactically sharp skipper of the *Zuikaku*, who departed at full steam into the concealment of a heavy rainsquall. American naval aviators fearlessly dived from 18,000 feet, down through bursting AAA shells to the *Shokaku*, pressing home their attacks to within a few hundred feet to ensure direct hits. Two 1,000-lb bombs hit the *Shokaku*, engulfing her in flame, smoke, and debris, and still the impressive damage control crew of the *Shokaku*, extinguished all fires to escape back to Japan to carry out repairs. While the Americans were attacking the *Shokaku*, Japanese torpedo-bombers using their *anvil* tactic attacked the *Lexington* from both sides. No matter which way Captain Sherman maneuvered the *Lexington* he saw a string of torpedoes. Minutes later, *Lexington* was hit on the port side by a torpedo, then another. Japanese dive-bombers scored two more hits, rupturing fuel tanks, starting four fires and as she listed 6° to port, signal flags flew up her mainmast: THIS SHIP NEEDS HELP. At 1707, Captain Sherman ordered *Lexington's* crew : Abandon ship. (*Yorktown* evaded all Japanese torpedoes).

The Japanese and Americans both lost one carrier. As the badly damaged *Shokaku* headed off for repairs, the Japanese were left with the carrier *Zuikaku* and her depleted complement of 47 serviceable aircraft. Decisively these two Japanese carriers would not participate at Midway. The score favoured the Japanese, but in strategic terms Admiral Shigeyoshi Inouye ordered the Japanese troop transports bound for Moresby to turn back, and the naval operation was postponed indefinitely. The Japanese empire had reached its outer limits. Japan's seaward invasion of Port Moresby had been stopped, as had their follow on plans to invade Fiji, Samoa, and New Caledonia. After their setback, the Japanese landed an invasion force at Gona, on the north coast of Papua on 21st July 1942, and embarked overland towards Port Moresby, (the main Australian base in New Guinea) with disastrous results. Fighting thru mountainous jungle in appalling conditions, the Japanese ground troops were defeated (for the first time) by a small force of Australians on the jungle pathway known as the Kokoda Track. Together they had saved Australia. The German Naval Attaché in Tokyo, Admiral Paul Wenneker; (who had previously commanded the *Lützow*, and was assigned to Japan to oversee blockade-running U-boats between Germany and Japan to exchange data, raw materials and optical equipment) said ; *In all the war, the greatest single shock to the Japanese was their defeat in the Battle of the Coral Sea because, they had been confident that they would completely conquer Australia by the end of 1942.*
This quote, again confirms that Japan intended to invade and conquer Australia.

FIGHTER TACTICS OVER THE PACIFIC

Apart from tactical surprise, the Japanese fighter pilot's main advantage during their first engagements over the Pacific was maneuverability. The Mitsubishi Zero had excellent acceleration, (over 5 mph per second at 15,000 feet through the speed range of 120 to 150 mph), a very low wing loading that enabled it to turn tightly, a great ability to sustain a climb at a very steep angle, and a reliable cannon armament. A Japanese fighter formation could stay in the air for six or seven hours (three times longer than British or German fighters) and usually flying at altitudes of 15,000 to 20,000 feet could operate efficiently at 27,000 feet or higher. Japanese naval pilots used a three aircraft *shotai* V formation perfected in China with 100-yard intervals between aircraft. Three *shotai* formed a nine aircraft *chutai*, with all pilots except the leader weaving loose and flexible, always ready for action or a ground attack role. Japanese tactics were mostly based on hit–and–run, with the *shotai* forming line astern for the attack, and pulling back upstairs after firing. Japanese fighter pilots generally avoided head-on attacks against Allied fighters, probably because most of their aircraft were unarmoured. *Head-to-Tail* attacks were favoured (except engaging U.S. bombers with rear guns), and the standard Japanese evasion was to go nose-up into a steep climbing turn to the left.

Early in the scrap I evaded the attack of one, and then jumped on his tail. He immediately resorted to the old Zero trick of zooming for altitude. I anticipated the pull-up and gave him a snap burst as he commenced his climb. My bullets hit him and he began to burn.   Lieutenant Noel Gayler

The Zero outclassed the United States Navy F 4F Wildcat fighter that was slightly slower at most heights, suffered initially from a much lower climb rate, and was unable to compete in a turning match at speeds below 200 mph. Although the Wildcat could not turn as quick or as sharp, it benefited from protective armour, self-sealing fuel tanks, and a stronger construction that made it more difficult to shoot down than the Zero. To begin, American flights tended to split into single aircraft when attacked and then often found themselves pursued by at least three Zeros. A pursued Wildcat pilot could only hope that a buddy would come and brush the Zero off his tail or *break* turning sharp across the Zero's flight path, which usually worked because the large nose of the Zero obstructed its pilots view. American Wildcat pilots came to rely on teamwork to hold their own against the agile Zero during 1942. The right tactics to defeat the Zero were to avoid turning dogfights with the more maneuverable Japanese fighters and to use the American fighter's greater speed for diving attacks. By the mid 1942, U.S. pilots realized that the weight and strength of their aircraft was their advantage.

The Cleanup Trio ~ Hiroyoshi Nishizawa transferred to the Tainan Kokutai Air Group in May 1942, flying Zero A6M fighter patrols from Lae in New Guinea, out over the Coral Sea and New Guinea with the Japanese *Gekitsui-O* or aces, Saburo Sakai and Toshio Ohta in a *chutai* commanded by Junichi Sasai. These Japanese aces become legends using swift pouncing tactics against unwary Americans. Flying sorties from Lae, they all scored triples in combat and Sakai once downed four opponents. Hiroyoshi Nishizawa chalked up six victories during one mission. Astonishing accuracy considering that a Zero had only 60 rounds per gun for its twin 20mm cannon and 500 rounds per gun for its two 7.7mm machine guns. Saburo Sakai who graduated first in his Naval Aviation Class of 1937 to be personally presented with a silver watch from Emperor Hirohito, wrote this of his friends:

Hiroyoshi Nishizawa was tall and thin for a Japanese, nearly five feet eight inches in height. He had a gaunt look about him and suffered almost constantly from malaria and tropical skin disease. He was pale most of the time, almost like a pensive outcast. Never have I seen a man in a plane do what Nishizawa could do with his Zero. Once on the air this strange man transformed into an extroverted, untamed genius. His aerobatics were at once breathtaking, brilliant, totally unpredictable, impossible, and heart stirring to witness. He also had a fine hunter's eye capable of spotting enemy aircraft before his comrades knew there was anything in the sky. Often we flew together with Ohta and were known to the other pilots as the Cleanup Trio.

The *Cleanup Trio* excelled at decoy tactics. Flying in circles one above the other at different altitudes, if the lower aircraft was attacked, the aircraft above it dived at the enemy fighters usually from behind or slightly below. Another signature tactic devised by the *Cleanup Trio* was used when they encountered a pair of Allied fighters. The right or left flanker of the *Trio* peeled off and dived, and if an Allied fighter followed, he became easy prey for the two other airmen of the *Cleanup Trio*. Another trick. Japanese fighter pilots attempted to draw off Allied stragglers into combat to exhaust their fuel supply by the time another flight of Japanese fighters arrived, to continue the fight and win. During their reign over the Pacific, the *Cleanup Trio* scored 233 kills.

THE CLEANUP TRIO

Hiroyoshi Nishizawa in his Mitsubishi Zero A6M3 over the Solomon Islands

Admiral Yamamoto, who developed a fatal passion for elaborate diversionary moves and for splitting his forces, had been thinking how to lure the American aircraft carriers that had escaped destruction at Pearl Harbour into a trap ~ by seizing two tiny U.S. island outposts of coral and sand known as Midway 1,770km N.W. of Pearl Harbour. Yamamoto devised a trap inspired by ancient samurai tactics with surprise as the essence, using a diversionary attack as the pivot to cause distraction while IJN airmen destroyed the carriers of the American fleet. It was Yamamoto's expectation that after Guam and Wake had been taken, the U.S. Navy would come to protect Midway and he would attack them with the most powerful fleet ever assembled : 11 battleships, 8 carriers, 23 cruisers, and 65 destroyers ~ 190 warships with more than 200 carrier aircraft.

Against the Imperial Japanese Navy, Admiral Chester Nimitz had 76 ships, no battleships, and three carriers to face Japan's eight. But, Nimitz had Lieutenant Commander Joseph Rochefort's code-breaking team in Pearl Harbour who reported that Yamamoto's strike against Midway was set for 4th June. In the subsequent action, where the key would be to find and attack carriers, the American Navy had one great advantage over the Japanese. Their ships were equipped with radar and were therefore far less dependent on flying standing patrols over their fleet, whereas the Japanese in the combat area, were obliged to keep at least one *shotai* of three fighters per carrier aloft, another *shotai* ready for immediate launch, and a third at a lesser state of readiness. Without radar the Japanese were dependent on visual sightings, and never did this weakness inflict more damage than during Midway.

Poised for total domination of the Pacific, Admiral Chuichi Nagumo steamed ahead with four of the six carriers from the fleet that had attacked Pearl Harbour. Before the blue dawn of 4th June 1942, Nagumo launched 108 aircraft, half his force to give Midway's defences a savage pounding, but his scout aircraft failed to spot the two sister ships, the carriers USS *Enterprise* and USS *Hornet* 320km to the northeast commanded by Rear-Admiral Raymond Spruance who, taking an intuitive risk, committed all his aircraft : 67 Dauntless dive-bombers, 29 Devastator torpedo-bombers and 20 Wildcat fighters to a desperate counterattack that by pure chance caught the Japanese fleet just as the their carriers were receiving their returning aircraft, and reloading for a second strike at Midway, when they should have been rearming (according to plan) to launch an airstrike against the lured in American aircraft carriers. This was the worst Japanese mistake of the war.

First to attack were the Devastator torpedo squadrons from the *Hornet*, but the superior Zero fighters defending their carriers, shot them to pieces. Eight times American airmen winged straight and low towards Nagumo's carriers and eight times they were blown out of the sky. Despite all their sacrifice not one torpedo ~ or *pickle* as they called them ~ hit a Japanese vessel. Only one survivor, Ensign George Gay who wounded in the arm and leg managed inflated his lifeboat was ever found, and he was rescued the next day. Next to attack was torpedo squadron VT-3 from the USS *Yorktown* escorted by the Wildcats of Lieutenant Commander Thach's VF-3 from USS *Saratoga* who engaged the Zeros, so when their Dauntless dive-bombers arrived over the scene, the Zeros had already been lured to low level to provide the dive-bombers an unopposed run in at the Japanese carriers.

THE BATTLE OF MIDWAY

Lt. Edward *Butch* O'Hare   LT-CDR *Jimmy Thach flying F1 with Lt. O'Hare in F4F-3A Wildcats ~ April 1942*

Wildcat taking off from USS Saratoga   *Jimmy Thach's Wildcat F1*

    USS *Enterprise's* VB-6 dive-bomber squadron searching for the Japanese fleet and near out of fuel, were directed by their skipper Lieutenant Commander C. Wade McCluskey to continue their search, and sighted the wake of the IJN destroyer *Arashi* steaming at full speed to rejoin Nagumo's carriers. Following the Japanese destroyers heading, McCluskey's squadron soon found the enemy carrier *Kaga* with its aircraft refueling hoses across its deck, making it a vulnerable target. McCluskey nosed his Dauntless into a screaming dive and his squadron followed, down to 600 yards above the waves before pulling their bomb releases over the *Kaga's* hardwood deck. Four huge explosions followed. McCluskey bleeding from five wounds made it back to the USS *Enterprise* with less than 20 litres of fuel. Even though some were lost in the ocean from fuel exhaustion, Admiral Nimitz credited ; *McCluskey's tactical decision to continue the search decided the fate of our carrier force and forces at Midway.* Next to attack, Lieutenant Dick Best shouted in the radio: *Don't let this carrier escape!* to his flight of dive-bombers swooping on Nagumo's flagship, *Akagi*. Dick Best's bomb landed adjacent to the bridge. The Japanese Admiral clambered out of a window to escape his burning flagship to see Lieutenant Commander Maxwell Leslie leading 17 dive-bombers from the patched up *Yorktown* to hit the *Soryu*. In less than ten minutes, three of the four Japanese fleet carriers : *Akagi, Soryu,* and *Kaga* were ripped apart by 1000-lb bombs and were visibly doomed flaming wrecks.

    The surviving fourth carrier *Hiryu*, undetected and steaming separate from the others, launched her planes toward the *Yorktown*. The radar officer on the USS *Yorktown* yelled *Bogeys, 32 miles, closing!* As a dozen of her own fighters circled overhead and her AAA guns started firing, *Hiryu's* airmen struck. As *Yorktown's* AA gunners blew apart one Japanese aircraft, its bomb exploded in an orange flash behind the carrier's bridge. Two bombs penetrated deep below and the *Yorktown's* bow erupted in flames. The Japanese airmen from the *Hiryu* had inflicted severe damage, but making another error, the final Japanese strike to finish off the *Yorktown* was postponed for returning Japanese airmen to eat. During their meal, 12 American dive-bombers from the *Enterprise* dived at the *Hiryu* out of the sun, scoring fatal hits on the *Hiryu,* and bringing the Pacific naval forces of the IJN and the U.S. Navy to parity. With that, Admiral Chuichi Nagumo withdrew the remnants of the Japanese fleet and Yamamoto seeing his plan for the Battle of Midway boomerang, sank into a chair staring into space…

Midway Atoll 24th November 1941 ~ looking west across the southern side of the Pacific atoll several months before the battle. Eastern Island with Midway's airfield is in the foreground, and the larger Sand Island is across the entrance channel to the west.

At the Naval War College in our Estimate of the Situation form we used to have : The enemy, his strength, disposition and - probable intentions. Later, probable intentions was changed to capabilities. We found that there had been a tendency to decide what an enemy was going to do and lose sight of what he could do. I have seen this happen in fleet problems at sea, and it is very dangerous. In war, we try to minimize rather than to avoid danger.

USN Admiral Raymond Spruance

Helldiver over Yorktown

THE THACH WEAVE

Lieutenant Commander John *Jimmy* Thach leading USS *Saratoga's* carrier fighter squadron VF 3 knew that the tactics of aerial engagements over the Pacific were as much about technology as about the skill and courage of airmen. If the technological advantage of the Japanese Zero could be neutered through the introduction of a new approach to fighter tactics, *Jimmy* Thach searched to discover such a winning tactic against the aerodynamic performance superiority of the Zero. After each day of flying, he worked night after night on his kitchen table using matchsticks to simulate opposing fighter flights, and devised a weaving tactic using a looser four-fighter formation, composed of two sections weaving back and forth across the sky to pick off any attacking Japanese fighters. In detail, he instructed his airmen to fly their fighters in pairs, spaced a distance apart dependent on speed, usually 300-400 yards apart. A cross look-out was kept on each other and the first section to see an attack coming, broke hard towards the other pair, and seeing the first pair break inward, the second pair would bank steeply turning in towards them, the pairs scissoring each other. A Zero dropping on the tail of one section would follow his evading target around to be faced by a head-on attack from the other two or break off his attack in a straight line, giving them an opportunity to reverse their turn and fire a fleeting shot against him. Thach had the satisfaction of seeing the first success of his tactic in action himself during the Battle of Midway, when a *chutai* of Zeros attacked his flight of four Wildcats. Thach's wingman Ensign Dibb, tailed by a Zero turned towards Thach, who dived under his wingman to fire a long burst into the incoming Zero's belly until its engine flamed, and every other time a Japanese Zero attacked one of Thach's Wildcats, the attacking Japanese fighter faced the guns of another American. This tactic countered the Japanese bounce and did not require radio warning. As this new tactic became known as the Thach Weave, *Jimmy* Thach wrote ;

Nobody can tell a fighter pilot how to meet all combat situations, but it was obvious from the beginning that this maneuver was working better than I had dreamed. The quick turn inward toward each other does two things to the enemy pilot. It throws of his aim and because he usually tries to follow his target, it leads him around into a position to be shot at by another part of our team. Follow the rules, utilize the elements of surprise, sun, and cloud cover, size up the enemy properly, choose the right approaches and attack smart, bracket, and make feints.

Island Hopping ~ After Coral Sea and Midway, the United States and its allies launched a sustained counter-offensive, beginning three years hard fighting towards Japan itself, pushing thru the jungles of N.E. New Guinea to the barricaded caves of Okinawa. The idea of island hopping was to capture well located and sometimes less heavily defended islands, one after another to be transformed into unsinkable carriers to isolate large Japanese bases (such Rabaul and Truk Atoll in the Carolines) and leave them to ~ *wither on the vine*. As each new island was captured, the first task was to construct an airfield on coral or carve one through dense tropical jungle to prepare for the next hop in a campaign of establishing strategic Allied airfields ever closer, and ever closer to Tokyo.

Guadalcanal ~ In August of 1942, the Japanese attempted to regain the initiative in the Pacific by constructing a large airfield at Lunga Point on Guadalcanal, a 90 mile long jungle island in the south Solomon Islands. The Japanese aim was to protect their stronghold at Rabaul, to sever the supply line across the South Pacific connecting Australia and New Zealand to America, and establish a staging base for further campaigns against the New Hebrides, Fiji, Samoa, and New Caledonia. On 7th August 1942, just nine days before the first Japanese aircraft was to arrive at their new airfield, the U.S. Navy began shelling Guadalcanal (and nearby Tulagi) beginning the Allied island hopping push to capture Japanese controlled islands in the Pacific. The U.S. First Marine Division then made their first amphibious landing of the Pacific war, advancing to the Lunga River to capture the near completed crushed coral runway, now so critical to the island's defence renaming it Henderson Field. Also seized were, vast stores of food and construction equipment of great use to the 11,000 U.S. Marines. For nearly two weeks the Marines worked around the clock, using useful captured Japanese machinery, to repair and extend the Hendersen Field runway to 6,000 feet, Soon afterwards, Marine aviators of the *Cactus Air Force* flying Wildcat and Dauntless craft from Hendersen Field achieved a kill ratio of four to one to gain air superiority over the Solomons with their frequent use of the Thach Weave. Japanese fighters including the *Cleanup Trio* were transferred from Lae to New Britain for long-range missions against U.S. forces on Guadalcanal, during which Marion Carl shot down Junichi Sasai. Although Saburo Sakai was injured (lost an eye) and eventually returned to duty, the Japanese had lost so many other experienced pilots that their pilot training had to be curtailed to keep pace with their demand for new airmen, with the inevitable consequence of a decline in the quality of Japanese pilots. In February 1943, the Japanese withdrew from Guadalcanal.

Flight of Black Cat Catalinas

Catalina PBY5 cockpit

B-17F Flying Fortress The Aztec's Curse cruising over Gizo Island in the Solomons after a striking their target behind ~ and cockpit below.

THE AKUTAN ZERO

U.S. Navy salvage team examining the Akutan Zero

The Akutan Zero ~ During the Japanese attack on Pearl Harbour, nine Zeros were shot down. From those wrecks it was learnt that the Zero lacked armour and self-sealing fuel tanks, but very little else. Then on 10th July 1942, the crew of a Catalina piloted by Lieut. William Theis returning from an overnight patrol flew over Akutan Island off the coast of Alaska, spotted a downed aircraft partially submerged in soft marshy soil. Gunner Wall called out *Hey, there's an airplane on the ground down there. It has meatballs on the wings.* The rising-sun insignia, and this is how it came to be there…

On June 3rd and 4th 1942, Japanese aircraft flew from the Japanese light carriers *Ryujo* and *Junyo* to attack the U.S. military base at Dutch Harbor in Alaska's Aleutian archipelago to draw units of the U.S. fleet north from Pearl Harbor and away from Midway, where the Japanese were setting a trap. During the raid of 4th June, 20 IJN bombers blasted oil storage tanks, warehouses, a hospital, a hangar, and a beached freighter, while 11 Zeros strafed at will. Chief Petty Officer Makoto Endo led a three-plane Zero section from the *Ryujo*, with pilots, Flight P.O.'s Tsuguo Shikada, and Tadayoshi Koga. Koga flew a light gray Zero, and as it had left the Mitsubishi aircraft factory a few months earlier on 19th February 1942, it was certainly their latest design ~ the Zero 52.

C.P.O. Makoto Endo led his section to Dutch Harbor, where it joined the other Zeros in strafing. It was then that Tadayoshi Koga's Zero was hit by ground fire. An intelligence team later reported; *Bullet holes entered the plane from both upper and lower sides.* One of the bullets severed the return oil line between the oil cooler and the engine. As the engine continued to run, it pumped oil from the broken line. Endo and Shikada accompanied Koga as he flew his oil-spewing Zero to Akutan Island 25 miles away that had been designated for emergency landings.

A Japanese submarine stood nearby to pick up downed pilots. The three Zeros circled low over the green, treeless island. At a level, grassy valley floor half a mile inland, Koga lowered his wheels and flaps and eased toward a three-point landing. Unfortunately, the long, flat field chosen by Koga for his wheels-down landing was deceptive. Under its inviting-looking grassy-green flat surface lay a treacherous waterlogged muskeg bog. As his main wheels touched, they dug in, and the Zero flipped onto its back, tossing water, grass, and clumps of mud. The valley floor was a bog with the knee-high grass that concealed water. Endo and Shikada circled. There was no sign of life. If Koga was dead, their duty was to destroy the downed fighter. Incendiary bullets from their machine guns would have done the job. But Koga was their friend, and they couldn't bring themselves to shoot. Perhaps he would recover, destroy the plane himself, and walk to the waiting submarine. Endo and Shikada abandoned the downed fighter and returned to their carrier, the *Ryujo* now 200 miles to the south, which would be sunk two months later in the eastern Solomons by planes from the carrier *Saratoga*.

THE AKUTAN ZERO

The Akutan Zero being loaded onto a ship

The Akutan Zero's secrets were revealed only after a remarkable series of events. Firstly, the Aleutians typically poor flying weather of low cloud with poor visibility greeted the Japanese aircraft over Dutch Harbor on 4th June. The *Aleutian Clag* persisted for weeks, so it was not surprising that, although only a short distance from Dutch Harbor, it was five weeks before a passing U.S. Navy PBY Catalina sighted the upturned Zero on 10th July 1942.

When the U.S. Navy sent a salvage team north to retrieve the fallen Zero, **Ensign Robert Larson** who was Thies's copilot was there when the aircraft was secured, and remembers reaching the Zero. *We approached cautiously, walking in about a foot of water covered with grass. Koga's strapped in body was upside down in the plane, his head submerged in the water. We were surprised at the details of the airplane. It was well built, with simple, unique features. Pushing on a black dot with a finger could open inspection plates. A latch would open, and one could pull the plate out. Wingtips folded by unlatching them and pushing them up by hand. The pilot had a parachute and a life raft.*

The salvage team eased the upside-down Zero onto a skid, dragged it out with a tractor, and righted it back in Dutch Harbor. The virtually undamaged Zero was crated and shipped to San Diego, where it was restored to flying trim for evaluation trials against the new U.S. Navy Grumman Hellcat. The first test flight exposed flaws of the Japanese Zero, which could be exploited with appropriate tactics. It was found that the Zero's ailerons froze at speeds above 200 knots, so that rolling maneuvers at those speeds were slow, and required tremendous force on the control stick. It rolled to the left much easier than to the right. Thereafter, Allied aviators were advised that if they were being outmaneuvered and unable to escape a pursuing Zero to go into a vertical power dive. Then at about 200 knots, roll hard right before the Zero pilot could align his sights. Further flight-testing confirmed that the Hellcat was 30 mph faster at all altitudes and could out-climb the Zero above 14,000 feet, and roll faster at speeds over 235 mph. Even so, the lighter Zero could still out-turn the Hellcat at lower speeds and maintain a better rate of climb below 14,000 feet. The official November 1942 U.S. Navy report of the Hellcat-Zero flight tests noted : *Do not dogfight with the Zero 52. Do not try to follow a loop or half-roll with a pull-through. When attacking, use your superior power and high-speed performance to engage at the most favourable moment. To evade a Zero 52 on your tail, roll right and dive away into a high-speed turn.*

With a Zero on my tail I did a split S, and with its nose down and full throttle my Corsair picked up speed. I wanted at least 240 knots ~ preferably 260. Then as prescribed, I rolled hard right. As I did this and continued my dive, tracers from the Zero zinged past my plane's belly. From information that came in from Koga's Zero, I knew the Zero rolled more slowly to the right than to the left. If I hadn't known which way to turn or roll, I'd have probably rolled to my left. If I had done that, the Zero would have turned with me, locked on, and had me. I used that maneuver a number of times to get away from Zeros. USMC Captain Kenneth Walsh

B-25 Mitchell *Ruthless Ruth* skip-bombs Japanese Kaibokan Type C escort ship No. 134 (6th April 1945)

As U.S. and Australian ground forces counterattacked strongly in the New Guinea campaign of early 1943, the Imperial Japanese Navy assembled a group of vessels to ferry thousands of ground troops transferred from China and Japan to reinforce Lae, Finschhafen, and Salamaua in New Guinea for a planned thrust that would ultimately secure the island. Eight Japanese transport ships escorted by eight destroyers and covered by 100 fighter aircraft set off from Simpson Harbour, Rabaul. Two tropical storms moved through the Solomon and Bismarck Seas in the last days of February 1943, then at 1500 hours on 1st March, the crew of the patrolling B-24 Liberator, *Cow Town's Revenge* found the slow Japanese convoy north of Cape Hollman. At 1000 hours on 2nd March, a second Liberator sighted the enemy convoy and flights of B-17 Flying Fortress bombers sank three Japanese merchant ships, including the *Kyokusei Maru* transporting 1,500 troops. After a Zero had seriously damaged a B-17, forcing its crew to take to their parachutes, its Japanese pilot shot at them as they descended and returned to strafe them in the water. Night-flying Australian *Black Cat* Catalinas from No. 11 Squadron RAAF kept the desperate Japanese convoy under close observation through the night.

The next morning, when the Japanese convoy rounded the Huon Peninsula a vengeful formation of 90 Allied aircraft from Port Moresby of USAAF B-25 Mitchells and RAAF Bristol Beaufighters caught them in the open and savaged them in a series of low-level attacks. Ideally, attacks were made down the length of the ship from either a bow or stern aspect at a calculated altitude of 200 feet at the time of drop. Other airmen were such strong advocates of masthead level attacks that they discarded their bombsights altogether as a useless piece of baggage. Others preferred the tactic of *skip bombing* ; flying low over the sea towards their targeted enemy ship and release their bombs to ricochet over the surface of the water to explode against the side of their target ship. Six transport vessels carrying 6,900 troops desperately needed by the Japanese in New Guinea with the destroyers *Shirayuki, Arashio, Asashio,* and *Tokitsukaze* were sunk 54 nautical miles south of Finschhafen. The eighth troopship was sunk by a motor torpedo boat. From 7,000 Japanese troops that set out from Rabaul, only 1,264 made it to Lae. Japanese destroyers and submarines rescued 2,427 survivors who were returned to Rabaul. The Japanese also lost 20 fighters for the loss of three U.S. Lightning aircraft, setting the pattern for the rest of the Pacific campaign where U.S. and Australian aircraft continued to dominate the skies. The triumphant sweep of Japanese forces across the Pacific had been halted. Soon their supplies became so scarce that in December 1944, General Hatazo Adachi as the CO of the residual Japanese 18th Army in New Guinea ordered : *While our troops are permitted to eat the Allied dead, they must not eat their own.*

BOSTON ~ TOPMAST & SKIP BOMBING DIAGRAMS

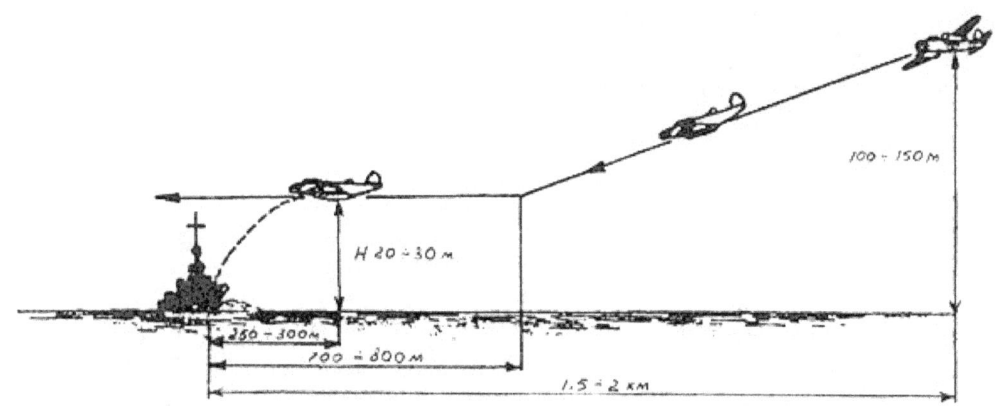

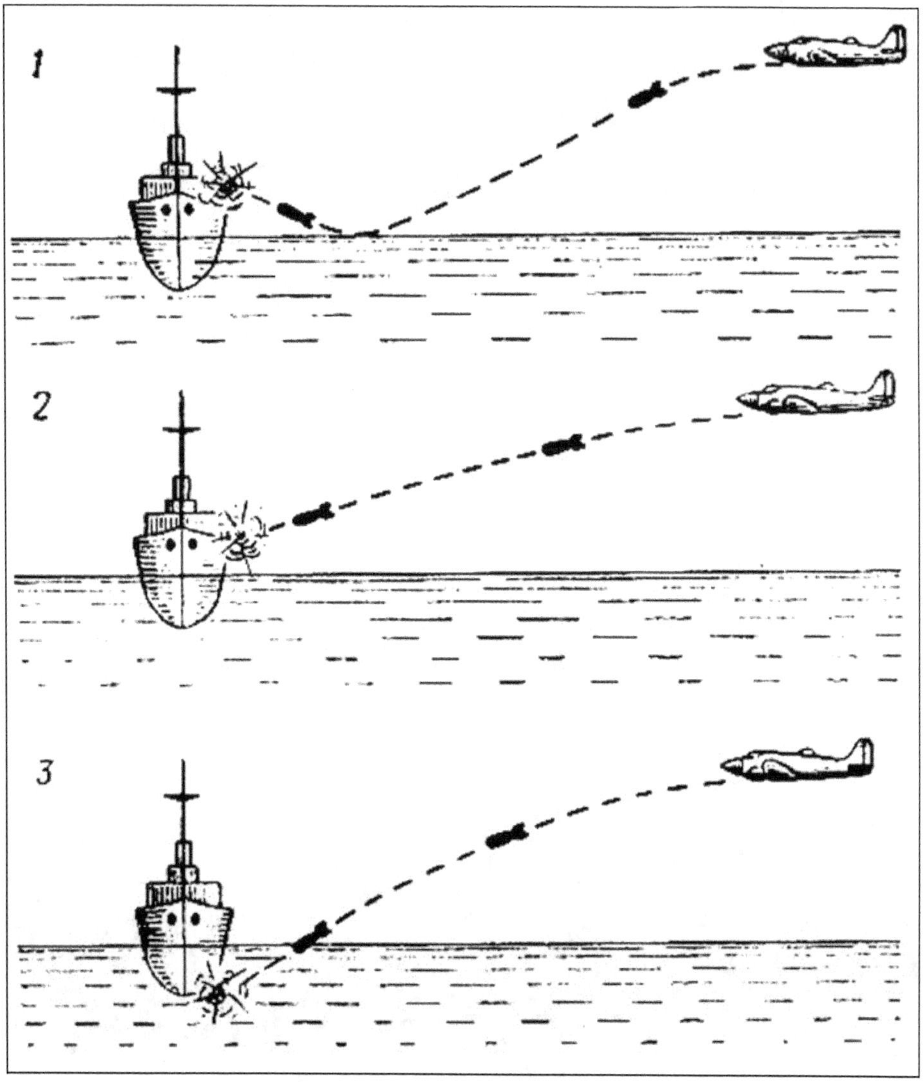

Dauntless over Tanahmerah Bay, Dutch New Guinea, as the US landing craft go in, April 22nd 1944

Dauntless over Wake Island

Avenger over Iwo Jima

Dauntlesses from Lexington hit Param Island, Truk Atoll

Marine Corsair at Cape Torokina on Bougainville ~ December 1943

*Major Gregory Boyington USMC
28 victories*

The Black Sheep Squadron wearing the ball caps donated to them by the St. Louis Cardinals, 1943 (National Baseball Hall of Fame)

Vella Lavella airfield in the Solomon Islands on 10th December 1943. U.S. Marine Corps Vought Corsairs of Marine fighter squadrons VMF-123 and VMF-124, Grumman F6F-3 Hellcats, a Douglas SBD Dauntless, and a RNZAF Curtiss Kittyhawk Mk.IV are visible on Vella Lavella which was seized in the summer of 1943 and served as a base of to support landings by Allied forces in the Treasury Islands and at Cape Torokina on Bougainville. The swift advance of Allied forces in the South Pacific soon bypassed Vella Lavella and the airfield ceased operations in September 1944 ~ less than a year after the first aircraft arrived.

THE BLACK SHEEP SQUADRON ~ USMC VMF-214

Formed in August 1943, the USMC *Black Sheep* Squadron VMF-214 consisted of a mix of highly experienced combat veterans and new substitute pilots who soon relied almost entirely on the training and skill of a seasoned and colorful United States Marine Corps Major, Gregory *Pappy* Boyington. But that name was invented by the press ~ in the Solomons, his pilots called the 30-year-old Major *Gramps*. The *Black Sheep* squadron, which started out as an impromptu group of replacements, was, in a matter of weeks, transformed by Boyington into one of the most elite groups of air fighters in military history. Their mascot was a black lamb named *Midnite*.

Ball Caps for the Black Sheep ~ In the fall of 1943, while war was on the minds of countless baseball fans, baseball was evidently on the minds of Pappy Boyington and his Black Sheep Squadron. In order to resolve a shortage of baseball caps, which were very popular among Marine airmen in the South Pacific, Boyington, and his men reportedly, made an unusual offer to Major League Baseball. The story goes that the airmen agreed to shoot down one enemy plane in exchange for every ball cap they received from players in that year's World Series. Sure enough, upon conclusion of the World Series, the St. Louis Cardinals sent 20 baseball caps along with a few bats and balls to the Black Sheep Squadron. In return for the caps, the Black Sheep made good on their end of the bargain. Instead of taking down just 20 enemy planes, though, the members pictured here accounted for 48 enemy planes destroyed. This photo opposite shows 20 original members of the squadron sporting Cardinals caps while posing on the wings of one of their Corsairs.   National Baseball Hall of Fame

Members of newly formed VMF-214 scramble from their Ready-Room to their Corsair fighters for an intercept ~ In Boyington's opinion; The Corsair was a sweet-flying baby if I ever flew one. No longer would we have to fight the Nip's fight, for we could make our own rules.

Some post-mission debrief reports from *Black Sheep* aviators included : Lt. Bob McClurg reporting his first kill in a head-on pass. *I just held the trigger down as we came at each other. I was scared to death.* Boyington's wingman Lt. Don Fisher. *I was right behind the Zero, and he blew ~ the wings went each way.* Lt. Bill Case fired 50' from a Zero's tail ~ too close for his Corsair's six wide-set guns to be effective. *I spent about 2,000 rounds figuring that out. I finally put the pipper up above his tail and about six to eight feet to the side… and hit him with three guns at a time.* Lt. John Bolt : *The first time I saw a meatball it was a full deflection shot, and he just zipped by. I was in a state of shock.* After Bolt flew a one-man unauthorized raid on Tonolei Harbor strafing troop transports and other vessels he reported to a concerned Boyington ; *I was only taken under fire from one gun… its 20mm tracers just floated by.* Lieutenant John Bolt received a congratulatory telegram from Admiral William Halsey, awarding him the Distinguished Flying Cross. After hearing such reports, Marine Corp HQ boosted VMF-214 Squadron pilot strength from 28 to 40.

After a US Navy squadron commander asked Major Boyington about his tactics, Boyington said ; *Tactics? Hell, you don't need tactics. When you see the Zeros, you shoot 'em down, that's all.* Another time, Boyington told his men ; *Don't worry about me. If you guys ever see me going down with 30 Zeros on my tail, don't give me up. Hell, I'll meet you in a San Diego bar and we'll all have a drink for old times sake.* They celebrated their achievements of 1943 on New Year's Eve in true *Black Sheep* style that night, by firing off so many pistol flares that the transport fleet offshore steamed off ~ fearing an air-raid! Thankfully, Major Gregory Boyington, skipper of *The Black Sheep* Squadron that shot down 94 enemy planes in 12 weeks of combat put pen to paper with his pilots to record their tactical repertoire.

AVIATORS OF THE BLACK SHEEP SQUADRON ~ USMC VMF-214

First Lieutenant Paul *Moon* Mullen
Guadalcanal June 1943
5 victories

First Lieutenant Alfred *Shorty* Johnson
Russell Islands October 1943

First Lieutenant Edwin Olander
5 victories

First Lieutenant Robert *Bob* McClurg
7 victories

HEADQUARTERS, MARINE AIRCRAFT, SOUTH PACIFIC, FLEET MARINE FORCE INTELLIGENCE SECTION

THE COMBAT STRATEGY AND TACTICS OF MAJOR GREGORY BOYINGTON, USMCR
COMMANDING OFFICER OF VMF-214

These reports, representing a digest of Major Boyington's ideas concerning, and contributions to, combat aviation were prepared by the squadron ACIO from information received from members of the squadron.

MAJOR BOYINGTON'S TACTICS
GENERAL OBSERVATIONS APPLYING TO ALL MISSIONS

Tactics in the air should be studied and developed in comparison with time-tried tactics on the land and on the sea. The principles of scouting, out-flanking, ambushing, etc., all provide a basis for the development of air tactics. Of course allowances and modifications must be made for our speed, for the additional dimension in which we operate. But land and sea experience are a starting point.

Fighter aircraft are designed, and fighter pilots are trained, to fight. If there are enemy aircraft in the air, and contact is not made, something is wrong. The only exception to this are those situations where we must stay close to something we are expected to protect; where to attack means that we have been lured away.

All missions must be preceded by thorough planning and briefing, with respect to the purpose of the mission and the purpose of any alternate missions, with respect to the rendezvous point and any alternate rendezvous points, etc. All possible contingencies must be considered in advance, particularly because all of our present operations are over enemy territory far from our bases. Success in the air is a lot of little things. Most of them can be taken care of before takeoff.

All missions must be flown as planned, and briefed unless there is real justification to the contrary ~ there must be discipline. Along with realizing the purpose of the mission, each pilot must realize fully his responsibility for its successful execution.

With proper planning and briefing, no use of the radio should (be) necessary except in emergencies and except in situations where tactical considerations require otherwise.

Every effort must be made to obtain weather information ~ and to make intelligent use of it.

All fighters must realize the critical importance of recognition, in order to distinguish our planes from those of the enemy, in order to identify the enemy's different types so that his particular points of weakness can be exploited and his particular points of strength respected.

Fighters must not go into combat feeling that the division leader or at most the section leader will answer for problems of tactics, navigation communications, etc. In these regards leaders and wingmen are the same ~ at any moment the customary leader may go down or may be required to return to base, or may become lost, or may be without a radio, and every one of the others must be ready to take over. And in particular, all pilots are equally responsible for spotting the enemy and for initiating immediate action either through their leader or by taking over the lead themselves.

Pilots must make steady careful observation a habit. They must have a system and a routine for scanning the air both above and below, behind, on the flanks, and ahead. The vigil must be unceasing.

FIGHTER SWEEPS

When bogies are called, the call must be worded that everyone will know the location of the division from which the call has come, and the location of the bogies either with respect to a geographical landmark or to the line of flight of the friendly force.

Surface bogies should not be called ~ unless it is certain that they really are enemy bogies. The comparative slowness of movement of friendly forces on the water makes any revelation of their presence or position undesirable.

MAJOR BOYINGTON'S TACTICS ~ FIGHTER SWEEPS

In all missions, all advantages of sun, weather, terrain, etc., must be exploited.

In danger areas high speed must be maintained. In less dangerous areas,
such a speed must be carried that all formations can stay together comfortably.

We must not climb into bogies. We must gain our altitude away in a position from which the action can be observed, and our climb must be made with a high forward air speed.

We must not pull up when closely and dangerously attacked. Speed is our defense.
With moderate loss of altitude and certainly without going all the way to the water and running for home, the enemy can be outdistanced, and then altitude and position recovered for further attacks.

Close attention must be given to the efficiency of firepower. Our guns' loading, their boresighting, their cleanliness, and their general performance at altitude as well as at the lower levels, must be perfect. As a result of tests of the comparative destructive power of tracer, armor piercing and incendiary ammunition, our squadron changed its belting from 1-1-1 to 2 incendiary - 1 armor piercing - 2 incendiary - 1 tracer. In combat we found this load much more satisfactory. We also found it desirable to use the ring sight entirely in determining range and establishing lead, using the tracer merely to check the bore sighting.

It must be remembered that with our armor plate, our self-sealing or purged fuel tanks,
and with proper evasive action, our rear vulnerability is not great.

On the other hand, our most successful runs against fighter opposition are from eleven to one o'clock ahead and from five to seven o'clock astern, from a level just above to a level just below.

The most common manoeuvre of the Jap fighter at present is a split-S,
which happens usually when he is approached from ahead or from astern.
In normal combat, if he cannot be hit before he has started down, he is usually gone.

In normal combat, in enemy territory it is not desirable (particularly for a single plane) to go below a base altitude, which might well be 10,000 feet. To go lower with a section of two planes may be desirable if both planes carry ample speed and are prepared to cover each other when necessary.

In normal combat, clouds may provide cover either for us or for the enemy,
and must be considered constantly in both connections.

All squadrons must keep the white star on their insignia freshly painted.
Cleaning of the fuselage causes them to become indistinct in a matter of days.

In all missions involving layers of aircraft stacked up through considerable air space,
all must remember the difficulty of maintaining visual contact through all the layers.

All pilots must check their oxygen equipment (as well as everything else) before takeoff.
To return to base early for oxygen reasons, indicates some negligence in preparation for the flight.

When a pilot decides that for some reason he must return to base, he should make a visual sign to the other member of his section that he must do so, and should also indicate by visual signal whether he is able to go back alone.

When a plane drops out, other member of the section should join up on another single plane,
if any is available. One-plane section as useless as a three-plane section.

Jap fighters approaching from angles ahead will usually turn away at the suggestion of a run
on them, or at any other suggestion that we mean business.

Fighters must hold their fire until within range, as indicated by the size of the target in the ringsight. Otherwise, the Jap will be warned by that first over-anxious burst, will split-S and be gone. On the other hand fire should be opened sooner in a head-on run because then we are closing faster and because the plane opening fire last usually turns away first and is a good target during that run.

MAJOR BOYINGTON'S TACTICS ~ FIGHTER SWEEPS

The larger a striking force, the greater its power, provided that it is not so large as to be unmanageable. At the present time it would seem desirable to restrict sweeps to 36 to 48 planes, upon the assumption of course that all or very nearly all of them will continue to the target and will participate in the action.

On each sweep the number of squadrons represented should be as small as possible :
each squadron, which does participate, should have a large number of planes.

On each sweep the number of different fighters should be small as possible.

The leader must fly in a position where he can be seen and followed easily. At the present time enemy patrols are not ordinarily airborne, or at least are not at altitude, before our arrival. Such being the case the sweep leader can fly satisfactorily in the bottom layer with the upper layers stepped back from his layer. However, if enemy patrols should be at altitude, it would be better for the sweep leader to be in the uppermost layer. By and large, therefore, the sweep leader should be at that level where it is expected that the first contact will be made.

The fighting should be kept in the same part of the air. It should not degenerate into small fights, some going in our favour and some going otherwise, with some planes out of the action altogether.

After the initial contact it is hardly possible to keep a division together, although it would be desirable if it could be done without throwing the division into a tail chase with only the leader doing any good. A section is a thoroughly satisfactory combat unit, and can be kept together more easily than a division. Every effort should be made to keep the section together, but the wingman should not be simply chasing his leader. If the efforts to keep together fail, the separated planes should join up as soon as possible on other friendly planes.

A rally point should be designated before hand for all planes on the sweep. If for any reason this is not done, those planes from any one squadron should have their own rally point. The rally point should be at a certain altitude, not too far removed from the probable center of action and preferably into the sun.

Our fighters must keep in the fight for the time specified unless lack of fuel, lack of oxygen,
or some other good reason require otherwise.

BOMBER ESCORTS

The responsibility for the success of a bombing mission rests partly upon the bombers and partly upon their escort. When the bombers do a good job (aside from hitting their target) they do the following : They keep good formation, with all units massed as closely as practicable. They select courses, which will allow sufficient clear air space above the bombers for the fighters to maintain visual contact. They select courses avoiding AA fire as much as possible. They use imagination in varying their times of attack, their altitudes of attack, their courses of approach and retirement, etc. They carry sufficient air speed for the fighters to stay together comfortably during the non-dangerous part of the approach and retirement as well as during the dangerous part.

The layers of fighter cover are ordinarily roving high, high, medium, low, and close.
On the approach all layers will tend to lean ahead of the bombers ; on the retirement they will be over the bombers. The higher the particular layer, the farther sideways its coverage should extend.

The roving high cover should fly as high as possible consistent with good visual contact with the bombers (not more than a 10,000 foot spread), but not at more than 30,000 feet. It should fly well ahead of the bombers on the approach and behind them on retirement. Its mission is the engagement and annihilation of enemy interceptors operating anywhere but particularly at the upper levels. It may leave the airspace over the bombers if necessary for the performance of its mission. It is a free unit.

The high and medium covers are ordinarily the second and third highest covers. Their positions are respectively 6000 and 4000 feet above the bombers. Unlike roving high cover, the high cover is not authorized to leave the air space above the bombers.

MAJOR BOYINGTON'S TACTICS ~ BOMBER ESCORTS

The fighters' air speed should be far greater than that of the bombers, for the safety of the fighters as well as for their great manoeuvrability. As a result they must weave, to keep their assigned positions. The low and close covers weave back and forth over the bombers' line of flight. The other covers ordinarily will put half of their strength on each side of the line of flight, each half then keeping to its side of the line of flight.

The low and close covers are respectively 2000 to 1500 feet and 100 to 500 feet above the bombers. They must hold those positions at all costs.

When the bombers are SBD's and TBF's (which usually approach in that order and a mile or a mile and a half apart), the low and close covers must go down with the bombers, leveling off at about 5000 and 2000 feet respectively. The upper layers must settle down proportionately.

Ordinarily the low cover must see that straggling bombers are covered.

It is desirable for the bombers to open up on the radio if the fighter cover is inadequate at any particular point. Someone in the bomber force should act as a fighter director or fire control officer. The fighters are often innocently unaware of enemy pressure at some particular point.

Fighter divisions should be able to keep together and fighter sections must keep together. The lower the layer in which a particular fighter happens to fly, the more vulnerable his position and the more prepared he must be to operate defensively with his section.

No fighter straggling can be permitted, on the part of single planes, sections,
or even divisions. Fighters in real trouble should dive under the bombers.

A bomber strike should be preceded by a fighter sweep,
timed to arrive at the target at least half an hour before the bombers.

STRAFING

The importance of thorough planning and briefing is particularly great in strafing. Every scrap of knowledge with regard to terrain, vulnerability of targets, location of AA defenses, etc., must be utilized. Each pilot must know exactly what his approach will be and where his targets will be found.

Probably no target is invulnerable to successful strafing. But strafing missions cannot be run off under just any circumstances. Surprise is absolutely essential. All cover of weather, darkness, etc., must be utilized. A mission which has succeeded largely because the attack was not expected certainly cannot be repeated immediately.

Strafing restrictions issued by intelligence and operations authorities must be strictly observed. But in the unrestricted areas there should be no hesitation or delay in destroying enemy targets which present themselves. Something which is wide open, can be gone the next.

High-speed runs are essential. Speed will reduce the number of rounds that can be delivered and will diminish the opportunity for observation, but it must be maintained. Each plane should make one run. If more firepower is desired the number of planes should be increased.

If the approach is made in line or in a flat echelon, the last mile or two of the approach should be made at a constant power setting so that all pilots will be able to devote full attention to the target.

Strafing must not be done in column. The greatest safety factor is achieved with a line or flat echelon. If the target is so small that all planes cannot get their guns to bear from a line or flat echelon, they should approach form different angles attacking as simultaneously as possible.

The approach must be as low as possible, with a momentary pull-up just before reaching the target, for the purpose of identifying the particular targets and getting the guns to bear.

An impulse to fire too early must be restrained. For effectiveness and for saving gun barrels, fire should be held until one is definitely within range.

Retirement must be low and very fast, with an eye for possible water spouts from heavy fire.

MAJOR BOYINGTON'S TACTICS ~ STRAFING

Upon the word to Scramble, it is important to get the fighters off the ground, and only second to get them joined up in their usual order. Any four planes can make a division if the take off has been mixed up. And if the take offs are unduly delayed, any two planes should proceed together as soon as possible. Planes should never proceed singly.

Since fighter direction by radio is imperfect, due to failures of radar or adversities of weather, it may be possible to locate the enemy only if our planes operate as a scouting force, i.e. It may be necessary to break the interception force into smaller units which should keep each other in sight but make their coverage wide as possible. On contact the force should be reunited.

When an enemy is approaching and the fighters go out, some of them should be required to remain over the area or object in danger of attack, in case the interception is not a complete success.

When it is expected that bombers are among approaching bogies, our interceptors should have as little as possible to do with enemy fighters. In such a case our mission is to prevent those bombers from doing any damage with their bombs. The bombers must be located, and shot down or at least their formation must be broken up and the individual bombers forced to jettison their bombs.

PATROLS - INCLUDING DUMBO AND TASK FORCE COVERS

Too often, fighters consider patrols and dumbo and task force covers a waste of time. However, along with performing the mission properly, the time involved can be utilized to good effect. Wingmen can develop their formation flying, making it perfect but effortless. Leaders can exchange lead with wingmen to see how well the wingmen can lead and to see how well they can fly wing. Divisions can perfect their teamwork and section manoeuvres by practicing tight turns, violent scissors, etc. Leaders can select patterns for flight the basis of which will be readily apparent to those following them, so that a minimum of concentration will be necessary in keeping the planes together. All pilots can practice their own systems of observation particularly in focusing upon very distant objects. In the case of a dumbo cover, the fighters can fly as if escorting bombers. Quite frequently, upon completion of the mission, the divisions can go into tail chases which will release energy and improve technique. The average tour of combat duty involves too much straight and level flying.

As much altitude should be held as is consistent with good visual contact with the object or area being covered. The position should be into the sun and generally between the object or area being covered and the enemy's territory.

ACCOMPLISHMENT RECORD OF THE *BLACK SHEEP SQUADRON* UNDER MAJOR GREGORY BOYINGTON USMCR

TWO TOURS : 12 September 1943-24 October 1943 & 27 November 1943-8 January 1944

- 94 Enemy planes destroyed in aerial combat (92 of them fighter planes ~ 91 over enemy territory)
- 32 Enemy planes probably destroyed in aerial combat.
- 50 Enemy planes damaged in aerial combat.
- 21 Enemy planes destroyed on the ground.
- 197 Total enemy planes destroyed, probably destroyed or damaged.
- 23 Japanese barges destroyed (3 of them loaded with troops destroyed).
- 125 Japanese bivouac areas and AA positions on New Ireland, New Britain, Buka, and Bougainville.
- 4 Japanese airfields strafed : Kahili, Kara, Ballale, and Borpop.

Successfully intercepted enemy formation attempting to attack task force reinforcing Barakoma. Assisted Bougainville ground troops in an untenable position by strafing enemy mortar positions. Participated in over 200 combat missions, flying 4,195 combat hours.

This record was established with the loss of 12 pilots missing in action, and six wounded in action. Only one pilot was injured and none killed in operational accidents.

(Transcribed by S/Sgt. Cyril Astolfi, French Air Force)

* * *

JAPANESE AIR TACTICS

The U.S. document, *Japanese Air Tactics up to May 1944*, noted some fundamental developments. At the end of 1943, The Japanese were observed to have adopted the tactical unit of flights of four aircraft composed of two pairs, but some of the slower fighters still use the original three aircraft formation, which has been effective when all pilots were experienced and capable of breaking formation and go into combat independently. Deceptive tactics continue to be used extensively in efforts to lure Allied aircraft out of formation. Decoys with indifferent camouflage are used, which fly straight and level at about 15,000 feet while a patrol of well camouflaged aircraft waits higher up. Other examples of Japanese deceits and feints are:

* The staging of fake dogfights.
* The showing of aerobatics to attract attention while the attack comes in from another direction.
* A Japanese fighter on the flank of an Allied bomber formation feints an attack while a second dives from ahead and above; then the attacker takes up the flank position and the feinter moves up ahead to repeat the same tactic.
* Japanese fighters fly on either flank to draw attention while one was always up in the sun ready to attack.
* Three Japanese fighters would position themselves on one flank while three were up in the sun; one would dive out of the sun and after his attack would take up a flank, then one of the wing fighters would move up into the sun and the manœuvre would be repeated in constant rotation.

Head on attacks against Allied fighters are generally avoided, probably because the Japanese armour is still inadequate against the firepower of Allied fighters. Attacks are usually made from high astern or from above and the side. Japanese fighter pilots attempt to draw their opponents into a steep climb and into a stalling position and then do a quick stall turn or loop back onto their opponent's tail. In order to make the most of their superior maneuverability, the Japanese fighter pilots often prefer to be jumped, and then execute a quick turn to get off their opponent's tail, rather than initiate the attack. Their favorite evasive tactics are half rolling, or steep climbs.

Against Allied bomber formations, the Japanese have a preference for the cautious head on attack. He can often be turned by a long burst at maximum range: his so-called Banzai suicide mania is not in evidence in the air and there is no report of his having rammed any of our bomber aircraft or having shown any inclination to do so. He respects the fire of heavy bombers and a concentration of tracers usually affects him. This is not to imply that he lacks courage. He often presses attacks to very close range in the face of a well-flown formation, but he does this less often than against individual aircraft or against weaknesses in guns. Against medium bombers: Attack from the upper rear or from the beam. Use a frontal attack only in an unavoidable situation, in which case attack from above or below. Avoid a cone of 30° behind the tail. Against heavy bombers: Keep out of the field of fire of revolving turrets on Allied aircraft. Attack from the front, below and towards the side or above and towards the side. In attack, one shotai alternating with one chutai will select targets as indicated by the formation leader and hamper the bomber's evasive tactics. In the case of attacks made from left and right when one side has finished, the other shotai will immediately repeat its attack.

Japanese fighter pilots with model aircraft discuss tactics against B-29s

BATTLE OF THE PHILIPPINE SEA

After the U.S. landing on Saipan on 12th June 1944 that wiped out a third of the Japanese aircraft based there, the Japanese fleet commanded by Admiral Jisaburo Ozawa comprising five large carriers, four light carriers, five battleships, and support vessels staged an offensive, code-named *Plan A-Go*. However, Ozawa's Japanese fleet was spotted coming into the Philippine Sea on the on 17th June by the submarine USS *Cavalla*. The following day, on board his flagship the carrier USS *Lexington*, Admiral Marc Mitscher arrayed Fast Carrier Task Force TF-58 off Saipan to await Ozawa's warships. At 1000 on 19th June 1944, Japanese aircraft were plotted 140 miles away from TF-58 and in a series of air engagements known as the *Great Marianas Turkey Shoot* the Japanese lost 433 carrier aircraft.

During a U.S. counter air strike against Admiral Ozawa's flagship, the newest IJN carrier *Taiho*, Japanese Zero pilot Sakio Kontatsu dived at a U.S. torpedo about to strike his carrier. Kontatsu died in the explosion but his sacrificial tactic saved the *Taiho*. Hours later, two Japanese carriers were torpedoed by U.S. submarines. The *Cavalla* fired three into the *Shokaku*, and the *Albacore* finally put a torpedo into the *Taiho*. The torpedo detonation ruptured two of *Taiho's* aviation fuel tanks and a series of catastrophic fuel vapor explosions killed 1,650 of *Taiho's* 1,751 crew.

Through growing darkness, TF-58 steamed west through the night, and at dawn, search patrols were launched at first light to find the Japanese. American searches eventually found the Japanese fleet at 1540, but the details were so garbled, that Mitscher had to wait until 1605 for a clearer report. With only 75 minutes until sunset, a furious 550 aircraft American air-strike against the Japanese Fleet began at 1830 that damaged several vessels and sank three ; the carrier *Hiyo* plus two oil-tankers, plus a further 65 Japanese aircraft. With night approaching, Admiral Marc Mitscher ordered his carrier's lights be turned on at 2030 hours to guide home his pilots. Already 50 aircraft had been lost in aerial engagements with the Japanese and the American aviators still had to fly thru the night to their carriers on near empty fuel tanks. Despite the risk from Japanese submarines, and night-flying aircraft, the U.S. carriers were fully illuminated ~ even with searchlights beaming up into the dark as the screening destroyers fired star shells into the sky to help the returning aircraft spot TF-58. Deck crashes wrecked 16 and 76 ditched at sea from empty fuel tanks, yet most U.S. airmen were rescued. Admiral Ozawa received orders to withdraw, and the battle was over. The Japanese fleet including its few remaining carriers had escaped serious damage, but having only 36 aircraft left from the 430 they began with, such loss of such irreplaceable airmen left them without effective air cover. Thereafter the Japanese relinquished their control of the Marianas and the central Pacific to the Allies.

THE MOTION OF ITS PROP CAUSES AN AURA TO FORM AROUND THIS F6F HELLCAT ON THE USS YORKTOWN. THE RAPID CHANGE OF PRESSURE AND DROP IN TEMPERATURE CREATED CONDENSATION OR DYNAMIC STATIC ROTATING WITH THE BLADES, THE HALO MOVES AFT, GIVING DEPTH AND PERSPECTIVE. U.S. NATIONAL ARCHIVES.

THE BATTLE OF LEYTE GULF

General Douglas McArthur's planned invasion of the Philippine Island of Leyte had the objective of forcing the hand of the Japanese Navy by strangling their sea supply routes to New Guinea, Malaya, and particularly, Borneo and Sumatra by which petroleum was shipped to Japan. Expecting such a move, the Japanese sought a decisive naval battle to halt the unrelenting Allied advance towards Japan, and an action amongst the Philippine Islands gave them the opportunity to inflict such a heavy blow against the Allied invasion force that they might even attempt to seek a negotiated truce. Two powerful forces comprised more that 700 ships including 120 warships, and more than 500,000 men converged on the Philippines. Royal Australian Navy forces with the U.S. 7th Fleet included five warships, three landing ships, and five auxiliary vessels. Then on 20th October 1944, the U.S. 7th Fleet, commanded by Admiral Thomas Kinkaid carried out the landings on the Philippine Island of Leyte.

The Japanese plan to defend the Philippines called *Sho-1* (*Sho* ~ Japanese for victory) required the formidable Centre Force of Admiral Takeo Kurita and Vice Admiral Shoji Nishimura to approach in a pincer maneuver to destroy U.S. forces after they began landing their troops on the beaches of Leyte. To achieve this, the Japanese first needed to lure U.S. 3rd Fleet carriers away from Leyte by convincing the impulsive Admiral William *Bull* Halsey to perceive the Japanese carrier force as the principal threat to the U.S. troops landing on Leyte. Vice Admiral Jisaburo Ozawa would command the Japanese carriers of Northern Force, and Admiral Shima's cruisers would assist.

Daybreak on the 24th October 1944 saw Admiral Ozawa's undetected force off northern Luzon preparing to strike U.S. warships with his few remaining aircraft, to lure Halsey's carriers away from the invading landing area, in a prelude to the Japanese trap. At 1115 Ozawa's search planes spotted a few American carriers, and launched a 76-plane strike against the Americans; but intercepted by American Hellcats their mission was a failure. However, in war it is often one man with good tactical sense that wins. A single Japanese airman flying a D4Y *Judy* dive-bomber followed the returning U.S. fighters ~ and did so unobserved. He singled out the light carrier USS *Princeton* and attacked. The AA gummers on the *Princeton* fired furiously, but the *Judy* pilot dropped his 250kg armour-piercing bomb through the middle of her flight deck. Inside the *Princeton*, the Japanese bomb wrecked several armed Avenger aircraft and ignited horrific fires within her hangar deck. Her aft magazines heated by the blaze in her hangar deck exploded in all directions killing 230. With her aft deck blown away and hangar deck fires spreading forward, Captain Bill Buracker gave the order to his crew, *Abandon ship!*

American patrol aircraft found Admiral Kurita's Centre Force, and Halsey ordered an airstrike sinking their battleship *Musashi*. Kurita now *seemed* to retreat west, while Japanese Northern Force carriers ~ the bait for *Sho-1*, were instructed by Ozawa to flood the airwaves, so Admiral *Bull* Halsey had no doubt that his IJN carriers were approaching. Halsey took the bait and steamed off searching for Ozawa's carriers; without leaving a single ship in the Strait or properly informing Admiral Nimitz. So far the Japanese plan was working. Kurita reversed his heading and around midnight steamed through the San Bernadino Strait undetected. During the early hours of the morning, Admiral Kurita turned down along the east coast of Samar, and headed for Leyte with nothing to stop him, except for the light U.S. escort carrier group *Taffy 3* commanded by Rear Admiral Clifton Sprague, comprising six light escort carriers (each carrying 30 planes), three destroyers, and four destroyer escorts.

In the early hours of 25th October 1944, the warships of *Taffy 3* were heading up the east coast of Samar. At 0630 the crow's nest lookout on Admiral Kurita's flagship, the battleship *Yamato*, reported seeing carrier mast tops over the horizon. Kurita ordered the guns of Centre Force to open fire. *Yamato's* guns spoke first, thundering 18-inch shells toward *Taffy 3*. Huge columns of water erupted between the U.S. flattops and Sprague's flagship carrier, *Fanshaw Bay*. Asking if the approaching ships were Japanese, the answer came back from his lookout: *The ships have pagoda masts!* Kurita's Japanese force of four battleships, eight cruisers, and 11 destroyers were 15 miles away, closing in for the kill. Sprague headed east for a rainsquall, launching every aircraft to stall the Japanese, and using the wind to his advantage, directed his destroyers to make a smoke screen to obscure the Japanese lookouts from clearly referencing the retreating carriers of *Taffy 3*. As American airmen attacked Kurita's ships form all points of the compass, Sprague ordered his destroyers to attack the Japanese warships. As the USS *Johnston, Hoel* and *Heerman* steamed through the carrier formation, four destroyer escorts USS *Roberts, Raymond, Dennis,* and *Butler* followed. The remainder of *Taffy 3* turned south retiring through shellfire, and then turned in a more southwesterly direction to place themselves between Kurita's warships and the invasion beaches of Leyte.

THE BATTLE OF LEYTE GULF

Eight small warships, steaming at full speed to their own probable destruction fired everything they had: guns and torpedoes, and began scoring such serious hits on the Japanese that they thought that they were being targeted by cruisers. USS *Johnston* pressed forward within torpedo range and fired a brace of deck torpedoes into the water, blasting the bow off the heavy cruiser *Kumano*, and also taking the cruiser *Suzaya* out of the fight as she stopped to assist. In return the gallant destroyer was shot to pieces by 14-inch shells from the *Kongo*. Kurita's men had never seen an escort carrier before, and thought that the Americans would never sacrifice two destroyer flotillas in a frantic action, if those carriers were not the cream of the U.S. Navy, the fast fleet carriers of Admiral Halsey's 3rd Fleet. The desperate destroyer attack persuaded the Japanese skipper of the heavy cruiser *Chikuma* to turn away east. Next the battleships *Yamato* and *Nagato* steered north away from the U.S. destroyers, after seeing their torpedoes coming at them. Kurita, becoming addled, ordered his fleet to retire. Sprague had saved *Taffy 3*, but at a terrible cost. The gallant vessels, *Hoel*, *Roberts*, and *Johnston* had all been sunk. USS *Johnston's* skipper, Captain Ernest Evans was posthumously awarded the Medal of Honor. Japanese salvoes of heavy shells from the *Tome* punched the escort carrier *Gambier Bay* full of holes, sinking her. Because of communication errors, many of *Taffy 3* survivors off Samar, who had abandoned ship, were not rescued for days and more good men were lost to thirst, sharks or madness.

The surviving battered ships of *Taffy 3* seeing Kurita's masts go over the horizon, then identified a flight of nine Zeros coming in, Vice Admiral Takijiro Onishi's *Shikishima Special Attack Force*. Onishi's courageous Zero airmen had volunteered to fly their 550-lb armour piercing bomb carrying fighters through the wooden flight decks to wreak destruction deep within the American escort carriers. Weakened by combat losses, Sprague's remaining vessels were unable to put up sufficient defence when three Zeros struck. Two human bombs smashed into *Kailin Bay* killing them self without serious consequence. The third Zero suicide craft flown by Lieutenant Yukio Seki slammed brutally into the flight deck of *St Lo*. The bomb pierced the deck, punching through to the hangar deck where aircraft were refueled and rearmed. A fire erupted, then six explosions smashed through *St Lo's* torpedo and bomb magazine sending huge flames in all directions. *St Lo* sank half an hour later. Elsewhere, HMAS *Australia* was hit on 25th October for a second time and forced to withdraw for repairs, marking the ascension of the age of the guided missile, for these *Kamikaze* were no more than that.

After receiving a report of Sprague's unencoded calls for help, Admiral Nimitz in Pearl Harbour needed to know what was occurring off Samar; what Halsey was doing; why he had failed to properly communicate his leaving the invasion area unprotected, and how to rectify the situation. Admiral Nimitz at CINCPAC HQ sent Admiral Halsey a radio message : TURKEY TROTS TO WATER GG ~ WHERE IS RPT WHERE IS TASK FORCE 34 ~ RR THE WORLD WONDERS. The first four words and the last three were *padding* used to confuse Japanese cryptanalysts with the beginning and ending of the true message marked by double consonants. But, a signals clerk in USS *New Jersey's* coding room who stripped the header first four words ~ and assuming the trailer *padding* THE WORLD WONDERS contained some relevant information, encoded the message and mistakenly sent :
GG WHERE IS TASK FORCE 34 RR THE WORLD WONDERS.

Halsey threw a tantrum and had to be bought under control by his officers. Three hours later he turned to change course for Samar ~ still searching for Ozawa's carriers. Ozawa had succeeded in his mission to lure Halsey away, and awaited his fate. Eventually Halsey's 3rd Fleet caught Admiral Ozawa's decoy carrier fleet of Cape Engano, (Engano being the apt Spanish word for deception). Halsey launched his aircraft to attack the Japanese carriers. With each Japanese warship firing different coloured AA fire for improved targeting, the U.S. pilots flew through multi-coloured flak explosions of blues, reds, whites, violets and yellows to sink the *Zuikaku*, the last surviving carrier of the raid on Pearl Harbour, plus the IJN carriers, *Zuiho*, *Chitose*, and *Chiyoda*. Admiral Ozawa had accomplished his mission, but without carriers the Imperial Japanese Navy was even more vulnerable.

After winning the Battle of Leyte Gulf, the land objectives on the Philippines were achieved much quicker. The Japanese were alarmed after the fall of the Philippines, at how quickly the flow of oil from its East Indies Empire dried up. While the Battle of Leyte Gulf confirmed that naval aviation would dominate the future of naval warfare, it also provides an excellent example of how decisions ; such as that of Clifton Sprague and the men of his heroic destroyers who saved the Leyte landing or the temptation of William Halsey, often have greater weight in war, than the size of their fleets.

ROYAL NAVY DEFENCE TACTICS AGAINST THE KAMIKAZE

In 1281, the Mongol emperor Kublai Khan launched a massive invasion fleet against Japan. With a navy that was too weak to oppose such an overwhelming force the Japanese realized that they were defenceless. Unable to fight, they offered prayers to their god-emperor to intercede on their behalf, to prevent the enemy barbarians from landing. Within hours a tremendous typhoon struck at the Mongol fleet, and Kublai Khan had lost so many ships and soldiers that he was forced to cancel his invasion plans. Interpreting the typhoon as a manifestation from heaven and an answer to their prayers the Japanese revered it as the *Divine Wind* or *Kamikaze*. Appropriately this was the name given to Admiral Takijiro Onishi's volunteer *Special Attack Corps*, formed in 1944, who like the great typhoon of 1281, were called upon to save the Celestial Empire from the enemy, and bring destruction down from the sky riding on the wings of the Divine Wind. Most of his *kamikaze* pilots who had attacked *Taffy 3* and HMAS *Australia* in the Philippines were only partially trained pilots, since the few remaining experienced Japanese naval aviators were far too important to throw away in expedient suicide attacks. Each Japanese airman, going into action wore a traditional scarf of white cloth around his forehead, the traditional *hachimaki* of the ancient Samurai, to keep any perspiration from their eyes.

> *I will do my utmost to dive head-on against an enemy warship, and fulfill my destiny in defence of the homeland... congratulate me. I have been given a splendid opportunity to die.*   Cadet Jun Nomoto

The British Pacific Fleet's first encounter with these, oriental *Knights of the Divine Wind* occurred on 29th January 1945 during Operation Meridian, a carrier strike against Japanese controlled oil refineries on Sumatra. The British commander, Admiral Phillip Vian knew some form of retaliatory attack was inevitable, so he maintained a continuous Combat Air Patrol above his Task Force, and kept his AA gun crews at Action Stations throughout the morning. Just before noon a wave of seven twin-engined *Sally* suicide aircraft led by Major Hitoyuki Kata were detected by radar, and the high-flying Seafires of the British CAP were vectored towards them. As these *Kamikaze* pilots attempted to crash onto HMS *Illustrious, Indefatigable, King George V,* their ship's AA gunners and the prowling Seafires kept them at bay, shooting down all seven Japanese hopefuls.

The Battle of Okinawa ~ Early in the morning of 1st April 1945, a deadly confrontation between the British and the Japanese occurred off Okinawa. A raid of 15 Japanese aircraft dived down towards the deck of the battleship HMS *King George V*. One of the Zekes machine gunned the battleship, and then roared towards the fleet carrier HMS *Indefatigable*... in a huge exploding *kamikaze* fireball he smashed into the base of the carrier's island superstructure and detonated its 500-pound bomb killing eight sailors. The big advantage that the carriers of the Royal Navy had was that their armoured decks made them far less vulnerable to *kamikaze* attack. HMS *Indefatigable's* damage control parties had extinguished the fire within four minutes and cleared the smoldering debris so that 38 minutes after, the first Seafire was able to land. But if a wooden-decked U.S. flattop was hit (and survived) it could be moored in harbour for up to six months being repaired. Late that afternoon, a lone Zeke targeted the carrier HMS *Victorious*. Leapfrogging over the destroyer screen, the Japanese pilot came in from astern and braving the hail of AAA fire hurled at him, lined up as if making a deck landing. Captain Denny's skill with the helm spoilt his aim, and clipping its wing on the port edge of the upper deck, the Zeke cartwheeled into the sea where it's bomb exploded 80 feet from the side of HMS *Victorious*, throwing tons of water, fragments of the *kamikaze* aircraft, and the dead pilot over the flight deck.

On 9th May, the *kamikazes* again targeted the carrier HMS *Victorious*. The standing patrol of Seafires shot one down as the others climbed to 3,000 feet. The first Zeke swooped down and took aim at *Victorious*, but Captain Denny once again proved too quick for his opponent. Thrown off balance by the carrier's sharp change of course the *kamikaze* pilot found himself on HMS *Victorious'* beam and not, as he intended, astern. The Jap pushed his throttle wide open trying to regain height for a second approach but his wheels grazed the flight deck. His under-carriage disintegrated and on its belly the flaming Zeke slid across the deck harmlessly and over into the sea. During the fighting for Okinawa, all four Royal Navy carriers were hit by *kamikaze,* but none took any serious damage, not so vulnerable as the Japanese thought, thanks to their 3-inch armoured flight decks, the seamanship of their skilled captains, the accuracy of their gun crews, and efficiency in battle of damage control parties. Future U.S. aircraft carriers would all incorporate the British armoured flight decks into their designs. During the terrible battle for Okinawa, the U.S. Navy lost 36 warships sunk, and some 368 damaged. The Marine Corps lost 7,600 men, but the Japanese lost over 172,000 soldiers during the *tetsu no ame* or rain of steel of the Battle of Okinawa, and Admiral Takijiro Onishi committed *hari-kiri* with his sword.

THE BIG BLUE BLANKET

Big Blue Blanket ~ At Ulithi, Commander *Jimmy* Thach convinced Admiral William Halsey that the only effective response to the *kamikaze* menace was to maintain what he called ; *The Big Blue Blanket*. In time it proved itself and became the standard tactic for opposing *kamikazes*, as *Jimmy* Thach wrote ;

I developed a system of combined offense and defence in depth. Defensive fighters were to patrol as far as 60 miles from the carrier task force. The carriers established a three-strike system to keep fighter patrols over enemy airfields almost continuously. In daylight one strike would orbit above an enemy airfield, strafing or bombing to disrupt activity on the ground. Meanwhile another strike was readied aboard a carrier and a third was en route to or from the airfield. At nightfall, fighters, and torpedo bombers would raid Japanese airfields strafe bombing, to dissuade takeoffs. This relentless raiding made it difficult for the Japanese to join up in large formations for an attack, in spite of the fact that there were some 6,000 enemy aircraft within range of our carriers.

On the island of Kure, the Japanese formed a unique squadron of 40 Kawanishi NIK2-J Shinden Kai *George* fighters designed to vanquish the Grumman Hellcat fighter. This specialist squadron's pilots were all practiced veterans, and included the few surviving aces of the Imperial Japanese Navy. After downing the Hellcat of U.S. Marine Captain Harvey Carter, Japanese ace Takeo Taminizu with a sense of humanity took off his own life preserver, flew low and threw it to Carter struggling in the sea, who gave a wave of thanks to Taminizu. Another memorable Japanese pilot, was Shoichi Sugita.

Shoichi Sugita plummeted like a stone. Coming out of his dive, he rolled in against a Hellcat and snapped out a burst. The blast from Sugita's four cannon set the American fighter's engine ablaze, and the Hellcat careened wildly through the air, out of control. Sugita rolled away and came out directly behind a second Hellcat sending his cannon shells into the fuselage and cockpit. The Grumman skidded crazily and plummeted into the ocean. After a third and a fourth kill Sugita landed.   Saburo Sakai

Air to Air Bombing ~ The Japanese aerial use of aerial bombs was first reported in May 1942 in the Solomon Islands, and they persisted throughout the war with air to air bombing against bomber formations after developing specific time-fuse bombs. The resultant explosion scattered a number of steel pellets filled with phosphorus in a downward cone like the pellets from a shotgun to penetrate the skin of enemy bombers, and the phosphorus igniting on contact with the air would set alight anything inflammable it contacted. Allied airmen reported seeing such bombs explode five to ten seconds after leaving a plane with a vivid yellow-red hue, followed by white smoke streamers (indicating phosphorous) in a waterfall effect. Attacks were made from head-on with bombs released between 1,000 and 2,000 feet above bomber formations. The damage caused to Allied aircraft by the rare Japanese use of aerial bombs was minimal and Japanese intention in persisting with this type of attack was more likely from its potential effect of disturbing a formation so their Zero fighters could pick off any stragglers. As reported in the 7th USAF Intelligence Summary of 7th October,1944;

On 19th September 1944, 13 B-24 Liberators bombed North Moen airfield, at Truk from a height of 19,000 feet. A twin-engine Jap fighter attacked the formation and dropped a phosphorus aerial bomb. One minute after bombs-away our planes were intercepted by four Japanese Zekes and a possible Nick. One two-plane coordinated attack was made from one o'clock high. Both of these planes dropped aerial bombs, after coming out of the sun. Five individual passes from between 12 and two o'clock followed this attack. These fighters also came in high and released three aerial bombs and made two shooting passes. The bombs hit low and wide.

One of our planes had a feathered engine and was subjected to two fairly aggressive attacks from five to eight o'clock high. However, the formation protected this plane by slowing down and keeping him well covered. The twin-engine fighter came in from the nose, high and out of the sun, and pressed his attack to 250 yards. He broke away to the right at two o'clock, exposing the belly of the plane. A phosphorous bomb was dropped, bursting approximately 300 yards at three o'clock. Although twin-engine fighters have been seen on many previous missions, this aerial bomb attack is the first reported from this type of fighter.

Both aerial bombs released in this coordinated attack were a new type. One of the bombs was observed prior to bursting and was described 1½ feet in diameter, six feet long that seemed to flutter down spinning to the left as it fell (indicating a mechanical fuse). The burst was orange-red and shrapnel was thrown out which looked like tracers. Both bombs were accurate as to altitude. One burst to the left and one to the right of the formation. One burst was close enough to one plane in B-Flight for the blast to jar loose lighting fixtures in the cockpit. Other passes were very unaggressive. No damage to our aircraft resulted from enemy fire.

*   *   *

Pilot and navigator of a Short Sunderland enjoying some refreshments...

Two gunners in Short Sunderland Mark I, sit at their positions with mg's mounted in the upper fuselage hatches.

Cocos Islands 1944 - Short Sunderland Mk V J of 205 SQN RAF moored off Direction Island to refuel.

Navigators desk on board a Sunderland.

JAPANESE FU-GO WAR BALLOONS OVER NORTH AMERICA

General Sueyoshi Kusaba and Major Toshiro Otsuki of the Imperial Japanese Army's Number Nine Research Laboratory after discovering a strong current of winter air blowing at an altitude of nine kilometers above Japan, hoped to use it to attack North America, in the same way the Austrians had bombarded Venice in 1849. Toshiro Otsuki's concept was to design and release bomb carrying paper balloons into the winter jet stream, to cross the Pacific in three days to set fire to the forests, farmlands and cities along the west coast of America. The initial order was issued for 10,000 balloons. Kusaba expected ten percent of the balloons released to reach some target. The researchers fabricated their *Fusen Bakudan* or *Fu-Go* balloon bombs from *washi*, traditional roadmap sized sheets of paper derived from mulberry leaves that impermeable and tough, were then glued together in laminations of three to four sheets using potato flour paste. Many of the *washi* workers who gathered in large indoor spaces throughout Japan, such as sumo halls and theatres were nimble fingered girls, who appropriated the paste and ate it. Inflated with hydrogen, the ten metre wide mulberry paper balloons could lift 800-lbs at sea level and were armed to carry either: one 12kg Thermite incendiary bomb, four 5kg incendiaries or one 15kg HE bomb. As U.S. B-29 bombers began raiding the islands of Japan, the first *Fu-Go* balloon was released on 3rd November 1944. Major Teiji Takada watched as their first balloon rose up and wafted out to sea ~ *The figure of the balloon was only visible for several minutes following its release until it faded away in the blue sky like a daytime star.*

Fu-Go war balloons released from Otsu, Inchinomirya, and Nakoso during the winter months when the jet stream is strongest, rose to jet stream altitudes between 20,000 to 40,000 feet, to travel east over the Pacific Ocean at an average of 200 mph to cross the Pacific Ocean in three days. During the day they cruised along at their maximum altitude, however at night they collected dew, became heavier, and descended. The non-rigid hydrogen balloons also contracted when cooled at night and fell further. So, when the *Fu-Go* balloon descended below nine kilometers, the altimeter electrically fired a set of blow plugs (charges to release ballast) to discard two sand bags at a time. This loss of weight would cause the balloons to rise up to the optimum cruising altitude, and continue until all sand ballast bags were gone. Similarly, when the *Fu-Go* rose above 11.5km or 38,000-feet, an altimeter activated a valve to vent hydrogen. Hydrogen was also vented if the balloon's pressure reached a critical level. This simple control system guided the *Fu-Go* through three days flight. The final blow charge dropped the balloon load of Thermite bombs from the final position on the ballast ring. Anti-personal bombs were also used. After the bomb load was gone, a 64-foot picric acid fuse was lit. Connected to a charge on the balloon itself, 84 minutes later the fuze fired a flash bomb to ignite the hydrogen and air mixture within the balloon envelope to explode in an enormous orange fireball.

By the beginning of 1945, Americans became aware that something odd was going on. Balloons had been sighted and explosions had been heard from California to Alaska. A balloon had been seen over Santa Monica, and fragments of *washi* paper were found in the streets of Los Angeles. Balloon envelopes and apparatus were found in Montana, Arizona, Saskatchewan, in the Northwest Territories, and in the Yukon. In Yerington, Nevada cowboys cut one up to use as a useful hay tarp. Another was found by a prospector near Elko who carried it into town on the back of his donkey. Eyewitnesses observed something with the appearance of a parachute descend over Thermopolis, Wyoming. A fragmentation bomb exploded and shrapnel fragments were found around the crater. On 5th March 1945, a minister's wife and five Sunday school children on a church fishing picnic in Bly, Oregon found one. Trying to pull it thru the forest back to their camp, it exploded and killed them all.

On the West Coast of the U.S.A. a secret operation was conducted to defend against the *Fu-Gos* called *Fire Fly* and included fighter aircraft to intercept the balloons, but as the balloons flew so high and fast, the airmen had little success, destroying fewer than 20 balloons. A troop of airborne fire fighters was assembled to put out the fires. These *Smoke Jumpers* were the first to jump to a forest fire and fight them in the way Americans still do today. They were called the *Triple Nickle* or the 555th Parachute Infantry Battalion. Incendiary bombs might have caused forest fires, but the *Fugos* launched at the beginning of autumn were less menacing at that time of year when American forests were too damp to easily catch fire. In North America, the authorities were far more worried about the severe possibility of the known infamous Japanese Bio-Warfare *Unit 731* commanded by Shiro Ishii, based in Pingfan, Manchuria, (where Chinese and Russian prisoners were infected with the bubonic plague, cholera, anthrax, and other hideous diseases) getting involved in the balloon project. Some Japanese generals had proposed loading *Fu-Go* balloons with cattle plague virus, and smut to wipe out crops.

JAPANESE FU-GO WAR BALLOONS OVER NORTH AMERICA

No one in North America considered that the balloons had come directly from Japan, thinking that landing parties from submarines had launched them from west coast beaches, or they may have come from Japanese-American internment centres. Once evidence of the ballast bags dark sand was recovered from crash sites and taken to the Military Geology Unit of the U.S. Geological Survey for investigation, researchers began their microscopic and chemical examination of the ballast sand to determine the details of any microscopic sea creatures, and its mineral composition. They confirmed that the sand had not originated from any North American or even mid-Pacific beach. It had come from Japan. In time the geologists were able to determine that the dark ballast sand had come from Ichinomiya beach, northeast of Tokyo on east coast of Honshu. Although this alleviated any fears of Japanese submarines lurking off North American shore, the intent of the Japanese balloons remained.

Censorship ~ The Japanese *Fu-Go* fire balloons had so far caused little damage, but their potential for fires and destruction was enormous. After *Newsweek* ran an article in their 1st January 1945 issue titled *Balloon Mystery*, the U.S. Office of Censorship decided to block the Japanese from learning any details of the effectiveness of their balloons, sending a message to all newspapers and radio stations asking them to refrain from mentioning any balloons or balloon bomb incidents, to ensure that the Japanese remained ignorant of their weapon effectiveness. If sufficient information had been disclosed concerning dates, locations, and descriptions of balloons sighted and bombs dropped, the Japanese may had been able to improve or correct their methods. Had the Japanese known any news of frightened civilians or panic, undoubtedly they would have pressed their campaign and further developed their *Fu-Go* fire balloon operations. The Japanese monitored every American broadcast and read all the American magazines and newspapers they could find, searching for clues. They read about unexplained forest fires on the west coast, and an unexplained fire in Boston, but still didn't know if any bombs had reached there or not. More bombs were released, and they learnt nothing; except for a single report of the tragedy in Oregon. Months later, the Japanese heard of a balloon bomb landing in Wyoming, which had failed to explode. The FBI and the army seized the balloon and quickly clamped a tight lid on the whole affair. Reports from all states showed little damage, due in part to the extreme cold of the high altitudes at which they travelled that froze the batteries preventing the proper operation of their mechanisms and bomb loads. With no subsequent evidence of effect, in April 1945, General Sueyoshi Kusaba was ordered to cease operations and again they missed their chance of dropping incendiaries during the dry season on the forests of America. The main response that affected the Japanese decision to drop their whole fire balloon campaign was the fact that North American newspapers and radios working under voluntary censorship, had kept quiet.

On 10th March 1945, in a quirk of history, one of the few remaining mulberry *washi* paper balloons to be released came down in the vicinity of the Manhattan Project's atomic production site at Hanford, Washington landing on a power line that fed electricity to the building containing the reactor producing plutonium for the Nagasaki bomb, and shut the reactor down. One *Fu-Go* was shot down by a U.S. fighter west of Alturas, California on 10th January 1945 and re-inflated for study at Moffet Field, California. In all the Japanese launched 9,300 balloons, of which 285 were found in North America. The balloons that came to rest often impressed their more curious finders who found that the paper was so durable that it could not be pulled apart by two men. After the last Japanese *Fu-Go* balloon was launched in April, there was a murmur of resumption of the *Fu-Go* campaign on 4th June when Lieutenant Colonel Shozo Nakajima, the propaganda spokesman of Imperial Japan forewarned :
Thus far these attacks have been on an experimental scale, and when the actual results of the experiment have been obtained, large scale attacks with death defying airmen manning the balloons will be launched. The balloon bomb is one of Japan's unique originations and it is especially significant in that by the use of this method of bombing we can attack the enemy mainland directly from Japan ~ something that the enemy cannot boast of.

Toshiro Otsuki, the inventor of the paper *Fu-Go* balloons and his secretary committed suicide in a forest in Izu Peninsula 100 miles southeast of Tokyo. Years later, a group of Japanese women who as high school girls had been required to make the balloons to which clusters of bombs were attached to drift across North America and kill, sent their heartfelt apology and a gift of two lacquered Ouchi dolls to the relatives of the Oregon children. With the symbolic dolls was a children's book in Japanese about their making the war balloons and the ensuing tragic church picnic accident, with their letter ;
We the undersigned, humbly offer our prayers from the bottom of our hearts for the souls of the six who lost their precious lives due to one of the balloon bombs we helped to build. We send our earnest resolve never again to go to war and our fervent hope to bring the people of the world together to live in peace and friendship among us all.

* * *

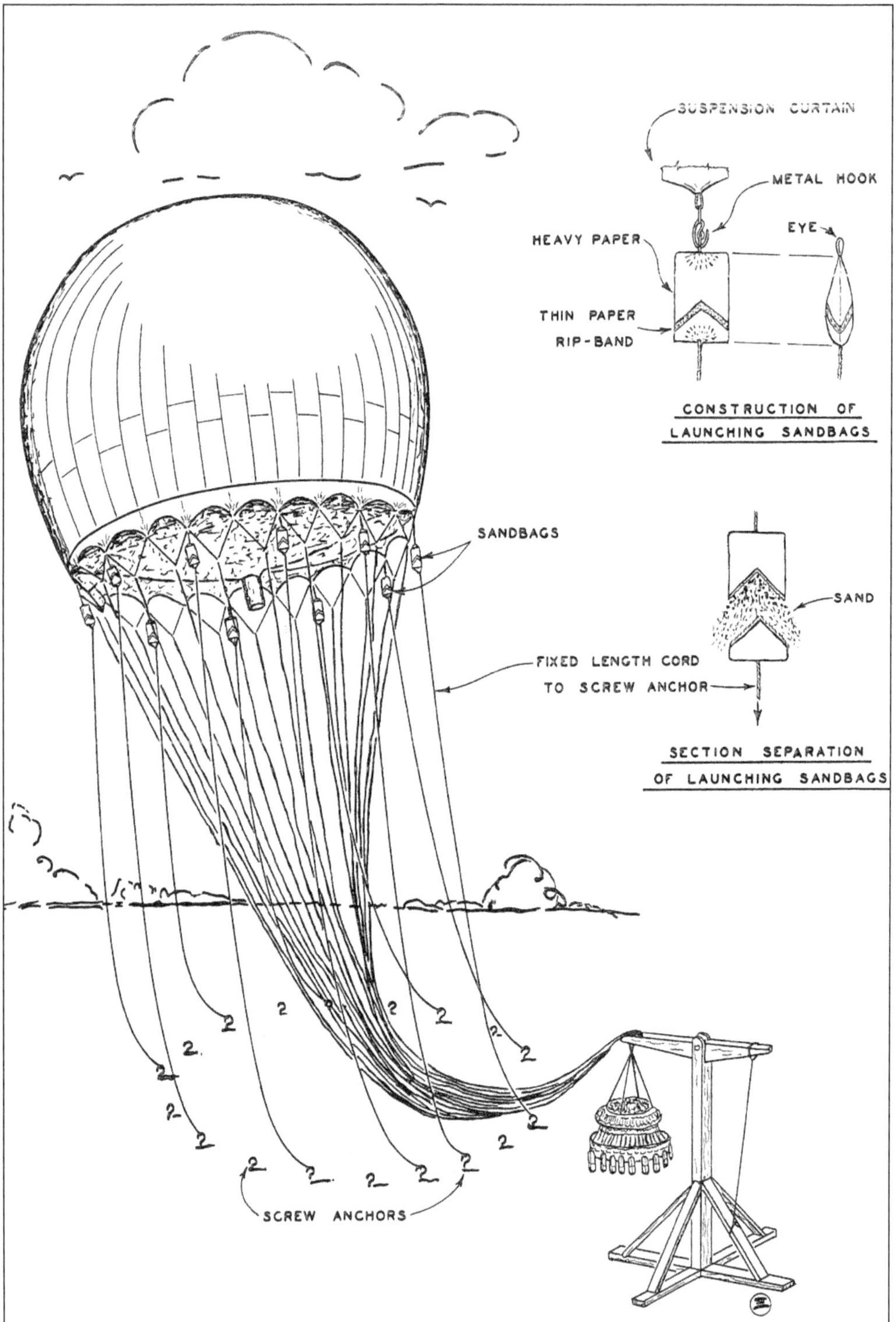

Above : Method used for launching Fu-Go balloons in winds above three miles per hour.

Opposite : Japanese Fu-Go balloon incident locations.

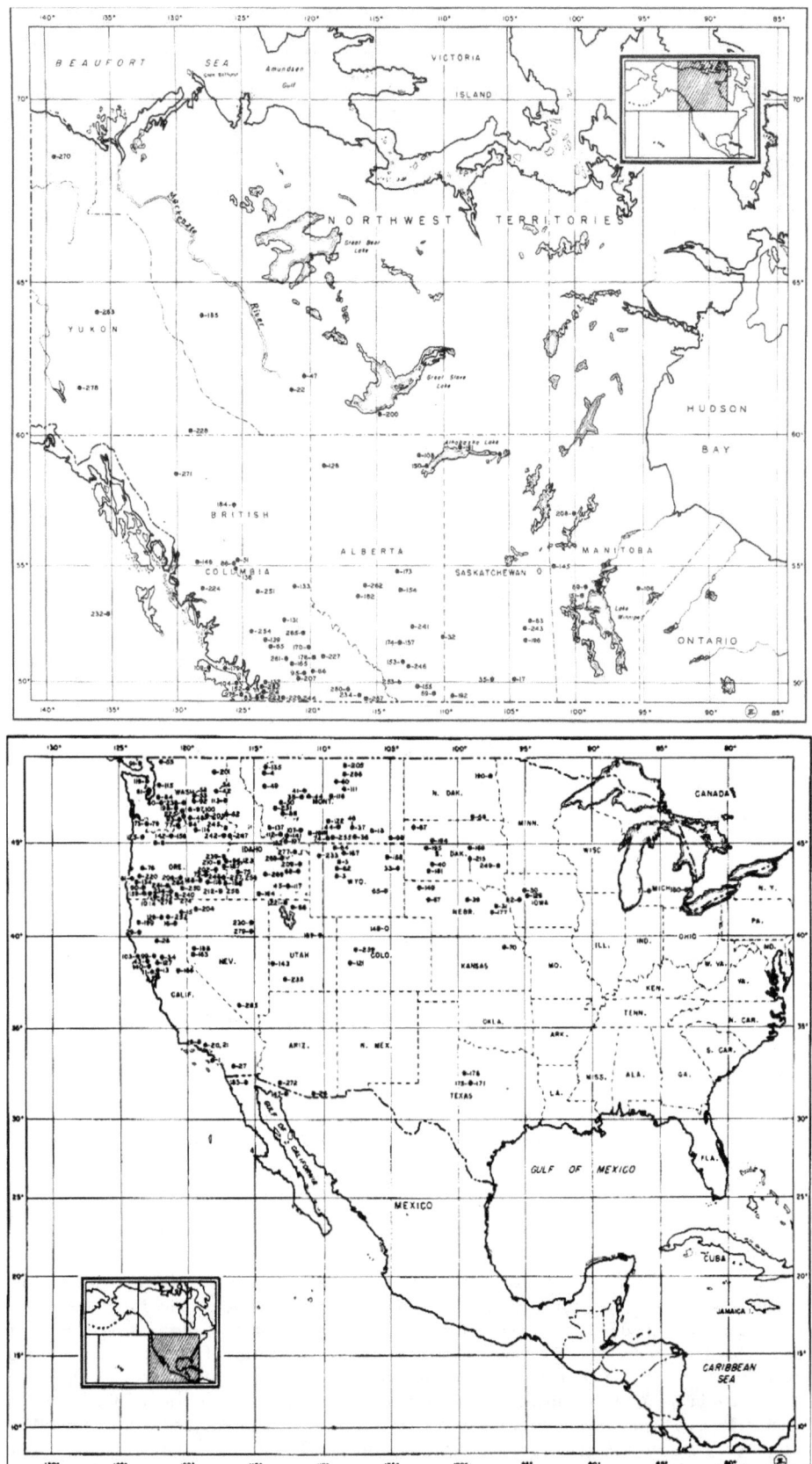

During a B-29 Superfortress raid against Singapore on 2nd March 1945, it was reported that 45% of Japanese fighter attacks were broken off before 500 yards, and only 26% pressed to less than 250 yards. Japanese fighter pilots preferred high 12 o'clock passes against the frontal aspect of B-29 bombers, but a new variation in the *12 O'clock Express* was then experienced. With a high degree of skill on the part of a Japanese Oscar pilot sighted two miles out, at two o'clock high, he approached within one mile, wagged his wings, and turned towards the B-29 from 12 o'clock in a dive. When 1,000 yards above the Superfortress he rolled over on his back through a vertical pursuit curve and opened fire at 500 yards. The Oscar roared to within 25 yards of the B-29's tail with exceptional judgment, raked the B-29 with one 20mm shell piercing the left wing root of the Superfortress.

On 20th January 1945, Major General Curtis Le May improved the combat efficiency of his Marianas bombers by altering their tactics from daylight high altitude raids with HE bombs to night low and medium altitude raids with a combination of incendiaries and HE. This led to the most destructive raid in the Pacific when 334 B-29s dropped 1,667 tons of bombs on Tokyo on the night of 9th March 1945. 16 square miles of the city was destroyed, 83,000 people were killed in the firestorm, another 100,000 were injured, and one million were left homeless. Thereafter, the B-29s roamed at will over Japan, destroying cities, industries, power stations, transportation routes, and food supplies. There was nothing the Japanese could do to impede the bombers, even aerial *Kamikaze* tactics failed. Superfortress crews were despised for the grim results of their raids, and captives were often beaten. Then, at Imperial GHQ in Tokyo, orders were signed for the murder of all POWs. On 5th May 1945, after an American B-29 bomber was shot down over southern Japan and its eight aircrew captured, these POWs were made available for medical experiments at the anatomy department of Kyushu Imperial University where they were all vivisected, organ by organ, while they were still alive.

During 1945, *kamikaze* pilots flew close to the sea level and Allied radar often failed to detect them. Although Allied intelligence estimated that three out of every four *kamikaze* planes were destroyed, one out of every four found his target and one out of every 33 *kamikaze* attacks sank a warship. Appalled by the terrible losses inflicted by the *kamikazes* in the Battles of the Philippine Sea and Okinawa and living in dread of the horrendous casualties that could follow in the wake of Operation Olympic : the invasion of the Japanese island of Kyushu. President Truman rushed the top secret Manhattan Project, the development of the atomic bomb to save Allied lives. And so the Pacific War, which had begun with a devastating air attack, similarly ended when U.S. bombers subjugated the spirit of the Rising Sun with atomic air strikes against the Japanese cities of Hiroshima and Nagasaki.

Nakajima Ki-43 Oscar

AMERICAN ACES OF THE KOREAN WAR

Captain Joseph McConnell ~ 16 kills

Major Frederick C. *Boots* Bleese ~ 10 kills

Major James Jabara ~ 15 kills

Major John F. Bolt USMC with his F-86 Sabre jet in Korea
Flying with *The Black Sheep* squadron he claimed 6 Japanese Zeros
to become the only U.S. Marine to achieve ace status in two wars.

*Photo dated 13th July 1953, two days after Major Bolt shot down his fifth and sixth MiG-15s to become the only U.S.M.C. ace of the Korean War. He achieved the aerial victories while flying with the 5th Air Force as an exchange pilot. The original caption states: *Major Bolt, who shot down six Japanese planes during World War II, has flown 89 jet fighter-bomber missions with the 1st Marine Aircraft Wing, which has been assigned only close support and interdiction missions in Korea. He has flown 37 Sabre jet sweeps with the 5th Air Force, which is carrying the air war to the MiGs.*

CHAPTER IX

THE KOREAN WAR

Working deep within the U.S. Manhattan Project, a lurking Soviet spy ring were able to lift enough top secret atomic data for their Soviet scientists at home to create a primitive fission bomb, and when they detonated their first atomic bomb in August 1949, the Soviets wiped out America's monopoly on nuclear weapons. Most spies were caught and imprisoned, but Ethel Rosenberg for her part in recruiting the nuclear espionage network was sentenced to death in the electric chair. U.S. President Dwight Eisenhower wrote to his son that if he had spared Ethel, it would have encouraged the Soviets to recruit more female spies. After the first 57-second jolt of electricity failed to kill Ethel Rosenberg, she was restrapped to the chair and given two more powerful jolts, becoming the first woman executed in America since Mary Surratt was hanged for her role in the assassination of Abraham Lincoln. While the two superpowers never engaged in open warfare, they ardently pursued their enmity through propaganda, espionage, and proxy wars for the ensuing four decades.

In June 1950, North Korean dictator Kim Il-sung, with the approval of Stalin and Mao Zedong, sent his army of across the dividing border of the 38th Parallel to invade South Korea. From the air, little could be done to stop the onslaught of enemy troops, as U.S. and South Korean troops withdrew to the southern tip of the peninsula to set up a defensive perimeter around the port of Pusan. Still in danger of being pushed into the sea, the small force defending Pusan were soon joined by United Nations contingents from Great Britain, France, South Africa, Australia, New Zealand, and ten other countries. From airfields in Japan, pairs of U.S. Far East Air Force, F-80C Shooting Stars took off every 15 minutes to pound the enemy. However their long flight times meant they could only remain a short time over the front, until the improvisation of extra capacity wingtip fuel tanks for the Shooting Stars.

The Americans formed a joint operations centre to cover their communication network of tactical air control parties who travelled in radio-equipped jeeps. However when the patrolling jeep crews scouted around the defensive perimeter calling down fighter-bombers on support missions, high mountains limited the range of their radio sets. Stinson L-5 *Mosquito* light aircraft were flown in to fill the gap, with their pilots being able to talk to both central operations HQ and the fighter-bombers. *Mosquito* airmen searching for the enemy, indicated targets to the circling fighter-bombers by lobbing smoke grenades, and when alone often hurled HE grenades down at the North Koreans.

Tank Busting ~ High velocity Shooting Stars displayed their skills on 10th July 1950, when they strafed and fired 5-inch *Holy Moses* rockets destroying 38 Soviet supplied T-34 tanks and 117 enemy trucks. Their method was to power dive at 450 mph through an angle of 45° at the enemy tanks, aiming at their vulnerable rears. Time delay fuses caused the 5-inch rockets to shatter against enemy armour before exploding, creating a shrapnel effect against the crews inside the enemy tanks. Such rocket detonations were so ferocious that attacking Shooting Star pilots needed to veer off more than 500 feet above a target, to be clear of the debris from a rocket detonated T-34 tank. Shooting Stars were responsible for more than 70% of the damage inflicted on the advancing North Korean forces and without their skills, the defenders might have been forced to evacuate the Korean Peninsula.

During August, the flow of troops and war material into the Pusan perimeter stabilized the situation, while fighter reinforcements flew in from Okinawa and the Philippines. In the air though, near-sonic speeds made it so frustrating for squadron leaders to hold patrolling formations together that smaller flights of four Shooting Stars and Thunderjets were flown. Arriving over the frontline at regular intervals, flight leaders clocked in at the joint ops centre, and directed to a tactical air control party supporting United Nations ground troops. Often the forward controller would indicate the target with coloured smoke or a *Mosquito* pilot might fire smoke rockets at an enemy held position. Then the warplanes attacked with bombs, rockets, machine guns, or their deadliest weapon Napalm. The flaming jellied fuel from a 110-lb Napalm bomb would splash along its line-of-throw engulfing enemy tanks within 30 yards of impact or flow flaming over the ground into enemy positions. Often flights of four B-29 Superfortress strategic bombers from Guam were asked to load up and help. Nearing Korea, their flight leaders contacted the operations centre for target information detailing their massive weapons load. Each carried 40 bombs : a standard 20,000-lb load out for a B-29.

MIG ALLEY 1950-1953

During September, the Americans were ready to fight back. From the carrier, USS *Valley Forge* in the Yellow Sea, Vought Corsairs and Douglas Skyraiders flew suppression sorties against North Korean ground forces, who changing their tactics moved at night to survive by day in camouflaged positions. Flying both jets and piston-engined aircraft, it became noticeable that pilots preferred jets. The airmen of No. 77 Squadron RAAF, who joined the United Nations force in July, were admired for their spirited flying of their Mustangs, but they might have preferred to fly jets. With fewer moving parts, jets are able to withstand more AA damage, and having no propeller gives a clearer view ahead. The cockpit of a Shooting Star was quieter than a Mustang, which made jets far less fatiguing to fly. U.S. Navy aviators flew from their aircraft carriers to support the troops below, and were often assisted by the British contribution to the air war over Korea, Hawker Sea Furies and Fairey Fireflies operating from the Royal Navy aircraft carriers, HMS *Glory, Ocean, Theseus* and *Triumph*.

Breaking out of their defensive perimeter in mid-September, for two days UN aircraft dropped napalm on the enemy held port city of Inchon, then on 19th September the men of the UN mounted a daring amphibious assault in the MacArthur manner, to take and hold Inchon on the northwest coast of Korea, 30 miles south of the 38th parallel, and then advance north to the South Korean capital of Seoul. From there they regrouped to march north to link up with airborne forces and consolidate.

The G-suit ~ One advantage defending U.S. pilots enjoyed, was their G-suits. Used for the first time in Korea, an airman's G-suit automatically constricted the lower part of their body during maneuvers to contain the flow of blood downwards. Although the maximum G force loading from which most pilots could aim and shoot with accuracy was 4G, pilots now could turn to attack or evade through 9G turns, endowing them with more capability than the previous limit of 5G to 6G, beyond which they risked losing their faculties and blacking out. The few obsolete Russian supplied Yak-9 fighters of the North Korean Air Force were poorly flown and soon shot from the sky by superior Shooting Stars. Yak-9 airmen were often seen spinning off out of control during a tight turn, after the centrifugal force of severe turn drained the blood from their brain, causing them to lose consciousness.

When the first and potentially most combustible conflict of the Cold War appeared over, hordes of well-supplied Chinese volunteers rekindled the fight. Forbidden to fly over Chinese airspace, U.S. B-29 bombers targeted the Korean ends of all Chinese roads and rail bridges spanning the Yalu River, the natural border between North Korea and China across which all Chinese reinforcements had to enter the warzone. A very old rule of war was demonstrated to the enemy. When damage is focused on the front line during an offensive, reinforcements might still add to the momentum of an attack and enable it to persevere. However, UN air flights made such determined long range attacks against the communists that caused such a lack of replacement ground troops and dire poverty of war material at the front, that their attacks were halted.

Over the Yalu River, a U.S. pilot reported seeing a swept-winged fighter takeoff from Antung in Manchuria, just north of the Yalu River. In November, confirmed sightings came of a Mikoyan-Gurevich Mig-15 over the Yalu and Chongchon River area, which soon became known as *MiG Alley*. The Soviet MiG-15 was armed with two 23mm cannon (80 rounds) plus one 37mm cannon (40 rounds) and just one or two hits from these shells were enough to kill. When the leader of four Shooting Stars spotted some MiG-15s below, the Americans zoomed after them in a tail chase toward forbidden Manchurian airspace. When the time came for the U.S. flight to break off their pursuit, the MiG-15s soared toward the sun, split into pairs, recrossed the river, and pounced. The sleek MiGs powered by an engine copied directly from the British Rolls Royce *Nene*, proved that they were 120 mph faster, and could out climb a Shooting Star in a tight turns. This time, there were no casualties.

Lieutenant Russell Brown flying a Shooting Star on 8th November 1950 over Sinuiju near the Yalu River made three strafing runs against ground targets. Then sighted two MiG-15s. Brown banked his Shooting Star into a firing position, head to tail after a MiG-15 in a climbing turn to the left. Brown fired four short bursts. All missed. The MiG-15 rolled over the top, and dived straight down. Brown snapped his Shooting Star nose down thru a hard dive. The heavier Shooting Star gained momentum. Brown began closing on the MiG-15, and thought… *Damm, I'm going to get him.* Firing three short bursts, crimson tinged flames, and dirty smoke poured from the right side of the stricken MiG… Then it erupted in a fireball of debris. Lieutenant Russell Brown had scored the first jet vs. jet victory.

* * *

Top & Middle : F-80C Shooting Stars   Bottom : F-84 Thunderjet ~ (from Operation Teapot)

MIG ALLEY 1950-1953

As North Korean radar operators plotted approaching UN air formations, flights of MiG-15s waited on the Chinese side of the Yalu River with almost full loads of fuel to be vectored across and intercept enemy aircraft, then streak back to Manchuria to loiter in anticipation for another fast strike. The MiG airmen had every advantage. If a MiG pilot found him self in strife, he simply headed for the northern side of the Yalu where the false cloak of Chinese neutrality protected him and because of their Chinese sanctuary MiG pilots remained in control of the time, place and capacity of their actions. The MiGs fought over their own territory while the F-80Cs flew from Kimpo and Taegu Airfields in the far south, and UN radar did not extend up to cover the Yalu-Chongchon River area of *MiG Alley*.

Another advantage favouring the enemy from October to March was the prevailing winds at 30,000 feet over Korea that regularly blew more than 100 mph from west to northwest. As fights drifted away from the MiG airfields, a fuel thirsty UN pilot could break off combat and head south. In the U.S. the new F-86A Sabre entered service equipping the famous, *Hat-in-the-Ring* 94th Fighter Squadron. With swept-wings to avoid the effects of atmospheric compressibility and armed with six 0.5-inch machine guns harmonized to converge 1,000 feet, the Sabre was fitted with a radar ranging gun sight that was wonderful when it worked, except it didn't that often. Three squadrons of Sabres arrived in Japan during December on the carrier USS *Cape Esperance* all showing salt air corrosion. Within two weeks, the Sabres flying from Kimpo Airfield had destroyed eight MiGs for the loss of one of their own. During combat, high stick forces and wide turning circles made aiming the Sabre tricky, and if speeds increased beyond Mach 0.95, the Sabre became even heavier on the controls. Early Sabres could not outturn the MiG-15 ~ also superior in ceiling (50,000 vs. 42,000 feet), acceleration and rate of climb, but the F-86 could always evade the MiG through a hard diving turn. Slow firing MiG cannons fitted for bomber interception had such a low muzzle velocity and subsequent longer time of shell flight that effective shooting at an evasive jet became incredibly difficult. The subsequent F-86E version of the Sabre with an all-moving tail plane gave U.S. pilots further advantage over the MiG-15.

High altitude stratospheric fighting ~ As height increases, air becomes thinner, provides less lift, and contains less oxygen for engines to burn, so an aircrafts available power diminishes. Hard turns at high altitude were not possible, and as the stalling speed of a Sabre at 40,000 feet pulling 4G was only 580 mph, attempting anything more than a gentle turn risked stalling in combat. Stratospheric tactics evolved with the winner being the pilot who could tempt his adversary into a stall inducing hard turn, to acquire an easy shot at him. Poor high altitude visibility in air so clear that the sun reflected every detail inside the cockpit canopy was another difficulty. Climbing to 50,000 feet, the sky turned dark blue and aggravated how a cloudless sky gave a pilot nothing to focus on. The sun sparkling on polished aluminum airframes or Perspex canopies raised the thorny issue of recognition. Due to the lack of oxygen, above 25,000 feet MiG-15s were hit and often went down leaking fuel without igniting until lower altitudes were arrived at. Survival was no longer found in tight turns. Sabre leaders flying near the speed of sound, could never let their speed lapse in swirling high-speed engagements or they would fall to the MiGs. Their wingman had to maintain his position, and keep his eyes open, so his leader could be the gun. In the air, Sabre pilots flying between 27,000 and 30,000 feet, below the contrail height could see those made by higher MiG-15s. Some MiG formation leaders, aware that their contrails could be seen far south of the Yalu, peeled off a flight before contrail height was reached to lure the Sabres into fixating on the contrails of higher MiGs to surprise them with a lower flight. This ruse was revealed one day when a flight leader, suffering a pressurization problem flew his flight at a lower altitude. Spotting a flight of low MiG-15s the Americans closed in, took aim, and downed three. Thereafter one Sabre flight flew below contrail height to respond to this MiG trick. It was said that the most skilled MiG airmen were Soviet pilots (called *honchos*) who rarely flew over UN territory. This was confirmed, when blond pilots in civvies were seen ejecting from shot up MiGs.

Helicopter Rescue ~ On 10th October 1950, Lieutenant David Daniels flew his Sikorsky helicopter 90 miles or 145km over enemy lines, to rescue a downed Hawker Sea Fury pilot of the Fleet Air Arm who had been holding off Communist troops with relays of Sea Furies until help arrived. Helicopter airmen successfully penetrated enemy territory to rescue hundreds of airmen from behind enemy lines or from the sea off the North Korean coast. After it became apparent that communist airmen disliked flying over the sea, a helicopter detachment was established on the island of Chodo, whose controller, call sign DENTIST coordinated his Sikorsky helicopters to extract downed UN airmen from the waters of the Yellow Sea. Helicopters also collected agents from behind enemy lines and flew medical evacuations delivering more than 8,500 wounded to MASH units.

MIG ALLEY 1950-1953

The Sabre handled well and it had a good radius of action. At height, its 5,200 pounds of thrust pushed it along at a top speed approaching that of sound to become supersonic in a steep dive. Finger-four flights staggering their takeoffs to arrive at five-minute intervals over the Yalu to support each other, and if a fight was imminent, let go their drop tanks to increase speed and fight in fours or pairs for up to 20 minutes, until their diminished fuel level forced them to return to Kimpo Airfield. This method of operation proved itself on 20th May 1951 in a swirling engagement which made Captain James Jabara the first ace of the Korean War, when a flight of 12 Sabres were pounced on by 50 MiGs, but reinforced by two Sabre flights, shot down three MiGs, and sent five home damaged.

Operation Strangle ~ At the end of May 1951, the UN air offensive had begun to strangle the flow along the Communist road and rail system by repeated attacks on specially selected choke points. Initially the operation went well, but once surprise was lost the Communists countered by concealing their supplies in caves and tunnels. To avoid further attacks, they repaired their roads, railway tracks, and bridges with cunning and convincing camouflage. The North Koreans sited many AA guns around obvious targets to claim 81 UN aircraft destroyed.

The Yo-Yo ~ Flying in sections of two, Soviet MiG pilots flew the air combat manœuvre (ACM) known to most airmen as the *Yo-Yo*. Whenever a flight of MiGs were pounced by two Sabres, *Honcho* airmen split — the lead *Honcho* pulled up so high thru a turning climb that the Sabres could not follow while his wingman flew a level or slight diving circle to hold the Sabres in play. As the Sabre pair below chased the lower MiG wingman, the high MiG *Honcho* positioned to pounce on the Sabres still pursuing his wingman. The *Yo-Yo* became the precursor of three manœuvres, which remain apart of a fighter pilots repertoire today, and all three moves saw their tactical début over *MiG Alley*.

Basic Manœuvre #1: the Defensive Split: When the question arose for an attacking pair ; which one to follow? The standard response of today is to thrust after the higher opponent. But in the sky over Korea, if you were flying a Sabre, your only option was to chase the lower MiG. The MiGs doing the defensive split forced the Sabres to chase one, while the higher MiG manœuvred against the Sabres.

Basic Manœuvre #2: the High Yo-Yo: While the MiG wingman lured the Sabres around in a full throttle circle, the MiG leader climbed fast with his aileron turned to keep the pair of Sabres in view. As the high MiGs speed lessened, he pulled hard over the top, diving through a firing position against the low Sabres. This *High Speed Yo-Yo* was used against a horizontally turning target when the speed of an attacker precluded any possibility of following its target through a turn.

Basic Manœuvre # 3: the Low Yo-Yo: Seeing a fast moving enemy turning wide, if an attacker cut the corner and failed, if he trailed far out and if his opponent had the advantage of speed, the hunter might become the hunted. Better to close on an opponent through a diving turn, converting altitude into airspeed to cut inside the defender's turn, soaring up on the far side of the circle to acquire a good firing position, using the same principle as the *High Yo-Yo*, just approaching from below.

Sabre pilots such as (future astronaut) Major John Glenn who vanquished three MiGs over Korea, flew at near maximum thrust in the warzone with a trained sense of alertness to neuter the ability of MiG-15 pilots to catch and surprise them. The refined Sabre F-86F which first flew in March 1952 with an engine that delivered 6,090 pounds thrust was even faster than the new MiG-15 bis. The new Sabre with 15% more thrust than the F-86E capably contested altitudes up to 48,000 feet from the MiGs.

As armistice talks droned on through the winter of 1952-53, the United Nations who were never going to commit more troops needed to win a ground war, continued to pulverize communist ground forces into some form of containment. From the communist perspective, they knew that even with unlimited Chinese volunteers on the ground, they were never going to win without air superiority. During early 1953, as the UN continued to attack North Korean bridges and supply routes, enemy air activity remained quiet. When reports did come in, during May it appeared that the MiG-15 pilots had lost their edge. Sabre pilots that month claimed 56 MiG kills for the loss of one F-86, and sensed the Soviets might have gone home. To spur on sluggish armistice negotiations, the Far East Air Force smashed the North Korean irrigation dams to deny them a year's supply of rice. First hit was the Toksan dam. Other dams were bombed, but Hwachon dam proved impregnable to bombs. Eight U.S. Navy Skyraiders from USS *Princeton* wrecked the dam, air launching torpedoes into the structure.

AIR COMBAT MANOUEVRES ~ YO-YOS & BARREL ROLLS

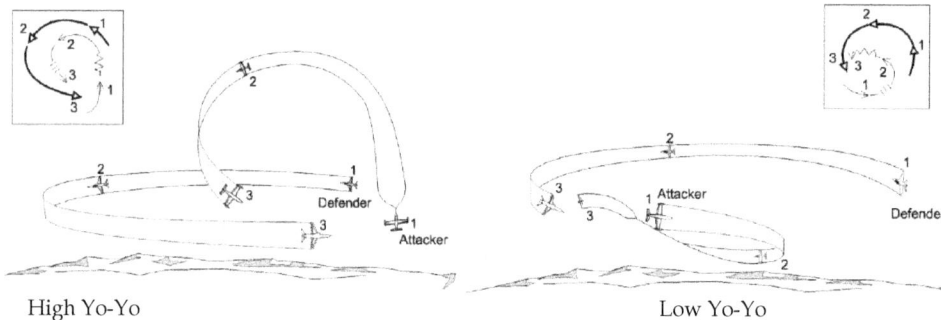

High Yo-Yo   Low Yo-Yo

The High Yo-Yo is a very effective maneuver and very difficult to counter, used to slow the approach of a fast moving attacker while conserving the airspeed energy. The maneuver is performed by reducing the angle at which the aircraft is banking during a turn, and pulling back on the stick, bringing the fighter up into a new plane of travel. The attacker then rolls into a steeper pitch turn, climbing above the defender. The trade-off between airspeed and altitude provides the fighter with a burst of increased maneuverability enabling the attacker to make a smaller turn, correct an overshoot, and to pull in behind the defender, having restored the lost speed while maintaining extreme energy.

The Low Yo-Yo is a most useful maneuver that sacrifices altitude for an instant increase in speed. This maneuver is done by rolling with the nose low into the turn, dropping into a steep slicing turn. The attacker can quickly decrease range and improve the angle of the attack, literally cutting the corner on the opponent's turn. The pilot then pulls back on the stick, climbing back to the defender's height to slow the aircraft and preventing an overshoot, while placing the energy back into altitude. A defender spotting this maneuver may try to tighten the turn in order to force an enemy overshoot. The Low Yo-Yo is often followed by a High Yo-Yo to help prevent an overshoot or several small low Yo-Yos can be used instead of one large maneuver.

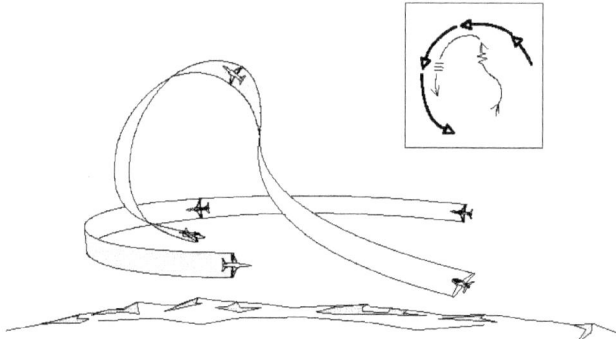

Barrel Roll : Spotting an attacker approaching from behind the defender will usually break turning sharply across the attacker's flightpath to increase AOT (angle off tail). The counter to a break is a barrel roll attack consisting of performing a roll and a loop at the same time. The barrel roll attack rolls away from the defender's break, completing the roll pointed in the direction of the defender's travel.

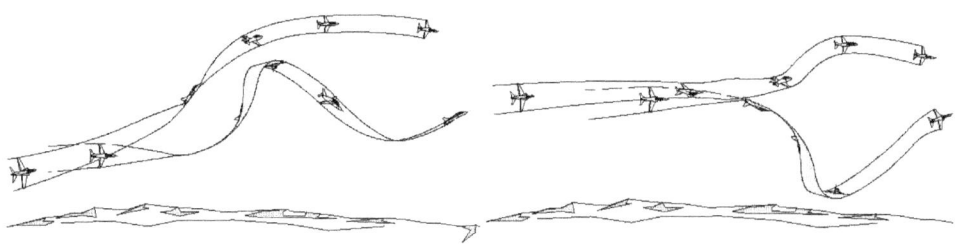

A high-G barrel roll performed over-the-top…   …and underneath.

AIR COMBAT MANOUEVRES ~ SCISSORS & DISPLACENENT ROLLS

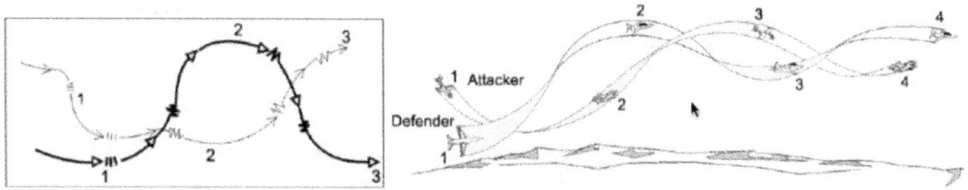

The scissors are a series of turn reversals and flightpath overshoots intended to slow the relative forward motion of the aircraft in an attempt to either force a dangerous overshoot on the part of the defender or to prevent a dangerous overshoot on the attacker's part. The defender's goal is to remain out of phase with the attacker, trying to prevent a guns solution, while the attacker attempts to get in phase with the defender. The advantage usually goes to the more maneuverable aircraft. **Flat scissors** or horizontal scissors usually occur after a low speed overshoot in a horizontal direction. The defender reverses the turn, attempting to force the attacker to fly out in front and spoil their aim. As the attacker reverses to remain behind the defender, the two begin a weaving flight pattern. Rolling scissors or the vertical scissors below, tend to happen after a high speed overshoot from the above flat scissors.

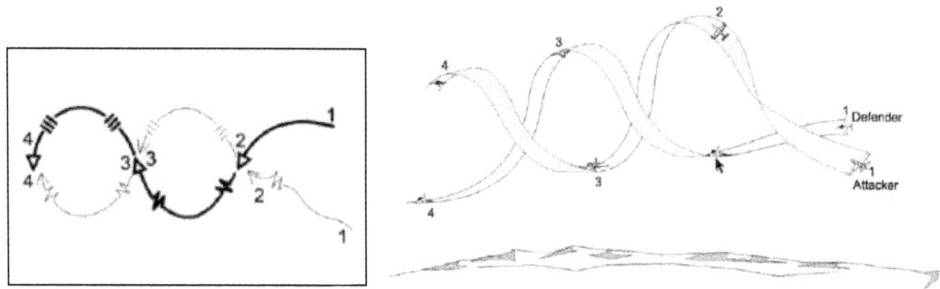

The Defensive Spiral (Below) ~ If a defender fails to outmaneuver an attacker and in time becomes ; *out of airspeed and ideas* the Defensive Spiral is a maneuver often used by defenders when all other options can not be used. The maneuver consists of dropping the nose low during the turn and going into a spiral dive, using gravity to supply the energy needed to continue evasive action. The defensive spiral becomes a rolling scissors performed straight down. The defender's focus is to stay out of phase with the attacker until the ground is precariously close. The advantage usually goes to the aviator and aircraft that can decelerate quicker and the defender can cut power, and extend the speedbrakes to force an overshoot. If this attempt is unsuccessful, the defender will usually pull out of the dive at the last possible second, hoping to cause the attacker to crash into the ground

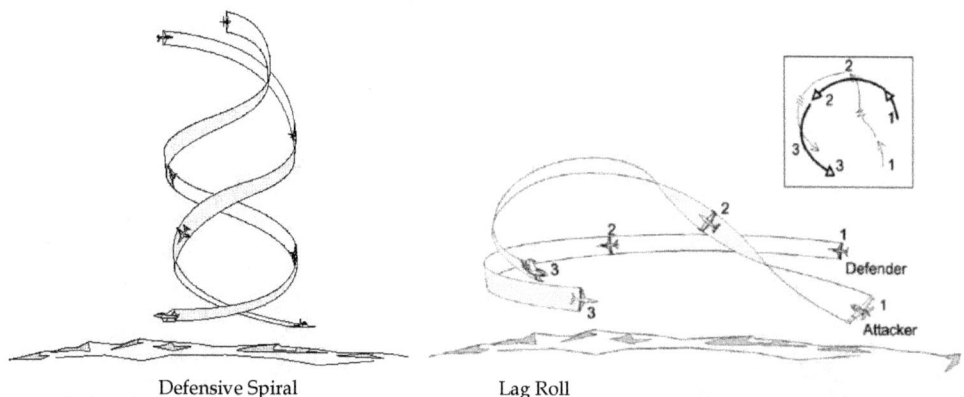

Defensive Spiral   Lag Roll

The Lag Displacement Roll ~ or *Lag Roll* is used to reduce the angle off tail by bringing the attacker from a position of lead pursuit to lag pursuit. The maneuver is performed by rolling up and away from the turn ~ then, when the aircraft's lift vector is aligned with the defender, pulling back on the stick, bringing the fighter back into the turn. This maneuver also helps to prevent overshoot caused by the high AOT of lead pursuit, and can be used to increase the distance between aircraft

Source: United States Navy ~ Naval Air Training Command ~ Flight Training Instruction Manual

Top : Sabre F-86A armaments ~ Below : Captain Joseph McConnell 39th FIS, 51st Fighter Interceptor Wing.

Top : Sabre F-86E in flight.   Below : Captain Joseph McConnell talking with his crew chief, 9th March 1953.

MIG ALLEY 1950–1953

Captain Joseph McConnell ~ Captain Joseph McConnell USAF vanquished eight MiGs before his Sabre was hit by North Korean ground fire. Heading south, over the Yellow Sea with his cockpit filling with smoke, McConnell trimmed his Sabre at 11,000 feet and ejected. He was only in the water minutes, before his rescuers arrived. Pulled from the water by a rescue helicopter flown by Don Crabb and Bob Sullivan, McConnell flying again the following day joked, *I barely got wet!* Captain McConnell wore a monocle to sharpen his eyesight, and his aim was magnificent. On the morning of 18th May 1952 he shot down two MiG-15s. That afternoon, while heading north McConnell's four-bird Sabre sweep was surprised by four MiGs attempting their repetitive ruse. The first pair of MiG-15s overshot the Sabres as targets to tempt. McConnell's flight broke, expecting to find more MiGs behind. McConnell found and targeted one through his monocle giving it a long burst with his machine guns. Shooting it to shreds, it burst into flames becoming Captain Joseph McConnell's third kill for the day, and his 16th aerial victory, making him the highest scoring ace of the Korean conflict.

The Fluid Four ~ Sabre pilots, finding their jets less agile than propeller-powered aircraft, developed a formation far more reactive than the finger-four formation of two pairs, based on interchangeable leader/wingman roles, known as the *Fluid-Four* or *Fighting Wing*. Involving two fighters flying 300 yards apart with a second pair 500 yards behind, 2,500 yards to the side, and 1,000 yards above, their spacing was derived from the time one pair could come to the support of the other.

On 27th May, the Korean Armistice was eventually signed, and the first conflict of the Cold War ended indecisively at 2200 hours on 27th July, with no clear winner or loser, with the 38th parallel continuing to divide two very different versions of Korea. The victory of the negotiated truce was only possible because UN airmen had flown aggressively. Keen UN military and naval pilots had ventured as far north as the line of the Yalu River, hundreds of miles from their southern airfields and aircraft carriers to contain the larger enemy fighter force. United Nations pilots flying F-86 Sabres during the Korean War claimed 817 MiGs for the loss of 58 Sabres ~ a kill ratio of 14 to 1. Considering that the U.S. Sabre and the Soviet MiG-15 were such closely matched aircraft it was evidence that in the jet age, as before, the quality of the airman and his tactics, as much as the quality of the aircraft that they flew, would continue to determine the level of victory or defeat.

After the truce, NKAF pilot No Kum-Sok defected and landed at Kimpo Airfield to receive $100,000 cash for delivering his MiG-15, and he confirmed that the Communists had lost over 800 MiGs (including two wiped out Russian units), and that the Chinese could not train pilots fast enough to replace those shot down. Without their airpower, the outnumbered but not outsmarted UN ground force would not have been able to survive the onslaught of the Chinese People's Volunteer Army.

* * *

Captain Frederick C. Boots Bleese ~ Captain Bleese flew two combat tours over Korea, including 121 missions in F-86 Sabres, in which he scored 10 aerial victories (nine MiG-15s and one YAK-9). Flying an F-86 Sabre in 1955, he won all six individual trophies on offer at the USAF Gunnery Championship (an achievement that still stands today). During 1955, Captain Frederick C. Bleese wrote his highly respected fighter tactics pamphlet, *No Guts, No Glory* (applicable more to pre-missile engagements). Among his thoughts, Captain F.C. Bleese's *Basic Principles of Offense* advised;

Never continue turning with another aircraft after you are unable to track him with your sight. Pull up immediately and keep your nose behind your tail. If he pulls up, you'll always end up on top because of your attacking airspeed. If, by using speed brakes, you can drift into the radius of turn of the aircraft you are attacking, do it in preference to the YO YO manœuvre. It takes less time to get your kill and you don't run the risk of being out-manœuvred by the aircraft you are attacking. Cruise at a high Mach. Look around; you can't shoot anything until you see it. Keep the aircraft you are attacking in sight. One glance away is enough to make you kick yourself for the next ten years. Attack from low and behind whenever possible. That's a fighter's poorest visibility area. Know the performance data on all aircraft you are apt to be fighting. Assume every pilot you meet is the world's best (you can swallow your pride that long) and manœuvre your aircraft accordingly until he shows you he is not. Don't shoot unless you're positive it's an enemy aircraft. When it's time to fire, you'll know if it's an enemy aircraft or not. If you can't tell you are out of range. Play on the team ~ no individualists. The quickest way to become a leader, is to be the best wingman in the Squadron. When in doubt ~ attack!

Gun camera footage of a MiG-15 being shot at by a Sabre circa 1953

Defector North Korean pilot No Kum Sok's MiG-15 at Kimpo Air Base

If you end up in front… your only salvation may lay in how quickly you get your nose down and get into this Diving Spiral. Maybe you can get some airspeed before you get hit too badly. Become familiar with this manœuvre for it is useful regardless of the type of aircraft being flown. Your best protector with an aircraft at six o'clock is G force. You need speed to get G's ~ thus, the diving spiral. Keep the G's on and keep the nose down so you can keep enough airspeed to make it rough for the shooter. Watch the aircraft behind you. If you can see the belly of his aircraft, he's beginning to pull lead and may be about to fire. Tighten up your turn immediately.

From: *No Guts, No Glory)*

From: *No Guts, No Glory)*

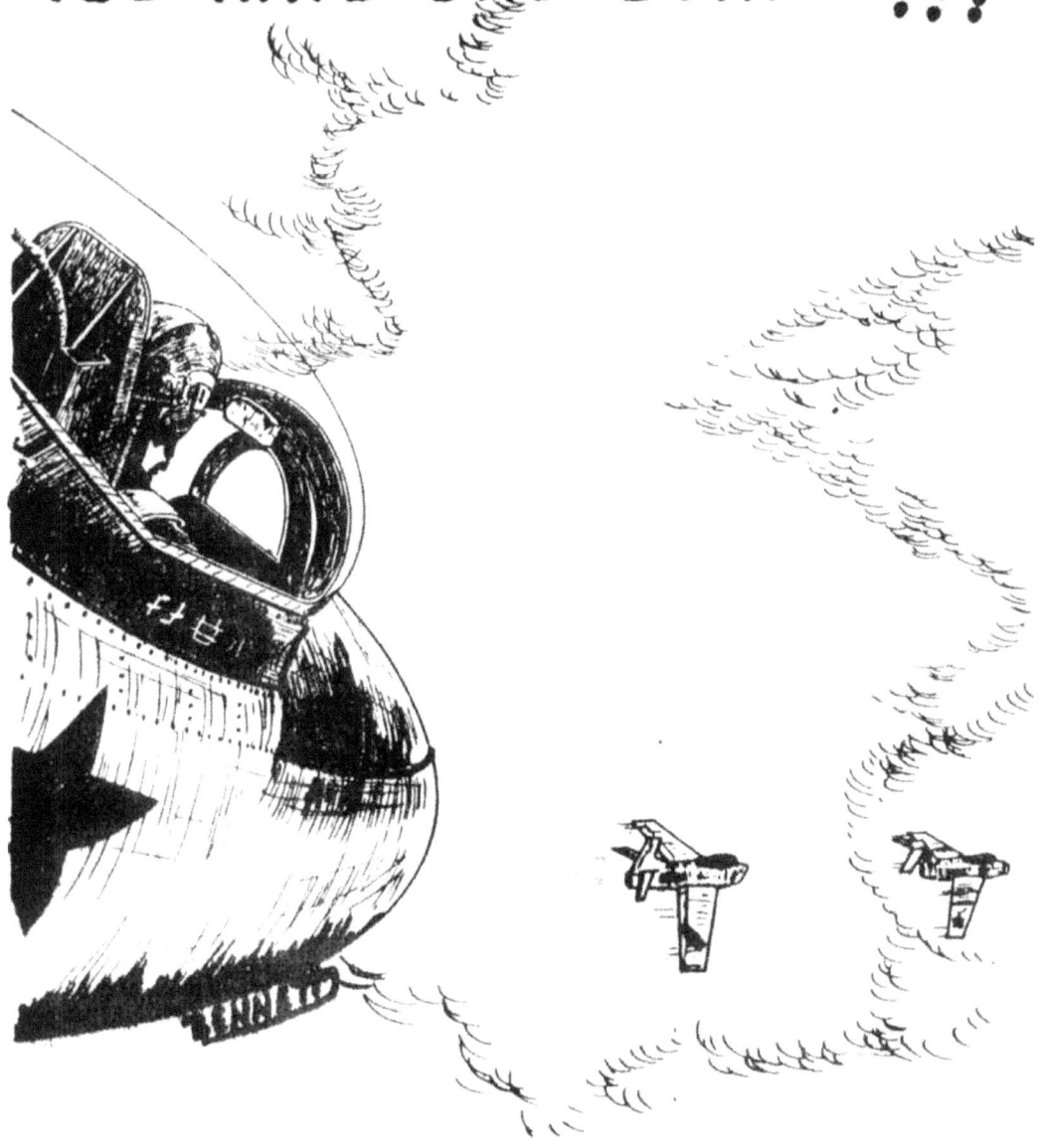

BASIC PRINCIPLES OF DEFENCE

1: *If you slow down, have an element high and fast for support.*
2: *Except at extreme ranges always turn into the attack.*
3: *If there are enemy aircraft anywhere in the area, get rid of external tanks and get your Mach up.*
   It's too late after you spot him.
4: *Keep your attacker at a high angle off.*
5: *Keep your airspeed up when patrolling.*
6: *Don't ever reverse a turn unless you have your attacker sliding to the outside of the radius of your turn.*
7: *If you have a hung external tank, leave the combat area.*
8: *If you lose your wingman, both of you should leave the combat area.*
9: *Know the low speed characteristics of your aircraft. If you are fighting aggressive pilots,*
   you will need all the know-how you can lay your hands on.
10: *Have a last ditch manœuvre, and practice it.*
11: *Keep a close check on your fuel.*
12: *Best defence is a good offense ~ is good most of the time ~ but know your defensive tactics.*
13: *Don't play Russian Roulette! When you're told to break ~ DO IT!*
14: *Avoid staring at contrails or the only aircraft in sight. There are a dozen around for every one you can see.*
15: *Watch the sun ~ a well- planned attack will come out of the sun when possible.*
16: *The object of any mutual support manœuvre is to sandwich the attacker in between the defending aircraft.*
17: *In any dogfight, the objective for the defender should be lateral separation. When this is achieved, a reverse*
    and a series of scissors will. If properly executed, put your attacker out in front. The rest is up to you.
18: *Place yourself in your attacker's shoes. How would you like to find an enemy flight positioned?*
    Be smart and avoid this formation for your flight.
19: *Don't panic – panic is your most formidable enemy!*
   (From: C^apt. F.C. Bleese's *No guts, No Glory*)

* * *

The Advent of the Sidewinder. Dr. William Burdette McLean of the U.S. Navy Ordnance Department with an idea to endow a standard dumb rocket with the function of a predator's eyes and brain, assembled a team of researchers to work on air-to-air rockets. McLean designed a missile with a photovoltaic eye that could sense (through a transparent seeker head) infrared radiation from the exhaust of a jet that reflecting off a gyroscopic mirror moved to track the heat source, trimmed the missile with proportional navigation to anticipate where the target jet was going, and remained on a constant bearing toward the target. Once fired, it reached Mach 2.5 in two seconds and sharing several qualities with another heat seeking predator of the Mojave Desert, the Sidewinder rattlesnake, William McLean's missile assumed its name to become the Air Intercept Missile or AIM Sidewinder, and his research team's rocket-powered Sidewinder missile changed air combat forever.

* * *

Taiwan Strait Crisis ~ In August of 1958, Red China tried to force the Nationalists off the disputed islands of Quemoy and Matsu by naval blockade and shelling. Over the Formosa Strait, air fighting began with Chinese Nationalist airmen flying surplus Sabre F-86Fs against Communist Chinese pilots in MiG-15 and MiG-17s. In response, the U.S. Navy fitted 100 Taiwanese Sabres with twin AIM-9B Sidewinder missiles. As in Korea, communist Chinese airmen flying MiG-17s retained their higher ceiling advantage over the Sabres, preferring to fly high above Nationalist Sabres to attack from a favourable position. During an engagement on 24th September, the Sidewinder changed that. Launched from up to three miles away, 14 Sidewinders downed 11 MiGs without loss to themselves. During a month of hostilities over Quemoy and Matsu, the Sabres downed 31 MiGs plus eight probable's for the loss of only two F-84G Thunderjets. After a faulty Sidewinder pierced a MiG-17 but failed to detonate, it was removed and delivered to the Soviets for examination and re-engineering. Working from stolen plans and their souvineered Sidewinder, the Soviets copied it so closely that their Vympel K-13 (Vympel being the apt Russian word for Trophy) shared the same part numbers.

Image series showing the launch of a AIM-101 Sparrow III missile from a McDonnell Demon in 1958

THE INDO-PAKISTANI WARS

Conflict in Kashmir ~ As Britain was dismantling its Empire to create the new countries of India and Pakistan, the Maharaja Hari Singh, ruler of the princely states of Jammu and Kashmir, whose lands were wedged between the two countries wished to remain independent. As he was Hindu, and his subjects were predominantly Muslim, the Prince signed a standstill agreement of convenience with Pakistan so that trade, travel, and communication could continue as normal. India declined to sign such an agreement. After persistent violence against Muslims in Kashmir, Pashtun tribesmen from the northwest frontier of Pakistan invaded Kashmir. The armed tribesmen created such strife and religious tension that a mutiny of Muslim regiments rose up in Gilgit, impelling Maharaja Hari Singh to request armed assistance from India. India refused to help unless he acceded his lands to India. Governor-General, Lord Louis Mountbatten proposed to resolve the situation if Kashmir temporarily acceded to India, with an agreement to hold a referendum plebiscite election. Clouds of confusion and controversial conflict followed Hari Singh's signing over of key powers to the Indian Government in return for military aid ~ and the promised referendum. Although Maharaja Hari Singh wanted Kashmir to remain independent, Pakistan argued that the Kashmiri should be allowed to decide the issue in a referendum on their future, while India pointed to the signed Instrument of Accession. These events detonated the 1965 Indo-Pakistani War.

As the outnumbered Pakistani Air Force of 90 machine-gun armed Sabre F-86F fighters prepared to tackle the Indian Air Force of 118 Hawker Hunter and 80 Folland Gnat cannon-armed fighters, the Pakistanis held the advantage of up to date technology. Some 25 of their Sabres were armed with AIM-9B Sidewinders having a two mile reach. In terms of performance, this redressed the balance, forcing all Indian Air Force pilots to assume that all PAF Sabres could be missile-armed opponents. The IAF Hawker Hunter was both quicker accelerating and faster than the Sabre, but accelerating away from a Sabre preparing to launch a Sidewinder was not the best option. Hunter pilots often decided to turn and fight, but in a fast turn, the high-drag Hunter lost speed more than the Sabre. Neither the Hunter or the Gnat carried air-to-air missiles, but IAF Gnat pilots could catch a PAF Sabre by snapping their nose into a hard climbing turn and firing a burst from their twin 30mm cannons : thump, thump, thump into an enemy PAF Sabre to destroy it ~ and be fêted in India as a *Sabre Slayer*.

At 0574 (PST) in the morning of 7th September 1965, two pairs of Sabres patrolling around their home base of Sargodha Airfield missed an approaching low-level incursion by six IAF Hunters. Pakistani Squadron Leader M.M. Alam and his wingman Flight Lieutenant Akhtar eventually spotted them short of Sargodha Airfield coming in fast. Alam decided on a missile shot against the last Hawker. Firing his Sidewinder in a dive at low altitude it wasted itself in the ground. The best way of firing an early Sidewinder at low altitude was to get below the target and fire with the cooler sky as a background to ease the missiles heat discrimination problem. The IAF Hunters flew over the treetops and as one nosed up to clear a high-tension cable, Alam heard a loud growl in his headset signaling a positive heat source. He fired his second Sidewinder that shot off to detonate near the Hunter in a proximity explosion rupturing its fuel lines. The Hunter flashed in the sky, and fell away in flames.

Alam and Akhtar lost contact. Soaring towards the Indian border, Akhtar's wingman called out ; *Contact. Hunters one o'clock.* The Hunters flying at 100–200 feet flew in a loose line abreast. Alam closed the combat range nearer and nearer to the IAF Hunters who had forgotten to jettison their drop tanks. Spotting Alam and Akhtar, they climbed steep to the left in line astern formation. Alam turning his Sabre inside their arc pulling 5G watched the Hunters lose speed at over ten mph per second, enabling him to close the gap while they turned in the same direction, lining up like decorative ducks on a wall.

The Hunter loses speed faster than the Sabre in a turn because of its higher drag. So in the turn I steadily closed up on the Hunters. As it was, they just slid back into my sight, one by one. I developed a technique of firing very short bursts ~ around half a second or less. The first burst was almost a sighter, but with a fairly large bullet pattern from six machine guns. It almost invariably punctured their fuel tanks. As we went round in the turn, I could just see, in the light of the rising sun, the plumes of fuel gushing from their tanks after my hits. Another half-second burst was then sufficient to set fire to the fuel, and as the Hunter became a ball of flame I would shift my aim forward to the next aircraft. The Sabre carried about 1,800 rounds for its six 0.5 inch guns, which can fire for about 15 seconds. Every fourth round is an armour-piercing bullet and the rest are HEI or high explosive incendiary. My fifth victim started spewing smoke and then rolled on its back at about 1,000 feet. The next time I fired was at very close range ~ at 600 feet or so, his aircraft blew in front of me.   M.M. Alam

THE INDO-PAKISTANI WARS

In three weeks, Pakistani Sabre pilots had fired 33 Sidewinders to claim nine Indian aircraft destroyed for their loss of 19 to capable IAF Gnat *Sabre Slayers*. In January 1966, both sides grew weary and when the Soviet Union hosted peace talks in Tashkent, Uzbekistan, both India and Pakistan agreed to recognize the prewar border in Kashmir. In reality though, tension continued, with neither India nor Pakistan satisfied with their borders. Both sides updated their inventories of war equipment. India purchased eight squadrons of Soviet MiG-21s and six squadrons of Sukhoi Su-7s, while Pakistan acquired 28 French-built Dassault Mirage-IIIs. Indo-Pakistani relations worsened again after Pakistan failed to accommodate the Bengali ethnic call for autonomy in East Pakistan during 1970, which led to secessionist demands. On 25th March 1971, a Pakistani Army convoy including tanks arrived at Dhaka University and encircled the premises to begin a selective massacre of the Bengali educated classes of East Pakistan. Many Bengali professors, artists, doctors, journalists, writers, cultural identities, and intellectuals were hunted down, and thousands were assassinated in a campaign to suppress the Bengali secession movement. Nearly ten million Bengali refugees fled to sanctuary in the adjacent Indian state of West Bengal. The U.S. prohibited the sales of Sabre aircraft and parts to Pakistan, but Pakistan maintained their inventory purchasing Iranian Sabres and Chinese Shenyang F-6s (MiG-19 copies). In August, the Soviet Union sympathizing with the Bengali independence movement signed an Indo-Soviet Friendship Treaty to also weaken the positions of its rivals, the U.S.A. and Red China.

The Indo-Pakistani War of 1971 ~ Serious fighting erupted again at 1745 on 3rd December 1971, when the Pakistani Air Force launched pre-emptive air strikes against 11 Indian airfields in N.W. India to push the IAF to operate from the disadvantage of long range airfields far from the frontline. The attacks, mounted with small numbers of Sabres failed to achieve their objective, and unwisely timed at dusk, gave the Indians all night to repair the runway at Amritsar Airfield, where the only serious damage occurred. India invaded East Pakistan to support the Bengali people and similar to the 1965 war, little air action occurred other than interception of low-level incursions. The IAF wiped out the PAF in the east in two days giving India control of the skies, while the Indian Navy blockaded East Pakistan. The long-range PAF Mirage IIIs were not flown ~ they could have reached New Delhi.

On 6th December 1971, when the demise of Pakistan in the east appeared certain, U.S. President Richard Nixon ordered a naval task force with the carrier USS *Enterprise* (carrying Phantom aircraft) and the amphibious assault ship USS *Tripoli* into the Bay of Bengal to evacuate U.S. nationals from Dhaka and to distract and disrupt Indian naval and air operations. The IAF flew 200 sorties on the 6th including an early morning MiG-21 mission to strafe Kurmitola Airport at Dhaka. A ceasefire was called at 1000 to allow the UN to evacuate foreign nationals but perhaps in error, a Canadian Air Force Hercules aircraft full of civilians was attacked, damaged, and then flew on to reach Bangkok. After the Jessore Airfield was attacked by IAF Su-7s and a bomb dump destroyed, the city was taken by Indian ground forces the following day. As East Pakistan convulsed, there came more atrocities. More than 300 Bengali physicians, professors, lawyers, landowners, writers, and teachers were found murdered in a field outside Dhaka. The victim's hands were all tied behind their backs, before they were bayoneted, garroted or shot. Then on 16th December, less than two weeks after it had begun, Pakistan striped of half its population, surrendered. India had accomplished its goals of a free Bangladesh, and the prospect of an early return of ten million Bengali refugees. When they came home they shouted proudly ~ *Jai Bangla! Victory to Bengal!* East Pakistan promptly seceded as independent Bangladesh.

Ever since, separatist militancy and cross-border firing between the forces of India and Pakistan through the mountainous region of Kashmir has seen a growing death toll of tens of thousands. In extreme conditions on the northern Saichen Glacier, the two rival forces have faced off since 1984 in such bitter cold that it still claims more lives than their skirmishing. The Kashmir dispute remains one of the most dangerous, as India and Pakistan are both nuclear powers. The bone of contention is now so suffused with ancient religious, cultural, and racial issues that there are times when even the Indians and Pakistanis do not seem to understand it all. India has recently suggested that, although the Kashmiri border cannot be redrawn, it can for the peace of a populace traumatized by timeless fighting and fear, and for the convenience of all concerned, perhaps be made irrelevant.

CHAPTER X

THE VIETNAM WAR

Vietnam becomes a French colony ~ In Vietnam during the late 18th century when its vast rice fields were controlled by squabbling feudal lords, frustrated peasants led by the Tan Son brothers revolted to unify Vietnam under their rule. Later, the Tan Son Army was defeated by the southern forces of the aristocratic Nguyen clan led by Nguyen Anh, who in 1802 adopted the name Emperor Gia Long in 1802, Gia derived being from Gia Dinh (Saigon) and Long from Thang Long (Hanoi). Aided by the interventionist French missionary, Pierre Pigneau de Behaine, who travelled to French Pondicherry in India to obtain the services of two ships, and a regiment of French officered Indian troops that sailed back to support Nguyen Anh's seizure of the Vietnamese throne. The French in return, had expected the new Emperor to favour them with trading and religious concessions. En contraire ~ Nguyen dynasty dealings with France through their traders, missionaries, diplomats and naval personnel, became a serious issue for the court and scholar officials. And more intriguing, some French officers chose to remain in colonial Vietnam and assumed the role of prominent mandarins. After persecuted missionaries and Vietnamese converts were tortured and executed in increasing numbers, religious factions in France pressured the government in Paris to consider their Imperial commercial and military interests; to persuade Emperor Napoleon III in 1858 to send a naval expedition to subjugate the Vietnamese into becoming a French protectorate. The first French assault commanded by Admiral Rigault de Genouilly attacked Da Nang harbour, but after tropical diseases decimated his troops, he changed his line of attack, sailing south to capture Gia Dinh. French troops acquired the six provinces of the Mekong Delta and the Vietnamese court in Hue ceded the Mekong as Cochin China. In 1893, France having acquired Vietnam, Cambodia, and Laos thus formed the colony of French Indochina.

The expulsion of the French ~ During 1930, Ho Chi Minh founded the Indochinese Communist Party to strive for reform after seeing how peasants in the countryside struggled under heavy taxes and high rents, while those on the rubber plantations labored hard for low wages. After Germany overran France in 1940, Japan invaded Indochina retaining the French colonial administration as a puppet régime. In 1941, Ho Chi Minh returned to Vietnam from exile to organize the Viet Minh Front to prepare an uprising by emphasizing reforms and independence; never harsh communist objectives. In March 1945, when a U.S. pilot was shot down over Vietnam and was rescued by Viet Minh troops; he was personally escorted back by Ho to the American 14th Air Force base at Kunming. Ho was offered money, but he only asked for the honor of meeting Major General Claire Chennault. During their meeting, Chennault thanked Ho, who promised to help any downed U.S. pilots and aircrew, requesting only an autographed photo of the general. He planned to use it later. Ho would show the photo to other Vietnamese nationalists as implied proof of good relations and support from the U.S. On 27th April, Captain Patti approached Ho to insert his OSS Deer Team agents under the command of Major Allison Thomas to work with the Viet Minh. Ho greeted Patti, *Welcome, my good friend*. The Americans had no idea that Ho was an ardent communist who secretly spoke fluent Russian and had been trained in the Soviet Union. Having allied itself with Ho Chi Minh and his Viet Minh troops to assist downed American pilots, and to harass Japanese troops in the ground, Major Thomas who was meant to indicate targets to the USAAF, spent most of his time with Ho and General Vo Nguyen Giap. The naïve Thomas was properly duped thinking that Ho and Giap were simple agrarian reformers. One agent noted that Ho did not know how to use a spade and Giap did not know how to milk a cow. Before Major Allison Thomas departed from Vietnam he asked Ho, *Are you a communist?* Ho replied, *Yes, but we can still be friends, can't we?*

After Japan surrendered, the Viet Minh forces rose up, captured Hanoi and Ho quoting from the American Declaration of Independence given to him by OSS agents, declared Vietnam independent. Ho Chi Minh agreed to allow French troops to return to Hanoi temporarily, in exchange for French recognition of the Republic of Vietnam, however the returning French bombarded Haiphong harbour, reoccupied Hanoi, and forced Ho, and his revolutionaries back into the jungle. After years of fighting under the Geneva Accord of 1954, Vietnam was divided along the 17th parallel with the communist Viet Minh, and the French in the south agreeing to hold free elections within two years. However, in the southern capitol of Saigon anti-communist prime minister, Ngo Diem refused to hold elections, and predicted another more deadly war. Ho Chi Minh returned to Moscow to seek Soviet aid. Radical communist land reforms in North Vietnam during 1957 forced thousands of landowners before Ho's treacherous *People's Tribunals* to be executed or sent to camps. Peasant unrest following the land reforms was suppressed with a Viet Minh campaign of terror, and the deaths of 7,000.

VIETNAM

After South Vietnam requested urgent assistance from the United States, President John F. Kennedy authorized the sending of maintenance experts and aircrews to keep U.S. loaned helicopters flying plus 400 Green Beret advisors to refine South Vietnam's counter-insurgency tactics. After President J.F. Kennedy was assassinated in 1963, Lyndon B. Johnson assumed the presidency of America, and in his subsequent 1964 presidential campaign declared : *We are not about to send American boys nine or ten thousand miles away from home to do what Asian boys ought to be doing for themselves.* Weeks later, the destroyer USS *Maddox* was cruising through the night in the Gulf of Tonkin, keeping an eye out for communist gun runners. Out of the darkness, lookouts on the USS *Maddox* sighted three Soviet P-4 torpedo boats coming at them at 45 knots. From 5,000 yards they launched torpedoes that all missed, passing to starboard. USS *Maddox* opened fire with its twin-mounted 5-in. guns damaging all three enemy vessels and stopping one dead in the water, helpless and aflame. President Johnson ordered the carriers USS *Ticonderoga* and USS *Constellation* to retaliate ; directing their aircraft to carry out air strikes against the North Vietnamese port of Vinh, and its nearby oil depot.

Operation Rolling Thunder ~ Beginning in 1965, more than 100 U.S. fighter-bombers began attacking targets in North Vietnam in an onslaught planned for eight weeks. Because the Viet Cong were such an elusive enemy, air power would focus its attention on the main enemy supply line to South Vietnam via a winding route over mountains and through jungle along the Laotian border that came to be known as the Ho Chi Minh trail to reduce their ability to fight. However, after each U.S. air attack, female Viet Cong construction units quickly repaired any bomb damage. USAF chief of staff, General Curtis Le May became so frustrated with Robert McNamara's policy of overseeing limited air strikes and having to order his airmen to fly multi-million dollar aircraft to blast cheap Chinese trucks transporting rice thru the jungle while being prevented from sinking enemy shipping in Haiphong harbour offloading 100 trucks at a time with weapons/supplies that he retired in disgust, delivering the apt comment ~ *Instead of swatting flies, we should be going after the manure pile.*

Soviet SAMS ~ In February 1965, Soviet Premier Andrei Kosygin visiting Hanoi was persuaded by the North Vietnamese to provide military aid to counter growing American involvement. Within weeks of his return to Moscow, Soviet surface-to-air SAM missiles began arriving in Hanoi. When a Lockheed U-2 identified the first SAM site, Lyndon Johnson asked his Secretary of State Dean Rusk: *What about the SAM sites? Question is, whether we let the clock tick or whether we take them out now.* Finally on 26th July 1965, President Johnson decided. *Take them out!* The next day, Operation Spring High opened with ten electronic warfare aircraft jamming the North's radar as 46 F-105s targeted the SAM sites with napalm pods and cluster bombs. The North Vietnamese responded by constructing more SAM sites than they had systems for, constantly relocating them to avoid becoming confirmed targets. Soviet SA-2 SAM launchers typically operated six missiles on launchers with a control van scanning the sky with Spoon Rest acquisition radar and Fan Song guidance radar to hold up to four aircraft. If a U.S. pilot ever heard a crackling sound in his headphones, he had just been acquired by enemy radar. To evade SAM acquisition, the radio alert ~ *take it down* ~ advised flying at treetop level. But, if a pilot was auspicious enough to see puffs of smoke and dust that indicated a SAM launch at his warplane, the evasive tactic was to turn hard and fly straight at the bright yellow of the streaking SAM. Then in the last possible instant, pull thru a high-G turn. SA-2 missiles could sustain an 8G turn at Mach 2, but flying a tighter arc gave a fighter time to escape. SA-2 missiles with ranges of 18 miles generally burnt out chasing well-handled fighters ~ and being over long, snapped apart in sharp turns.

Disco and Red Crown ~ The early warning radar detection equipment carried onboard the U.S. Lockheed Super Constellation AWAC aircraft known by call sign *Disco* that took off from Saigon to fly over Laos was supported by the radar warning and control ship *Red Crown* in the Gulf of Tonkin. Together they scanned the skies. The only all MiG-21 killing ace, Captain *Steve* Ritchie said that on every occasion that he shot down a MiG-21, he had received advice from either, *Disco* or *Red Crown*. From the opposing perspective, North Vietnamese early warning radar was assisted by a Chinese radar installation on Hainan Island, and by Russian spy ships also in the Gulf of Tonkin, which often shadowed U.S. Navy warships. Once a U.S. air strike was plotted coming north, NVAF MiGs rose to form a Lufbery Circle (transposed five decades into the future) to give them all-round cover and enable the NVAF airmen to wait… until sighting an American air strike, then peel off and attack. American airmen soon learnt that the NVAF pilots ~ who had received their basic training in the Soviet Union and advanced training over China ~ preferred sneak attacks to swirling engagements.

VIETNAM

The airmen of the NVAF were also equipped with MiG-17 trans-sonic, cannon armed fighters that while slower in speed and their rate of climb, still out-turned any U.S. fighter. A preferred tactic of MiG-17 pilots was to prowl low, looking for the heavy smoke trails of U.S. Phantoms low on fuel returning from missions. Phantom pilots, unable to out-turn a MiG-17, fired up the after burners to outrun their problem 700 to 1,500 mph. More potent was the supersonic MiG-19 armed with three 30mm cannon and a selection of missiles. MiG-21s flew in pairs to seek and strike approaching American air formations from behind, accelerating beyond supersonic Mach 2 supersonic speeds to fire their two Vympel K-13/Atoll missiles, then streak off. However, every time a MiG-21 was caught low, the structural weakness of the MiG-21 restricted its speed at low altitudes to give the stronger built American fighters a minimum of a 100 mph speed advantage during low-level engagements.

The two Soviet air-to-air Atoll missiles carried by MiG-21 were not as fine the Sidewinder from which they were derived. They were ineffective fired from a turning MiG-21 pulling more than 2½ Gs. Atoll acceleration times were weaker, their rocket fuels burned faster, and they were easier distracted by heat sources other than their intended targets. American fighters at maximum thrust could often outrun the Atoll in its final flight stage. Neither the Atoll, nor the Sidewinder could follow an evasive fighter through a tight turn, but for a F-105 Thunderchief fighter-bomber to achieve maximum evasive speed, it had to jettison its load. For this reason, the MiG-21 threat to Rolling Thunder became so potent that whenever a flight of MiGs were detected coming in, the Thunderchiefs had to jettison their ordinance, and this was the precise objective of the MiG-21 airmen : to prompt U.S. Thunderchief formations into wasting their loads before they reached their assigned target destinations.

Operation Bolo ~ In late 1966, veteran U.S. ace Colonel Robin Olds (with 12 *Luftwaffe* kills) then commanding the 8th Tactical Fighter Wing based in Thailand's Royal Thai Air Force Base Urdorn approved Captain John Stone's concept of luring their elusive NVAF MiG opponents into a trap. Based on the premise that intelligence analysts in Hanoi had become so adept at identifying the more vulnerable Thunderchiefs from the Phantoms by their radio frequencies and strike wing call signs, Captain Stone and a group of junior officers at Urdorn, with contributions from Colonel Olds planned a new version of the *Bait and Switch* ruse, code naming it Operation Bolo. The Phantoms would assume the subterfuge signatures of a typical Thud mission, using the same routes, altitudes, and call signs as Thunderchiefs to catch any MiG-21s vectored towards Olds' formation expecting to find F-105 aircraft. Colonel Robin Olds, call sign *Wolf 01* and his airmen knew that their Phantoms greater thrust margin and lower wing loading (ratio of gross weight to total wing area) would perform far better than the Thuds with the sustained turning rate of their Phantoms more on par with enemy MiG-21s. The fighter sweep would involve 56 Phantoms flying seven flights in two waves of four aircraft, with Robin Olds' 8th TFW in the lead. The flight names of the 8th were auto derived. Robin Olds' first flight was call-signed *Olds*. Captain Daniel James led second flight *Ford*, and Captain John Stone who had initiated the Operation Bolo fighter sweep led *Rambler* flight. Other Phantoms were assigned diversionary missions to infuse confusion, and to range over North Vietnam to block the MiGs escape.

On the morning of 2nd January 1967, Colonel Robin Olds' Phantoms took off from Urdorn base in Thailand to skirt the NVAF airfields of Phuc Yen and Gia Lum and fly on towards Hanoi. A radio warning came in from Super Constellation *Disco* that MiGs were taking off from both and heading to intercept them. As expected, the silver MiGs were coming. The Phantom airmen knew their enemy was under the clouds, seeing them on their radar. Olds led them through such a masterful turning maneuver that when the first MiG-21s nosed through the cloud, *Olds* flight were on their six and locking on. Their bait and switch ruse had worked. After a swirling aerial engagement, the Phantoms flew home after suckering seven NVAF MiG-21s to their fate ~ one of which was nailed by *Wolf 01*.

The first MiG zoomed away and I engaged the afterburner to get in an attack position against this new enemy. I reared up my aircraft in a 45° angle, inside his turn. He was turning to the left, so I pulled the stick and barrel-rolled to the right. Thanks to this maneuver, I found myself above him half upside down. I held it until the MiG finished his turn, calculating the time so that, if I could keep on turning behind him, I would get on his tail with a deflection angle of 20° at a distance of 1,500 yards. That was exactly what happened. He never saw me. Behind and lower than him, I could clearly see his silhouette against the sun when I launched two Sidewinders. One of them impacted and tore apart his right wing. Suddenly the MiG-21 erupted in a brilliant flash of orange flame. It fell twisting, corkscrewing, tumbling lazily toward the top of the clouds.   Colonel Robin Olds

F-4 PHANTOM ACES ~ COLONEL ROBIN OLDS

Colonel Robin Olds ~ 17 aerial victories ~ and his men celebrate the legenary fighter pilot's 100th combat mission over Vietnam

Captain Steve Ritchie ~ 5 kills. Admired by Colonel Robin Olds who described him as *brilliant*, Ritchie was a gifted and dedicated flyer maintained his skills by flying every two or three days to achieve a thorough understanding of the weapons systems he used to place himself in the forefront of USAF fighter pilots, where he was well known for his *intelligent aggression*.

VPAF

VPAF pilots with their MiG-17 aircraft

North Vietnamese VPAF ace Nguyen Van Coc ~ 9 kills
Shot down during Operation Bolo

Nguyen van Coc meeting with Ho Chi Mihn

North Vietnamese VPAF MiG-21PF on a landing roll, with its drag chute deployed

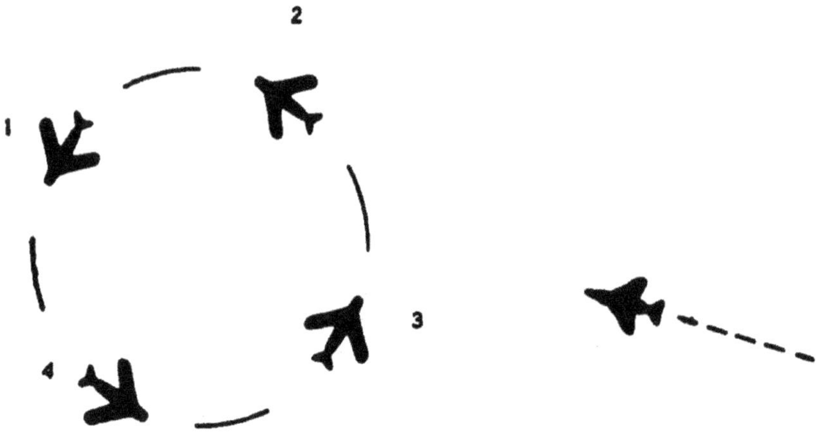

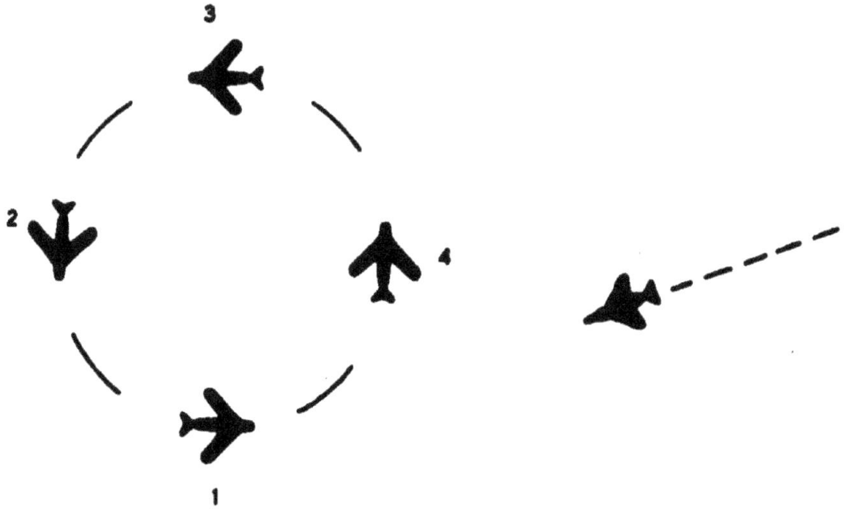

WAGON WHEEL FORMATION ~ The Wagon Wheel was a very significant tactic devised by the North Vietnamese Air Force in 1967 for MiG-17 defense. This tactic was in fact a modification of the Lufbery Circle. The Wagon Wheel was composed of a group of MiG-17s operating from orbit. Whenever they came under attack, they would enter an orbiting formation to provide 6 o'clock coverage for each other enhancing mutual defense. Through the use of the Wagon Wheel, the MiG-17s could effectively utilize their superior turning capability to force an overshoot by USAF aircraft while still providing 6 o'clock coverage for the preceding MiG in the orbit.

The wheel formation was used in one of two ways: (1) the circle could tighten to prevent the faster moving, heavier U.S. aircraft from getting into the turn, or (2) each time a USAF aircraft engaged an orbiting MiG, another MIG would cross the circle at full power to gain a firing position on the attacker.

Among the methods introduced for attacking MiG-17's in a Wagon Wheel formation was one in which U.S. aircrews initiated a tangential attack from outside the periphery of the Wagon Wheel to gain position for an AIM-7 shot. When lock-on and positive identification were secured outside of minimum missile range, this attack was effective and presented little threat to the attacking aircraft. However, the low altitudes of the Wheel created excessive noise problems on the attacking aircraft's scope, and radar lock-ons were the exception rather than the rule when it was extremely difficult through ground clutter or to attain a full system lock-on. When this tangential attack was initiated for an AIM-9 or gun attack, the high angle-off of the attacking aircraft made it relatively easy for the MiG to force an overshoot before a tracking solution was achieved.

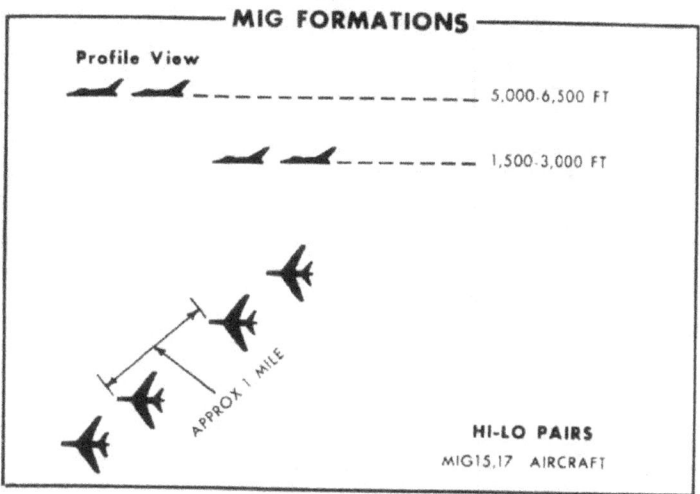

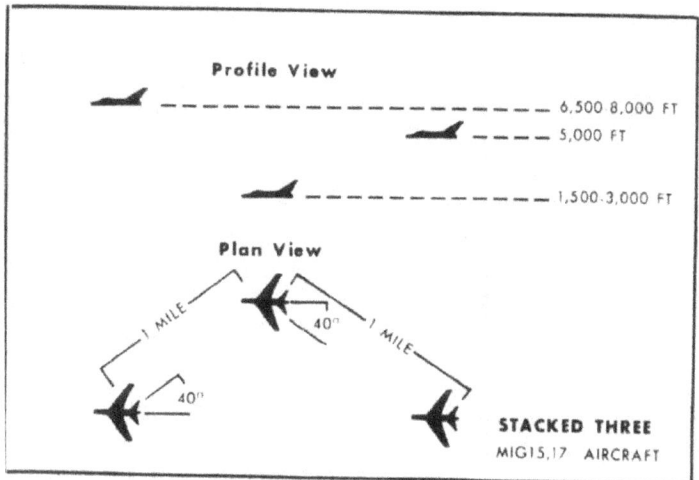

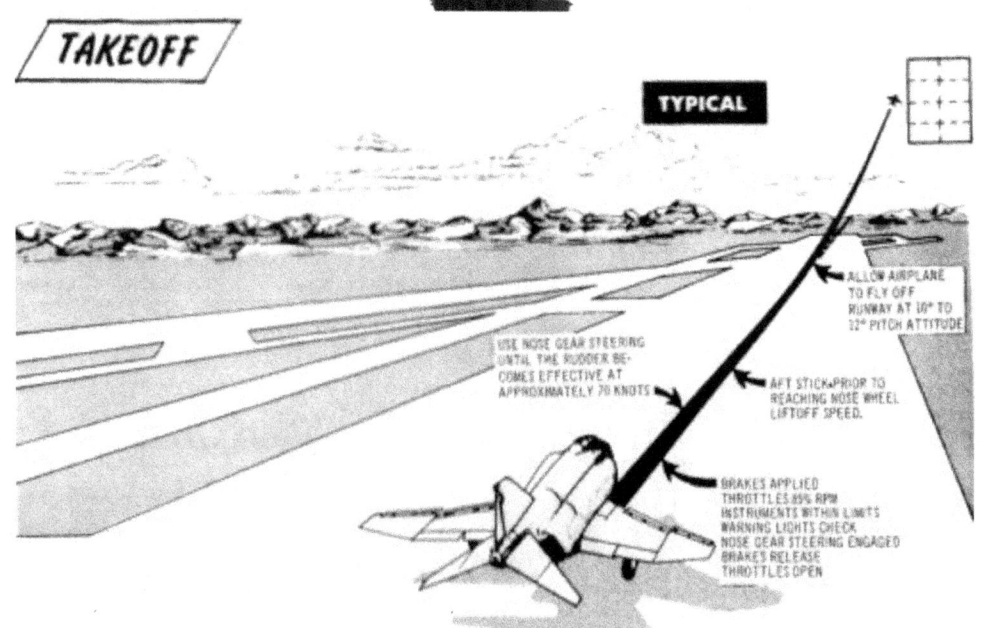

Vertical Rolling Scissors ~ a defensive, descending, rolling maneuver in the vertical plane. The purpose of the maneuver is to gain an offensive advantage if the enemy overshoots a flight path in a vertical plane. The maneuver is used when the enemy cuts off in the vertical plane during the defender's zoom maneuver. When it is observed that the enemy is cutting off, the USAF aircraft would turn down into him to increase the angle of over-shoot. Once the overshoot has been achieved, the nose of the USAF aircraft will be low and the enemy's nose high, so to press the attack, the enemy must pull his nose down also. At this point, when the enemy has been committed nose-low, the USAF attacker would roll 180" toward the defender's flight path and pull into him. If the timing is right the enemy will not be able to match his attacker's attitude, and will overshoot in front of the attacker's flight path ~ the rolling maneuver is then continued around to the enemy's 6 o'clock position.

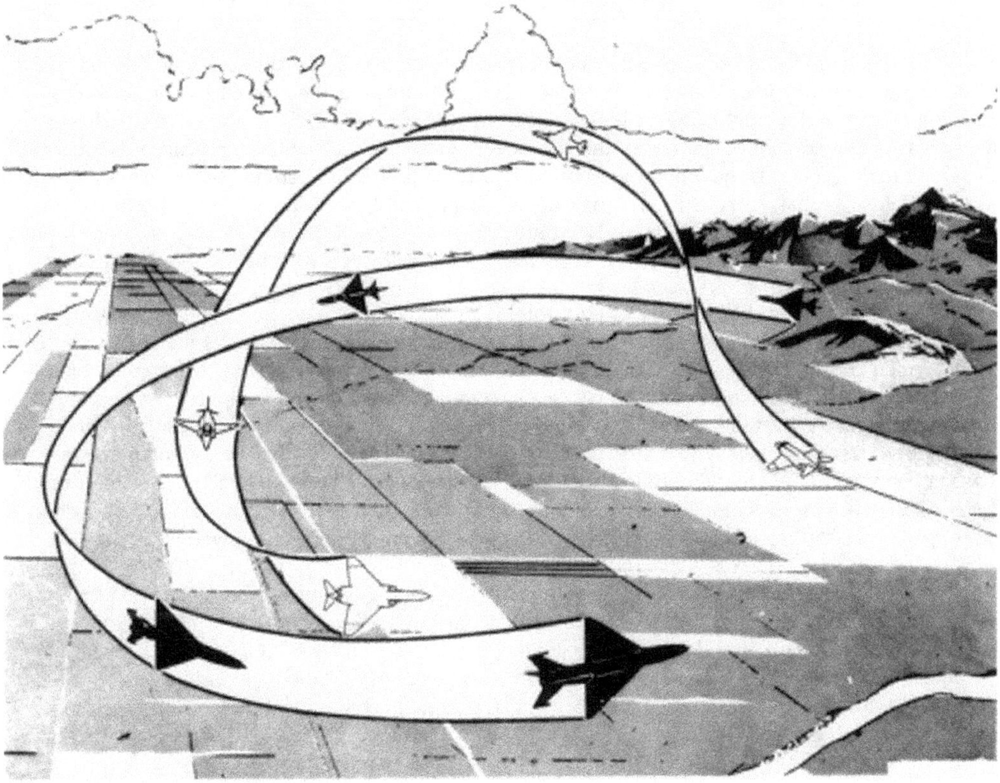

AS PORTRAYED IN PERSPECTIVE SKETCH

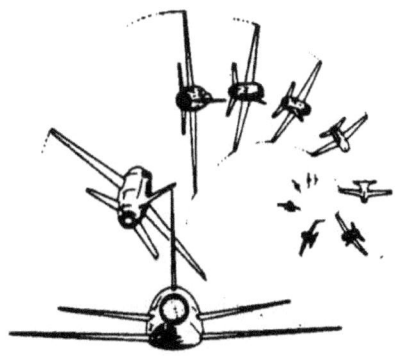

Barrel Roll Attack ~ *The offensive barrel roll is a three-dimensional maneuver used to reduce a high angle-off while maintaining nose-tail separation. Its purpose is much the same as the high-speed yo-yo. It is used instead of the high-speed yo-yo at a large angle-off in order to lower the apex of the attack. The range at which the maneuver is begun varies greatly and depends primarily on overtake and angle-off. Generally, it is initiated at a range of one to three miles, but the attacker must be flying at a relatively high-calibrated airspeed. The maneuver is initiated by rolling to match the defender's angle of bank, then loading the aircraft. As soon as the aircraft is loaded, the pilot rolls it in the opposite direction ~ over the top.*

Air-to-air Sidewinders scored one hit out of every seven fired in action over Vietnam. In the cold air of high altitudes they invariably locked on as intended, but nearer the ground in rain or through contrasting clouds and sunlight, the Sidewinder became indiscriminate about the source of heat to home on, and often strayed off target. Without a direct heat source to home on a head-on shot was futile. These early version IR heat homing missiles had to be fired from behind. The other victory of Operation Bolo was that no NVAF airman was seen again until late March 1967. When they did appear 36 enemy MiGs were shot down in 12 months. Olds' 8th TFW destroyed 23 of them. Bob Hope described Olds, with four confirmed victories as ~ *the largest distributor of MiG parts in South East Asia!*

The capacity of U.S. airborne radar offered minimal possibility of detecting low-level aircraft, as any radar echo from an enemy aircraft was simply lost in ground reflections. In August 1967, NVAF MiG-21 airmen took advantage of this. Flying in pairs as low as they dared, till abreast of an approaching American air strike. The MiGs attacked, soaring above and behind the Americans. Diving from behind they fired their Atoll missiles then either zoomed away or climbed high for another diving pass with their twin 23mm cannons. When it was realized the enemy had come from below, flights of Phantoms carrying four Sidewinders were positioned to cover the tail of highflying formations. In seven months of using this tactic, NVAF airmen, making the best of the MiG-21 lack of combat persistence hauling just two missiles, shot down 18 U.S. fighters for their own loss of five.

In November 1968, as the number of U.S. aircraft lost over Vietnam neared 900, the kill ratios achieved by USAF and U.S. Navy pilots in combat were far lower than expected. For the US Navy, it was just 3.7 enemy aircraft shot down for every American plane lost. Navy analysts determined that their pilots, both in F-8 Crusaders and F-4 Phantoms, were not skilled enough in air combat maneuvering (ACM) and lacked experience against dissimilar types enemy aircraft types ~ in particular the smaller, faster Soviet-built MiG-17 *Fresco*. Consequently, the U.S. Navy asked USN Captain Frank Ault to compile a classified full scale Air Weapons Study. In the *Ault Report* he revealed as missiles had given fighter pilots the option of fighting at longer range, U.S. airmen were losing engagements because they hadn't been properly trained or learned how to use their missiles. Ault recommended that ACM skills be restored to prominence during fighter pilot training, using different aircraft as near in performance to enemy fighters as possible. After the *Ault Report*, the study's recommendations were approved by the U.S. Navy established its *Post-graduate Course in Fighter Weapons and Tactics* at Miramar Naval Air Station in California, which became Top Gun or TOPGUN.

Miramar opened in March 1969, and projects undertaken at the inauguration of TOPGUN known as *Have Drill* and *Have Doughnut* tested acquired MiG-17 and MiG-21 aircraft against USAF and USN planes and pilots. The results were alarming. Every pilot so underestimated the MiG-17 that they were all *shot down* in the engagements. The extraordinary performance of the MiG-17 below 450 kts and below 10,000 feet was largely unappreciated, pilots were overconfident, and training was thin in understanding the true capabilities of the enemy's principal dogfighter. To correct these deficiencies, the TOPGUN program recruited the best pilots from each Carrier Air Wing to attend. Captain Ault's findings were critical in regard pilots and how they misused air-to-air missiles. As the majority of pilots were unfamiliar with the narrow limitations of the air-to-air missiles and radar systems, they tended to launch against enemy aircraft from outside the envelopes. TOPGUN therefore focused on maneuvering against dissimilar aircraft, bring them within the envelope and simulate firing a missile, and after every flight, carefully evaluated whether the missile launched within the narrow parameters achieved success. To do this, they would fly from NAS Miramar and use the U.S. Marine Corps' missile range at nearby MCAS Yuma. TOPGUN also taught pilots how to exchange airspeed for altitude by using manœuvres such as the Immelmann turn and the Split-S, to put them behind an enemy fighter where an opponent was vulnerable to a heat-seeking Sidewinder missile which could accelerate to Mach 2.5 (1,840 mph) or three times faster than most enemy jets, and use its IR heat sensor to lock in a targets tailpipe for a homing flight from 2.6 miles.

At the end of each TOPGUN class, graduating pilots returned to their squadrons and to teach the others the techniques they had learned, thus multiplying the program's effectiveness across the fleet. As U.S. pilots were usually killed within their first ten combat missions, if they could survive that, their chances of surviving an entire tour of 100 missions were very high, therefore **Ault's plan was a masterclass** to teach a needed skill set that was being lost to time ~ that of the modern fighter pilot.

Miramar's first class included Lieutenant Randall *Duke* Cunningham of VF-96 *Fighting Falcons* from the carrier, USS *Constellation*. On 19th January 1972, Lt. Randall Cunningham flying a Phantom (with his radar intercept officer Lt. *Willie* Driscoll in the backseat) spotted a pair of MiG-21s far below. Cunningham dived to attack. Switching on his Sidewinders, he heard them growl in his headphones as they locked on. The MiG detecting the tuned Sidewinder sounded a high-pitched warning to its pilot that an enemy had targeted him. Cunningham fired as the NVAF airman engaged his afterburner and pulled his MiG through a hard right turn, to shake off the Sidewinder. Top Gun had taught Cunningham that a MiG-21 could outturn a Phantom, so instead of directly chasing the MiG, Cunningham dropped down to treetop height to widen his missile firing range. The MiG-21 pilot banking left came out of his turn to search for his enemy. Far below, Cunningham in one fluid motion, nosed his Phantom up slightly to tune another Sidewinder missile onto the MiG-21 at 1,000 yards. The fired Sidewinder blazed across the sky like a streak of lightning, smashing through the MiGs tail. The training of Miramar had proven itself. There was celebration onboard the carrier *Constellation*. Cunningham's kill had also been the U.S. Navy's first kill in almost two years over Vietnam.

On 10th May 1972, the carrier USS *Constellation* launched its VF 96 Squadron *Fighting Falcons* to target the rail yard and warehouses of Haiphong Harbour. As Phantom *Showtime 100* flown by Cunningham and Driscoll was climbing away after dropping their bombs on Haiphong, they saw a Phantom hit by AA fire and disappear in a dark fireball. Then they were bounced. Driscoll in the backseat of *Showtime* reported a MiG-17, *Seven o'clock low and climbing*. Driscoll saw another four MiGs split into two pairs and come after *Showtime*. The first pair turned left, closing fast. Cannon fire from the MiG leader shot by as the second pair zoomed behind. Cunningham turned tight to the left, knowing from Top Gun that the MiG-17 devoid of hydraulically boosted controls became difficult to pilot at high speeds. Cunningham broke hard left, and the MiG unable to match the Phantoms responsiveness in a high-speed turn, over shot. Cunningham slashed to correct his turn as the MiG flew on higher, slightly to the right. Perfect firing position and the MiG pilot knowing it, engaged his afterburner, pouring raw fuel into his exhaust to gain speed and out fly Cunningham's Phantom.

The MiG-17 pilot had made his worst possible choice — widening the airspace between them. Cunningham with no guns on *Showtime* knew he was within minimum range for a Sidewinder launch. He selected his Sidewinder, acquired the MiG with a good growling tone in his headphones, and launched a missile that slashed through the sky to destroy his opponent. Cunningham then soared up to 15,000 feet to scan the scene below. Four MiGs appeared from behind. Eight more were in a turning engagement with three more Phantoms. Suddenly a Phantom broke off in a quick left turn, pursued by two MiGs Cunningham dashed to his six, targeting one enemy MiG-17 with a Sidewinder, but unable to fire from the risk it might equally lock on the Phantom as on his opponent. Cunningham: *Showtime, break right, break right.* (No response) *Showtime, reverse starboard.* ~ *If you don't, you're dead.*

The pursued Phantom cut right to clear his arc of fire. Cunningham selected a missile for launch with only the MiG to his front. Releasing his growling Sidewinder it tore after the MiG-17 spearing up its tailpipe for Cunningham's second kill of the mission. To confirm his six was clear, Cunningham rolled through 360° for his back-seater Driscoll to scan the sky behind and below. Another MiG was seen in the distance, lining up on their six to nail them in a hail of 37mm Soviet cannon shells. Cunningham snap rolled on full power, and through a sequence of deft rolls, left the MiG far behind.

Cunningham and Driscoll flew *Showtime* east to the coast and headed for the USS *Constellation*. From head on, a lone MiG-17 with a tempest of cannon fire flaring from its nose came straight at them. Luckily none hit *Showtime*. With no guns to reply, Cunningham full-throttle snapped the Phantom's nose into a hard vertical climb ; matched by his NVAF enemy in seconds, to be canopy-to-canopy. Cunningham's twinjet Phantom's faster rate of climb would only put it in front of the MiG-17 so he pulled over the top. Tracer flashed past. Cunningham decided to take advantage of the MiG-17s momentum as before. Another vertical zoom, Cunningham cut throttle, extended his air brakes, and the MiG overshot. The MiG pilot dived straight down to maximize his speed, knowing he was set for Sidewinder, but hoping the American missile might not home if he could confuse it with background heat emissions. Cunningham locked and loosed a Sidewinder at the MiG that detonated to send the MiG and its fireball of debris tumbling in flames. The capable NVAF pilot perished in his wreckage. Three kills on one mission. Cunningham and Driscoll became the first aces of the Vietnam War, and the first dual aces to score their kills by missile.

SHOWTIME 100 ~ CUNNINGHAM AND DRISCOLL

U.S. Navy McDonnell Douglas F-4J Phantom II NG-100 assigned to fighter squadron VF-96 *Fighting Falcons* Carrier Air Wing 9 aboard the carrier USS *Constellation*. Cunningham and Driscoll flew *Showtime 100* for three MiG-kills on 10th May 1972. However, their Phantom was later hit by a NV SA-2 surface-to-air missile and this crew had to eject over the Gulf of Tonkin.

Lt. Randall Cunningham, Lt. William Driscoll   Phantom II firing pairs of *Zuni* 5-inch rockets

VICTORIES OVER LAOS

American HQ received reports that MiG aircraft airborne below 20" North latitude increased from a daily average of five flights in December 1971 to an average of 10 per day in January of 1972. By March, the North Vietnamese fighter inventory included 120 MiG-15s and MiG 17s, 33 MiG-19s, and 93 MiG-21s ~ and these MiG-21s could be vectored by radar with swift timing against U.S. strikes and support forces, with the North Vietnamese determining the structure of most strike forces soon after U.S. aircraft left the ground (although no more than 180 of their MiG aircraft were combat ready).

American forces had the advantage of chaff-dispensing flights to degrade the enemy SAM and AAA gun-laying radars. This de-gradation was further supplemented by EB-66 electronic jamming, U.S. Navy jamming, and jamming pods installed on all strike aircraft. In the Gulf of Tonkin the U.S. Navy continued to operate the early warning radar ship *Red Crown* while the U.S.A.F. still deployed their airborne counterpart codenamed *Disco,* to provide their forces with MiG warnings. *Red Crown* was more effective along the coastal zone, and *Disco* for inland areas. B-52 strikes flew *over the beach* into Laos, and U.S.A.F. fighters flew combat air patrols and escort flights. From early 1972, MiGs increasingly began to penetrate Laos to check these strikes, but USAF F-4 Phantoms were on hand to greet them. So for a brief period of time in 1972, F-4s engaged MiG's in air-to-air combat over Laos. The first U.S.A.F. aerial victory in four years at night, took place on 21st February over N.E. Laos, 90 miles S.W. of Hanoi. Major Robert A. Lodge was aircraft commander and 1st Lt. Roger C. Lecher was his weapon systems officer in an F-4 flying MIGCAP from 555th TFS ~ Major R.A. Lodge reported:

Red Crown called out bandits ~ MiGs at our 060"position and proceeded to vector us on an intercept. I descended to minimum en route altitude, and at approximately 2123 local, my WSO detected and locked on a target at the position Red Crown was calling Bandit. The target was level at zero azimuth and closing, with the combined velocity of both aircraft in excess of 900 knots. I fired three AIM-7's, the first at approximately 11 nautical miles, the second at 8 nautical miles, and the third at 6 nautical miles. The first missile appeared to guide and track level, and detonated in a small explosion. The second missile guided in a similar manner and detonated with another small explosion, followed immediately by a large explosion in the same area. This secondary explosion was of a different nature than the two missile detonations and appeared like a large POL (petroleum, oil and lubricants) explosion with a fireball. The third missile started guiding in a corkscrew manner and then straightened out. No detonation was observed for the third missile. We had no more AIM-7s left, and broke off and egressed at low altitude. Two other MiG-21s then attempted to pursue us. We were low, over 500 knots airspeed, and the MiGs broke off after a 30-nautical mile chase and continued to drop back. Another F-4 was flying radar trail during the entire flight and was about 5,000 feet higher than us on the final attack.

Linebacker Operations ~ On 8th May 1972, President Richard M. Nixon announced the resumption of bombing of North Vietnam and the mining of entrances to its ports, and began *Operation Linebacker* for renewed and generally unrestricted air strikes against key military targets in North Vietnam. Throughout April and early May, additional US. Navy attack carriers joined the line in the Gulf of Tonkin, large numbers of B-52 heavy bombers were deployed, and more fighters based in Thailand. During an air strike in the Hanoi area on 8th May 1972, two MiGs fell to USAF Phantom F4D aircrews. Major Barton P. Crews and his WSO, Capt. Keith W. Jones, Jr., downed one MiG-19 ~ the first enemy aircraft of this type destroyed by an USAF crew. Major Barton F. Crews described his skirmish :

On 8 May 1972, a flight of four F4Ds was fragged to provide MIGCAP for strikes hitting the Hanoi area. I was scheduled as number three, with Capt. Keith W. Jones as my weapon systems officer. After the flight arrived at the preplanned orbit point the flight proceeded north of Yen Bai airfield and then made a 180 right turn heading south. After crossing the Red River, the lead aircraft called ~ Bogies, 12 o'clock ~ I immediately acquired them visually and identified them as four MiG-19s. I called over the radio ~ They're not friendly. The lead aircraft commander confirmed that, and directed the engagement. I set up my attack on the northernmost element of MiG-19's and started a closure on what appeared to be the number two man. My WSO stated that he couldn't get a lock-on so I pulled the pipper up to the MiG and fired one AIM-7. I estimated the range was less than 3,000 feet. I did not see the missile impact as I directed my attention to the lead MiG. Captain Jones stated he saw a yellowish chute go by. As I was trying to get my pipper on the lead MiG he did a break and ruined my tracking solution. At that time my number four aircraft said over the radio ~ That's a kill. Shortly after that my number four WSO, Lieutenant Michael Holland, called ~ Bandits at 6 o'clock. I then broke off my engagement, went into the clouds, and lost the MiG's. Later on the ground, 1st Lt. Michael T. Holland, the aircraft commander and weapon systems officer on my wing, confirmed seeing a chute and observing the MiG do a slow roll to inverted position and start down.

VICTORIES OVER LAOS

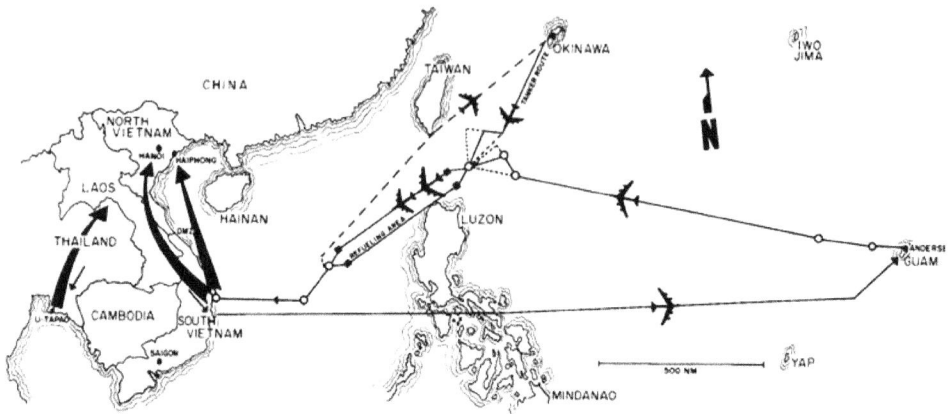

Western Pacific. The westbound route shows the flightpath taken by B-52 formations to arrive over the S.E. Asian mainland, from where they dispersed on a variety of routes into North Vietnam. Most sorties returned over the South and back to Guam. (Diversions from these basic routes were often made to meet timing requirements or for supplemental inflight refuelings).

Three MiG-21s became the prey of U.S. Air Force aircrews on 8th July, with two MiG-21s destroyed being downed by the same aircrew ; Captains Steve Ritchie and Charles De Bellevue, who flying the lead F-4E in a flight of four from the 555th TFS. The flight was on MIGCAP in support of a *Linebacker* strike flying west of Phu Tho and south of Yen Bai. Ritchie reported the double MiG victory:

Disco and Red Crown advised our flight of bandits southeast of our position, approximately 35-40 nautical miles. The flight headed toward the threat in patrol formation and crossed the Black River on a southerly course. Red Crown and Disco shortly thereafter advised that the bandits and our flight had merged.

The flight then turned to the north, met two MiG-21s at 10 o'clock, made a slight left turn, and passed the MiGs head-on. I then unloaded and executed a hard left turn, as the MiGs turned right. I maneuvered to a 5 o'clock position on the number two MiG, obtained an auto-acquisition boresight radar lock-on, and fired two AIM-7 missiles. The first missile impacted the number two MiG, causing a large yellow fireball as the MiG broke into parts. It continued to disintegrate until impacting the ground.

I then unloaded again for energy and turned hard right in pursuit of the lead MiG-21, who was now in a rear-quarter threatening position on aircraft 4. I maneuvered into a similar position on the lead MiG as was achieved on his wingman previously. Another radar auto-acquisition lock-on was obtained and one AIM-7 missile fired. The missile impacted the MiG, resulting in a large yellow fireball. This MiG also broke into parts and began to disintegrate. The front of the aircraft was observed impacting the ground in a large fireball. The flight remained in tactical support formation throughout the flight and egressed as a flight of four.

These two aerial victories increased Captain Steve Ritchie's score to four, and the U.S. Air Force had its first ace of the war when Ritchie had his fifth MiG victory confirmed for 28th August 1972. Ritchie was flying lead Phantom of a MIGCAP flight, with Capt. Charles B. De Bellevue as his WSO.

We acquired a radar lock-on on a MiG-21 that was head-on to us. We converted to the stem and fired two AIM-7 missiles during the conversion. These missiles were out of parameters and were fired in an attempt to get the MiG to start a turn. As we rolled out behind the MiG, we fired the two remaining AIM-7's. The third missile missed, but the fourth impacted the MiG. The MiG was seen to explode and start tumbling toward the earth. It was an entirely different situation. The MiG flew at a much higher altitude than any of my other MiG kills and at a much greater range. I don't think the MiG pilot ever really saw us. All he saw were those missiles coming at him and that's what helped us finally get him. The kill was witnessed by Captain John Madden, aircraft commander in number 3.

The most important thing is for the crew to work well together. They have to know each other. During the Linebacker strikes on 9th September, I knew what Steve was thinking on a mission and could almost accomplish whatever he wanted before he asked. I was telling him everything he had to know when he wanted it, and did not waste time giving him useless data.
Captain Charles B. De Bellevue

VIETNAM

Three air-to-air missiles were responsible for MiG kills : the AIM-4, AIM-7, and AIM-9. The AIM 4 Falcon used on Phantoms downed five MiG's (four MiG-17s and one Mi G-21). Falcons were capable of speeds from Mach 2 to Mach 4, and all models had an effective range of more than 5 miles.

The AIM-7 Sparrow used by Phantoms, accounted for 50 MiG kills ~ more than any other missile (eight MiG-l7's, four MiG-19s, and 38 MiG-21's). Sparrow was a solid fuel, radar homing, air-to-air guided missile with a HE warhead for use against high-performance aircraft under all-weather conditions and from all angles (including head-on). The AIM-7 used in S.E. Asia had a supersonic capability and aircraft flying at either subsonic or supersonic speeds could launch the missile.
The AIM-7 Sparrow later became part of the primary armament on USAF and USMC fighters.

AIM-9 Sidewinders air-to-air missiles downed 33 MiGs (14 MiG-l7s, 2 MiG-19s, and 17 MiG-21s). Sidewinder was one of the simplest weapons, with few electronic components and few moving parts. It required little training to handle and assemble. Powered by a single-stage, solid-propellant rocket, Series B, D, and E of this missile used a passive guidance system, which homed in on the engine exhaust of a target aircraft. Series C utilized a semi-active radar guidance system. Sidewinders were approximately nine feet long, capable of Mach 2.5 and with an effective range of more than 2 miles.

Counter-insurgency tactics ~ In the Malayan Emergency, the British correctly assessed that their best hope of defeating communist insurgents lay in isolating them from their supplies, and tracking them down in the jungle. Australian soldiers using similar tactics in Vietnam constrained Viet Cong contact with the local population to deny them their source of sustainment and supply reinforcement through unrelenting infantry cordon, search, and destroy operations. Foot patrols by troopers of the 3rd Squadron SAS operating with the 1st Australian Task Force in Phuoc Tuy Province, proved themselves able to give chase and surprise the Viet Cong guerillas around Nui Dat, Long Tan, Binh Ba, and other areas without exposing themselves to the ambushes that had claimed so many previous casualties. Australian infantry patrols avoided jungle tracks and clearings, instead finding quiet ways through bamboo thickets and tangled foliage. Their method was to move forward a few steps at a time, then stop, listen… and proceed again, taking as much as nine hours to sweep a mile of terrain. Apart from jungle terrain, there were also dark rubber plantations to sweep, dangerous areas of evenly spaced trees that offered any hidden enemy, very clear fields of fire. The primary SAS role was reconnaissance, to detect and document Viet Cong routes. During their first tour of 134 patrols, 3rd Squadron SAS commander Major Reginald Beesley kicked down the kill boards put up by previous units. He made the point, advising his unit ~ *We are not here to kill people, but to gain information.* Australian and New Zealand SAS teams killed 598 Viet Cong, losing only two men to the enemy. After such a result, the Viet Cong were so weakened that they were forced to operate in small groups, engaging the Australians only if it was essential to their survival. In the words of David Hackworth, the most decorated U.S. soldier in Vietnam ; *The Aussies used squads to make contact… and brought in reinforcements to do the killing* ~ *they planned in the belief that a platoon on the battlefield could do anything.*

* * *

Although the fighter pilots of the USAF favoured the *fluid-four* to cover their lead pair from enemy surprise attacks, U.S. Navy pilots mixed their traditional Thach Weave with the Korean *fluid-four* to create the naval *loose-deuce*. Either pilot was free to adopt the lead role, while the other covered as his wingman. Naval flight leaders accepted that it was essential to pass the lead to the pilot who had the enemy in sight. With TOPGUN refining the skills of U.S. aviators, the kill ratio over Vietnam soon improved from a mediocre 2 to 1 to their superb 12.5 to 1.

FLUID FOUR

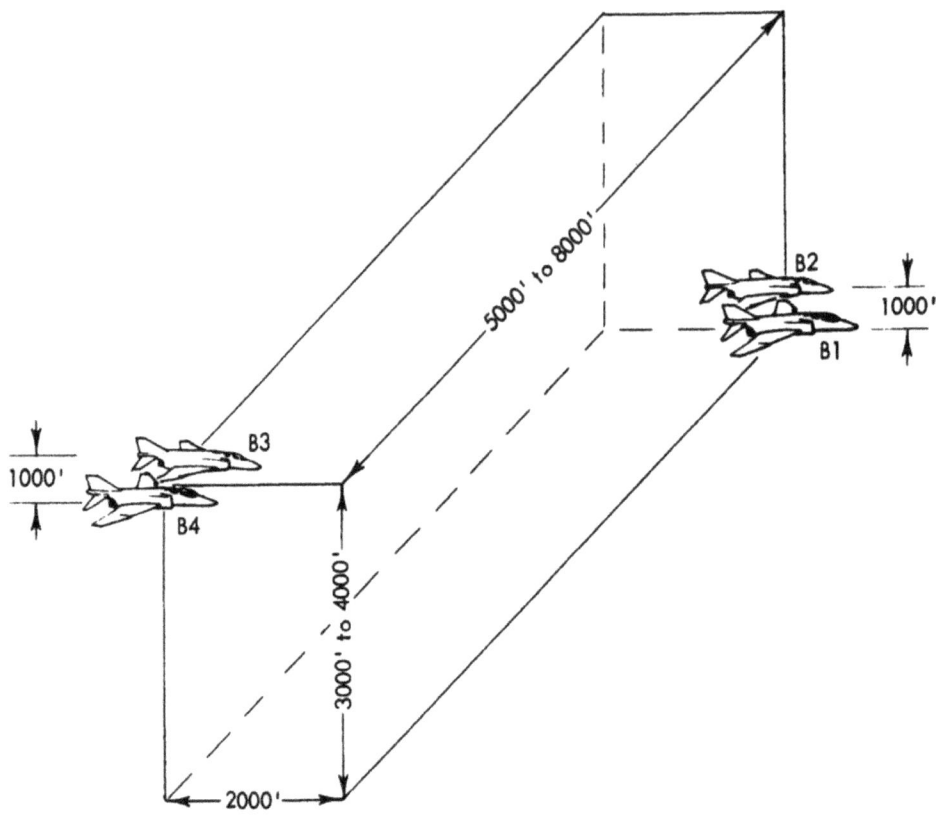

FINGERTIP OR FINGER FOUR
(All at Same Elevation)

The Fluid-Four formation consists of a four-ship formation, now an optimum for air-to-air combat. It is offensive, maneuverable, and has good mutual support. It was employed during daylight hours only when counter-air activity was anticipated. The lookout capability with four aircraft allows sizing the formation to cope with any expected threat ~ even in some surface-to-air missile environments the formation can be sized to provide ECM coverage and yet retain air-to-air capability. The fluid-four formation consists of two elements. The second element maneuvers off the flight leader's element to provide mutual support both from positioning and lookout. The wingmen fly off the flight and element leaders, and they position themselves to provide the best coverage for the entire formation. The fulfillment of each individual's responsibility allows the flight to conduct offensive operations with security from any potentially lethal 6 o'clock attack.

BASIC ESCORT FORMATION

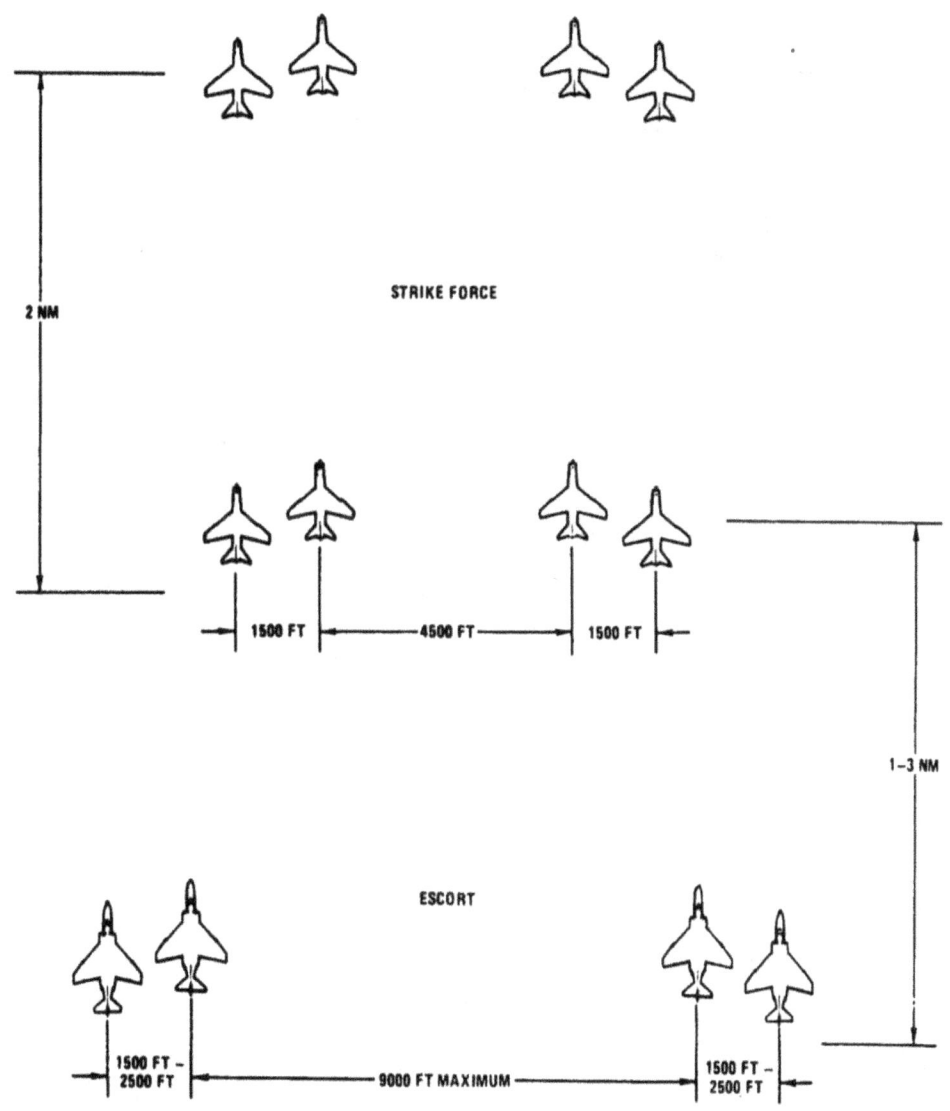

Escort Formations ~ purpose was to provide protection for escorted aircraft as well as the escorts. The tactics employed in S.E. Asia depended on the size and speed of the escorted force plus the anticipated tactics of the enemy, with the tactics for each mission tailored to fit specific requirements. Fighter escort formations essentially were dictated by the strike force formations. When escort aircraft had equal or better performance characteristics than the force being escorted, a variety of escort formations could be developed. Generally, the escort would prefer to fly a fluid formation approximately two miles behind the force. To obtain most protection from a four-ship escort, the force formations would spread the elements no more than 4,500 ft. apart with wingmen 1,500 ft. out. The strike flight would spread 7,500 ft. wide and would be managed by an escort flight utilizing a maximum of 9,000 ft. between elements, maintaining line abreast with wingmen 1,500 to 2,500 ft. In case of chaff flights, escorts placed themselves high or outside during chaff bomb delivery to avoid falling canisters. Weather conditions dictated different positions for better visual coverage.

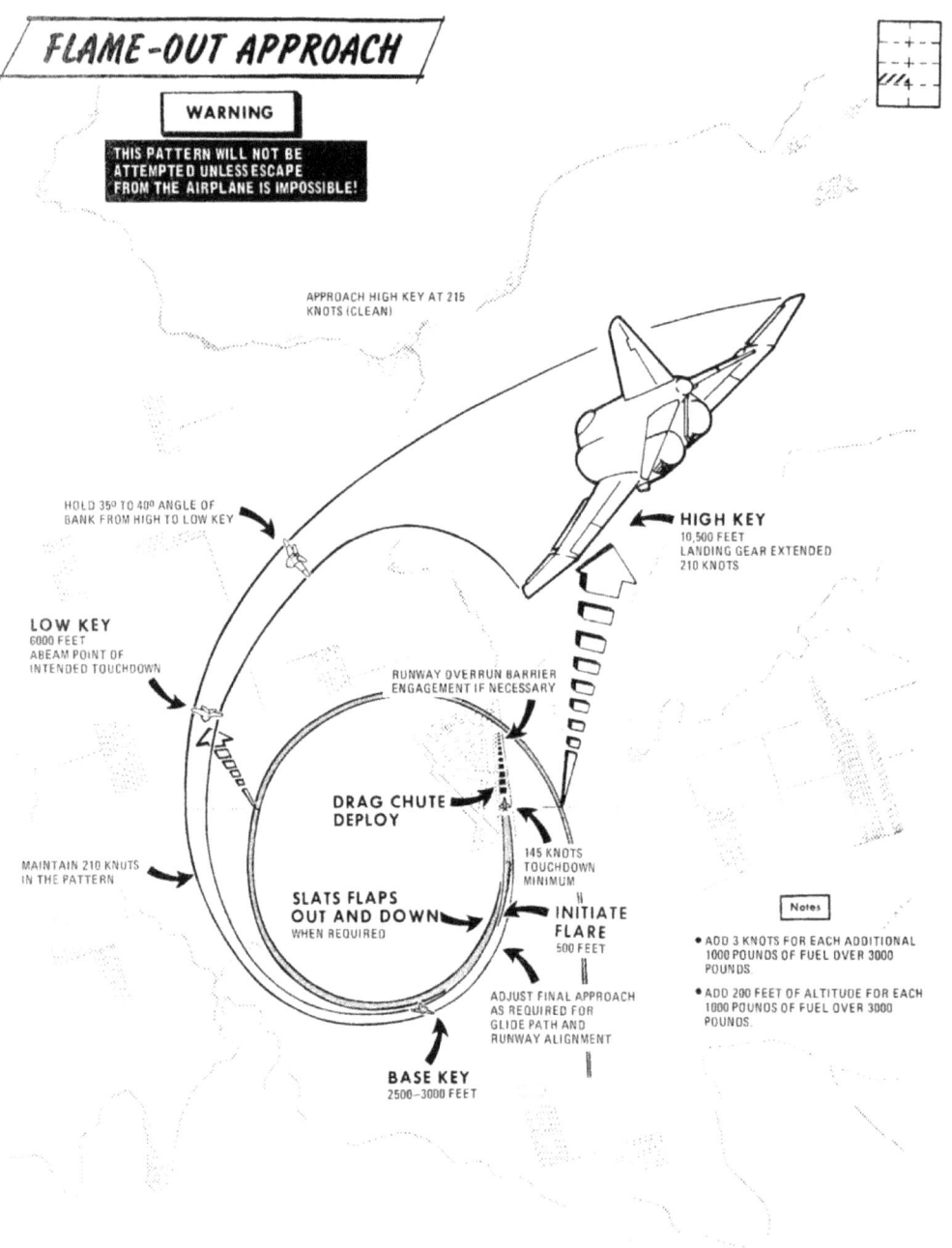

Figure 3-10

VIETNAM

The Paris Peace Talks ~ Speaking from the White House in a televised speech to the nation, President Richard Nixon appealed to the public. *And so tonight – to you, the great silent majority of my fellow Americans, I ask for your support, for the more divided we are at home, the less likely the enemy is to negotiate at Paris. Let us be united for peace. Let us also be united against defeat. Because let us understand: North Vietnam cannot defeat or humiliate the United States. Only Americans can do that.* Hanoi intended to use the televisions of democracy to turn the people against their government, sending letters of praise to the leaders of the anti-war movement (who Nixon called *bums blowing up campuses*) praising them, *may your offensive succeed splendidly.* Spiro Agnew described the anti-war campaigners as *ideological eunuchs and communist dupes of an effete corps of impudent snobs who characterize themselves as intellectuals.*

While the Americans hoped for a ceasefire agreement to emerge from the Paris peace talks, it became clear to them that their hardy opponents would not let them disengage on acceptable terms. After the communists walked out on 13th December 1972, President Nixon unleashed the use of maximum air power to force the communists to comply, authorizing Operation Linebacker II to take the required action that U.S. generals had called for, back at the beginning in 1965 : the unrestricted bombing of military targets in North Vietnam. Two days after the walk out of the North Vietnamese 200 B-52 bombers began shattering targets across Hanoi and Haiphong, then continued for 11 days and nights until all priority targets in Hanoi were rubble. Ten airfields and the northern rail network ceased to exist. The communists answered by firing fusillades of up to ten SAMs toward each B-52 heavy bomber. Three B-52s were lost on the first night. Six more fell on the third night.

The North Vietnamese delegation, bombed back to the Parisian peace table on the 29th December resumed talking while the unrelenting pressure of the bombing continued to destroy their homeland. On 15th January 1973, the North Vietnamese agreed to the terms of a 23rd January ceasefire, under which the last U.S. troops would leave in March 1973. America had won the war, but at the cost of 2,000 plus aircraft. The question remains. *If there had been a severe application of air power back in 1965, would it have curtailed America's longest war?* Nobody knows. The Vietnamese experience was unique. In jungle terrain with the impediment of monsoonal weather, the tactics and logistic methods of the Viet Cong in a warzone almost devoid of decisive military targets, made worthwhile air interdiction a huge challenge. The North Vietnamese knew that the régime in the south would be vulnerable without the presence of U.S. air support, so they signed the Paris Treaty.

As the last U.S. soldiers departed from South Vietnam, in 1973, President Richard Nixon declared. *The day we have all worked and prayed for has finally come.* Two years after the Americans and Australians went home, the rejuvenated North Vietnamese Army violated the Paris Treaty in March of 1975 and turned against the south. Without foreign air support, the South crumbled. After the communist triumph, thousands of refugees trying to leave Vietnam set off in small boats out into the South China Sea. Many never made it, perishing in storms as their leaky boats sank beneath them or were robbed, raped, and murdered by Thai fishermen turned pirates.

*   *   *

With Russian SAM missiles threatening their survival, U.S. airmen listened with attention to any proven tactician who could advise them how to evade the SAMs and Soviet Atoll air-to air missiles. Colonel Robin Olds after he returned home from Vietnam was promoted to Brigadier General and appointed Commandant of Cadets at the United States Air Force Academy in Colorado Springs to restore morale in the wake of a major cheating scandal. Robin Olds offered them this advice ;

Anybody who doesn't have fear is an idiot. It's just that you must make the fear work for you. Hell, when somebody shot at me, it made me madder than hell, and all I wanted to do was shoot back. Younger guys have to think before they start a maneuver. With me, its instinct ~ here come the SAMS! The trick is seeing the launch. You can see the steam. It goes straight up, turns more level, then the booster drops off. If it maintains a relatively stable position, it's coming right for you and you're in trouble. You're eager to make a move but can't. If you dodge too fast it will turn and catch you; if you wait too late it will explode near enough to get you. What you do at the right moment is poke your nose down, go down as hard as you can, pull maybe three negative Gs at 550 knots and once it follows you down, you go up as hard as you can. The SAM can't follow that and it goes under. But I never got over confident. I mean, if you're one or two seconds too slow, you've had the schnitzel!!

CHAPTER XI

THE MIDDLE-EASTERN WARS

It all began with a false report from Soviet intelligence. As the speaker of Egypt's parliament, Anwar al-Sadat who was in Moscow during May 1967 for talks with the Soviet PM Andrei Kosygin was boarding his return flight to Egypt he was ushered aside by the Soviet foreign minister for a sharp warning ~ *Israeli troops are massing along the Syrian border*. Sadat reported it to Egyptian President, Gamal Abdel Nasser, who ordered Egyptian Air Force patrols to fly over and photograph the border area, but after their recon aircraft photographs were studied, nothing extraordinary was observed. Soviet diplomats spread more fiction, to change the balance of power by provoking a limited war.

Egypt made the opening move on Sunday 14th May 1967, when President Nasser ordered the forward deployment of 130,000 ground troops across the Suez Canal and onto the Sinai Peninsula. On 17th May, Egypt coerced UN forces on the Egyptian-Israeli border to withdraw. Days later, Nasser ordered an airborne battalion to occupy the old fortress of Sharm El-Sheikh overlooking the Straits of Tiran and closed the Gulf of Aquaba to Israel's eastern port of Eilat and waited. King Hussein of Jordan flew to Cairo on 30th May 1967 to sign a defence pact with Egypt. At the beginning of June, when President Nasser called on the Arab nations to destroy Israel, Soviet diplomats again interfered, conveying to the Egyptians the clear understanding that they would only support Egypt, if they were seen to be refraining from being the aggressor. Thereafter the Egyptians adopted a defensive posture, and remained politically prepared to absorb any Israeli attack.

The Moked Project ~ About five years before these events, the commander of the Israel Air Force, Ezer Weizmann ordered his chief of operations, Yak Navo to report back with a decisive plan of how to achieve air superiority in a hypothetical future war. Over a terrain of barren rock and desert, the new state of Israel on the edge of the Mediterranean, was bordered by the hostile lands of Egypt to the south occupying the Sinai Desert, Syria to the north of Israel looked down from the Golan Heights. Jordan was also in the north. It was assumed that all three of these immediate neighbours would engage in war against Israel. Israeli surveillance detachments began listening to Egyptian Air Force communications to detect and document their early morning activities to reveal any patterns: when Egyptian patrol aircraft landed low on fuel, and when they change shifts, to learn when the Egyptians were least prepared to meet an attack. Soon they knew more. Where the real aircraft were parked, and where the decoys were parked. Navo's plan called for the determination of targets, specification of weapons, and coordinated flight plans for a minimum of five waves redeploying the same aircraft.

The final concept was, for Israeli warplanes to destroy their neighbour's runways, grounding their aircraft long enough for them to destroy their parked fighters. To cause maximum damage, French delay-fused Durandal 1,200 lb anti-runway bombs would be used to penetrate the surface of a runway then, detonate beneath to crater the enemy runways. To achieve air superiority on the first day, Israeli airmen were expected to fly five sorties a day. Much depended on how quickly their French built Vautors, Mirage IIICs, and Super Mystére fighters could be refueled and reloaded, and sent out on later attacks. Israeli ground crews were able to achieve turn around times of ten minutes. Without the men or material to fight a protracted war, the IAF plan framed by Yak Navo, called for tactical surprise through a pre-emptive air strike against the Egyptian Air Force to attain air superiority with a holding operation to the north. Israeli commanders told their chief of staff General Yitzhak Rabin, *We have to mobilize and launch an attack within 72 hours. If we give the Egyptians more time they will pack the Sinai with more divisions*. Israel mobilized its reserves and waited… poised to attack on two fronts. On the night of 5th June the Israeli cabinet authorized *Operation Moked*.

Operation Moked began just after dawn on 5th June 1967. Israeli warplanes streaked north low over the waves and below the radar, before heading over the Egyptian dunes west of the Nile Delta to carry out pre-emptive bombing and strafing air strikes. The fine turning and fast Mirage IIICs capable of 1,470 mph armed with two 30mm cannon, two Sidewinders, and carrying two 1,200-lb bombs would crater the runways of Cairo and beyond. The swept-wing Mystére IVs would attack the airfields along the Suez Canal where the most modern MiGs were based. The slower Ouragans would target the airfields on the Sinai Peninsula. The stout Vautors that could carry several tons of bombs would fly deep into Egypt to attack the Egyptian air bases around Gardaka and Luxor. The Mirage IIICs returning from attacking the airfields in the Cairo area, lighter and more maneuverable without their bombs would support the attack of other Israeli jets over the Suez Canal area.

THE SIX-DAY WAR ~ 1967

We observed total radio silence. We flew at the height of the waves for about 15 minutes. We flew low over the sand dunes. We crossed the Suez Canal at Kantara, and entered the Nile Delta. As we flew over the delta, farmers waved at us – they probably thought we were Egyptian. At exactly 0745 we pulled up to 6,000 feet. I looked down and saw the MiGs glinting on the edges of the runway. I knew we had caught them by surprise.
                                                          Squadron Leader Ran Peker IAF

Israeli aircraft flying below the 2,000 feet minimal operational altitude of Egypt's SAM-2 batteries roared over ten Egyptian airfields across the Sinai Peninsula. Climbing to 6,000 feet, they rolled inverted to dive bomb the runways with *switches on*: all cannon and missiles armed to attack the Egyptian runways with rocket-assisted bombs. On a day with no wind, the Israelis delivered their bombs with superb accuracy, gouging deep craters in enemy runways to prevent the Egyptian MiGs from taking off. Mirage and Mystére airmen strafed the MiGs with short three-second bursts. Setting the Egyptian MiGs alight, high-octane fueled fires spread to other aircraft.

The IAF first wave had destroyed ten Egyptian airbases, cratered their runways, and shot up the stranded MiGs where they were parked. The first Israeli attack wave lasted 75 minutes and destroyed 197 aircraft plus eight radar stations for the loss of ten airmen. One specific Egyptian runway in northern Sinai had only been strafed, but not bombed. The Israelis were saving it to fly in supplies, and evacuate their wounded.

The first wave returned to their airfields to be quick-turned, refueled, and re-armed in eight minutes. They took off at 0930 as the second wave to target residual Egyptian airfields, destroying 107 more EAF aircraft for no loss, and returned. Quick-turned again, the third wave departed at 1215 to destroy another 48 Egyptian aircraft, plus their first targets in Syria and Jordan. King Hussein's single squadron of Jordanian Hawker Hunters were intercepted and destroyed. In the afternoon the Syrian Air Force were targeted and devastated. Any MiGs that had taken off were simply overwhelmed. The drag of the delta-winged Mirages that enabled them to lose speed in a tight turn, was understood and exploited by Israeli pilots, who forced the MiGs to overshoot, in their version of the *let him pass* ruse. The slower the fight became, the more lethal the Mirages became, shooting down 79 MiGs for the loss of 3 Mirages. All of Israel's air-to-air victories had fallen to the 30mm cannons of Israel's Mirage IIIs. The IAF after having achieved air superiority on three fronts, were now free to support their advancing ground troops on three different fronts against three different Arab armies ~ one at a time.

Mitla Pass ~ Egyptians troops in prepared defensive positions on the Sinai Peninsula, could have blocked the Israeli troops advancing on the Suez Canal. But, on the third day of fighting, their commander Field Marshal Abdel Hakim Amer ordered his formation on the Sinai to evacuate, and be back on the Egyptian side of the Canal within 24 hours. Feeling predisposed to a defeat, the Egyptian soldiers drifted into disorganization as they retreated west across the Sinai in loose convoys.

In a scene reminiscent of the massacre in the passes of 1918, Egyptian troops moving in their thousands through the narrow Mitla Pass in an elongated convoy were found and strafed by Israeli aircraft that returned to napalm hundreds of enemy vehicles. After the IAF air strike on the Mitla Pass, the surviving Egyptians staggered through an area strewn with burnt out wrecks and charred corpses. Other retreating Egyptians were also caught and strafed in the Gidi Pass. Thousands of Egyptian soldiers trundled 200km east through terrible heat to reach the Suez Canal, but many perished.

Despite an official UN ceasefire, Israel pressed on to solve their problem of strategic depth on three fronts ~ by acquiring the natural barriers of the Suez Canal in the west, the Jordan River in the east and the Golan Heights to the north. Then the ground commanders gave their troops the order to halt. In six days, Israel had quadrupled its territory. They had acquired the Sinai Peninsula, the Gaza Strip, the remainder of historic Palestine and the West Bank of the Jordan River with the city of Jerusalem, and from Syria, Israel had captured the strategic position of the Golan Heights. Egypt, Jordan, and Syria were content (at the time) to accept a cease-fire proposed by the UN Security Council, which spared them from further military defeat, and loss of territory.

SOVIET MiG-21 GROUND ATTACKS

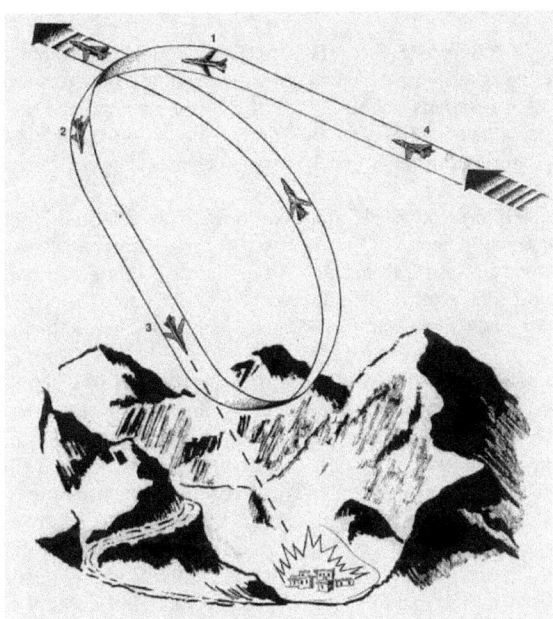

From the Soviet MiG-21F manual ~ Moscow 1960

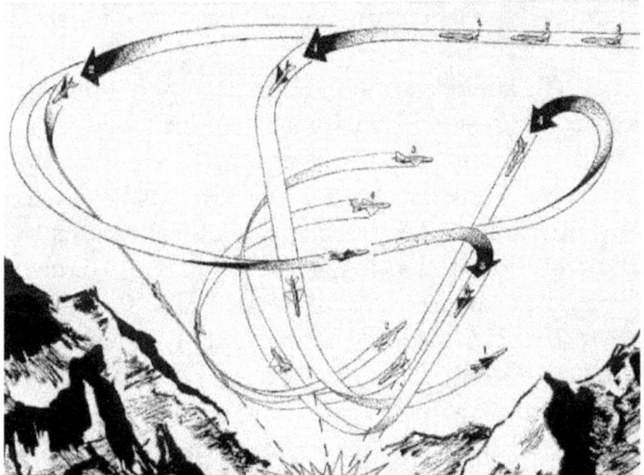

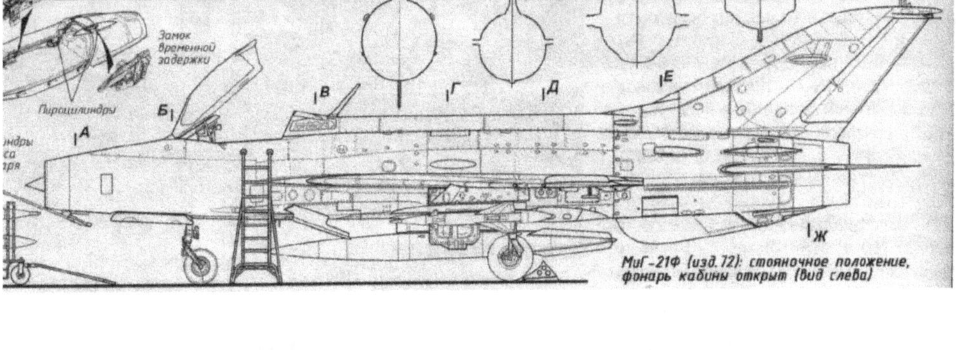

THE ELECTRONIC SUMMER ~ 1970

Operation Rooster 53 ~ Hoping to recapture the Sinai Peninsula from Israel, during 1969 the Egyptians requested and received vast supplies of military equipment from the Soviet Union. From the photographs of an IAF reconnaissance mission, the Israelis recognized a new Soviet P-12 radar unit sitting on the Egyptian beach of Ras-Arab. Planning began on Christmas Eve for a mission to capture it, and carry the entire radar system back to Israeli territory, using Sikorsky CH-53 helicopters.

Two days later at 2100 hours, Skyhawk, and Phantom aircraft attacked Egyptian targets along the Suez Canal to create a distraction, to cover the approach of three Aérospatiale Super Frelon helicopters carrying paratroopers towards Ras-Arab beach. As they landed, the Israeli paratroops surprised the Egyptian guard group and took control. Israeli paratroopers in part stripped the P-12 radar system, and called in two Sikorsky *Jolly Green Giant* CH-53 helicopters from across the Red Sea.

One Sikorsky lifted the sling-loaded communications and command caravan off the beach while the second Sikorsky lifted the acquisition radar across the Red Sea toward Israeli controlled territory. The four-ton radar, heavier than the helicopter hauling it, soon endangered the mission. The strong cables from the radar caravan to the *Jolly Green Giant* warped the structural ribs of the helicopter so intrusively that one cable slowly ruptured an internal hydraulic pressure hose. The anxious helicopter pilot, tracing his options of whether to release the radar or risk a loss of control and crashing, chose to fly on, and with the last of his hydraulic pressure crossed the Red Sea, setting his helicopter down with the retrieved radar. The second Sikorsky set down its lighter load and headed back to retrieve the radar. Locating and lifting the radar into the air, it nearly rendered another helicopter inoperative. The new radar system was delivered to waiting intelligence specialists. After the Israelis had studied the radar it was passed on to the U.S. (as much other captured Soviet equipment had been previously).

After taunts from Egyptian skirmishers in early 1970, Israel responded sending two Mirages through Egyptian airspace at low level over the rooftop of President Nasser's house in Heliopolis, shattering many widows. Faced with Israeli air superiority over the Suez Canal area and over his own home, Nasser turned to the Soviets for help. The Soviet Prime Minister, Andrei Kosygin was happy to oblige, arranging for an airlift of Soviet air defence equipment, including several Soviet technicians to instruct in its operational use. The Soviets offered further support, redeploying five squadrons of their 135th Mig-21 MF Air Regiment to defend Cairo, Alexandria, and the Aswan Dam. U.S. President Richard Nixon countered, informing the Soviets that the U.S. would supply the IAF with 50 Phantoms, equipped with anti-SAM AGM-45 Shrike anti-radiation missiles, plus 128 Skyhawks. Israeli airmen circumvented the complication of Soviet involvement by avoiding areas where the Soviets patrolled, and the Russian airmen reciprocated, by not flying within 30 km of the Suez Canal. With eerie prescience, in March of 1970, Israeli Defence Minister, General Moshe Dayan predicted ;

This summer is going to be an electrifying one, an electronic one. There will be some fighting, but its accent will be probably revolve around the deployment and setting up of new weapon systems. There will be incursions.

The situation on the Syrian front escalated quickly after the Syrians sent supersonic MiG-21s over Haifa shattering many of their southern neighbours widows. The IAF answered, sending low flying Phantoms over the Syrian capital of Damascus. Israel was in good shape for Moshe Dayan's envisaged *Electronic Summer* confrontation between Egypt's Soviet-built SA-2, and Russian crewed SA-3 systems designed to hit low-level IAF aircraft, and the opposing jets of the IAF loaded with U.S. supplied electronic countermeasure (ECM) devices. The electronic summer got off to an early start when Israeli Phantom airmen practiced how to bomb from such low altitudes they could scarcely miss, and like U.S. pilots in Vietnam, learned how to recognize the *SAM Song* – the change of pitch in tracking signals indicating a high flying SAM was pursuing them, and that the time for escape and evasion had come. The electronic war was now humming at full frequency. In June, Egyptian SAM crews along the Canal Zone locked on to Israeli Phantoms flying too close, and in six days three IAF Phantoms fell to SAMs fired in a *ripple* or sequenced pattern that proved difficult to evade. The Egyptians then shelled specific Israeli outposts across the Suez Canal to lure the IAF into flying retaliatory air strikes against particular antiquated and expendable artillery positions, only to find batteries of SAMs waiting for them. In the wake of more Phantom losses, the IAF commander Mordecai Hod described Egypt's reinforced air defences as ~ *the Russian fist covered by an Egyptian glove.*

P-12 radar captured during Operation Rooster 53

U.S. translated Soviet SA-2 site diagram

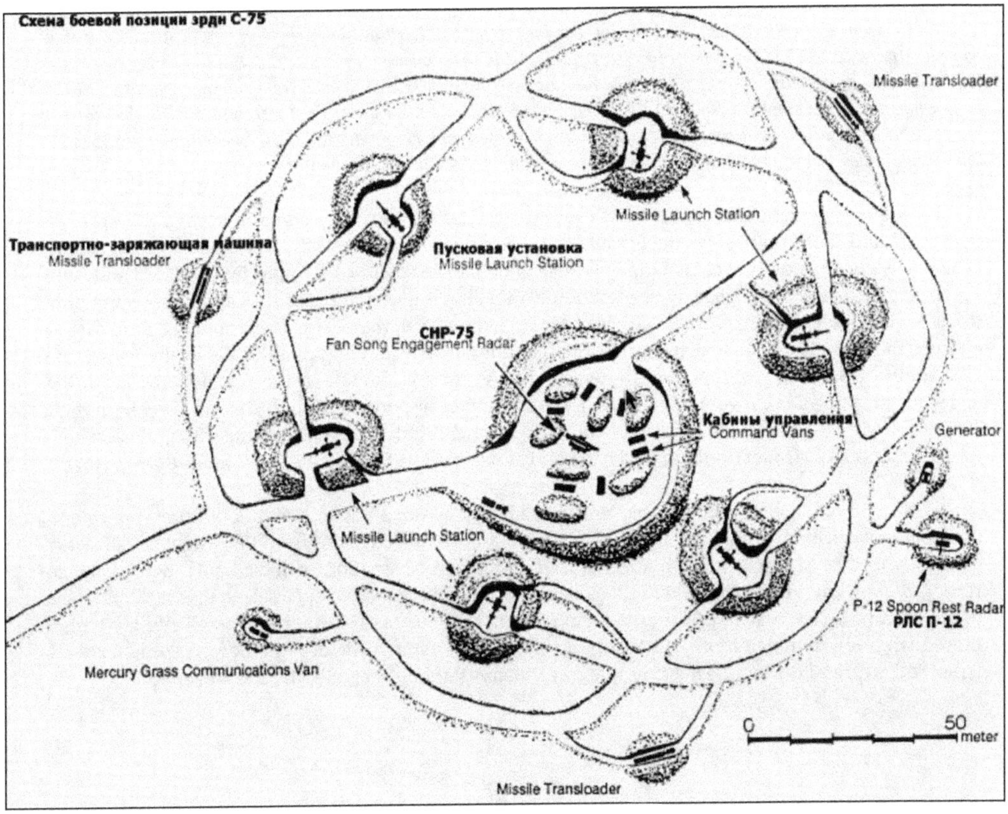

THE ELECTRONIC SUMMER ~ 1970

Operation Rimon 20 ~ On 25th July, a flight of Soviet flown MiG-21s intercepted a flight of Israeli Skyhawks during a ground attack mission over the Sinai Peninsula. One Soviet airman locked and launched an Atoll missile at a Skyhawk, damaging it and forcing it to land at Rephidim Airfield. The Israelis had to react. The IAF who had planned and carried out aerial ambushes before under the code name *Rimon* (Pomegranate) simply updated their proven plan and redesignated their trap : *Rimon 20*. The operation called for a flight of four Mirage IIICs to fly toward the Suez Canal at dune crest level, then as they penetrated Egyptian airspace to the south near Hurghada, to soar up to 35,000 feet, turn north and pose as a routine reconnaissance flight. As soon as the Soviet MiG-21s rose to intercept the lure of the Mirages, they would turn west toward an aerial snare of waiting Phantom F-4Es.

On June 30th at 1400 hours (Israeli time), four Mirage IIICs each armed with a pair of Sidewinders flew low across the Gulf of Suez. As they entered Egyptian airspace, they soared up to 35,000 feet in a tight formation to reduce the usefulness of enemy radar surveillance. By construing the composition of their force to appear on Arab radar screens as single aircraft or entity on a reconnaissance flight their true intention could not be reasoned by the enemy. Ten minutes after acquiring the climbing Israeli Mirages as a radar plot, the Soviets scrambled 20 MiG-21s as five flights of four. As the Soviet quartets vectored toward the Mirages, two flights of four Israeli Phantoms rose up from below.

Flight leader Avihu Ben-Nun fired two long-range Sparrows at the MiG-21s downing his first and breaking up the Soviet formation. As they separated into mutually protective pairs, Mirage pilot Avraham Shalmon sighted two MiGs astern of a pair of Phantoms. Warning them, Shalmon thrust closer firing a Sidewinder that shattered the lead MiG. The third kill of the day came from Asher Snir, who shot a Sidewinder into a MiGs fuselage, setting it on fire. The Soviet pilot ejected but his chute opened too soon. Soviet parachutes designed to open automatically at 10,000 feet, enabled their airmen to plummet. Not to freeze or suffocate at high altitude, as this airman perished during his descent. A pair of supersonic Phantoms flown by Avihu Bin-Nun and Aviam Sela separated to hunt alone. As the MiG-21 flown by Soviet pilot Captain Georgy Syrkin pulled through an arc to emerge facing the Phantom flown by Avihu Ben-Nun, the Israeli swept through an Immelmann turn that positioned him above, and behind Syrkin's MiG-21 ready for the kill. Then in Ben-Nun's words ;

By this time I'd realized the Russian pilot was inexperienced; he didn't know how to handle his aircraft in a combat situation. At 15,000 feet he proved this fact by trying to escape in a steep dive down to 7,000 feet. All we had to do was follow him, lock our radar onto him, and fire a missile. There was a tremendous explosion, but the MiG came out of the smoke apparently unharmed. That made me mad, and I fired a second missile that turned out to be unnecessary. The Russian aircraft had, in fact, been severely damaged by the first missile; suddenly, it burst into flames and fell apart. By the time the second missile reached it, it wasn't there.

As the engagement entered its fourth minute, the Israeli ground commander gave the order to disengage and withdraw before the Soviets could scramble, and more MiG-21s into the engagement. (The Soviets were known to be making good use of data from satellites: Cosmos 596, 597, 598 and 600). The Israeli airmen flying *Rimon 20* headed home after shooting down four MiGs to report seeing only three Soviet pilots eject from their stricken MiGs. Hours after the event, Israel making an effort to obscure the identity of their victims, reported shooting down four Egyptian MiGs in the El-Sukhna area south of Suez City. Egypt rightly denied the loss of any aircraft in the Canal Zone. From the engagement, three tactics were seen: MiGs flying into their opponents at high speed to break up their formation: Mirages luring the Soviet flown MiGs toward flights of IAF Phantoms flying underneath the radar, and IAF airmen dividing into pairs and single aircraft to tease the MiGs apart even further.

With neither Israel nor Egypt able to demonstrate any advantage, international pressure to end the pointless attrition compelled Israel to agree to a ceasefire effective from 7th August 1970. Arab leaders who met to discuss the predicament of the Israeli occupation of their lands, agreed not to recognize Israel, not to negotiate with Israel, and not to make peace with Israel, so it would only be a matter of time before war would erupt again. Prior to the 1967 Six Day War, no Israeli city lay within more than a few minutes flying time from a potentially hostile airfield. Now however, the territory Israel had acquired during the Six Day War gave them useful buffers of both time and space.

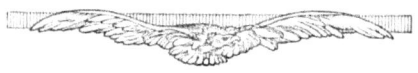

THE OCTOBER WAR ~ 1973

When President Nasser of Egypt died of a heart attack in September 1970, he was succeeded by Anwar Sadat, who was determined that the Egyptian Army would take their revenge against Israel. Israeli intelligence failed to take Anwar Sadat seriously. Even in September 1971 when Sadat offered ; *If Israel committed itself to the withdrawal of its armed forces from Sinai and the Gaza Strip under the provision of UN Security Council Resolution 242, Egypt would then be ready to enter into a peace agreement with Israel.* Following Israel's rejection of his peace offer, Anwar Sadat flew to Moscow to ask the Soviet Premier Leonid Brezhnev to re-equip the Egyptian Air Force. Sadat asked for the latest Soviet fighters to contest Israeli air superiority and re-conquer the Sinai Peninsula. After the Soviets refused the Egyptian request, Sadat said in frustrated English : *To be or not to be.* The Egyptians with initiative then created a substitute war plan requiring no new weapons. In an extension of the electronic summer, Egyptian SAMs would defy the established air superiority of the IAF by injuring any attempts to carry out ground support roles in the first days of the war. The original Egyptian plan to attack Israel in concert with Syria, code named *Operation Badr* (Arabic for Full Moon) aimed for a limited advance of 12 miles into the Sinai to gain a foothold on the far bank of the Suez, then consolidate under the protective cover of their SAM missiles to inflict a limited defeat on Israel, and alter the status quo.

To keep their innovative war plan secret from Israel, the Egyptians had to move their vast army to the Suez Canal without alarming Israeli lookouts on the eastern side of the Suez Canal. The Egyptians could not hide their ground forces, but they could hide their intention. Between October 1972 and October 1973 the Egyptians mobilized their forces 23 times to attune the Israelis into seeing Egyptian ground forces moving back and forth. Israeli commanders remained convinced that Egypt would not attack until they had the prerequisite air power. King Hussein of Jordan worried that his kingdom might become involved in another futile war, secretly flew to Israel to warn them, but the Israeli commanders thought he was mistaken. During an early October movement of Egyptian troops, the Israelis failed to react to two events. Egyptian troop movements under radio silence (something new) while a convoy of 300 Egyptian ammunition trucks headed out of Cairo towards the Suez Canal. Then on 3rd October, Israeli reservists manning a line of outposts along the Suez Canal reported seeing Egyptian troops hammering in poles, perhaps boat moorings they thought, indicative of Egyptian measures to cross the Suez Canal. The Egyptians had already buried their pontoon bridge sections in the dunes near the waterway. Then, on 5th October, Israeli agents reported the Soviet evacuation of their civilian and nonessential personnel from their embassies in Cairo and Damascus. Classic signs that hostilities were imminent. The next day was Yom Kippur, the Holiest day of the Jewish calendar.

Israel's border with Egypt, along the Suez Canal was mostly defended and patrolled by reservists doing their annual service. War began at 1400 hours on 6th October, when their defensive positions quaked from an Egyptian bombardment. An overconfident Israel had been caught unprepared. Egyptian commandos in rubber dinghies paddled across the canal to the far shore. Reaching the opposite bank, they scrambled up the sand ramparts to attack Israeli outposts. Israeli ground forces on the Suez line were stranded and surrounded as 23,000 Egyptians achieved their first day objectives of crossing the canal, pressing forward, and holding their captured ground. After the first wave of Egyptians crossed the Suez, tank-hunting detachments armed with Soviet AT-3 Sagger anti-tank guided missiles moved inland to shoot up 53 Israeli tanks from concealed positions. Four hours later, the first Egyptian tanks were ferried across the Suez Canal on a pontoon bridge. Pushing forward, on the first day, the Egyptian ground force advanced nine miles east across the Sinai Peninsula.

As the first flights of IAF Skyhawks and Phantoms screamed low over the desert, Egyptian SAMs and AA gunfire shot ten from the sky in the first 30 minutes. Rather than concentrating on the missile sites, Israeli airmen had been ordered to first attack the Egyptian mobile bridges, all hidden in smoke. Unable to penetrate the severe Egyptian missile cover, more were shot down. After their loss, the IAF forbid flights within ten kilometers of the canal, and waited for the Egyptian force to make an error. On the morning of 14th October, the Egyptians made their mistake, in resuming their offensive. Leaving their protective SAM air defence umbrella, the Egyptians advanced and began losing tanks to IAF air strikes. Israeli tanks waiting in the distance, moved up to exploit the open Sinai terrain, taking advantage of their long-range accuracy and unhampered maneuverability. As the Egyptian offensive stalled, Israeli General Ariel Sharon counterattacked through the seam of two Egyptian divisions. Israeli ground forces that crossed the Suez Canal with supporting tanks, destroyed 40 SAM launchers, then, called in air strikes to attack the Egyptian ground force. The Egyptian Army became encircled, and by 23rd October the road to Suez was captured, cutting off two Egyptian divisions of 45,000 men.

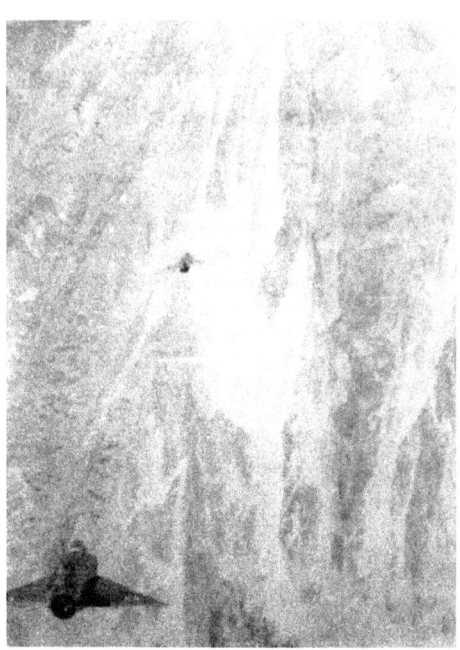

EAF MiG-17 targeted by an IAF Phantom F4   *MiGs in the gun sights of the IAF*

THE OCTOBER WAR ~ 1973

To the north, across the Golan Heights, Syrian forces with surprise on their side now attacked. With aircraft flying top cover, and spearhead by a force of 1,200 Syrian tanks, 60,000 troops overwhelmed the Israeli border outposts in attack timed for 1400 hours to coincide with the Egyptian assault to the south. Meanwhile, Syrian helicopters airlifted commandos onto Mount Hermon to capture a key Israeli observation post that dominated the surrounding area. As Syrian tanks continued trundling through Israeli defences, and crossed the fine farmland buffer of the Golan Heights, two Israeli brigades facing odds of five to one, heroically sacrificed themselves being wiped out by nightfall as the Syrian tanks retook the southern half of the Golan Heights. Then the Syrians blundered, when they paused 12 miles shy of Israel. This gave enough Israeli units time to recover, and within days they launched an armoured counter attack to recapture the Golan Heights. The Israeli tank crews fired more accurately and reloaded faster than the Syrians and on 10th October, only four days into the war, were able to force the Syrian tanks to withdraw.

After acquiring a broad inventory of Soviet air defence weapons, Syrian ground forces were protected by Soviet SA-2s, new SA-3 systems effective at altitudes as low as 350 feet, and SA-6 SAMs mounted on armoured cars to follow their ground forces into battle. White spirals of SA-6s climbing in the blue sky struck many Israeli Phantoms in the early hours of the Syrian offensive. Disappearing in puffs of grey smoke, they emerged as falling flaming hulks of shot down Phantoms. Syrian troopers on the ground said that it was rare to see parachutes of Israeli pilots who ejected in time. The Syrians also had SA-7 *Strella* (Russian for Arrow) shoulder-launched missiles, plus four-barreled radar-guided 23mm AA cannons mounted on ZSU-23 Shilka tracked carriers that fired up to 3,500 rounds a minute through the wide horizons of the Golan Heights. As the SAMs attacked, Israeli pilots flying over the Syrian front developed an evasive tactic of weaving in pairs, of crossing their jet effluxes to create even hotter spots in the sky to distract the heat-seeking missiles.

It was a hell of a fight. Wherever I turned I saw a Phantom behind a MiG, and a MiG behind a Phantom. These Israeli pilots were really good; it was not the standard of performance we saw at the start of the war. They had lost the new, inexperienced ones against our forest of missiles along the Suez Canal. I pulled behind a Phantom and attacked with my gun... my cannon shells hit the Phantom and it exploded like the sun. I had my own problems. I wanted to make a forced landing to save the plane but that was crazy. If I had tried it I would have been killed because other Phantoms had hit the runway that was now full of holes. At 500m I ejected. Then I went back to the Squadron, but I couldn't fly for the rest of the war. Qadri el-Hamid EAF

After the October War, Israel admitted that SA-6, SA-7 SAMs, and Shilka AA gunners had downed 33 Phantoms and 53 Skyhawks or around 20% of the Israeli Air Force inventory. The lesson from the October War for aviators was that a wise and auspicious enemy had proven that it could outfight a technologically advanced opponent in a limited offensive, by using a wide range of air defence weaponry to rebalance an inferior air force. After a series of great mistakes, in that they had failed to account for Egyptian changes in operations and tactics, the Israelis were nearly beaten in the first 48 hours of the Egyptian and Syrian October offensive. After the Israeli intelligence failures of the October War, they established the Office of the Devil's Advocate to look at specific military situations and ask : What if? What next? Then come up with an analysis to challenge the accepted wisdom.

Operation Nickel Grass ~ The outcome of the October War might have been very different had King Hussein of Jordan been prepared to join the fight and support the Syrian attack, especially when it is appreciated how Israel's resources had declined after just a few days of fighting. Israel having been caught by surprise then came under considerable logistical stress from their heavy consumption of all types of ammunition, in particular of Sidewinders. The United States authorized *Operation Nickel Grass* on 13th October to airlift a rapid resupply to Israel. The Arab states responded, declaring that they would limit or even halt all oil shipments to the U.S. if they continued to support Israel in the conflict. A complete oil embargo was placed on the U.S. that saw fuel prices rise, and become scarce during the 1973 Oil Crisis. Israel under global economic pressure heeded the call for a ceasefire. Four years later, on 19th November 1977, Anwar Sadat made a historic visit to Jerusalem to negotiate the peace he had previously offered the Israelis, and Egypt became the first Arab country to recognize Israel. Drifting further away from the Soviet sphere of influence, Anwar Sadat reopened the Suez Canal in 1975. More peace talks in America in 1978 led to Anwar Sadat and Menachim Begin signing a peace agreement at Camp David that resulted in normal relations. Israel soon after returned the Sinai Peninsula to Egypt.

MiG-23 ~ in Libyan markings ~ wings extended

Fast Eagle 102 ~ F-14 Tomcat flown by CDR Henry Kleemann   USN airmen Kleeman & Music who fought two LAF Su-22s

Sukhoi Su-22 M3 *Fitter* ~ in Libyan markings

THE GULF OF SIDRA INCIDENT ~ 1981

The U.S. Navy's new Grumman F-14 Tomcat variable sweep-winged fighter aircraft, which had been designed for short takeoff and landings from the flight deck of a carrier did so with its wings spread wide, and then to achieve its top speed of Mach 2.3 or 1,560 miles per hour at 36,000 feet, its wings were swept back through a delta shape similar to the French Mirage. With 25% of this twinjet two-seater aircraft constructed of titanium, a metal more suited to the heat of supersonic flight than aluminum, the F-14 Tomcat achieved magnificent agility. Armed with a six-barreled rotary cannon, (with a firing rate of 6,000 20mm shells per minute) and also equipped with radar-guided Phoenix and Sparrow missiles, the F-14 Tomcat also carried the new Sidewinder Lima with its *all-aspect capability* that could lock and home on enemy aircraft from any angle, which removed the historic tactical need to manœuvre behind a target. Projected at eye-level in a head up display (HUD) the advanced avionics of the Tomcat offered its pilot the combat data of : distance to target, closing speed, what weapon was primed to fire, and how many missiles or 20mm cannon shells were remaining.

In December of 1979, a mob of Libyan agitators attacked and burned the U.S. Embassy in Tripoli and their dictator, Muammar Gaddafi violated international law by claiming the Gulf of Sidra as Libyan territorial waters, with entry to the Gulf crossing Gaddafi's *Line of Death*. It was known that 2,000 Soviet air defence technicians maintained Gaddafi's SAM sites and 490 aircraft. On 19th August 1981, a pair of F-14 Tomcats from the carrier USS *Nimitz*, flying over the Gulf of Sidra, detected two radar blips climbing from Gurdebayeh Airfield, and then soar out over the sea. The two Tomcats, flown by U.S. Navy Commander Henry *Hank* Kleeman and Lieutenant Lawrence *Music* Muczynski monitoring their radar contacts, banked into position for a head on pass toward what appeared to be two Soviet-supplied Sukhoi-22s piloted by Libyans now on a direct course for the USS *Nimitz*. At a combined closure rate of 1,000 knots, the U.S Navy aviators saw a flame from one Su-22 as it launched an Atoll missile straight at them from a range of 1,000 feet. Kleeman and Muczynski turned sharp to port, to dodge the missile as the two Sukhois shot past. Each U.S. pilot targeted one Libyan. Commander Henry Kleeman had tone to launch his Sidewinder Lima, but the Libyan flew toward the sun to confuse any heat-seeker. As the Libyan Sukhoi veered away from the sun, Kleeman fired, striking the Libyan warplane. Lieutenant Muczynski launched a Sidewinder at the second Sukhoi that rocketed up the Su-22s tailpipe, detonated, and tore the Sukhoi in two.

Baghdad 1981~ At Osirak, 18 miles south of Baghdad, a French-built 70-megwatt Iraqi nuclear reactor, constructed for the purpose of atomic research for peacetime uses, was near completion. Despite assurances from France, the suspicious Israelis considered the reactor capable of producing the specific kind of isotopic uranium required for atomic bombs. After Iraq's official newspaper was so immodest to report that it had been Iraq's aim to acquire the means to build such a weapon, Israel resolved not to allow a potential enemy to attain a weapon of such potency. In the spring of 1981, the Iraqi reactor was weeks away from becoming operational. But at the moment it had not been stocked with nuclear fuel so there was no danger of a leak. However, when it became active, bombing it any time thereafter, would expose all of Baghdad and beyond to atomic radiation.

At 1640 on Sunday 7th June 1981, six missile-armed F-15s took off from Etzion Airbase, and flew east, with eight F-16 Falcons following their leader who carried two video-guided *smart* bombs. The other seven F-16s each carried a pair of delayed-action fused 2,200-lb unguided Blockbuster bombs. The Israeli formation flew east, zigzagging over Saudi Arabian desert nearing the Jordanian border. Then over southern Jordan, three F-15s were detected by JAF radar, however one IAF pilot responded in such perfect Arabic that he convinced the radar operator that he was part of a Jordanian air patrol. Over the Gulf of Aquaba, the IAF formation flew over the yacht of King Hussein of Jordan on holiday. King Hussein bi Talal, who deduced that they were heading for the Iraqi reactor, sent a warning but his message was never received. Half an hour before the Israeli warplanes arrived, the Iraqi soldiers manning the AA defences left their posts for their afternoon meal, and switched off their radars.

As the Israel pilots spotted the rooftop cupola of the reactor at Osirak, F-15 Falcon flight leader Ze'ev Raz led his formation to 2,100 feet, to swoop through a 35° dive at the reactor complex, aiming his video-guided smart bomb at the rooftop cupola, smashing a gaping hole through the containment dome of the reactor. The rest of his flight followed, releasing their bombs at five second intervals through the hole to detonate inside with a series of quaking explosions that collapsed the thick concrete roof burying the Iraqi reactor core under hundreds of tons of rubble. The attack had been timed for a Sunday, to spare the French workers at the site who always took the day off.

THE BEKAA VALLEY ~ LEBANON 1982

Israeli Ambassador shot in London ~ On 3rd June 1982, as the Israeli Ambassador to Great Britain, Sholmo Argov was leaving a diplomatic affair at the Dorchester Hotel in the centre of London and entering his car around 2300 hours ~ a young man who had been loitering outside the building took out a pistol and fired two shots. The first bullet narrowly missed Sholmo Argov's protection officer. But the second projectile went straight through Argov's head. The protection officer pursued the gunman into nearby South Street, and shot him. Police checking a Brixton flat, later arrested two other uninjured assassins who had fled the scene in a car. The detained gunmen, who confessed to being members of the radical Abu Nidal group of Palestinian terrorists (with ties to Syria, Libya and Iraq) were convicted of the shooting and sentenced to a total of 95 years imprisonment (both went insane). Argov, who remained lucid after the shooting, survived but could no longer move his hands or legs.

The gunshots of the Palestinian terrorists in London, served Israel with proximate cause to order ferocious air strikes over southern Lebanon against all known terrorist targets in and around Beirut. That in turn triggered the militants of the Palestine Liberation Organization (PLO) to retaliate with artillery and rocket attacks in a 24-hour barrage against Israeli farmers and settlers in northern Galilee. Between July 1981 and June 1982, PLO terrorist raids and shelling from southern Lebanon killed 25 Israelis and wounded tenfold more. Defence Minister Ariel Sharon used these developments to plan an incursion into southern Lebanon to destroy the bases of the PLO and remove the artillery threat to Israel. After Israel prudently notified Syria, that they wished to avoid any unpleasantries with Syrian forces located in southern Lebanon, Damascus reinforced its Lebanese detachment, deploying 19 SAM batteries in the Bekaa Valley as a deterrent gesture against Israel. The Israelis decided that the SAMs had to be suppressed, and the inept SAM operators invited disaster upon themselves.

The Syrians used mobile missiles in a fixed configuration; they put the radars in the valley instead of the hills because they didn't want to dig latrines ~ seriously. The Syrian practice of stationing mobile missiles in one place for several months allowed Israeli reconnaissance to determine the exact location of the missiles and their radars, giving the IAF a definite tactical advantage on the eve of battle. Even so, the Syrians might have been able to avoid the complete destruction of their SAM complex had they effectively camouflaged their sites; instead they used smoke to hide them. That made them easier to spot from the air. The Syrians also chose to ignore maneuver and camouflage. Alternate firing positions, defensive ambushes, regular repositioning of mobile SAMs to confuse enemy intelligence, and the emplacement of dummy SAM sites are fundamental considerations for the effective deployment and survivability of ground-based air defences. General Leonard Perroots USAF

On 9th June, Israel attacked through the Bekaa Valley ridges up to 6,500 feet high that framed the agricultural plain of lower Lebanon, some 10 miles wide, and 25 miles long. Having experienced loss from Soviet-supplied SA-6 systems previously, the Israelis had found the answer to defeating them. The Israeli anti-SAM missions began. As a decoy flight of pilotless drones were flown over southern Lebanon to bait the Syrians into switching on their SA-6 engagement radars. The duped Syrian SAM operators did so, preparing to initiate a ripple-fired fusillade of SAMs against the approaching enemy, the pilotless IAF drones. As positive SAM electronic radar emission signatures were detected, SA-6 radars close to the Israeli-Lebanese border were targeted by informed Israeli artillery moving forward. Syrian SAMs beyond the range of the Israeli gunners, were attacked by low flying flights of Phantoms, which howled in from different directions to fire long range Shrike anti-radiation missiles and drop cluster bombs on Syrian radar vans and their associated SAM launchers. The IAF had destroyed 17 of Syria's 19 SAM batteries and their radar sites ~ even after the Syrian SA-6 operators had launched 57 SAMs against them on the first day. The following day, the two surviving SA-6 launch sites along with some SAM sites that had been replenished overnight were targeted, and all destroyed.

As soon as attacking IAF Phantoms were detected, Syrian MiGs were scrambled to intercept. The Israeli planners had expected this, and flights of F-15 Eagles and F-16 Falcon fighters were waiting in the sky west of the Bekaa Valley to be vectored toward any Syrian fighters attempting to disrupt the SAM suppression raids. Up to 90 Syrian MiGs rose to challenge 60 IAF aircraft. Syria lost 35 aircraft. Israel suffered no loss. The main Syrian fighter over the Bekaa Valley was the MiG-21, plus some swing-wing MiG-23 all-weather interceptors, and a few Su-20s. It is of note, that the Soviet-supplied Syrian aircraft were stripped-down export models. Even so, Syrian pilots showed little skill in the air.

The IAF flew McDonnell Douglas F-15 Eagles and General Dynamics F-16 Falcons, as well as their older, but still very effective F-4E Phantom IIs. The F-15s and F-16s, in addition to having better acceleration and maneuverability at combat speeds, had superior radar and cockpit visibility that often resulted in early detection of the enemy, and in the delivery of unexpected shots.

THE BEKAA VALLEY ~ LEBANON 1982

These lethal shots came in the form of 20mm cannon shells during close range engagements, or from a mix of radar-homing AIM-7F Sparrow missiles, and improved AIM-9L *all-aspect capable* Sidewinder Limas, now so sensitive that they could home on the friction of an enemy fuselage thrusting through the air, and these wonders would account for the majority of air-to-air kills over Lebanon. Radar-homing Sparrows that could be launched beyond visual range enabled Israeli airmen to fire missiles at their Syrian opponents, very often undetected from launch till impact to deny the Syrians any opportunity to either escape, evade or to return fire. The Syrians had no comparable missiles, having to rely on their 1960's era Atoll AA-2 air-to-air missiles.

Jamming ~ The IAF working to obstruct Syrian command networks while enhancing its own, made effective use of a modified Boeing 707 electronic intelligence (ELINT) platform equipped with stand-off jammers capable of disrupting several enemy frequencies at once with little out-of-phase disturbance, in so doing minimized any self-jamming of friendly frequencies. Using other airborne jammers aboard *Jolly Green Giant* helicopters, a thorough jamming of Syrian communications and radar systems isolated Syrian MiGs from their ground controllers, leaving them even more vulnerable to the attention of Grumman E-2C Hawkeye AWAC (airborne warning and control) aircraft orbiting off the coast of Lebanon.

With a 200-mile surveillance range, the Hawkeye AWAC monitored Syrian radar emissions and transmitted any threat data to attacking Israeli fighters or relayed intercept vectors to patrols of IAF F-15 Eagles and F-16 Fighting Falcons if any Syrian MiGs were detected taking off to intercept approaching Phantom on fighter-bomber missions. Israeli jamming caused such chaos that Syrian MiG airmen were seen flying figure eights with no idea what to do after finding themselves in a communications blackout, deprived of any contact from their ground control centre.

By the end of July 1982, the Israeli Air Force, with the advantages of the F-15's look-down radar and the all-aspect capability of the AIM-9L Sidewinder Lima missile had downed 84 Syrian aircraft, for the loss to themselves of a few helicopters, one Phantom and one Skyhawk downed by PLO fired SA-7s. Part of the disproportionate Israeli score was attributed to their adoption of the same hardware, training, and concepts of operations (such as constant retention of initiative) that had been developed in America by aviators such as Robin Olds during the 1970s. Another reason, was the ability of IAF fighter pilots to fly surprise enhancing blind-side tactics against Syrian MiGs made possible by having Hawkeye AWAC aircraft vector F-15s and F-16s into side attacks against Syrian MiGs from where their Soviet radar systems were known to be, next to useless.

Roof Knocking ~ The code name the IAF used to warn the inhabitants of buildings, to give them time to flee an air strike on a suspected PLO or Hamas terrorist weapons cache stored in or under mosques, schools or other targets formerly considered off-limits by the IAF. During air operations over southern Lebanon and Gaza, Israeli intelligence officers used the ruse of telephoning civilians in good Arabic, pretending to be a sympathetic Egyptian or Jordanian expressing horror of the war, asking about their family and local conditions, being supportive ~ *Oh, God help you, God be with you.*

The caller then asked whether the family supported Hamas and if there are any fighters living near, this time gathering intelligence to preplan future targets. Because Hamas built its tactics around the use of human shields, Israel strived to distinguish between the Hamas terrorists and those innocents living in the middle, Israeli security agents continued to telephone the residents of specific buildings to give them up to 15 minutes of warning time before an airstrike.

The war in Lebanon, which had been sparked by the assassination attempt on Sholmo Argov, resulted in the expulsion of the PLO from Lebanon and an 18-year Israeli military presence in southern Lebanon that ended with Israel's complete withdrawal in May 2000. Sholmo Argov in 1983, from his hospital bed in Jerusalem (where he had been ever since his return from London) dictated these comments to a friend, who passed them to a columnist on the Israeli newspaper *Ha'aretz,*

If those who planned the war had also foreseen the scope of the adventure, they would have spared the lives of hundreds of our best sons. They bought no salvation. Israel should go to war only when there is no alternative. Our soldiers should never go to war unless it is vital for survival. We are tired of wars. The nation wants peace.

* * *

Sea Harrier ~ Falkland Islands

Flight-Lieutenant David Morgan RN ~ DSC 4 kills   Commander Nigel *Sharkey* Ward RN ~ DFC AFC ~ 3 kills

CHAPTER XII

THE FALKLANDS WAR

The 1982 Falklands War ~ The old adage of war, that the crisis in which you find yourself is never quite the one you expect, applied well to this campaign. Sovereignty over the Falkland Islands, a windswept British dependency 300 miles east of the Strait of Magellan has been disputed by Britain and Argentina since 1690, when Captain John Strong first set foot on the Falklands and Commodore John Byron raised the Union Jack over West Falkland. The British returned to the Falklands in 1833 with two warships, HMS *Tyne* and HMS *Clio* led by Captain James Onslow to evict Argentinean settlers, and reclaim the islands for Britain. Since then, most islanders have been of British extraction. Then in March 1982, near the old whaling station of Grytviken on South Georgia, the Argentine naval vessel, *Bahia Buena Suceso* arrived to disembark a party of scrap metal merchants with the manner of soldiers who raised the Argentine flag. When a British Antarctic Survey team found the Argentine ship in the harbour, and 30 salvagers ashore with the Argentine flag flying, the incident was reported to Governor Rex Hunt in Stanley, who demanded they take down their offensive flag.

The initial international reaction was, that the whole thing was too much like a comic opera plot to be taken seriously, however the Argentine Navy were at sea and they were not going to back down. Their comic opera was in danger of becoming *Hamlet*, particularly in London where it was viewed that South Georgia and the Falkland Islands were British territory, and British citizens were under threat. Urged on by the British Prime Minister Margaret Thatcher, U.S. President Ronald Reagan intervened with a personal telephone call to the leader of the repressive and brutal Argentine junta, General Leopoldi Galtieri on the evening of 31st March. Reagan got nowhere. Many inept and self-delusory officers beyond his control, who had convinced themselves that Britain was a softhearted democracy, and that the British had no stomach for a fight, were sweeping General Galtieri along. The U.S. Secretary of State, Alexander Haig who flew down to Buenos Aires in Argentina to negotiate said: *I might make better progress if I can get to Galtieri when he is sober.* Haig often got to Galtieri or found that his Foreign Minister Costa Méndez could agree on something, only to have another junta member appear on the scene to say it was unacceptable or that the invasion fleet had already been committed. Admiral Jorge Anaya, the main advocate of the Falklands invasion, went so far as to call Haig a liar. Haig facing the real tragedy of a comic opera government told President Reagan that hostilities were inevitable and that the U.S. should back Britain. The U.S. Government authorized a secret shipment of AIM-9L Sidewinder Lima missiles to give British airmen the edge in their coming air-to-air combat. The British nuclear-powered submarine HMS *Conqueror* set course for the Falklands, and with its low noise signature would operate at minimal risk from the modest Argentine anti-submarine capabilities.

On 2nd April 1982, as Argentine forces invaded the British territory of the Falkland Islands, Governor Rex Hunt informed the Royal Marine officers commanding the Falkland defenders with the profound words: *It looks like the buggers really mean it.* As the Argentines came ashore on East Falkland the 80 Royal Marines and sailors of Naval Party 8901 opened fire, causing loss to the Argentines. When six armoured personnel carriers rumbled within anti-tank missile range Marines Gibbs and Brown fired, wrecking one enemy vehicle. After holding off an encroaching Argentine battalion for three hours, Governor Hunt asked the small force of bold Royal Marines to lay down their weapons. The Argentines then humiliated the British prisoners of war, by publishing pictures of Royal Marines surrendering, and lying on the ground. The Argentine CO General Mario Menendez began deploying 2,000 troops on West Falkland, 1,000 around Goose Green, and 9,000 on the heights west of Stanley. The island's main airfield outside of Stanley consisted of a single 4,100 feet runway with one taxi strip leading from the runway to a small maintenance hangar. Menendez overlooked the idea of extending the runway when it should have been his highest priority. When the U.S. Marines captured the island of Guadalcanal in 1942, their Seabees repaired and extended their crushed coral runway in two weeks. The terrain at the eastern end of the Stanley runway was level enough to be graded to an operational length of 6,000 feet. If the Argentine Air Force (*Fuerza Aérea Argentina*) were able to operate attack aircraft from Stanley, they may have forced the British carriers further away from the Falklands zone. Instead of this unrealized potential, the FAA dispersed a few Pucaras and Iroquois helicopters on small grass airfields. The Argentine defenders set up 35mm and 20 mm AA gun positions around their airfields, and airlifted in supplies using Hercules C-130 flights, sometimes landing as often as every two hours, with average turnaround times of 30 minutes.

THE FALKLANDS WAR

When Argentina invaded the Falkland Islands, they initiated a conflict that the British had never made any contingency plans for. So, they dusted off an old plan for Norway and simply amended it. Margaret Thatcher announced to the House of Commons that a naval task force would be sent to the South Atlantic with orders to retake the Falklands. All Argentine forces had to withdraw, and the wishes of the Falkland Islanders were paramount. The British attempt to recover the Falkland Islands would be hazardous, in the face of an absence of any real intelligence reports on the Argentine fleet. Britain's gathering Royal Navy task force, commanded by Rear Admiral Sir Sandy Woodward included the carriers HMS *Hermes* and HMS *Invincible*, escorted by four destroyers, five frigates, and three nuclear powered submarines. On Monday 5th April 1982, the British task force steamed south towards the Falkland Islands 7,500 miles away. Argentina deported the Royal Marines of Naval Party 8901 who had defended the Falkland Islands, and as part of the task force, they were now returning.

Ascension Island. The Royal Navy first settled this mid-Atlantic island when it established Georgetown as Napoleon Bonaparte was exiled to St Helena. During 1942, U.S. Army engineers constructed Wideawake Airfield with dispersal facilities for 24 large aircraft as a useful staging post between the U.S.A. and the theatres of war in North Africa and Southern Europe. However, after U.S. troops were withdrawn at the end of WWII, the airstrip on Ascension fell into disuse. After some improvements, Ascension Island became the forward airbase supporting the Royal Navy task force as they steamed south toward the Falklands. Relocated RAF Lyneham Hercules C-130s flying south from Ascension were the only way of delivering urgent/vital supplies to the carrier group, flying down to drop packages by parachute into the sea, with flotation to permit the stores to float until picked up by the ships. The airdrop technique always worked well, but once a ships boat sent to drag a large load alongside a destroyer for pick up, took three hours to do the job. A killer whale had fallen in love with the parcel the Hercules had dropped, and chased away the naval boat every time it tried to get near. Eventually the destroyer itself had to chase off the amorous orca and recover the essential cargo.

Vulcan operations. The RAF worried that Argentine aircraft operating from Stanley on East Falkland posed a threat to British ships (particularly the carriers), considered how to deny enemy aircraft the use of the airfield. During early 1982, RAF Vulcan bombers were being withdrawn from service, and three had just been presented to various USAF museums. Then someone discovered the hard truth that no Vulcan in the UK had an air-to-air refueling (AAR) probe. The AAR probes were needed in a hurry, and seemed unobtainable. What followed for the British was quite embarrassing. *The Vulcans we just gave to the USAF museums. They had probes on them didn't they? Yes they did.* So a team of RAF technicians flew across the Atlantic, arrived in civilian clothes, and went around various USAF museums to remove the necessary Vulcan probes from the recently arrived museum exhibits.

The first Vulcan bomber raid, refueling en route to the Falkland Islands came on 1st May when the lone Vulcan XM607 piloted by Flight Lieutenant Martin Withers flew from Ascension Island to bomb Stanley Airfield. Withers flew the Vulcan on a high-low-high profile approach from 25,000 feet, descending to avoid Argentine radar, then climbing to 10,000 feet for his bomb delivery. Through the darkness at 0446 local time, Withers released his 21 1,000-lb conventional iron HE bombs, scoring at least one hit on the runway. Turning for home, Withers radioed ~ *Superfuse* ~ the codeword to convey the success of his mission. A few days after this first Vulcan strike, the Argentines redeployed a squadron of their Mirage III fighters north to protect their capital, which reduced their ability to escort anti-shipping flight missions against the approaching British naval task force. Five Vulcan missions were flown, three against Stanley Airfield, after which the Argentines made the runway appear damaged by heaping mounds of earth up overnight, shaping mock bomb craters. This deception, mislead the British as to the condition of the airfield and the success of their raids. The fourth Vulcan air raid carried two AGM-45A Shrike anti-radiation missiles to destroy the U.S. designed and supplied Westinghouse surveillance radar located near Stanley ~ a real pain for the Royal Navy with its ability to track their carriers and provide targeting data for FAA anti-shipping attacks. However, when this Vulcan strike came, the Argentines resorted to the tactic successfully used by the North Vietnamese. After detecting any inbound aircraft on their radar, the Argentines simply switched off their radar. After each Vulcan air strike, Argentine airmen took off from Stanley within 24 hours of any attack. The runway remained open to FAA Pucara flights and their Hercules aircraft flew 33 daring night missions into Stanley bringing in men and stores (and flying out 263 wounded plus one British POW).

THE FALKLANDS WAR

The Sea Harrier ~ In the British war plan to retake South Georgia and the Falkland islands, the British task force needed three types of aircraft: a naval interceptor to protect the fleet, a ground-attack warplane to pound Argentine ground defences, and an agile infantry support aircraft to cover British ground forces as they advanced from their landing beach toward the main enemy garrison at Stanley. The versatile British Sea Harrier, a vertical short-takeoff and landing jet with such advanced avionics that allowed its pilot to fly day or night in any weather, filled all three of those roles. With a maximum speed of 690 mph, the Sea Harrier would face Argentina's faster Mirage fighters over the Falklands, but the Mirages had to sacrifice their speed margin, to come in low under the radar heavily loaded. The Sea Harrier, designed to fight low and slow, would use its maneuverability to stop midair, hover or veer off sharp in fresh directions. The same exhaust nozzles that gave the Sea Harrier its ability to take off vertically also enabled it to outmaneuver conventional warplanes by using a tactic known to Harrier airmen as *viffing* (Vector In Forward Flight). By adjusting their exhaust nozzles to reverse thrust, Harrier airmen caused their mounts to decelerate and veer away to the side. In coming hostile engagements, a sudden VIFF would see pursuing fighters overshoot and fall victim to the Sea Harrier's twin 30mm Aden cannons or two Sidewinder Lima missiles. Facing a frontal opponent, the Sea Harrier pilot could use his forward-and-down scanning *Blue Fox* radar to plot an adversary up to 40 miles away, until a graphic display on his windshield flashed tracking data indicating to the pilot, when to fire. Using the AIM-9L Lima, the pilot aimed within the general direction of his target, and pressed to launch. The IR sensor did the rest, homing in on the detected hot engine or exhaust nozzle.

The great concern of the Royal Navy during the Falklands War was their task force's vulnerability to Argentine air attacks, and the possible attrition rate of their Sea Harriers, not only from enemy action, but also from weather and other hazards of operating in the South Atlantic in the middle of winter. The British naval task force that set sail for the Falklands only had the 20 Sea Harrier aircraft of 800 and 801 Squadrons, Fleet Air Arm at sea with them. When it was appreciated how these few aircraft were expected to provide the task force's only air cover against some possible 230 FAA warplanes at the outset of hostilities, most minds thought they might need some backing up, so the navalization of No. 1 Squadron RAF's 18 Harriers became a top priority. The RAF Harriers received anti-corrosion weatherproofing and were updated for Sidewinder Lima operations. Nine aircraft of the second wave were also fitted with IRCM flare dispensers. At the same time, the Harrier pilots of No 1 Squadron RAF practiced carrier ski-jump and deck landings, and received intense training in CAP missions and armed reconnaissance tactics. When complete, the Harriers flew south to Ascension, then onto the aircraft hide on the container ship MV *Atlantic Conveyer's* deck between three walls of containers. Next, the Harriers flew to the task force, landing on the carrier HMS *Hermes*.

Operation Paraquet ~ Several Royal Navy warships led by the destroyer HMS *Antrim* detached from task force to repossess South Georgia Island. In the morning of 25th April, London warned that the WWII era Argentine submarine *Santa Fé* was on her way toward South Georgia with men and supplies. The British ships distanced themselves, then launched armed helicopters to hunt the enemy submarine. Lt-Cmdr. Ian Stanley took off from HMS *Antrim* flying a Wessex named *Humphrey* on an anti-submarine search. Detecting a radar blip, Stanley spotted *Santa Fé* cruising on the surface outside Cumberland Bay. *Humphrey* attacked the submarine, dropping two depth charges detonating close to her port casing that forced the damaged *Santa Fé* to steer for Grytviken. A Lynx helicopter took off from HMS *Brilliant* to air-launch a torpedo at *Santa Fé*, but missed. Next a Wasp helicopter, piloted by Lt-Cmdr. J.A. Tony Ellerbeck DSC from HMS *Plymouth* fired two AS12 missiles that exploded inside the submarine's large fin. HMS *Plymouth's* Wasp fired a third AS12 missile at the Argentine submarine while *Brilliant's* Wasps strafed her hull with machine gun fire. The British warships now closed in at full steam to finish her off. Captain Bincain, commander of the disabled *Santa Fé* unable to submerge, leaking oil, and streaming smoke beached his stricken vessel alongside the British Antarctic Survey jetty at King Edward Point. Listing and apparently on fire, her crew ran off toward Grytviken. *Antrim* and *Plymouth* far out in Cumberland Bay fired a spirited 235-round barrage of 4.5-inch shells. None touched the Argentine troops defending King Edward Point. The point being, that this is what will happen if you do not surrender! In a brisk operation, *Antrim's* Wessex *Humphrey*, and *Brilliant's* Lynx helicopters, flew relays to land a detachment of Royal Marine/SBS/SAS troops at Hestesletten. After forming up, they advanced through the abandoned and derelict whaling station at Grytviken. At 1715 hours as British ground troops neared King Edward Point, the Argentines after just 23 days at the rough settlement ~ without firing a shot ~ sang their national anthem and hoisted three white flags. With South Georgia safely back in British hands, the Battle for the Falklands could begin.

THE FALKLANDS WAR

Sinking of the ARA General Belgrano ~ The British declared a 200-mile Maritime Exclusion Zone around the Falkland Islands, warning that they would attack any Argentine vessel entering the zone. The submarine HMS *Conqueror* (Captain Christopher Wreford-Brown) approached to patrol the zone. On 28th April, HMS *Conqueror* detected a large vessel at long range nearing the Falkland Islands. It was an old cruiser that had been launched in 1938 as USS *Phoenix*. She had survived Pearl Harbour, to be sold to Argentina, and re-named *General Belgrano* was commanded by Captain Hector Bonzo : *Our mission wasn't just to cruise around on patrol, but to attack. We were anxious to pull the trigger.* Threatening the task force, the British war cabinet sanctioned the destruction of the old cruiser. *Conquerer* launched two 21-inch torpedoes… after 50 seconds one hit amidships, seconds later another hit, tearing off her unarmoured bow, sending the *General Belgrano* to the bottom of the South Atlantic. With her perished 322 sailors. The remaining Argentine Navy lost their resolve, and turned for home.

HMS Sheffield ~ On Tuesday 4th May, two FAA Super Étendard aircraft took off from Rio Grande, each armed with an Aérospatiale Exocet AM 39 anti-ship missile, to be fired from a stand off position. The Argentine airmen flying a circuitous route flew nearer to launch a low-level surprise attack. When 25 miles away from the British warships, the two FAA airmen rose to 120 feet, switched on their radars for a target check, then air-launched their Exocet anti-ship missiles, and nosed down to 50 feet to turn for home. Rear-Admiral Sandy Woodward had positioned three destroyers, HMS *Glasgow*, HMS *Coventry*, and HMS *Sheffield* well forward on radar picket duty. The two approaching Argentine aircraft were detected by the radar watch onboard HMS *Glasgow* who reported it to their sister ships: *Coventry and Sheffield this is Glasgow. We're holding a passive radar contact. We'll maintain our warning red. Suggest you do likewise.* However, HMS *Sheffield* some 100 miles south of Stanley only received part of the warning from HMS *Glasgow* and far worse : HMS *Sheffield's* radar frequency at that moment was blotted out by a satellite transmission from her communications equipment. The two Exocets fired by Captain Augusto Bedacarratz and Lt. Armando Mayora, skimmed toward the Royal Navy destroyers.

The first Exocet missed HMS *Glasgow*, but the second missile rocketed into HMS *Sheffield*, piercing her amidships and punching deep inside the British destroyer. The warhead did not explode, but the impact and its unburnt fuel, sparked fires of acrid smoke that soon surged out of control. With 20 dead sailors, 26 wounded and a damaged warship ablaze with minimal power, Captain Sam Salt's crew fought to save their sinking ship, but the order went out in the afternoon to abandon *Sheffield*. HMS *Arrow* took off 260 survivors, and the *Sheffield*, gutted and deformed by her still burning fires, drifted for four days before it sank. Had the Royal Navy some form of airborne early warning aircraft, it would have been able to detect enemy aircraft before they ever came within firing range. The only prior warning of an air attack came from four Royal Navy submarines operating at periscope depth off the Argentine coast in the vicinity of FAA airfields or from picket ships in advance of the Fleet.

After the sinking of HMS *Sheffield* on 4th May, and until mid May, things seemed relatively quiet. However, further Vulcan raids targeted Stanley Airfield, RAF Hercules aircraft dropped spares to the Fleet, and the two flights of No. 1 Squadron RAF Harriers had transited 3,000 miles from Ascension Island to the Royal Navy task force. Sea Harrier attacks and naval shore bombardments softened up the Argentines. To avoid the threat of enemy ground-to-air weapons and small arms fire, Sea Harrier pilots simply flew outside or below their respective engagement zones. Additional threats came from plentiful Argentine shoulder-launched Blowpipe and Russian SAM-7 missiles. The British airmen flew fast and low to negate this threat. What hit the Sea Harriers was infantry small arms fire with 25% of Sea Harriers returning from missions with holes. Apart form one Sea Harrier that suffered a serious fuel leak all others returned safely where RN engineers affected some ingenious repairs, so no Harrier spent longer than 48 hours dormant in the hangar before it was flying again.

Landing at San Carlos ~ After United Nations talks failed on 20th May, the Royal Navy carried out an unopposed night raid through the sheltered deep waters of San Carlos Bay in the Falkland Sound. Then, before sunrise in the early morning of 21st May, British troopships approached the landing beach in San Carlos Bay. Out of the range of enemy artillery, the surrounding high hills shielded the off loading British landing ships against any attacking Argentine aircraft whose radar was useless near the obscuring hills. The first British troops landed in a gale, spread out, and took all the high ground around San Carlos without a casualty. The bridgehead was secured by 3rd Commando Brigade (Brigadier Julian Thompson), and broadened by the men of 2nd Parachute Regiment, known as 2 Para who advanced to the top of Sussex Mountain to dig in, and secure the overlooking heights.

ARGENTINE AIR TACTICS

Because of their limited flying time over the Falkland Islands, FAA pilots flew predictable attack routes that gave an immense advantage to the Sea Harrier pilots searching visually for them. British airmen just waited for incoming Argentine aircraft at points over the north and south coastline of West Falkland, at landfalls where low-flying Skyhawk pilots routinely checked their navigation. Invariably the Argentine airmen flew close by these points, and the Sea Harrier pilots intercepted. Even if Sea Harrier aircraft were able to operate in weather conditions impossible for conventional carrier aircraft, the threat of Exocet missiles kept the two carriers at ranges limiting the effective flying time of Sea Harrier patrols. After the hard experience of their early attacks on the British fleet, FAA pilots needed to fly in fast and low to survive the combined threat of RN ship-to-air missiles, and the Sea Harriers. Above 50 feet, Argentine Skyhawk pilots believed they would be shot down, but below 50 feet salt spray obscured their ability to see detail through their gun sights. So, for anti-ship missions, Skyhawk airmen drew two lines on their windscreens, to bracket a targeted ship between the lines on ingress, and release their bombs as the ship passed under the nose of their Skyhawks.

The Argentine FAA tried in vain to fly Mirage III escorts for Skyhawk missions. Because the Mirage maneuvered best at high altitude, they flew top cover for Skyhawks approaching through lower altitudes more favourable to the Sea Harrier. The British took full advantage. Pairs of Sea Harriers flying efficiently below 10,000 feet waited to intercept the Skyhawks and their escorting Mirages. Such an outcome transpired on 1st May, when an Argentine radar crew on the Falklands transmitted to escorting Mirage pilots, Captain Gustavo Cuerva and his wingman, Lt. Carlos Perona that two Harriers were on their 12 o'clock. Cuerva spotted the Harriers, and fired two Matra missiles. One misfired and wobbled into the sea. The other failed to lock on and flew off without success. Cuerva and Perona rather than engage the British, dumped their ordinance and veered off to escape. As the Harriers banked around ~ in their standard flight pattern one mile apart so that each pilot could cover the unseen aspect behind and below his brother airman's aircraft ~ the Harrier flown by Flight Lieutenant Paul Barton launched a Sidewinder at Perona's Mirage that broke it in two. The next Harrier piloted by Lieutenant Steve Thomas launched a Sidewinder that shattered Ceurva's Mirage. The sixth Mirage III sortie that day flown by Lt. José Ardiles was interrupted by Lt. Tony Penfold, who fired a Sidewinder that detonated near Ardiles' Mirage, causing it shed parts and crash in the sea.

Air Battles of 21st May ~ The first Argentine reaction to the British landings at San Carlos came from their Falklands based Pucara aircraft on Goose Green, one of which was shot down over Sussex Mountain by a D Squadron SAS fired Stinger SAM missile. Sea Harriers shot down another Pucara, plus nine FAA Daggers. HMS *Broadsword* fired a Sea Wolf missile, ripping apart another Pucara. Fusillades of AA fire from elite British infantry ashore and Royal Navy vessels offshore compelled FAA pilots to zoom in low and fast, but their bombs, often well aimed and on target, failed to explode. The main Argentine air raids came at 1030, 1300, and 1500 hours. At 1300 eight Argentine Skyhawks near-missed the frigate HMS *Ardent* with two bombs. Four more Skyhawks in a tight V-formation came over Chatres in West Falkland, but were sighted by Sea Harriers flown by Lt-Cmdr. Mike Blisset and Lt-Cmdr. Neil Thomas. The Skyhawks turned low to escape and were shot out of the sky by Sidewinders near Christmas Harbour. Then ten Daggers streaked over West Falkland. HMS *Brilliant* vectored two Sea Harriers at them. Lt-Cmdr. Fred Frederiksen fired a Sidewinder at 1435 hours into one that went down near Teal River inlet. HMS *Brilliant's* radar watch, plotted a flight of three Mirages coming over West Falkland and vectored two Sea Harriers at them. Commander *Sharky* Ward and his wingman Lieut. Steve Thomas waited north of Port Howard flying a racetrack pattern. At each turn they looked to check each other's six and spotting them, rolled out of their turns on full thrust to lock on. The Argentine pilots spotted the Harriers and firewalled their throttles for home. Lieut. Steve Thomas got behind them, and fired a Sidewinder Lima that shattered the first Mirage. Thomas locked onto the second Mirage, thrusting off on full afterburner and fired a Sidewinder that exploded beneath its wing disintegrating it. The third Mirage fired at Ward, and snap dived to escape.

I looked round and there was this Mirage passing underneath me, beautiful colours. Going ahead, all I had to do was pull down hard because I was only at about 300 feet. I pulled down hard and he was right on the deck. He didn't stand a chance because I got in behind him and fired my missile. It hit, a full motor burnout and the whole Mirage exploded, the right wing cart wheeling through the air and I thought that guy didn't get out… But he did. He must have seen the missile coming at half a mile and ejected. I didn't see that being directly behind. All I saw was a great flame, then pieces, so we got three confirmed kills. Commander Nigel Ward

ARGENTINE AIR TACTICS

HMS Ardent ~ 30 minutes later, Captain Alberto Philippi was leading his flight of three FAA Skyhawks toward the Falklands at 27,000 feet. From 100 miles off, they descended to 100 feet and nearing Falkland Sound sighted HMS *Ardent* off North West Island. The Skyhawks slashed toward the stern of HMS *Ardent*. Captain West turned HMS *Ardent* at full speed to bring his AA guns to bear, but whichever way he turned, there was a Skyhawk lined up for an attack run against *Ardent's* stern.

At 1000m from the target, I ascended from 30m to 100m and concentrated to arrange the crosshairs of my sight on the stern, without thinking of anything else, it was aligned. When the center of the crosshairs was superimposed on the target, I pressed the button that launched the bombs. I increased speed and commenced a snap turn to escape to the right, descending once again to fly just above the waves... Meanwhile I listened to the voice of my wingman José Cesar Arca shouting ~ Well done, sir! ~ I looked over my shoulder and saw the frigate with much smoke in the stern.
Captain Alberto Philippi

HMS *Ardent's* rear flight deck looked as if a giant can opener had opened it up. The two trailing Skyhawks targeted and bombed *Ardent's* stern. Their hits failed to explode, but still disabled her missiles. HMS *Ardent* was on fire aft, and flooding fast. With 22 fallen sailors and many casualties, Commander Alan West signaled to HMS *Yarmouth* ~ Take off my men. HMS *Ardent* sank the following morning. As the Argentine Skyhawks followed the same route out as they arrived, two Sea Harriers flown by Lt. Clive Morell RN and Flt-Lt. John Leeming RAF intercepted them near Swan Island.

I immediately ordered external cargoes jettisoned and to escape with the hope of reaching refuge in the clouds that were in front of us. But I felt an explosion in my tail and the nose of my warplane elevated uncontrollably : I needed the support of both hands on the stick that was unresponsive. I looked to the right and saw a Sea Harrier at 150m coming in for the kill. I reduced speed, and ejected. I felt a forceful explosion when the canopy popped, and immediately there was a forceful pain in the nape of my neck. My final thought before passing out was ~ I am falling like a rock.
Captain Alberto Philippi

Philippi's Skyhawk had been hit by a Sidewinder launched from Lt. Clive Morell's Sea Harrier. Captain Philippi landed in the sea 100m from the coast and swam to shore. He walked for two days before being rescued by an FAA helicopter. Clive Morell's wingman, Lt. John Leeming fired short bursts from his twin 20mm cannons at Lt Marquez's Skyhawk, low above the water, all just missing. Marquez broke right. Leeming fired. A long burst thumped into his Skyhawk that exploded in a flash.

Missile trap ~ On 23rd May, Argentine air attacks resumed. HMS *Broadsword* and HMS *Coventry* positioned as radar picket vessels off Pebble Island posed a threat to FAA aircraft as a missile trap. Under the watchful eyes of HMS *Broadside's* radar room, a flight of four Argentine Dagger aircraft approached low over Pebble Island. Two Sea Harriers were vectored toward them to intercept. Firing three Sidewinders, they downed three Daggers. In the morning of 25th May, Argentine National Day, the first flight of Skyhawks came over. HMS *Coventry* launched a Sea Dart missile, shooting down one. The rest of the flight abandoned their mission and turned away. Near midday, HMS *Coventry* launched her second Sea Dart of the day, shooting down another Skyhawk. At 1400, four Skyhawks flew low over the folds, and contours of the hills to hide from British radar ~ then emerged to attack.

As we began our run (one minute out at top speed) they fired a missile. I don't know what kind of missile it was; perhaps a Sea Dart because the frigate was of the Type 42 series, which turned out to be the Coventry. The missile passed over us at about 900 to 1200 feet. Once I framed the ship in my sight, I saw nothing else for about 30 seconds but I felt something like the impact of cannon fire before releasing my bombs. As we were coming toward the frigate, we noticed that it quickly began to change course 90°. By the time I released my bombs I was 40° off from the ship. We passed over and noticed three bomb hits on it. Flight-Lieut. Jorge Neuvo

HMS *Coventry's* desperate evasion was her ruin. She had steamed in front of HMS *Broadsword's* bow negating her Sea Wolf missile system's target solution on the four incoming Argentine Skyhawks. Her desperate move blocked any chance for HMS *Broadsword* to fire, and HMS *Coventry* was lost. Three of four bombs dropped penetrated HMS *Coventry's* port side killing 19 sailors and wounding 25. With her power all gone, HMS *Coventry* listed heavy to her port and began to sink. Slowly capsizing, RN helicopters and HMS *Broadsword* came to retrieve survivors from the doomed HMS *Coventry*. Two Skyhawk airmen targeted *Broadsword*. One bomb missed. Another hit the sea, bounced up through her quarterdeck destroying the Lynx helicopter then fell over the side, exploding harmlessly in the sea. The Royal Navy had suffered, but they had protected their troopships. Scores of soldiers were ashore.

THE FALKLANDS WAR

Ship losses ~ Argentine attack priorities were inexplicably poor, often targeting screening picket warships, while rarely attacking the decisive vessels that these warships were actually protecting. Reports of ships lost were always received with great concern. Had the Argentines persisted with air strikes after 25th May, the outcome of the campaign might have been different or worse if the Argentines had actually fused their bombs properly. Many bombs went straight through the ships, while 14 others that did hit failed to explode. On 25th May, the two British carriers with the container vessel *Atlantic Conveyer* were 90 miles northeast of Port Stanley and heading for San Carlos Water. Captain Roberto Curillovic, and *Tenenté de Navio* Julio Banaza flying Super Étendards detected them. The Argentine pilots fired their Exocets from 25 miles away that targeted HMS *Ambuscade* that detected the missiles on her radar and warned the RN warships, which all fired their defensive chaff. The Exocet broke lock and veered off to retarget the *Atlantic Conveyer*, pierced her port side, killed 12 sailors, and ignited an unstoppable fire. The *Atlantic Conveyer* sank with three Chinooks, six Wessex, and one Lynx helicopter, tentage for 4,000 men, and most British vehicles. The Argentines who thought they had hit HMS *Hermes* had still struck a serious blow to the British plan. Brigadier Julian Thompson's troops now had to walk 60 miles to Port Stanley. Daring 2 Para preferred to fly, and hijacked the existing Chinook *Flying Angel* on 2nd June for a surprise advance to Bluff Cove.

The standard of tactical operations, planned and performed by the FAA was poor. The Argentines always had the potential to attempt a variety of feint attacks, spoof raids or mass strikes against a priority target, but at no time tried anything innovative to overwhelm the British defences. Such initiatives were always within the capacities of the FAA and would have been difficult to guard against. Even with up to eight Argentine aircraft in the Falklands area of operations at times, no more than five Argentine aircraft were ever used in a single attack. Often when FAA airman did attack they followed the same flight path, and this oversight helped the Sea Harriers pilots focus against them. One intruder, a FAA Hercules flying maritime detection, popped up from low level north of Pebble Island to radar sweep a survey of the British Fleet. Detected by HMS *Minerva*, a Sea Harrier piloted by Cmdr Nigel Ward from HMS *Invincible* detected the enemy FAA aircraft on his radar and descended through the clouds to engage. With his target in sight at a height of 200 feet, Ward launched a Sidewinder that sparked a fire on its port wing, and then finished it firing 240 rounds from his 30mm cannons that perforated off the port wing. The C-130 then went into a descending right hand turn. Its starboard wing sliced into the ocean, cart wheeled and broke up. Its seven aircrew all perished at sea.

Exocet missiles ~ Both HMS *Sheffield* and MV *Atlantic Conveyer* had been sunk by Exocet missiles air launched from FAA Super Étendard aircraft, a aircraft-weapon combination seen by the British as the main threat to their carriers. The operating range of the Super Étendards flying from Argentina extended a little beyond the Falklands, compelling the British carriers to remain well to the east of the islands, which in turn imposed a corresponding range issue for the Sea Harriers. The British knew that the Argentines had begun the war with only five Exocet anti-ship missiles. After three detected air launches, was assumed that two more remained. What the British did not anticipate was an enemy ability to adapt and launch an Exocet from a land vehicle. Supporting the British ground advance with her naval gunfire, HMS *Glamorgan* was hit by a shore based Exocet missile through the port side of her stern hangar deck killing 12 sailors and wrecking her fully fueled and armed Wessex helicopter. On 23rd May, HMS *Antelope* was hit and later sunk. British troops called San Carlos Water, *Bomb Alley*.

Bite and Hold tactics ~ The British proposed a land campaign of a series of limited offensives, envisaging forward movement, then biting and holding key terrain features such as mountains or ridgelines then consolidating their newly won position with more troops, and bringing up supplies. By biting and holding, the British could cut through the Argentine defences and reduce their opponents ability to acquire intelligence reports from their foot patrols or forward observation posts. By giving British troops frequent triumphs in the field, it would also bolster British morale, while cowing that of the Argentines. The surviving helicopters would help the troops move forward, carrying their artillery, stores, and sometimes them into battle. The invasion could not have proceeded without the aircraft carriers and their Sea Harriers that offered operational protection. The Sea Harrier had faced its first combat over the Falklands, and British airmen had proved that even in the microchip age, that the tactics of mobility, flexibility and surprise were still as important as ever. As aircraft speeds continued to increase, reaction times diminished, and only higher levels of pilot training could compensate for this. The Sidewinder Lima showed its fangs in the Falklands campaign where British Sea Harriers scored 25 kills from 27 launches against faster aircraft in marginal weather.

THE FALKLANDS WAR

Goose Green ~ On Wednesday 26th May, Colonel Herbert Jones known as H, led the 500 troops of 2nd Battalion Parachute Regiment down Sussex Mountain toward their next objective, Darwin Hill. At the top was a sniper. Men were falling with headshots. With his sub-machinegun fully loaded, Colonel H. Jones went forward alone into the heart of the battle straight at an enemy position on Darwin Hill. Up the ridge H charged, shouting to his men. *Come on A Company ~ get your skirts off!* Then the heroic commander of 2 Para fell mortally wounded. H was awarded a posthumous Victoria Cross. Major Chris Keeble took over and restored the situation. With the exhausted men of 2 Para he pressed forward, firing Milan anti-tank missiles at the dug in Argentines. The first Milan shot over their heads. Corporal David Abols launched his Milan AT missile to blast apart the Argentine command trench. The remaining Argentine defenders, perhaps next to be killed by such a terrifying weapon, found their will to fight broken and surrendered. Corporal Abols won the DCM for his action, and 2 Para consolidating their position, peered over the brow of Darwin Hill to see their next *bite and hold* objective. Goose Green was their mission. To free the villagers locked up in the Community Hall the following day, and to capture the alternate Argentine airstrip.

As the sun rose, C Company 2 Para emerged from a forward slope to find themselves pinned down in daylight, being shot at by a 35mm AA gun at 2000m range to which they had no answer. Suddenly, from the sky came a flight of three Sea Harriers to hit the enemy at Goose Green with cluster bombs and rockets and save the day. Major Chris Keeble, to prevent casualties on both sides, sent a message to the Argentine CO that the threat of British airpower would become even more apparent. *The red berets are the best soldiers in the world. Surrender or face the consequences.* Suggesting he would direct a flight of Sea Harriers onto a nominated target. Not to attack, but to demonstrate his ability to delivery firepower onto a position. *We'll show you what we can do, and unless you surrender, we'll come and do it to you.* In the morning, the Argentine commander, Commodore Pedroza surrendered his pistol and 1,500 men to Major Keeble. The villagers of Goose Green were thrilled to be released by 2 Para. This is a fine example of an inspired officer in the field, focusing on psychological warfare to hold the threat of available and recently demonstrated airpower over an opponent, then using it to win in a memorable manner, while saving the soldiery of both sides the loss of many men.

The weather ~ The British knew that the war would become progressively more difficult to win as the harsh South Atlantic winter approached. The weather throughout the war had been atrocious and they had been under great pressure to end the war. Between May and June, there had been 17 days when aircraft could not participate in the conflict because of rain and low clouds, indicating that wild, windy, wet, and cold weather would soon become the norm. British ground troops also facing the impediment of hideous weather who were now pressed by a need for expediency, moved the landing ship LSL *Sir Galahad* towards the Falklands in the hope that the poor flying weather might continue to provide their ship with this hoped for additional protection. But as many soldiers and aviators have experienced in prior battles, reliance on the predictability of weather can be precarious.

On 8th June, as the skies cleared over Falkland Sound, the frigate HMS *Plymouth* was sailing to West Falkland to shell an Argentine position, when she came under attack from FAA Dagger aircraft that dropped four UXBs (unexploded bombs). Later that same day, the landing ships LSL *Sir Tristram* and LSL *Sir Galahad* were sailing up the East coast of the Falklands (without an armed escort) to disembark the 1st Battalion Welsh Guards at Fitzroy in Bluff Cove. In broad daylight, packed with ammunition and troops, *Sir Galahad* was still attempting to disembark troops. These landings, devoid of air cover, took place in full view of Argentine observation posts that would have reported it. The LSL *Sir Tristram* had almost completed her unloading when three Argentine Skyhawks targeted her, dropping bombs and scoring hits with two, which happened not to explode. The Skyhawks then targeted LSL *Sir Galahad* and having time to aim well, dropped two bombs into the still occupied landing vessel killing 43 Welsh Guards, seven sailors, and injuring 115 more. If the troops had landed the instant *Galahad* had arrived, the casualties would have been mercifully fewer.

Four more Argentine Skyhawks appeared over Choiseul Sound that afternoon. Two Sea Harriers flying overhead piloted by Lt. David Morgan and Lt. David Smith dived from 10,000 feet to target the Skyhawks. Morgan locked and fired one Sidewinder shattering the rear Skyhawk. Morgan flew on, fired a second Sidewinder and sent another Skyhawk into the sea. Smith next locked onto the last fleeing Skyhawks and fired, to spear the third Argentine up the tail in a massive flash that split the exploding Skyhawk. The fourth Skyhawk dropped his tanks, and was extremely lucky to escape.

THE FALKLANDS WAR

What he could have done was to have flown troops forward, and harassing, patrolling and mounting local attacks on our perimeter, which would have slowed us down. Brigadier Julian Thomson

Nine days after the British landing in San Carlos Bay, Major General Jeremy Moore assumed command of the land campaign, and moved his ground forces up to capture the hills around Stanley. To save his men from being caught advancing over open terrain in daylight (as happened to 2 Para), Moore decided to conduct most ground approaches and assault operations by night, to reduce the risks of Argentine defensive fire from the ground, and perhaps from the air. The British planned to take the Argentine positions on the string of hills centered on the craggy heights of Mount Longdon. After success on the night of 11th June, the 2nd Battalion Scots Guards were given the toughest objective of the campaign : to take Tumbledown Mountain in yet another night action. The Argentine 5th Marines were dug in up there, and they outnumbered the Scots Guards two to one. An undetected forward platoon of the Scots Guards climbed onto a rocky ledge of a higher peak to pour fire into the exposed Argentine flank, as support platoons of the Scots Guards advanced to capture and control the heights of Mount Tumbledown. From the top, they could see a few lights on in Stanley, some five kilometers away. All weekend, the Sea Harriers pounded Argentine positions around Stanley, and on Sunday the first successful laser-guided bomb attack was made.

As the sun rose, on Monday 14th June, 2 Para on Wireless Ridge saw a fleeing mass of Argentine infantry trundling back into Port Stanley. The men of 2 Para, who were eager to enter Stanley first, turned off their radios and ran off, following them down the road to Stanley. The first British unit to reach the outskirts of Stanley was 2 Para only to be halted and ordered not fire on the enemy by Brigadier Julian Thompson, as negotiations for the surrender of the Argentine forces had began. General Moore gave 2 Para clearance to advance further toward Stanley, but to halt at the racecourse. With the Antarctic winter fast approaching, there was some urgency to get to Stanley. Even though General Mario Menendez still had 9,800 well-supplied troops around Stanley, his other soldiers had already surrendered the heights of Wireless Ridge to the Scots Guards on Mt Tumbledown, and the reputable Gurkhas were advancing on Mt William. Then the shelling stopped. Major Bill Dawson came out of his tent and announced ; *Gentlemen. I've just heard that the white flag is flying over Stanley.*

The next morning, 2 Para were the first into Stanley. General Jeremy Moore, commander of the British land forces arrived soon after to accept the surrender signed by General Mario Menendez. On the first morning of peace the Royal Marines of Naval Party 8901 who had defended Stanley against the Argentine invasion, proudly raised the Governor's flag over Government House. Around Stanley British troops found an impressive array of unused Argentine war equipment, ten potent armoured cars, artillery pieces, missile radar systems, and several serviceable aircraft ; including nine Iroquois helicopters and one Chinook helicopter, which were all taken to Britain to be used. The Argentineans had not been overcome by numbers, but by a military culture of British resolution and courage.

Three days after the Argentine surrender of the Falkland Islands, General Leopoldi Galtieri and his disgraced Junta were forced from power. Margaret Thatcher won an early British election with a majority, and many of the ill thought out British defence cuts proposed in the 1981 Defence Review didn't eventuate. British forces on the Falklands began constructing a comprehensive Sea Harrier facility ashore at Stanley and on 4th July, the Sea Harriers flew ashore, fully armed with Sidewinders. Despite vile weather, RN Sea Harriers remained at the 2,000 feet longer airstrip re-designated RAF Stanley until 1985, when Phantoms assumed their air defence role. A purpose built major airfield with its associated structures constructed near Mount Pleasant Peak was commissioned in mid-1985 to accommodate large long-range jets capable of reinforcing the Falkland Islands at very short notice.

The Falklands War was very much the forerunner of recent Expeditionary Operations, which require a strong joint services ingredient. The British at the time did not have a Joint Headquarters, and the need for one was one of the lessons of the Falklands campaign. The outcome of the war had been dependent on the use of sea power, to place elite soldiers in the right place at the right time. After the loss of the two destroyers, two frigates, two LSLs, and a container ship, much was learned about the vulnerability of modern warships. Three quarters of British fighting ships around the Falklands had been damaged in one way or another and the damage would have been worse if Argentina's ordinance had been effectively prepared. The failure of those 13 Argentine bombs that had hit British warships all had the potential to disable or destroy those blessed British warships.

THE FALKLANDS WAR

If the Argentineans had been able to hang on for say another two weeks, they would probably have won.
Rear-Admiral Sir Sandy Woodward

The real concern was the sustainability of the warships. Admiral Sir *Sandy* Woodward made it quite clear what his situation was. The ships were being battered to pieces by the South Atlantic winter; the harsh combination of ice and salt water was degrading the effectiveness of communications systems and his early warning systems. The time when Admiral Woodward would have to take his ships back to a warmer part of the ocean was rapidly approaching. His estimate was that perhaps they could stay on station for another five or six days at the very most. On 16th June, facing force 10 winds, high seas, and icing conditions, all flying was cancelled. So it was success at the 11th hour in terms of critical naval support. If the Argentines were to ask themselves : could they have fought differently, and changed the outcome? Considering the lack of success of their bombing raids and their ongoing, and unrectified problem with unexploding ordinance, the answer is yes.

The destroyer HMS *Glasgow* was hit by bombs, which didn't explode. The frigate HMS *Brilliant's* AA missiles destroyed four Argentine aircraft. The frigate HMS *Broadsword* evaded destruction with Argentine bombs bouncing off her stern, while she shot down three aircraft. HMS *Plymouth* destroyed five Argentine aircraft as she was hit by four UXBs. One inept Argentine pilot on a bomb run against HMS *Antelope* hit her mast after dropping his bombs (all missed) to be shot apart by a Sea Wolf missile launched from HMS *Broadsword*. Had these warships all been struck by exploding ordinance and lost, the defense capability of the British task force would have been severely weakened. Argentine losses would have not been so great, and their capacity to continue the fight would have been maintained. Lord David Craig the retired Marshal of the RAF once remarked. *Six better fuses, and we would have lost.* The bombs that the brave Argentine airmen carried were not designed for low altitude targets, so the bombs had a retarding trigger mechanism that tended to make the bombs go right through the ship. Seven British ships had been lost, however at least 14 times the Argentines had dropped bombs that did not explode ~ so that was once again, something for the British forces to be very thankful for.

The unfortunate loss of HMS *Sheffield* revealed that the defence of the Royal Navy was imperfect. The ships of the task force were vulnerable to attack by such modern low flying anti-shipping missiles that could not be : detected, engaged or destroyed ; because they were without airborne early warning aircraft, an omission resulting from the complexity of the task that could not be provided in time. Several years earlier, the Royal Navy had retired the last of its Fairy Gannet AEW Mk 3 aircraft from the carrier HMS *Ark Royal* that had been scrapped in 1978. Too late the mistake was realized, and two Sea King helicopters were fitted with Searchwater radars to fly as AEW radar pickets, however these Sea King AEW Mk 2 versions of this helicopter only became operational at the end of the Falklands campaign, when they sailed south onboard HMS *Illustrious*. And so, the vessels of the task force were required to operate under threat. The error of depriving the Royal Navy of its AEW capability by retiring its few Gannet aircraft was cause for much loss of life, and several ships.

Timing ~ The original plan of the Argentines was to invade the Falkland Islands in December 1982. They knew that the new government of Margaret Thatcher had made the decision to sell the two aircraft carriers of the Royal Navy, reducing it to a coastal defence force. If Argentina had waited until December for Britain to cut her navy, and to receive another shipment of Exocet missiles from France Britain would have been incapable of responding. But when Britain protested about the flag-raising episode on South Georgia Island, it provoked the Argentines into bringing forward their planned invasion of the Falklands, making it an imminent operation. In the uncontrollable momentum of war, the naïve Argentine Junta made their move far too early and their timing could not have been worse. Having committed the forces of Argentina to the Falklands, failure for the Junta became *not an option*. However within their defective decision, their failure became invisible, inconceivable, and inevitable. The carriers and warships of the Royal Navy were at sea on naval exercises and the active vessels of the Royal Navy were transformed into an operational task force overnight. Sometimes in life, sport and in the art of warfare, making fewer mistakes than that of your adversary wins the day, and in the course of the Falklands campaign, the path of history may have been altered by one less mistake by Argentina, or if perhaps if Britain had made one more mistake. British airmen and naval anti-aircraft missile crews with their skills and will to win, successfully protected their naval task force, and with their ground troops, achieved a swift victory in the South Atlantic.

* * *

CHAPTER XIII

DESERT SHIELD & DESERT STORM

In the early morning of 2nd August 1990, Iraqi tanks rumbled through the streets of Kuwait City. In subsequent days, the invading Iraqi Army overran, and subjugated the entire Emirate of Kuwait. Iraqi President Saddam Hussein then boasted: *Kuwait has been returned to the fold of her motherland, Iraq.* His historical justification was that Kuwait had been part of the Ottoman Turkish Empire province that had formed Iraq. However, Kuwait had already achieved independence from the British Empire, (after having been a British protectorate since 1899). Iraq then concentrated their ground forces along Kuwait's southern border facing Saudi Arabia, putting half the world's oil reserves within their reach. There was little to stop the potential threat of an Iraqi invasion of Saudi Arabia, and turmoil loomed for global oil supplies. Although Iraqi ground troops showed no indication of menacing Saudi Arabia, they stood firm in their occupation of the Emirate. In Kuwait City itself, the Iraqi occupation force committed various outrages, including punishment beatings, torture, shooting civilians in the street, plus the widespread looting of houses, shops, and businesses. U.S. President George Bush (the senior) and British Prime Minister Margaret Thatcher called for an international military Coalition to respond in an expeditionary campaign of armed intervention. Troops preparing for the campaign, felt uneasy about Iraq's use of chemical and biological agents, knowing that the Kurdish inhabitants of Halabja had been massacred with mustard gas, and nerve gas. Saddam Hussein then linked any Iraqi withdrawal from Kuwait to the departure of Israel from occupied Palestinian territory, and threatened to use chemical weapons against Israel if the UN Coalition attacked Kuwait.

To deter further aggression, on 8th August 1990 President Bush ordered Operation Desert Shield: the expeditionary airlift of the U.S. 82nd Airborne Division to Saudi Arabia, and fresh orders for six U.S. Navy carrier groups (carrying 202 attack aircraft) to change course and head for the Red Sea. Britain, France, and Italy announced their intention to send forces, and within months the Coalition included the forces of 30 countries, including some welcome Arab contingents from Egypt and Syria. Within days, flights of U.S. F-15C Eagle fighters were carrying out air patrols along the Iraqi-Saudi border to counter the prospect of Iraqi forces moving south to threaten the oilfields of Saudi Arabia. All efforts of diplomacy expired, and on the 29th November 1990, UN resolution 678 called for the Coalition to use all possible means of force to reverse Iraq's aggression, and gave Saddam Hussein until 15th January 1991 to withdraw all Iraqi forces from Kuwait. After the deadline came and passed, President Bush signed the document that took the UN Coalition to war, which further authorized Operation Desert Storm ~ the armed expulsion of all Iraqi forces from Kuwait.

Task Force Normandy ~ Coalition pilots began flying high speed *Fence Check* sorties toward the Iraqi border, to buzz their frontline radar sites, and then turn away. The Iraqi radar operators had to choose whether to ignore such feints, (and risk being lulled into a false sense of security) or to switch on. However, switching on would give away their reaction times, procedures, and operating frequencies. Two forward Iraqi radar sites sited to give Baghdad early warning were slated for destruction. Apache helicopters were chosen for the mission for their ability to : fly at night, low infrared signature, reduced radar signature, and the accuracy of their standoff weapons. At 0100 (Baghdad time) on 17th January, eight U.S. Apaches of the 1st Battalion, 101st Aviation Regiment each carrying four Hellfire missiles pulled up from Al Jouf Airfield into a pitch-black sky, with all lights off. As the Apaches flew north over the Iraqi-Saudi Arabian border, they split into Red Flight and White Flight to locate the two forward radar sites near the border. Two miles from their targets, Red and White Flights split into four units to fly low over desert in line-abreast formation, with a mile of separation. Nearing missile range, the Apache helicopters slowed to 40 mph, settled into a hover, and waited for H Hour. At 0238, the Apache airmen heard their leader give the attack order : *Party in ten.* On cue, ten seconds later they released their Hellfires. One Iraqi emerged from a radar van, and seeing missile flashes in the night, hurried back inside. Others appeared from an opposite door, and ran off into the darkness. Hellfires rocketed through the air smashing electrical generators, communications vans, radar dishes, and all perimeter AA guns. In four minutes, a firestorm of 32 Hellfire missiles, 100 Hydra 70mm rockets, and 4,000 30mm chain gun rounds had killed 150 Iraqis on the ground and opened a 20-mile flight corridor for follow up Coalition aircraft to stream over Iraq with impunity. The Apache helicopter force transmitted : *California AAA* : The radar sites had been destroyed, followed by : *Nebraska AAA* : There had been no Coalition casualties. What the returning the Apache airmen didn't know, was that Nighthawk stealth aircraft had flown over undetected by Iraqi radar, and were already on their way to strike Baghdad.

THE PERSIAN GULF WARS

Stealth technology ~ At 0300, eight small angular F-117A Nighthawk aircraft that had been designed to be invisible to enemy radar, took off from King Khalid Airbase in southwest Saudi Arabia to soar up into the night sky and head for Baghdad. The Iraqi SAM sites, and hundreds of AA guns firing from the rooftops of Baghdad put up a barrage of fire. The unseen and undetected Nighthawk aircraft flew on through streams of tracer arcing up through a firestorm of shell and missile airbursts. The radar invisible Nighthawks flew in pairs. As one Nighthawk defined the target with a laser beam, a second launched a guided 2,000-lb bomb toward the target. Major Gregory Feest delivered the first bomb of the war, from his Nighthawk through the roof of Baghdad's air defence HQ, severing its ability to communicate with its operational elements. Pairs of Nighthawks then reversed roles to attack their next target. On that night, Nighthawk airmen hit their targets with an accuracy of 86%.

Tomahawk cruise missiles ~ At 0305, coalition warships launched a wave of 54 Tomahawk missiles against targets in Baghdad. These wonders incorporated a ground mapping program that required specific terrain features to navigate ~ except couldn't over the featureless Iraqi desert. So, they were programmed to fly over the spine of the Zagros Mountains of Iran, then west into Iraq and up the identifiable main streets of Baghdad to knock out 28 of Baghdad's electrical power stations and plunge the city and most of Iraq into darkness. Precision bombing had become a reality. In the first 24 hours, more than 200 targets were hit with 1,000-lb warheads including airfields, railroad yards, bridges, communication towers, and all major freshwater reservoir dams. Other Tomahawks carried *Kit-2* warheads that dropped spools of carbon filament wire to short out 85% of Iraq's electric grid system.

After that, the Coalition turned to acquire air superiority against the Iraqi air force and air defence network of early warning radar posts and independent radar/SAM/AAA sites. After early Nighthawk missions destroyed the central Iraqi air defence HQ, most IrAF airmen decided to sit out the air campaign within 300 of their fortified aircraft bunkers, however 148 IrAF airmen flew off to internment in Iran. Iran retained their MiGs and Mirage aircraft, even though they were of little use to them, as their own aircraft were all American made. Iran had neither spares for them, nor properly trained technicians, so the interned IrAF fighters just sat unattended for years on the tarmac, decomposing in dust storms and the harsh desert weather. On 17th January, an IrAF MiG-29 Fulcrum became the first Iraqi aircraft of the fight to fall, after being detected and destroyed by an F-15C. Eagle.

I see that somebody's targeted on me. It's an Iraqi fighter who wants to target me and shoot me down. So, I remember firing a missile. I remember seeing this big blue flame coming across the sky, three to five seconds long and I remember seeing it, and then it goes away. I remember thinking to myself, that my missile hit him. The target is destroyed, but what the hell.
Major Jon Kelk

On January 19th 1991, four U.S. Eagles flying over a cloudbank above western Iraq heard an AWAC warn of bogeys 80 miles to the north. In seconds the Eagles detected multiple radar plots. A pair of MiG-25 Foxbats at 10,000 feet heading south while two MiG-29 Fulcrums veered away toward Baghdad. The two Foxbats keep coming. The American airmen switched their radars to lock on the Foxbats, but the Foxbat pilots pumped out chaff, dived low and thrust over the desert floor to blend with ground reflections and broke radar lock. Captain Larry Pitts swirled his Eagle through a tight Split-S and spotted a Foxbat. Pitts' arced his Eagle inside the Iraqi's turn ~ selected a Sidewinder AIM-9M with minimal smoke exhaust. Locked on and fired. As the Mike rocketed toward the Foxbat's burner plume, the Iraqi decoyed the heat seeker away with flares. Pitts opted for a radar-guided Sparrow that failed to explode. Pitts fired a second Sidewinder M, and the Iraqi saved himself again with decoy flares. Pitts locked on and launched his third Sidewinder : *I fired the missile from 6,000 feet behind him, and the missile goes straight up his tailpipe and blows up the back end off his plane*. Airmen preferred to fire longer-range Sparrows to reduce the need of tail chase attacks at closer Sidewinder ranges, and as most Iraqi pilots were reluctant to engage, they were targeted and taken out with six o'clock Sparrow shots. Improved long-range Sparrow missiles led experts to declare that dog fighting was obsolete, as they had with the coming of jets when air combat manœuvre training was minimised. Then the U.S. Navy had to establish Topgun. Because of the continuing need for winning aerial tactics, the U.S. F-14 Tomcat was armed with guns and designed as an agile supersonic dogfighter, as were the F-15 Eagle, F-16, F-18 Hornet, and F-22 Raptor. Within these advanced aircraft, the situation remains the same. As long as a pilot has a missile, it can be fired at an enemy beyond visual range. Otherwise similar tactics of the Great War are called for. The need to get behind your opponent with good fighting spirit ~ to be brave and face your opponent in the ultimate test for a gun kill.

DESERT STORM

Tornadoes and Buccaneers ~ During the first week of Desert Storm, the RAF lost four Tornado F 3 aircraft from low level flying (trying to avoid Iraqi radar) either from flying into the ground, being hit by AA fire or trying to avoid it. So, questions were asked : *Was this a necessary or sensible thing to do?* Apparently not, Coalition HQ respected RAF airmen, but were growingly anxious about their tactics. RAF airmen requested a change of tactics (after another F 3 was lost) and low level flying was halted. Older 1970's RAF Blackburn Buccaneer jets that the Tornadoes had been superseding were flown to the Persian Gulf. The Buccaneers carried a device to designate targets from high altitude and soon RAF airmen were flying safer above Iraqi ground fire. The Buccaneers, despite their age, began flying target designation for Tornado flights that carried the laser-guided bombs to crater Iraqi runways.

Targeting Saddam ~ A week into the war, Coalition aircraft controlled the skies and bombed at will. But on the 20th January, two F-16s were shot down over central Baghdad by intense AAA barrage fire. Even with much of Iraq's air force destroyed, Saddam Hussein remained defiant, refusing to withdraw from Kuwait saying : *It's no problem. We can rebuild everything, even the morale of the people.* Officially Saddam Hussein was not a target, but the Coalition flew 260 sorties against places where he might be hiding. Spies reported that Saddam operated from and slept in ordinary homes, changing his location nightly, driving alone around Baghdad, in an old taxi or plain truck to avoid being obvious.

Saddam Hussein retorted that he would turn the coming ground campaign into a nightmare for the coalition soldiers, promising them the *Mother of all Battles*. Saddam Hussein ordered his forces to retaliate and hit back at the Coalition of Western and Arab countries, authorizing mobile Scud missile launchers hidden in the desert, to aim and fire seven missiles at Tel Aviv. The Scud was an Iraqi modified version of the relatively inaccurate Soviet SS-1B Scud missile with a 400-mile range that carried a 550-lb warhead. Added to the threat, was the dread that the Iraqis might load their Scuds with chemical weapons. Saddam Hussein believed that Israel would retaliate and join the conflict. The Arabs in the coalition would then refuse to fight alongside Israel ; the Coalition would collapse, and with it, the war. The Israeli government reported traces of nerve gas in the debris of one of the missiles and prepared for the worse. If Saddam Hussein could provoke the Israelis into attacking him by firing almost ten Scuds a day at them, it would be difficult to hold the Coalition together. So, 20% of Coalition airpower was redeployed on missions over 28,000 square miles of desert to search for the 14 known mobile Scud launchers.

The UN airmen couldn't find them. Neither could they find the military transport bringing up fresh missiles to the hidden launchers. The Iraqis were transporting the Scuds in civilian buses. After taking out all the seats, they simply slid a missile in through the back, and if a Coalition airman spotted a bus, they overlooked it, assuming it carried only passengers. For 43 days, British SAS troopers patrolled the desert in what they called the Scud Box closest to Israel looking for Scuds. Two days after they arrived, they began cutting communication cables and destroying microwave towers to halve the number of Iraqi launches from the Scud Box. The U.S. Patriot air defence missile system, originally designed to shoot aircraft down was reset as an anti-Scud missile system to protect Israel. It is now known that the Patriot was a myth of good faith that failed to destroy one Scud, but the finer point of the gesture was what kept Israel out of the Gulf War.

Two weeks into the war, Saddam Hussein ordered units of the Iraqi Army across the Kuwaiti/Saudi Arabian border to provoke the Coalition into a ground war they were hoping to avoid. The Iraqi Army advanced on the abandoned Saudi border town of Ras Al Khafji to capture Coalition prisoners to use as human shields, to tie onto their tanks and overrun the Saudi Arabian oilfields, supposing that Coalition airmen would not target them. On 28th January, an unmanned U.S. Marine drone flew over Iraqi forces on the southern Kuwait-Saudi Arabian border, beaming back images of their advance across the border towards Ras Al Khafji. The Iraqis were able to continue their advance uninterrupted (for reasons unknown) and occupy the Saudi Arabian town. Within Ras Al Khafji, there were two six-man U.S. Marine recon teams hiding. That night U.S. artillery moved up, and at dawn the Marines hiding in Ras Al Khafji began giving precise instructions to their gunners where to fire. Two days later, Saudi Arabian forces moved up to retake their town, losing 19 soldiers. Coalition airstrikes using A-10 Thunderbolt tank busting aircraft firing Gatling guns and Maverick missiles, supported by Apache helicopters into the night, using FLIR forward looking infra-red radar and firing hundreds of guided missiles, destroyed two Iraqi divisions attempting to reach Ras Al Khafji.

DESERT STORM

During a briefing General Norman Schwarzkopf outlined his war plan to divisional commanders stating that Coalition airmen were required to attrit the Iraqi forces in Kuwait to the 50% level prior to any ground offensive. Airmen were also directed to isolate the Iraqi occupation force in Kuwait by striking all key bridges to prevent any Iraqi forces from escaping, and to hinder all reinforcements and supplies from entering Kuwait. During a coffee break that day, Major-General J.H. Binford Peay III, commander of the U.S. 101st Airborne Division made a clear comment on the strategy : *If we can bomb them to 50% in three weeks ~ why don't we take another three weeks, and get the other 50% for good measure?*

Iraq had 400,000 troops and hundreds of tanks lurking in Kuwait behind a barrier of minefields, trenches, and barbed wire. General Schwarzkopf relied at first on B-52 bombers to carpet bomb them, but a mistake in their navigation program meant that most of their 72,000 bombs fell on empty desert. Even when this was rectified, high wind blew most bombs off course to score a daily strike rate of just six to eight Iraqi tanks. Following the training exercise *Operation Night Camel*, aircrews of two F-111F jet aircraft found as they flew over the Iraqi frontline at night, that even the best dug-in and camouflaged Iraqi tank presented a sharp white rectangular IR image on their heat detecting displays. Particularly in the hours between sunset and midnight. The Iraqi tanks, still warm from the daytime sun dissipated their heat far slower than that of the surrounding desert sands. Thereafter, F-111F aircraft carrying laser-guided bombs took to the air to destroy 920 Iraqi tanks with an accuracy of 72%.

In retaliation, from 22nd February onwards, detachments of the Iraqi Army began the accelerated detonation of Kuwait's oil wells causing dense smoke to drift over the desert that reduced visibility to three miles. This infers that they may have been preparing to pull out of Kuwait City, and withdraw from Kuwait before the Coalition ground offensive began at 0400 on 24th February. So far, the war had been fought mostly in the air. Next came a 100-hour period of ground combat in which the resistance of the Iraqi Army was slight and when they did make a stand, they only displayed tactical ineptitude with large numbers of starving and startled soldiers, surrendering without firing a shot.

The Road of Death ~ On the 25th February, the soldiers of Iraq formed up to flee Kuwait City in any vehicles at hand. An F-15C crew, pilot Captain Merrick Krause and Weapons Officer Major Joe Seidl about to take off fully loaded with bombs were paused, and given a fresh mission. They had to attack, and stop the Iraqi military convoy leaving Kuwait City. As Krause and Seidl walked toward their aircraft, their CO Colonel David Baker spoke to them : *It is not just a retreating army, these guys are rapists, murderers, killers, and coincidentally a Scud has just hit Dhahran Airport and killed 60 Americans...* inspiring them to harden their hearts, and stop the enemy from getting out of Kuwait.

As we dove out of the clouds, the picture was absolutely astounding. There were thousands of headlights heading on every road that led north out of Kuwait City. We had 12,500-pound bombs and we elected to drop them three at a time. Now we drop our three bombs and it's a perfect delivery. The bombs impact in a string right across the highway with the center bomb impacting in between two trucks, causing both of the trucks to burst into flame. There was now a traffic jam beginning and I could see cars pulled off to the side of the road. Some headlights on, some headlights off. A lot of gunfire was coming from that area, but it was random, it was in all directions. So I picked a spot in front of us that looked like the highest concentration of that traffic jam and where the shooting was, and we dropped three bombs on that. After that, I came up very hard because the shooting erupted in front of us, even before the bombs exploded. Once the bombs exploded, other headlights went out for miles in every direction. And then to our disbelief, all of a sudden the headlights come back on. Joe Seidl said : Hey, it's getting pretty hot down there. We need to get out of here! I replied : Yeah, we've got one more pass and then we're out. We came in on that last pass and I dropped the bombs and went straight ahead, out towards the water, trying to get as much speed as I could.   Captain Merrick Krause

As Krause and Seidl bombed the lead vehicles on the road at Mutla Ridge west of Kuwait City, they halted the escaping Iraqi convoy and created a traffic jam, similar to the Turkish 7th Army that had been massacred in 1918 along Wadi-el Far'a or the 1967 Egyptian withdrawal west through the Mitla Pass. Other aircraft loaded up with cluster bombs and arrived. Coalition airmen calling it the *Road of Death* knew it had to be targeted as it was an armed military retreat of those who had raped and pillaged their way through Kuwait, and were now trying to escape before they were caught. Flights of Coalition aircraft slotted for 15-minute airstrikes over the killing zone left the scene after destroying hundreds of loot laden vehicles, and leaving charred and hideous corpses along the road. Reporting on this horrific sight, war correspondents later counted fewer than 300 dead Iraqis.

DESERT STORM

With the liberation of Kuwait, the Coalition had achieved their primary aim. However the hideous televised images of charred carnage on the *Road of Death* proved to be so evocative to the imaginations of the Coalition leadership in Washington that President George Bush became so apprehensive about the perception of U.S. forces continuing to bomb a retreating enemy that he ordered a ceasefire on 28th February 1991 that regrettably allowed Saddam Hussein to remain in power for another 12 years. After the war, the U.S. Defence Intelligence Agency estimated that Coalition airstrikes had set back the Iraqi regimes development of a nuclear potential for three to five years, and their chemical and biological development program by five to eight years.

* * *

Osama bin Laden said that the presence of U.S. soldiers in the Muslim holy land of Saudi Arabia had motivated him to launch a series of attacks against U.S. interests that reached their zenith with the hijacked passenger aircraft attacks against the twin towers of the World Trade Centre in New York. When President Bush's son, George W. Bush became U.S. President, he declared a global war on terror that still continues to this day. As part of this war on terror, in 2003 after George W. Bush was advised that Saddam Hussein supported terrorists and still possessed weapons of mass destruction, he ordered the commencement of Operation Iraqi Freedom, after being continually provoked by Iraq that had shot down 35 Coalition warplanes in the preceding months from now well coordinated AAA fire, and SAMs launched from no-fly zones.

British and U.S. forces invaded Iraq, with the principal aim of finding hard evidence of the Iraq's supposed weapons of mass destruction, even though there had been no tangible sighting of any WMD or WMD enabling equipment for years ~ only conversations of inference between Iraqis, and an assumed belief that the Iraqis had learned how to conceal their WMD elements away from U.S. surveillance assets. Coalition Special Forces tasked with hunting down the Scud missiles and Saddam Hussein's WMD arsenal searched in vain from just west of Baghdad to the Jordanian border, and between the Syrian and Saudi frontiers.

The absence of the WMD was one aspect of the supposed threat. Another ongoing pain was how to counter the very mobile, and easily concealed Scud launch vehicles or other potential launch vehicles of new WMD. During the 1991 Gulf War, nearly all reports of Scud launch vehicles turned out to be those of ruse Scud-like decoy vehicles of similar appearance and weapon sensor signature. At least 80% of all Scud missile launches were fired at night, displaying no preparatory pre-launch emission. Within five minutes, Scud launch vehicles then departed the scene, leaving a realistic decoy at the launch site, while the true launch vehicle escaped undetected into the desert to conceal itself. Coalition aircraft patrolling over known Scud launch zones, detected 42 Scud missiles being launched, but only eight pilots were ever able to intercept and destroy a detected Iraqi launch vehicle. The level of Iraqi ingenuity in their Scud operations in their deception, decoy, and concealment was exceptional. This explains why Coalition Intelligence continually underestimated the Iraqi Scud force by a factor of five. There were not 42 Scud launchers in the desert as they thought, but 225. What proved to be most effective in detecting Iraqi Scud launches flew in space, in the form of U.S. defence satellites designed during the Cold War to detect Soviet ICBMs.

The real lesson of the Scud episode was that the true worth of a weapon is in its power as a political device, regardless of its operational impact, and Iraq's use of these simple Soviet missiles demonstrated the potential during the opening phase of the Gulf War to politically tear the UN Coalition apart. American leaders in good faith applied the promise of the Patriot system to support Israel, and stabilized the situation.

John Boyd in 1952   Colonel John Boyd USAF

F-100 taking off from Nellis AFB circa 1959 ~ note checkerboard pattern on the vertical stabilizer and nose, indicating it was a *Hun* from the Fighter Weapons School, and it was in the *Hun* that Boyd became famous as *Forty-Second Boyd*.

Fighter Weapons School F-100s on the ramp at Nellis AFB

Epilogue

Lieutenant John Boyd, a young USAF pilot in the 1950s, issued a standing challenge to all comers. Starting from a position of disadvantage, he'd have his aircraft on their tail within 40 seconds or pay out $40. He never lost. His unfailing ability to win any air engagement in 40 seconds or less earned him his nickname, *40 Second Boyd*. During the 1960s, Boyd studied aerodynamics, and in the 1970s he helped design the F-16 *Fighting Falcon*. After that, from his cramped second floor office in the Pentagon, he focused on creating an expression of tactical agility ~ of how we make such decisions. Boyd's resultant OODA loop of : observation-orientation-decision-action (similar to Eric Hartmann's 1942 tactical quartet of *See-Decide-Attack-Break*) to operate inside a foe's OODA loop, acting swiftly to outthink and outmaneuver a foe *to make us appear ambiguous, and thereby generate confusion and disorder*. Colonel Boyd conceptualized focusing airpower on an adversary to defeat them and survive by the decisive use of the OODA loop ; by out thinking and out acting them, by knowing more and destroying the enemy capacity to think on the same level as you are. In John Boyd's notion of conflict, the target is always the enemy mind. It's all about rapid assessment and adaptation to a complex and rapidly changing environment that you can't control. He considered how to isolate an opponent, how to render their OODA loop inoperable, and he proposed three methods :

Physically, we can isolate our adversaries by severing their communications with the outside world as well as by severing their internal communications to one another. We can accomplish the former via diplomatic, psychological, and other efforts. To cut them off from one another, we should penetrate their system by being unpredictable. Mentally, we can isolate our adversaries by presenting them with ambiguous, deceptive, or novel situations, as well as by operating at a tempo or rhythm they can neither make out nor keep up with. Operating inside their OODA loops will accomplish just this by disorientating or twisting their mental images so that they can neither appreciate nor cope with what's really going on. Morally, our adversaries isolate themselves when they visibly improve their well being to the detriment of others, by violating codes of conduct or behavior patterns that they profess to uphold or expect others to uphold.   Colonel John Boyd

On the opening nights of both Gulf Wars, Colonel John Boyd's OODA loop was put into practice ; by jamming and targeting the Iraqi radar network, their command and control centres, by destroying the HQ of Iraq's air defence commanders, and scaring any survivors witless. The Iraqi capability to react had been targeted, shaken, and so disrupted that it was reduced to the antique simplicity of runners carrying messages by hand. Over Iraq, air superiority was won by retaining the initial element of surprise and exploiting a fast fluidity of action to repeatedly penetrate, splinter, envelop, and vanquish the disconnected remnants of the IrAF. Coalition aviators using the OODA loop, detonated the Iraqi ground forces comfortable view of their world, and with their familiar perspectives shattered, the confused defence of Iraq struggled under pressure as Coalition airmen continued to reduce their ability to carry on by further constricting their frames of reference.

* * *

All through the rich stream of aviation history, and whatever the level of technology applied, aerial warriors have always exerted a powerful influence on the outcome, As always, it will be the quality of those who fill the ranks that well motivated and well led, will provide the decision makers, the practitioners on the ground, and the warriors in the air who get the very most out of the weapons they operate. Both now and in the future, versatile fighter aircraft will continue to use radar, air-to-air missiles, and automatic cannons capable of high rates of fire. Aerial warfare has seen the emergence of maneuvers to break radar lock or to exhaust the kinetic energy of an incoming missile by changing the aircraft's course. Even so, aerial engagements with auto-cannons and missiles still adhere to many of the aerial tactics pioneered during the Great War and remain applicable today. As long as the missiles carried on a warplane are a finite resource, military and naval aviators will continue to need the tactics of winged warfare. One profound lesson of military history is that many battles have been lost rather than won. For example, the *Luftwaffe* foolishly forfeited the Battle of Britain because of their fatal change of focus from attacking the decisive RAF fighter airfields to targeting London. Therefore, if you want to win, dedicate yourself to your craft. Properly trained and well-read airmen rarely fall, however if they do it is often only from some misfortune ~ so keep your radars on sweep, and avoid any premature lock ~ a sure way to warn off your opponents before they come within missile range.

General alarm of the inhabitants of Gonesse, occasioned by the fall of the air balloon of M. Montgolfier.

On a clear winter's morning in November 1783, the Montgolfier brothers launched the first untethered manned balloon flight. The paper bag rose gently above the grounds of the Château de la Meutte on the western outskirts of in Paris. 30 minutes later, after reaching a height of 1,000m the slightly singed balloon then slowly descended among the windmills of Butte-aux-Cailles. Enough fuel remained on board at the end of the nine-kilometer flight, to have allowed the balloon to fly another 30km.

APPENDIX I

ANTIQUE APPLICATIONS OF AIR POWER

It came one November night during the cold winter of 1782, in the old French town of Annonay. The Montgolfier family of paper makers had been discussing the success of their *Papier de Montgolfier*, which they had recently refined so well, and had just been appointed as the providers of royal paper. Joseph, a practical and well-educated fellow resting in his parlour armchair, began to ponder France's old thorn ~ of how to assault the British fortress of Gibraltar impregnable from both land and sea. Gazing into his warming parlour fire, watching the smoke curl up the chimney, Joseph grasped the potential of hot air by gazing at his wife's chemise as it rippled by the fireside, and gently inflated as she draped it over the hearth to dry. Joseph drifted into a tactical vision... of airborne troops riding on the same force carrying the embers up his parlour fireplace ~ could Gibraltar be taken from the air?

Joseph funneled the hot air from his fireplace into a light paper bag. Smoke formed within, and the bag rose to the ceiling. Joseph wrote to Étienne, who was at the time studying architecture in Paris; *Get in a supply of Taffeta and cordage, quickly, and you will see one of the most astonishing sights in the world.* Constructing a three-metre cubed chamber from thin wood, the brothers covered the sides with lightweight taffeta. After setting a fire beneath their craft, it soared into the air on 4th June 1782. For ten minutes it floated through the air, before wafting toward a nearby field. Naïve peasants, seeing it fall from the clouds toward the green fields of Gonesse, imagined the coming of an aerian apparition... The rustic mob slashed into the descending craft with their scythes and pitchforks, tearing it to shreds.

Once news of the Montgolfier's achievement had travelled from the southern provinces to Paris, King Louis XVI and Queen Marie Antoinette in great suspense to see Joseph and Etienne's remarkable discovery invited them both to their Palace of Versailles. A booming cannon on 19th September 1783, announced the impending flight of the Montgolfier balloon, *Réveillon Aerostatique*. A roaring fire was built in a pit below the mouth as eight men held the sack while it filled and took the form of a sphere. As it ascended into the sky, three creatures could be seen suspended in a wicker cage, a sheep called *Motauciel* (climb to the sky), a wild duck, and a feisty rooster. After reaching a height of 480m, the *Réveillon* ended its two-kilometer flight in a safe descent. The sheep, which was feeding complacently, had trodden on the upset rooster, and the wild duck continued quacking with apparent satisfaction.

The Montgolfier's fabricated their next balloon, the *Globe Aerostatique* from silk, and lined it with three layers of thin paper in preparation for the first manned flight. King Louis XVI, anxious about the unknown affects of high altitude upon people offered two condemned criminals, mentioning that if they survived, they would be pardoned. Hearing this, a young physician Jean-Francois de Rozier asked his friend, Marquis François d'Arlandes to use his closeness with Duchess Yolande de Polignac to request King Louis XVI grant him the honour of being the first to rise into the sky. The King agreed, and Marquis d'Arlandes seeking recompense for his patronage, asked to accompany de Rozier on a future flight. A wicker balcony for passengers encircled the base of the *Globe* and slung inside of this by light chains was an iron brazier, for de Rozier and d'Arlande to add fuel for a wider ranging flight. To begin, the *Globe* was moored to the ground, then on 3rd October, from limpness into form, as it filled with hot air the *Globe* billowed, bucked, and tugged at its ropes. Jean-Francois Pilatre de Rozier became the first person to ascend in a captive balloon, rising in the *Globe* to a height of 30m.

Early on 21st November 1783, the Montgolfier's were ready to launch the first manned free flight. The Marquis d'Arlandes and de Rozier took their stand on opposite sides of the *Globe's* grand balcony, each equipped with bundles of fuel to feed the fire. Each also carried a large wet sponge, in case the *Globe* caught fire. As the *Globe* was released, with swiftness and majesty it rose while the balloonists waved their hats to the speechless crowd below. The *Globe* ascended to a height of 920m and floated over the rooftops of Paris for 30 minutes. Beyond Paris, and still carrying enough fuel for two hours flight, burning embers from the fire began to scorch the inner envelope of the *Globe*. In time, de Rozier dampened the embers with his sponge and took off his coat to subdue the fire. The *Globe* began a serene descent into a vineyard nine kilometers from their launch point. The Marquis d'Arlandes and de Rozier thanked the local farmers with a fine bottle of champagne for allowing them to land in their field. In December of 1783, King Louis XVI elevated the Montgolfier family to the nobility, for having developed the concept of Joseph's tactical vision into practical free flight ~ all in the span of one year.

THE ORIGIN OF MILITARY AVIATION

Nicholas-Jacques Conté   Jean Marie-Joseph Coutelle

In the flawed dawn of the French Revolution, both King Louis XVI and Queen Marie Antoinette were guillotined in 1793. During the same year, André de Villette, the third man to have flown in a Montgolfier balloon thought of using balloons *for an Angel's-eye view* of the field of battle to identify and report the positions and manœuvres of an enemy to a commander on the ground. The French Committee of Public Safety of the revolutionary régime, once persuaded of the idea, commissioned Jean Marie-Joseph Coutelle a competent physicist from the Dijon Academy to command the first balloon observation unit, *le Compagne D'Aérostiers*, and to establish it from men with skills relevant to military ballooning. Leading chemists and military artists were enlisted, plus carpenters and masons to construct the brick furnaces required to extract hydrogen, the gas recently isolated by Henry Cavendish, which weighed one tenth as much as an equal volume of air.

A fellow officer in the army, and chemist Nicholas-Jacques Conté (the inventor of the artists Conté crayon and the modern pencil) was appointed to direct the *Compagne D'Aérostier's* training academy and centre of manufacturing balloons at Chateau de Meudon that had been King Louis XVI's hunting field. Nicholas Conté led all research into the techniques of the balloon fabrication process, refining balloon shapes and choosing all materials. The process of isolating the hydrogen content of water, was improved inexpensively by passing steam over red-hot iron particles within cast iron tubes to generate enough of the lighter than air gas to fill a balloon in 15 hours. Coutelle's men dressed in stylish blue uniforms with black collars red braid and buttons inscribed *Aérostiers* soon finished their first balloon *L'Entreprenant* or Enterprise from silk; the lightest, strongest, tightest and finest fabric. It was then sealed with a flexible varnish containing India rubber developed by Conté so impervious in reducing the leakage rate of hydrogen that *L'Entreprenant* remained inflated for two months. During further experiments at Chateau de Meudon, hydrogen leaking from a glass flask exploded and Conté lost the sight in his left eye. Deemed a war wound, he was rewarded by promotion to brigadier.

In May of 1793, over the Seine in Paris, Coutelle ascended to the upper limit of his 500m mooring cable to evaluate the feasibility of using tethered balloons for military observation. During observation flights in action, two men would go aloft. One to devote his attentions to surveillance, while the other signaled reports and communicated with the ground. During two flights, Coutelle could see the seven bends of the Seine and observe details 29km distant through his telescope. His signaler communicated with their fellow *patriotes* on the ground by waving flags and placing written messages into weighted bags fitted with metal rings to slide down the mooring cable. Coutelle also noted that the strain on a tethered balloon could be appreciably lessened, by attaching the mooring cable to a team of horses with some give and take in their movement, rather than simply being tied to a rigid holdfast.

BALLOONS AT WAR

The Battle of Maubeuge ~ Coutelle's first lieutenant, de Launey was a master mason by trade, so when the seven new iron retort tubes of their hydrogen generator, which had been expressly made at le Creusot ironworks arrived at Maubeuge, it was de Launey who was assigned to set the iron tubes in a new reverberatory brick furnace. This involved the laying of 12,000 bricks in the College Court at Maubeuge. Once the generator was complete, it took 30 hours to inflate the envelope of *L'Entreprenant*. The quality of Conté's varnishes was imperative, and the gas-holding properties of his finest varnish was so successful that a balloon could remain inflated for six months - with an envelope of single silk. The recipe for this varnish applied to all French war balloons was subsequently lost, and never equaled since. After the *Aérostiers* reinflated *L'Entreprenant* at Maubeuge, Captain Jean Coutelle rose in *L'Entreprenant* to the cheers of his troops below, and the thunderous roar of a French cannon salute.

From the wicker basket of *L'Entreprenant*, rising above the battle lines on 2nd June 1794, Coutelle and his ordinance officer reported in detail and disposition on their opponents, the emigré noblemen with the royalist Austrian and Dutch troops. Brilliant in their aerial reconnaissance, they aroused a dread in the ranks of the enemy. The Austrians protested that the use of a balloon was against the rules of war, and as *L'Entreprenant* appeared on its fifth ascent, the Royalists retaliated and opened fire with two 17-lb howitzers from a concealed hollow. The first cannonball shot over *L'Entreprenant*'s envelope while the second passed between the balloon and the basket. Coutelle greeted their efforts with a shout of *Vive la République!* and then seeing that the Austrian gunners were acquiring his range, signaled his crew to feed out more cable from their windlass to reach an altitude beyond their range.

I raised myself to the full length of the cord, a height of 500m and at this height with the help of a telescope; I was able to count the number of cannon on the ramparts. The soldiers of the enemy all, who saw the observer watching them and taking notes, came to the idea that they could do nothing without being seen. C$^{apt.}$ Coutelle

The Battle of Fleurus ~ On 18th June 1794, Captain Coutelle received orders from General Jean-Baptiste Jourdan for the *Aérostiers* to join the army at Charleroi. From Mauberge, Coutelle's men formed two files and heaved their inflated balloon at such a height that it passed over the cavalry along the 48km route. Following their arrival, for three days Coutelle ascended to make preliminary observations. While the Battle of Fleurus was fought on the 26th June, Captain Coutelle and General Morlot remained aloft in *L'Entreprenant* for nine hours dropping a series of messages reporting the movements of the enemy to General Jourdan's forces below. By responding to written questions from the ground and being seen to slide their reports down the tether cable in a signals bag, their Austrian and Dutch opponents became demoralized seeing their positions reported for the attention of French artillery. In this new way of war, French tactics were directed entirely from their pioneering mastery of the air. The approach of enemy reinforcements was also observed and General Jourdan was able to force the opposing Austrian and Dutch troops to surrender before they became aware of their approaching support. The Battle of Fleurus was the first action where airmen contributed decisively to a victory. Their action had opened the gates of Brussels and for France's control of the Austrian Netherlands. Coutelle's gallant *Aerostiers* were the heroes of the hour, particularly with the ladies of the town who they took aloft in the basket of their balloon. After Lieutenant Alfred Selle de Beaucamp was hauled back to earth with one lady, he sought immediate permission to marry his new friend.

The Treaty of Campo Formio signed in October of 1797 ended the war between France and Austria, and gave the new French commander, Napoleon Bonaparte time to consider the conquest of England. Rumours of extraordinary new French weapons saw the British look to the sky and report in December, further rumours of a planned enemy airborne invasion; for the French Armée to fly to across the English Channel in balloons to attempt their conquest. More realistically, Brigadier Conté persuaded Napoleon to include the proven *Aerostiers* in the army for his Egyptian expedition. Unfortunately all their equipment was lost in the wreck of *La Patriote* in the naval Battle of Abukir in August 1798. After his adventure in Egypt, Napoleon Bonaparte with his old-world conservatism in matters of war and its conduct and with his distaste for novel ideas, decided against the continued use of balloons. After disbanding the two companies of *D'Aerostiers* and closing the balloon centre at Chateau de Meudon, Napoleon had the chateau demolished to salvage its marble for the construction of the Arc de Triomphe in Paris. So the progress of military aviation was suspended. Perhaps it was because Napoleon had such a capable genius for war that he depended alone upon himself and scorned assistance? Perhaps it was because if the benefits of balloons were widely known to be of great use, his future rivals might make equally effective use of them, to thwart his coming campaigns.

* * *

THE FIRST AERIAL BOMBING ~ VENICE 1849

The war between French forces of Napoleon Bonaparte against Austria in 1796 saw the French invade Lombardy and capture Milan. The following year, Napoleon thrust eastward to occupy Bergamo, Brescia, Verona, and moving thru the Brenner Pass, he questioned Venice's neutrality after the passage of Austrian forces through Venetian territory. After having remained independent for 1,070 years, on 12th May 1797 the army of Napoleon Bonaparte stood poised on the Veneto to attack Venice. The Great Council of Venice met to consider Napoleon's demand for their surrender to avoid battle, then dissolved the Venetian Republic, and surrendered the city. On Napoleon's authorization the city was tastefully looted ; for example taking to Paris such treasures as the *Horse's of Saint Mark*. Venetia, as inclusive of the city of Venice, Istria and Dalmatia, were all ceded by Napoleon Bonaparte to Austria in October 1797, and confirmed as Austrian possessions at the Congress of Vienna in 1815.

A revolt against the occupation of the Austrians who controlled much of Italy erupted in 1848, when the Venetians declared their realm a republic. Austrian forces gathered and retaliated by blockading Venice causing terrible hunger and outbreaks of cholera among the Venetian defenders. Although the Austrian forces of Field Marshal Joseph von Radetsky had surrounded Venice by both land and sea, they soon found themselves tactically impotent. Unable to get their artillery within effective firing range, they had been thwarted by Venice's coastal defences and shallow lagoons.

Then a perceptive Austrian artillery officer, *Leutnant* Baron Franz von Uchatius began thinking. Remembering the Austrian defeat inflicted by the French *D'Aerostiers* over Fleurus he pondered the potential of an aerial approach, suggesting the use of unmanned hot-air paper balloons. The paper balloons would need to be strong enough to carry an incendiary bomb and stay aloft for a set time, then drop to explode on contact. First attempted on the bright morning of 12th July 1849, during the Festa of the *Madonna della Salute,* a series of 20 balloons were released from the Austrian warship *Vulcano*, however the wind proved too strong and most balloons floated over the island of Venice and fell beyond, discharging their contents on the Austrian besiegers encamped on the mainland.

The balloons appeared to rise about 4,500 feet. Then they exploded in mid air or fell into the water or blown by a sudden southeast wind, sped over the city and dropped on the besiegers. Venetians, abandoning their homes crowded into the streets and squares to enjoy the strange spectacle ~ when a cloud of smoke appeared in the air to make an explosion, all clapped and shouted. Applause was greatest when the balloons blew over the Austrian forces and exploded, and in such cases the clapping Venetians added shouts of ~ Viva! Bravo! Buon Appetito!

For their second attempt on 22nd August 1849, the Austrians planned to release a string of 200 hot air balloons over the city of Venice to drop incendiary bombs onto the Venetian garrison below. Each Austrian balloon, governed by a 33-minute time-fuse device carried a single 30-lb bomb. A launch position was chosen on the windward side, and one trial balloon was prepared for release over a course laid out on a naval chart. An armed trial balloon, with a bomb attached had its timing fuse set, and this was released towards Venice. Fine adjustments were made, and the balloon bombs started dropping into Venice. One fell in the market place causing slight damage, but it had a critical psychological impact on the Venetians. The cumulative effects of the siege : exhaustion, disease and starvation, and finally the bomb carrying balloons of the Austrians were too much for the struggling Venetians who surrendered two days later. When the siege was over, Field Marshall von Radetsky wisely allowed the patriots to slip away, hoping to avoid any future commotion or ill will.

The Austrian Empire then set about administering their new acquisition by abandoning the traditional methods of the ancient *Venetocracy*. Any image carved in stone on the walls or columns of the old city to indicate the presence of Venice's power were chiseled away and destroyed forever. Austria then wiped out everything associated with patrician rule, beginning with their traditional title, *Nobil Uomo* (Nobleman). The Austrians required the ex-patricians to fit in with the aristocracy of their own Empire by accepting the simpler title *Nobil* (without the addition of *man,* which gave rise to some bitter humour) or they could purchase the title of count or prince of the Austrian Empire. One of the great successes of Venetian rule, had been the peace and harmony of the relationships between the Italian, Venetian and pro-Venetian elements, and the Slavs in Istria and Dalmatia. The Austrians, without realizing the damage they were doing (particularly to themselves) thoughtlessly rekindled old rancours. As the Italian community were drawn to the appeal of their national *Risorgimento* or Resurgence, the Slav community responded to the appeal of pan-Slavism, looking towards Serbia and thinking of a union of southern Slavs, which in time would bring about the tragic synchronicity of events that sparked the Great War, and the end of Austria's Hapsburg Empire.

FIRST AERIAL BOMBING - VENICE 1849

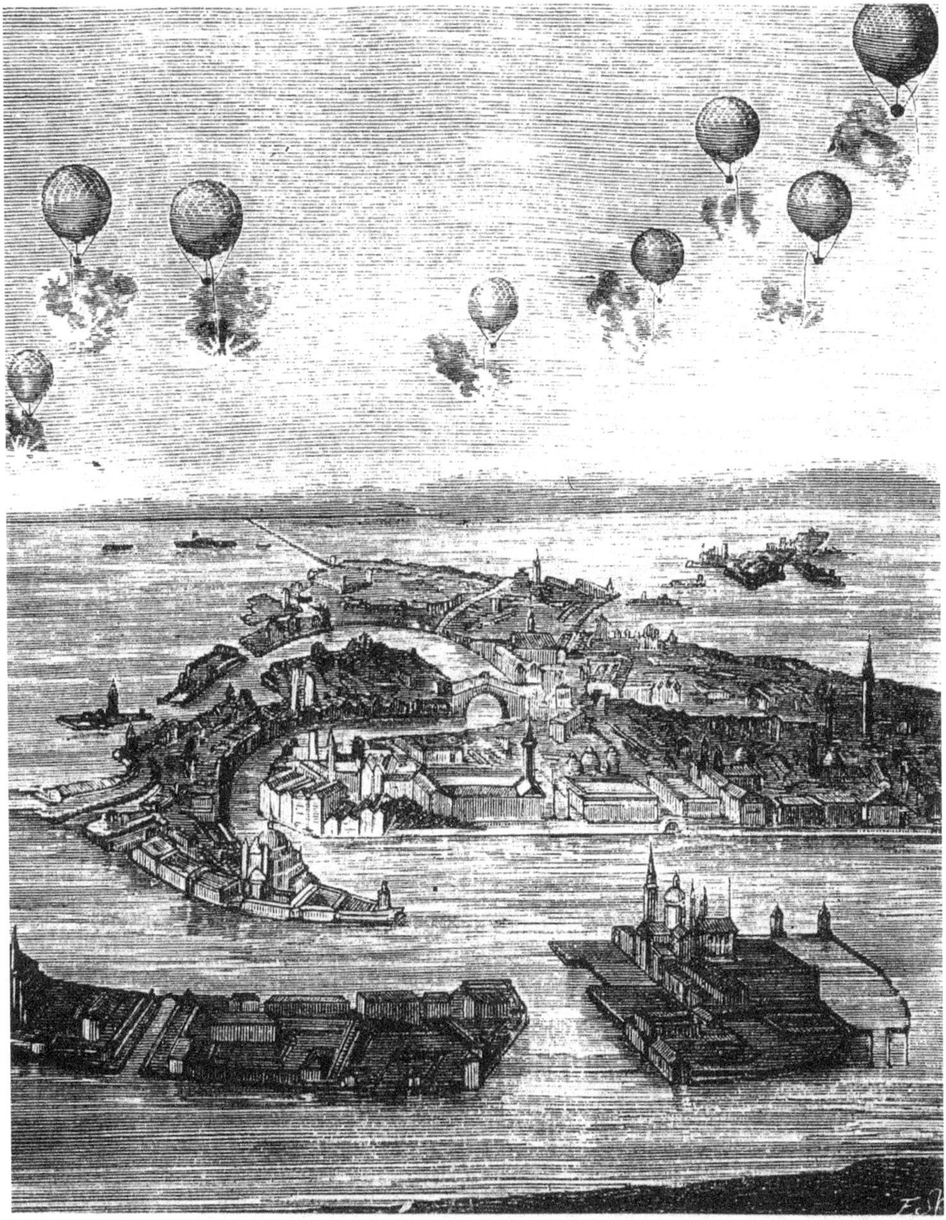

PROFESSOR LOWE AND THE AMERICAN CIVIL WAR

One week after civil war broke out between the Union and Confederacy, the 29-year-old balloonist Professor Thaddeus Lowe cast off in his version of the *Enterprise* from Cincinnati, Ohio, still wearing his top hat and frock coat from a farewell party. Lowe was on a test flight to discover if the prevailing upper winds might carry him directly east to assist in his planned balloon flight across the Atlantic Ocean. However, the winds blew him S.S.E. for nine hours and some 1,000 miles down to Unionville, South Carolina. A keen young lady, who first arrived on the scene, began assisting Lowe. After releasing the remaining gas and folding the balloon, she invited Lowe into her family's home for a meal of corn-dodgers. Other more suspicious residents arrived who took Lowe under armed guard to a local detachment of Confederate soldiers who arrested him on suspicion of being a Union spy.

Thaddeus Lowe was unable to persuade the soldiers that he was a man of science, until the local newspaper editor and innkeeper, A.W. Thompson recalled enjoying a balloon ride with him during 1860. Lowe was thus allowed to spend the night at the inn rather than the county jail ~ the first sleep he had enjoyed in 40 hours. In the morning, Lowe was released and given a celebrity's tour of the city. The mayor presented him with signed papers to ensure his unmolested return passage to Cincinnati.

On his arrival, he offered his talents to the Union. A meeting with U.S. President Lincoln was arranged, where Lowe hovered in the *Enterprise* 500 feet above Washington, and able to see clearly for 25 miles in every direction, demonstrated how balloons could serve the Union Army by transmitting telegraphed reports from a tethered balloon to a ground HQ. The first telegraphic message tapped out from the air was addressed to President Abraham Lincoln who appointed Thaddeus Lowe his Civilian Chief of Army Aeronautics with the pay of a colonel and gave his written authority to requisition equipment and personnel, to form the Union Balloon Corps.

Balloon Enterprise ~ In the Air
WASHINGTON, June 17th 1861
To the President of the United States

SIR, ~ The point of observation commands an extent of nearly 50 miles in diameter. The city, with its girdle of encampments, presents a superb scene. I have great pleasure in sending you this dispatch ~ the first that has been telegraphed from an aerial station ~ and to know that I should be so much encouraged, from having given the first proof that the aeronautic science can render great assistance.

First Battle of Bull Run, July 1861 ~ Lowe's soldiers of the Balloon Corps proudly wearing their new cap badges depicting a balloon with the letters B.C. smartly stamped on it, took part in their first action observing for Union General Irvin McDowell. They learnt that accurate observation of an opponent's position did not have to be done from 500 feet, but was possible from over 2,000 feet. During a free flight in his freshly revarnished *Enterprise*, when he was buffeted by strong winds and hit by infantry ground fire, Lowe was forced to land behind the Confederate lines. Dismounting his punctured balloon, he shot a few inquisitive Confederates and took cover in a nearby forest. His wife Leonine, seeing her husband's crash from a hill waited for the night. Then brilliantly masquerading as a farmer's wife, she rode a wagon in to rescue him, and returned thru the cover of darkness to safety.

Falls Church, Virginia ~ After a defeat near Manassas, Lowe made a free flight above 1,000 feet in the stars and stripes adorned *Union*, to determine the position of the Confederates using field glasses and a telescope to correct a report that the Confederates were making a general forward movement. Ascending on 24th December 1861, Lowe sighted an enemy camp three miles away at Falls Church and sent a report from his onboard telegraph down his telegraphic wire and attached mooring cable to General Smith's HQ. By using maps marked on an earlier flight, Lowe observed the Union gunners fall of shot from the air, and continued to correct their accuracy on targets, unseen from their position.

The Peninsula Campaign ~ In January of 1862, Lowe's seven balloons *Union, Intrepid, Constitution, United States, Washington, Eagle* and *Excelsior* accompanied General George McClellan's Union Army on their campaign up the James River Peninsula toward the Confederate capital of Richmond. Accompanying McClellan and Lowe, was the 25-year-old Count Ferdinand von Zeppelin on leave from the Prussian Army to observe the American Civil War and in particular to learn all he could on the subject of military ballooning from Thaddeus Lowe. The Union Balloon Corps now travelled with mobile hydrogen generators developed by Lowe that used sulphuric acid and iron filings to produce enough hydrogen that vented through pressure regulators, could fill a balloon in three hours.

THE AMERICAN CIVIL WAR

McClellan, an enlightened commander fully appreciated the value of balloon reconnaissance, and extended his military telegraph to all balloon sites, then up the mooring cables and into the wicker baskets so that all observers could report direct to his field HQ, any news regarding the configuration and strength of the opposing Confederate defences at Yorktown. The Union balloons were made of thick silk with a coat of varnish enveloped by a netting of Italian flax thread, while the baskets were made of willow and cane, and had an armored floor. During the siege of Yorktown, Lowe observed a heavy Armstrong cannon explode. He supposed from the extreme angle of elevation, its crew had at least double charged the piece in an attempt to shoot down his balloon. Lowe's report of 4th May 1862;

It was through the midnight observations with one of my war-balloons that I was enabled to discover that the fortifications at Yorktown were being evacuated, and at my request General Heintzelman made a trip with me that he might confirm the truth of my discovery. The entire great forest was ablaze with bonfires, and the greatest activity prevailed, which was not visible except from the balloon. At first the General was puzzled on seeing more wagons entering the forts than were going out, but when I called his attention to the fact that the ingoing wagons were light and moved rapidly (the wheels being visible as they passed each campfire), while the outgoing wagons were heavily loaded and moved slowly, there was no longer any doubt as to the object of the Confederates. General Heintzelman then accompanied me to General McClellan's headquarters for a consultation, while I, with orderlies, aroused other quietly sleeping Corps commanders in time to put our whole army in motion in the very early hours of the morning, so that we were enabled to overtake the Confederate army at Williamsburg, an easy day's march... Without the time and knowledge gained by the midnight observations, there would have been no battle of Williamsburg, and General McClellan would have lost the opportunity of gaining a victory, and the Confederates would also have gotten away with more of their stores and ammunition.

Confederate Balloons ~ The south's first balloonist, Captain John Randolph Bryan had volunteered for special service to distinguish himself. When all was revealed to Captain Bryan he told his friends, *My ardour to go on special service had been much cooled at the bare thought of being suspended in mid-air by what appeared to me as a mere thread under a hot air balloon.* Captain Bryan having just become the commander of the Confederate Balloon Service also acquired a new name, *Balloon Bryan.* His balloon could only be inflated with hot air from a field stove burning turpentine soaked pinecones, but it still gave him enough lift for a few hours aloft. Bryan thought of it, *nothing but a big cotton bag coated over so as to make it air tight and intended to be inflated with air, as gas was a thing not to be had in those places.*

During the Peninsula campaign, the Confederates used their first balloon in action near Yorktown on 13th April 1862. While observing his opponents, Bryan was shot at by Union artillery and had such a petrifying near miss that he requested an immediate discharge from the Confederate balloon service. His commander General Joseph Johnston expecting a better example quashed his request reminding Captain Bryan, *My dear sir, I fear you forget that you are the only experienced aeronaut I have with my army.* The Confederate method to protect Bryan and his balloon was to harness a team of six artillery horses to his mooring cable and have his ground crew standby for evasive action. The team of horses could then be ridden away at full gallop hauling the balloon away from enemy gunfire. Captain Bryan last flew at night, after one of his ground crew tangled his leg in the rising balloons mooring cable. The cable had to be cut, so Bryan was cast adrift in his balloon to waft helplessly towards the Union lines. Then for a moment it paused... slowly the wind changed, and Bryan's balloon floated back to the battle lines of the Confederacy where Southern soldiers mistook his craft in the dark sky for a Union balloon. Peppering his envelope with musket balls, they brought Bryan's balloon down in tatters.

Seven Pines (or Fair Oaks) 31st May and June 1st 1862 ~ Prior to this battle, Union balloonists were sent aloft to select a site for a supply bridge to be constructed across the Chickahominy River flowing full of spring rain. A map reference chosen was reported from the balloon, and the bridge built. Thaddeus Lowe then ascended in his tethered balloon *Intrepid* to a height of 2,000 feet, and spotted a column of Confederate troops and cavalry moving from Richmond toward the Union lines. Messages were tapped out from his balloon to alert HQ and Lowe was told to keep watching. By noon, Lowe wired McClellan that the Confederates were headed for Fair Oaks. At dawn the next day, Lowe reported the enemy less than four miles away from the Union lines and still advancing. General McClellan had been confident that the Confederates were feigning an attack, however Lowe seeing the situation clearly from above, continued sending reports, and saved the Union army from defeat. At the Battle of the Seven Pines, Confederate gunners developed the *Blanket of Fire* tactic, in which they created a field of fire through which an ascending or descending balloon would have to pass. They all missed Lowe's Union balloons, but some shells landed to damage the camps beyond.

Alexander Gardner's photograph ~ Studying the Art of War ~ taken at Fairfax Courthouse, Virginia, June 1863, depicts four officers in the Union Army from left-to-right : Major Ludlow, Colonel Dahlgren, Colonel Joseph Dickson Adjutant to Hooker (wearing a straw hat), Lieutenant Rosencranz and the foreign military observer, Count Ferdinand von Zeppelin of the Prussian Army who travelled with Union forces to observe the war. Zeppelin ascended in an air balloons, and became known for the massive combat airships that he later developed.

Confederate Dress Silk Balloons ~ The south's finest corps commander, General, James Longstreet wrote for the Century Magazine in 1886 ; *We longed for the balloons that poverty denied us. A genius arose for the occasion and suggested that we send out and gather all the silk dresses in the Confederacy and make a balloon. It was done and soon we had a great patchwork ship of many and varied hues.* The Confederacy found it simple shopping in Savannah and Charleston to purchase 40-foot lengths of multi-coloured dress silks over the counter. The procured silk was then sealed with a concoction a la Conté of rubbery varnish prepared from dissolving old rubber railcar springs in naptha. When complete the colourful balloon christened the *Gazelle* was deflated and delivered to Richmond in 1862. There it was reinflated with coal gas from the Richmond gas works, tied to a locomotive, and taken down the river railroad to any point on the map requiring observation ~ adhering to the timeless military adage, that advises ; *Time spent on reconnaissance is seldom wasted.* Colonel Edward Porter Alexander, who rode the *Gazelle* aloft to a height of 1,100 feet for seven hours during the Battle of Gaines Mill, used it on 27th June to report the moves of General Slocum's advancing Union division.

On the 4th July 1862, E.P. Alexander after having flown repeatedly in his balloon decided to have the *Gazelle* brought down, deflated and stored on the deck of the Confederate gunboat CSS *Teaser* on the James River. One day the tide went out and left *Teaser* stranded on a sandbar at Turkey Bend. Seen from the wheelhouse of the patrolling gunboat USS *Maratanza*, the skipper fired two shots into her, rupturing *Teaser's* boiler. The *Maratanza's* crew boarded the deserted wreck of the *Teaser*, captured the silken *Gazelle,* and sent the remnants to Lowe. A second dress silk balloon flown by Charles Cevor during the 1863 siege of Charleston was blown from its mooring and captured. Edward Porter Alexander who rose in rank to command the southern artillery at Gettysburg in July 1863, after seeing no further use of balloons by the Union wrote : *I have never understood why the enemy abandoned the use of military balloons early in 1863, after having used them extensively up to that time. Even if the observers never saw anything, they would have been worth all they cost for the annoyance and delays they caused us in trying to keep our movement out of their sight. Although we could never build another balloon, my experience with this gave me an idea of the possible efficiency of balloons in active campaigns.*

* * *

THE PARIS AIRLIFT ~ 1870

During the Franco-Prussian War, the advancing Prussian Army encircled Paris on 19th April 1870 stranding British journalist Thomas Gibson Bowles, correspondent of the London *Daily News*. After milk, cheese, and the usual meats began to run out, the besieged Parisians savored horsemeat then cats and dogs. As these became scarce, the butchers of Paris in the necessity of war looked to the zoo. Bowles wrote that he had dined on *Cuissot de Loup Sauce Chevreuil* (Haunch of Wolf with Deer Sauce), *Terrine d'Antilope aux Truffles* and a slice of elephant that he duly described as, *tough, coarse, and oily*. With the urgent stimulus of needing to send his dispatches to his newspaper in London, it was his idea to suggest an airmail balloon service to carry priority mail out of Paris.

His concept had merit and in September of 1870, the French Post Office announced that they would construct balloons under the spacious roofs of the Orleans and Northern Railway stations, and soon commence the regular dispatch of airmail balloons to maintain communications with the outside world and the provisional government at Tours. As there were two professional aeronauts in Paris at the time, Goddard and Yon, they were entrusted to oversee balloon operations at the Orleans and Northern Railway stations respectively. Silk as a fabric was barred on the point of expense alone. As a single journey was all that was needed to be calculated for each craft, simple calico would suffice. Slight differences in fabrication were adopted at the two stations. At the Northern station, plain white calico was used, sewn with a sewing machine, whereas at the Orleans station the red and white lengths were hand stitched. The varnish was a mixture of linseed oil with driers, laid on with rubber to render the material gas-tight and quick drying. The first Paris balloon, *L'Neptune* was released on the 23rd September with Bowles' dispatches and a sack of mail. The aeronaut Jules Durouf later wrote :

I left the Place Saint Pierre Montmartre at eight in the morning of September 23rd. A strong east wind was blowing. I rose to a height of 3,000m and was then driven in the direction of the Arc de Triomphe. Going still westward I perceived the Prussians in clusters below, and with a telescope, could distinctly see them pointing cannon at me. I saw the balls ascend almost perpendicular into the air, exhaust their impetus, and fall to the ground. Some of the balls arrived quite high enough to make the balloon vibrate perceptibly. Infantry fired at me with their rifles almost all the way from Paris to Mantes, but I was entirely out of their range. I distinctly saw the Prussian Army in the valley of the Seine, in seven lines, all flanked by cavalry. After a voyage of 45 minutes, I judged that I had gone far enough to come down in safety and descended in the park of a chateau near Evreux, belonging to Admiral La Noury. I was then able to take my dispatches on to Tours.

It must have been an impressive sight for both the Parisian defenders and the Prussian invaders to witness the departure of each balloon and their aeronauts carrying cargo to the outside world. To boldly ascend above the roofs of the capital, wafting away to an uncertain and hazardous fate, running the gauntlet of enemy gunfire, then catching the upper winds to be borne away for as long a period as their limited ballast releases could be maintained. In the first airlift out of a besieged city, vivid red and white striped balloons ascended almost daily until the Prussians deployed a mobile anti-aircraft gun designed by Alfred Krupp. Many balloons were targeted, but few were damaged. The Paris authorities then decreed from November, that the balloons would only depart by night, which merely substituted one hazard for another in the hands of enthusiastic amateur aeronauts. After night departures began, one craft sank in Arcachon Bay off the southwest coast near Bordeaux. *Le Jacquard* was blown over the English Channel and came down in the sea off Falmouth in Cornwall. Fishermen recovered the mail, but the brave aeronaut drowned. On 26th November, *Ville d'Orleans* set the new flight record of 15 hours, and touched down in Norway, some 600 miles north of Christiania. By the end of November, 30 escape flights had been made. In December, another 15 balloons escaped. In total, the Parisians successfully launched 66 balloons that carried 122 passengers plus more than two million letters or ten tons of mail. From a financial point of view, the cost of the venture was covered by 20 centimes postage for four grammes, on par with the famine prices of the siege. Also established, was a homing pigeon airmail service by Gabriel Maugin, who took off with a cage of 60 pigeons in the *Ville de Florence*, of which 57 birds flew back with thousands of microfilmed messages.

Although the defenders of Paris did not derive much actual assistance from aerial observation balloons as generally used in warfare, they ensured regular communication and important airlifts. Had the siege continued, an even more elaborate use of balloons would have been observed. When hostilities terminated in May 1871, there were six balloons in readiness at Lisle, waiting for a northerly wind to attempt to fly into Paris from the outside. The following year, after ceding the two border provinces of Alsace-Lorraine plus five billion gold francs to the Prussian invaders, the old ballooning centre at Chateau de Meudon was reopened, after laying in technological limbo for seven decades.

* * *

THE FIRST AIRSHIPS

The first decades of the nineteenth century were a time when ballooning became the sport of the rich, and when the idea of an airship was uppermost in the minds of those who visualized future air travel. The French inventor, Jules Henri Giffard was one of these visionaries, and in 1851 he patented ; *the application of steam in airship travel*. Others had flown, but Giffard was determined to build a steerable steam-powered airship, with a means of changing direction to fly wherever he wished. As all available steam engines were too weighty, Giffard needed to design a light one-cylinder steam engine with the best possible power to weight ratio. The creation and construction of Giffard's small steam engine had only one purpose : to apply the driving force of steam on board his airship. Henri Giffard's eventual engine weighed 180kg. There was also the additional weight of its boiler plus enough coke fuel and water for a one-hour flight. The steam engines three horsepower or 2,200 watts would still be adequate to turn the three metre propeller 110 times a minute to provide sufficient forward thrust. Giffard realized the danger of putting a furnace so close to his gas envelope, so he placed a piece of wire gauze in front of the stoke hole, and could pivot the chimney downward to direct the steam mixed with combustion gases through wire gauze, and away from the envelope. With its long keel evenly distributing the suspension load, Giffard calculated that it would all be light enough for his airship to lift. Finally, Giffard mounted a rear rudder of triangular sailcloth, and pressure-filled his 43m long airship envelope with hydrogen to acquire its aerodynamic shape, and to maximize its lift.

Powered and controlled flight ~ Henri Giffard's airship was ready to fly on the 24th September 1852. Taking off from the Paris Hippodrome, he rose to a height of 2,000m and at seven kilometres per hour flew 27km to Élancourt. Henri Giffard demonstrated for the first time that his powered airship obeyed his helm, as he turned and piloted his airship in slow circles, proving that controlled flight was possible. However, when it came time for Jules Giffard's return flight to Paris, a stiff breeze developed and against a gathering headwind, Giffard was unable to refuel and return to Paris.

The world had to wait 30 years until August of 1884, for the next significant step in aviation. Two captains of the French Army, Charles Renard, and Arthur Krebs were preparing their airship for its first test-flight at Chateau de Meudon. Their 50m long *La France* used Charles Renard's recent invention of the zinc-chlorine flow battery connected to an 8-hp Théophile electric motor to power its seven-metre, four-bladed propeller. Ready for take off, Renard and Krebs flicked their electrical switches and *La France* rose up. Flying a figure of eight they covered an eight-kilometer circular course in 23 minutes and returned to their starting point after completing the world's first return flight.

SANTOS-DUMONT CIRCLES THE EIFFEL TOWER

Alberto Santos-Dumont, the Brazilian heir of a famous coffee family came to live in Paris in 1891. Hoping to become the world's first *sportsman of the air* he hired a balloonist and learnt how to pilot. Learning fast, he had successfully designed and flown eleven balloons and five airships by 1905. His next project, the bold yellow airship #6 was fitted with a 20-hp Buchét engine that incorporated two significant innovations : joints of aluminium reinforced by piano wire to minimize weight, and suspension lines of wire to reduce the air resistance of #6 by half. The purpose of #6 was to claim the *Deutsch de la Meunthe* prize of 125,000 francs for the first controlled flight from Park Saint Cloud, around the Eiffel Tower and back again in less then 30 minutes. Before his attempt, he announced : *I am not interested in the money... I will give half to my team, and donate the rest to the poor of Paris!*

After lunch on 19th October 1901, he signaled to his crew ~ *Let go all!* His airship rose up, thrusting straight for the Eiffel Tower to make it in nine minutes. As he circled the Eiffel Tower at 50 metres, the crowd of cultured Parisians in the staircase called out, *He's coming back!* After miscalculating his turn, he only just missed colliding with it. The motor misfired, and without power... #6 drifted in the wind. Leaving the safety of his nacelle, Santos-Dumont walked along the airframe of #6 to revive his engine. He got #6 under control again, and succeeded in getting back to his starting point of Park Saint Cloud in 29½ minutes, to win the 125,000 francs of the *Deutsch de la Meunthe* prize.

THE BEGINNINGS OF AVIATION IN ENGLAND

Lord Garnet Wolseley, recalling Bonaparte's advice, *you must change your tactics frequently*, on his first balloon ascent said; *I believe in novelties, the more novelties the better! Had I been able to employ balloons in the earlier stages of the Sudan campaign, the affair would not have lasted as many months as it did years.* In this inspired British pictorial of 1878, a naval officer attacking an enemy coastal fort maneuvers a second bomb-carrying balloon over his target. The button beneath his finger will release the bomb.

BRITISH MILITARY BALLOONING

In England, experiments began at Woolwich Arsenal in 1878, and in 1879 a military school for balloon training was established at Chatham, under Captain Charles Templer of the Royal Engineers. British forces took their balloons on colonial expeditions to China where they were used for the preparation of maps and during an expedition sent to Bechuanaland in 1884 under Sir Charles Warren to expel raiders and reinstate the colony's natives. The Sudan Expeditionary Force used captive balloons during 1885 when ascents over El Tib and Tamai saved many. During the South African War, four balloon sections took an active part in the difficult campaign to support the forces of the British Empire on the ground. Suspended beneath each balloon a sturdy wicker basket carried two trained observers into the air to hover over the battlefield and act as the eyes of the British artillery, to seek out the enemy and report anything of interest through their connected telephone line to HQ.

Siege of Ladysmith ~ The 2nd Balloon Section with eight balloons under the command of Major G.H. Heath arrived at Ladysmith on the 27th October 1899. Days later the Boer enemy surrounded the town blocking any access to further supplies of bottled hydrogen. As long as their supply of hydrogen permitted, each khaki balloon was filled with 11,000 cubic feet of gas and tethered at a height of 3,000 feet. Over Ladysmith, balloon reports were flashed to the ground HQ using heliograph mirror signals.

Magersfontein ~ General Buller sent the 1st Balloon Section commanded by Captain H.B. Jones to join Methuen's advance on the Modder River. There they observed the Boer enemy at Magersfontein, (whenever their Boer opponents could be seen). A timely 1st Balloon Section report reliably informed Lord Methuen that a perceived 1,500 yard gap in the Boer line had closed up and no longer offered an opportunity for one his brigades to slip through. After that, a line of Boer reinforcements were seen coming from Spytfontein, with indications that they were preparing for a flanking attack against the British right. Only the steadiness of the Guards and the accuracy of their fire thwarted their purpose. It was by another observation from the air that the howitzers got the range of Boer horses concealed in a gully and accounted for 200 of them. Unfortunately, balloon observers were unable to detect the well-camouflaged Boer positions at the foot of the hills of Magersfontein. As advancing British infantry failed to spread out from a compact formation, Boer marksmen were able to inflict horrific casualties on them, contain their advance, and finally force them to withdraw.

At *Vaal Krantz*, the British war balloon came under heavy fire and had to be retired out of range by means of the rope, which holds it captive.

BRITISH BALLOONING DURING THE SOUTH AFRICAN WAR

Siege of Mafeking ~ The Boers surrounded Mafeking on 13th October 1899 and as Colonel Robert Baden-Powell's soldiers protected the city, he inspired the boys of the town to form the Mafeking Cadets to take on a support role within the town. Boer artillery, firing smokeless artillery shells to aid their concealment, hoped to contain the advance of the British 10th Division moving toward Mafeking at Fourteen Streams, north of the River Vaal. However, Lieutenant R.B.D. Blakeney sent observers of the 3rd Balloon Section aloft near the river to direct accurate gunfire from British 5-inch howitzers that thwarted several Boer attempts to hold the advance of the British relief column. Spotter corrections were shouted through a megaphone and the results were good. A week later, the British brought up a 6-inch gun mounted on a railway wagon and within two days, counter-battery fire directed from these balloonists had silenced the Boer artillery. While the attention of the Boers was focused on this, the main part of the British 10th Division outflanked the Boers, by crossing a river to the west. The experience of British ballooning during the South African War had shown that enemy infantry or artillery en route could be monitored from a hovering balloon from a distance of 11,000 yards.

Colonel Robert Baden-Powell's defence of Mafeking is one of the great feats of tactical deception in the history of warfare. The British who didn't have any barbed wire for their defensive perimeter, instead walked around their outer perimeter with theatrical high steps to give their observing Boer opponents the impression they were finding their way through wire entanglements. Baden-Powell's other great ruse was setting up his *secret factory* to fabricate land mines. Even if he only had a few sticks of dynamite, some of his troops filled wooden boxes with soil and boldly plodded out to bury their fake mines around the defensive perimeter. The pretence put on for the Boer spies within the town culminated with the notice of a test firing of one of the mines, requiring the restriction of the townsfolk from the test area. Baden-Powell and his artillery officer then went out to the advertised test area alone and detonated a stick of dynamite down an ant nest. Word soon spread that the perimeter was heavily mined. The Boer commander of 12,000 enemy troops, General Piet Cronje dispatched a note to Baden-Powell protesting against his use of land mines and demanded the town's surrender. Baden-Powell who only had 1,000 troops replied resolutely, that if he did not like mines he should find himself another battle. For seven months, the Boers remained duped that the Mafeking perimeter was strongly fortified with land mines and barbed wire. Piet Cronje's strong Boer force could have easily overrun Mafeking with one assault, but he never tried. British troops eventually relieved Mafeking on 17th May 1900, after Baden-Powell's soldiers had suffered 30% casualties and were almost out of ammunition : but they had beaten their opponents with bluff. Colonel Baden-Powell who had become England's principal hero of the South African War received a telegram, detailing his promotion to Major General at the age of 43 (the youngest general in the British Army) and after being so impressed with the boys of his Mafeking Cadets, he established the International Boy Scouts.

The British Army's first airship ~ a sausage-shaped contraption, almost 120 feet in length and 30 feet in diameter was fitted with a 50-hp Antoinette engine and christened with the Latin motto of the Coldstream Guards, *Nulli Secundus* or Second to None. With Colonel John Capper R.E. at the controls for her maiden flight, he took off from Farnborough at 1100 hours on 5th October 1907. *The Times* remarked on the day, *it was an open secret that London was their objective*. Winds from astern blessed the flight and by midday *Nulli Secundus* making 26 mph flew over Kensington Palace. Then she toured London, passing over the grounds of Buckingham Palace to salute King Edward VII. A circuit round the grand dome of St Paul's Cathedral crowned the flight after which Colonel Capper turned for home, however *Nulli Secundus* was not to make it back to her Farnborough base as planned.

The friendly southwesterly winds of the outward flight were now adverse… and strengthening. After creeping over the Oval, Colonel Capper realized that he could no longer make headway against the wind, so set course for the Crystal Palace where a ground party of Royal Engineers eased the airship down to a soft landing onto the centre of the Crystal Palace cycle-track to conclude the 50-mile flight in three hours. Capper said : *We should have no difficulty in keeping in the air for another six hours. France and Germany now have a worthy rival in the aeronautic accomplishments of the first dirigible to be designed for war purposes by the British Government.*

Still tethered to the ground five days later in strengthening winds the still inflated airship had some of her mooring ropes snap free. To avoid further damage *Nulli Secundis* was deflated, her hydrogen was vented off through escape valves and a slit through the envelope. Finally the flapping airship was packed up and returned on the back of a lorry to Farnborough. In June 1910, the smaller airship *Beta*, was reliably powered by a 30-hp Green engine, and flew with success for 3,000 miles.

* * *

THE WRIGHT MILITARY FLYER

WILBUR WRIGHT (From: *L'Aérophile* December 1908)

Although Wilbur and Orville Wright were not the first to achieve powered flight, on December 17th 1903 their aircraft took off from Kitty Hawk Beach, N.C. for an inaugural flight that lasted 12 seconds and covered 120 feet. Later that same day the *Wright Flyer* flew 59 seconds over 859 feet. The Wright brother's essential secret was their ability to morph their *Flyer's* wing tips in flight to give lift to banking, turning or to sustain the lateral balance of their craft. In 1908, the brothers flew an improved version of the *Flyer* over Washington and the U.S. Aeronautical Division issued the first specifications for a military aircraft: the ability to lift two men for a flight time of one hour at a speed of 40 mph. The following year in 1909, Orville demonstrated the brother's newest 30-hp two-seater *Military Flyer* version that enabled a pilot to carry an observer-gunner to qualify for the $25,000 agreed sale price in addition to a nice $5,000 government bonus for surpassing the requisite 40 mph speed. In France, when Wilbur Wright was invited to make a dinner speech, he wryly responded, *I only know of one bird, the parrot that talks ~ and it does not fly very high.*

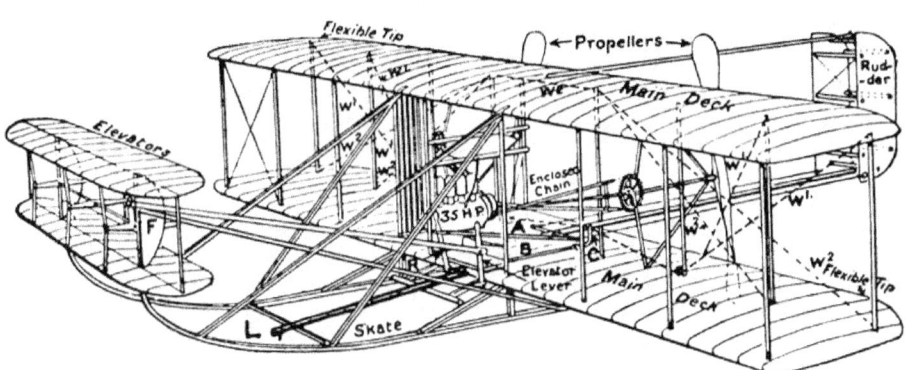

Diagram of the Wright Flyer assembled by Wilbur Wright in France ~ June 1908

The birth of American naval aviation came on 30th June 1910, when Glen Curtiss carried out a series of bombing tests. Flying his own biplane design over the outline of a battleship marked out with buoys on Lake Keuka in New York, he dropped 18 out of 20 mock bombs onto the outline of the mock battleship target. Curtiss then flew 137 miles along the Hudson River averaging 55 mph to New York, where he flew over Manhattan and circled the Statue of Liberty. In the winter of 1910, Eugene Ely made the first shipboard takeoff flying a Curtiss Model D Biplane from a ramped platform of planks raised above the bows of the cruiser *USS Birmingham*, anchored in Chesapeake Bay. Then on 18th January 1911, he landed on the stern of the *USS Pennsylvania* 13 miles away in San Francisco Bay. Using an arrestor of 22 wires with attached sandbags, Ely touched down half way along the landing area by hooking some N° 11 wire, to stop within ten yards. After lunch on board with the skipper, Ely took off and returned to California.

In 1912, the U.S. Navy successfully launched an aircraft from a ship by catapult, and while Lieutenant Milling piloted a *Military Flyer*, his observer-gunner Captain Charles de Forest Chandler test fired a fitted Lewis machine gun from a height of 250 feet and scored five hits on a target of six-by-seven-foot cheesecloth. The first aeroplane used in a military operation was employed by the United States in February 1911 to observe the Mexican frontier near Juarez during some revolutionary fighting in Mexico. The aviator was Charles Hamilton, and the machine a Wright *Military Flyer*. Then in April of 1914, as part of the U.S. military expedition against Mexico, flying boats designed by Glen Curtiss were sent aloft to search the seas ahead of a U.S. naval squadron steaming for Vera Cruz for mines. This was the debut of naval aviation engaged in supporting warships at sea.

THE HENRI FARMAN BIPLANE

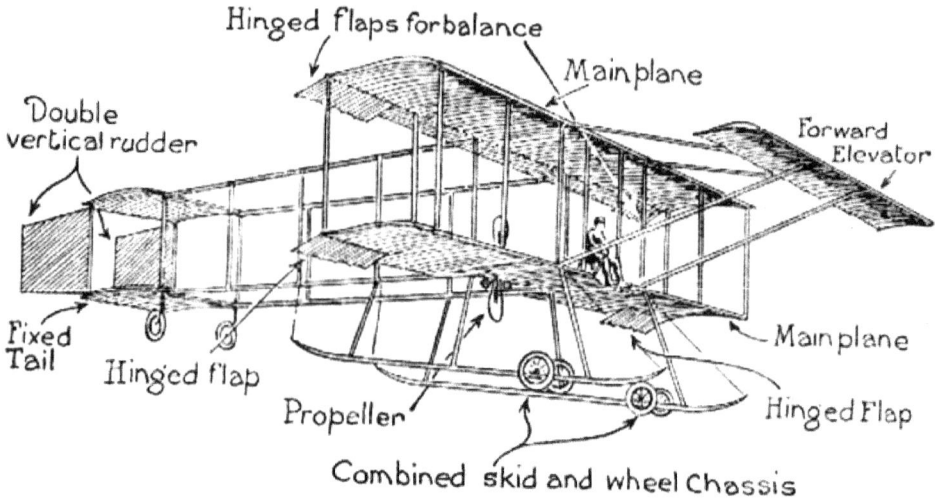

Meanwhile in Europe, the most memorable biplane at the great Rheims Aero Carnival of 1909 was that designed and flown by Henri Farman, whose light and efficient biplane represented a big step forward in aeroplane construction with several excellent features. With its lightly constructed tail of two horizontal planes, a vertical rudder was fitted between them. In front of the main-planes on light wooden outriggers was the horizontal elevating plane. Another of the innovative features of this aeroplane was his method of obtaining lateral stability. Henri Farman fitted small flaps or horizontal planes, hinged to the wings, and termed them *ailerons*. The controls of the Farman biplane that became standard, were a hand and a foot lever. The hand lever, when moved forward or backward, operated the elevating plane. When shifted from side to side, it actuated the wing *ailerons*. The pilots feet rested upon a pivoted bar, which he swung from side to side to move the tail rudder of the aeroplane.

Another feature of Farman's biplane was the landing chassis. Appreciating the disadvantages of the Wright brother's launching rail, Farman aimed for something lighter and more efficient. He devised a chassis combination of wooden skids and bicycle wheels. Below his biplane, on either side of two long wooded skids were two small pneumatic-tyred bicycle wheels paired by a short axle. The wheels were held in position on the skid by stout rubber bands, which passed over the axle. Normally, the skids were raised off the ground by the wheels on which the biplane actually ran. However in the case of a rough landing, the chassis was designed so that the wheels were forced up against their rubber bands, so the skids of the machine came into contact with the ground. Then, when the force of the shock had been absorbed, the wheels came into play again.

Apart from the constructional excellence of his innovations, Henri Farman fitted a 50-hp motor, which was to have a remarkable influence on the development of aviation. The seven-cylindered revolving *Gnome Omega* used on Farman's biplane in August of 1909 was initially regarded as odd. However in 1909, all other aeroplane engines, both air and lightened water-cooled types, after perhaps 20 minutes had a tendency to overheat, owing to their engines high turning speeds and either lose power or stop altogether. But, the *Gnome* had its seven cylinders revolving around a fixed crankshaft, with petrol and oil fed to the cylinders thru its stationary hollow crankshaft.

The advantages of this efficient engine were many. Its method of construction enabled the Farman to be built very light that enabled more fuel to be carried to extend its operational range, and for a pilot to carry a handgun, rifle or machine gun aloft : major advances for military aviation. The design of the seven cylinders revolving at near 1,000 revolutions a minute mechanically cooled itself and its flywheel effect gave a smooth even thrust to the propeller. The internal complications of this engine ran so well that Henri Farman remained in the air for more than three hours, winning the 1909 Rheims Aero Carnival prize for the longest flight, as well as the passenger carrying prize for carrying two passengers round a ten kilometre course in ten minutes. Louis Paulhan in a Farman III on 27th April 1910 won a £10,000 prize for flying the 183-mile aerial journey from London to Manchester. After Louis Bleriot's crossed the English Channel in 40 minutes, British and French military planners began to take all aspects of aviation seriously.

PREPARATIONS FOR WAR

A number of enthusiastic young British naval and army officers inspired to fly at their own expense helped establish the Royal Flying Corps (RFC) on 13th May 1912, with both a Naval Wing and Military Wing to meet the requirements of each service. The role of the Military Wing was to cooperate with the other branches of the Army and by flying over enemy rear areas to provide intelligence on enemy strength, troop dispositions, and traffic movements to locate enemy artillery, depots, dumps, and troop encampments. HQ using this and other sources of information, could then outline an opponent's main force and estimate their intentions. Squadrons were formed, consisting of three flights, each with eight pilots and four aircraft, plus the squadron commander and his aeroplane.

The Royal Navy required its aeroplanes and airships to search vast areas of ocean, to keep an eye on the position of an opponent's main naval forces, to preserve the tactical initiative, and avoid being surprised. The airships of the Royal Navy enjoyed the advantages of greater range and being able to carry a useful load, so wireless equipment was installed to send and receive across hundreds of miles. They could hover with slowed engines, to detect mines and submarines. To maintain extended operational efficiency their crews were accommodated in comfort to offer them a decent rest to ensure maximum alertness. Within months the Senior Service ceased to call its air arm the Naval Wing and unofficially substituted the title, Royal Naval Air Service. The RNAS then established several seaplane bases along the east and south coast to face the perceived threat from Imperial Germany, and in particular to protect the powerhouses, the naval harbour of Portsmouth and their fuel tanks

By the beginning of 1914, there were two approaches to military aircraft design. The *tractor* with its engine in front of the pilot powering a propeller in the nose or the *pusher* with the engine mounted behind the cockpit with its propeller reversed. The pusher offered its pilot and observer seated in the nose an excellent view over the ground below ~ and if the observer was armed, he had a clear field of fire over a wide forward arc. However, the tractor with its cleaner aerodynamic shape offered better aerial performance. Typical aeroplanes of 1914 were capable of almost 60 mph, with operational ceilings of 3,000 to 12,000 feet and flight duration times of between two to four hours. All Royal Flying Corps squadrons were assembled near Nether Avon in Wiltshire, in June 1914 for a month of special training. The mornings were devoted to experimental work and the afternoons for lectures. There were courses in meteorology, engine and aeroplane construction, plus how to use a compass. Then exercises in photography, reconnaissance, and navigational flying between various landing grounds.

Even to the most optimistic, there was a perceptive dread in the air… war was not far away. On the morning of 28th June on the streets of Sarajevo, seven conspiring *Union of Death* or *Black Hand* assassins armed with four pistols and six bombs (supplied from Serbian Army arsenals) were waiting at several points along the well publicized route of the motorcade of the Archduke Franz Ferdinand, nephew and heir to the Emperor of the Austro-Hungarian Empire, who was accompanied by his wife the Countess Sophie. The morning was sunny and warm. At 1010 an assassin threw a bomb at the second automobile, the vehicle flying the Hapsburg pennant, and carrying the Imperial couple. Franz Ferdinand seeing the hurtling black bomb flying through the air raised his arm to deflect it away from Sophie. The bomb glanced off Franz Ferdinand's arm ~ bounced off the convertibles cover, and descended to detonate under the next car, injuring at least 20 spectators. The inept assassin swallowed a vintage suicide pill, but its old poison only made him vomit. Trying to drown himself, he threw himself into the nearby river, but it was too shallow and the crowd pulled him out, giving him a severe thrashing. When the Archduke arrived at the Sarajevo Town Hall he interrupted the mayor furiously shouting; *Mr. Mayor, I come to Sarajevo on a visit and I get bombs thrown at me. It is outrageous!* All gathered, then waited for the Archduke's speech, for his papers still wet with blood, to be retrieved from the second bombed car. The days schedule was cleared for Franz and Sophie to visit the wounded at the hospital. Continuing on, the motorcade rounded a corner at 1045 just outside of Moritz Schiller's food store, when the seventh assassin emerged eating a sandwich. He pulled a pistol from his pocket. Cocking it, he took a step toward the car and fired two bullets. The first hit Archduke Franz Ferdinand, slicing through his jugular before lodging in his spine. The second bullet hit Sophie. She perhaps took a bullet for her husband who died pleading, *Sophie dear! Stay alive for our children!* Both were dead in minutes. Kaiser Wilhelm II of Germany, alarmed by the assassination of his friends, pressed Vienna to make appropriate investigative demands on Serbia, for its role in the assassinations. Serbia leaned toward Imperial Russia for assistance, but Russia tied by treaty to support Serbia, was also tied by treaty with France ~ and by extension to the British Empire. Mired in the impenetrable diplomatic squabbling of opposing alliances, all too soon the Empires of Europe drifted from the deadlock of the July Crisis, to the predicament of having to choose sides, mobilize and declare war…

More than one true thing may be said about the causes of the war, but the statement that comprises the most truth is that militarism and the armaments inseparable from it made the Great War inevitable. Armaments were intended to produce a sense of security in each nation ~ that was the justification put forward in defence of them. What they really did was to produce fear in everybody. Fear causes suspicion and hatred. It is hardly too much to say, that between nations, it stimulates all that is bad and depresses all that is good.   Sir Edward Grey

When the Great War erupted, people asked as they still do, what was the cause of the quarrel? Many believed it impossible that the continent of Europe could be plunged into war through a local squabble in the Balkans. But while the greatest events, as Sophocles pointed out many centuries ago, may seem to spring from trivialities, the causes are always profound. The truth of this is abundantly clear in the origins of the Great War. They can be traced in various deep and different ways, and as far back as in the primitive days when the Teutonic warriors sallied forth for a place in the sun which they regarded as theirs by *Right of Conquest* ~ and if we examine the various episodes that lead to the Great War, we soon become conscious of a tide of affairs that gathers force and carries all before it.

Acting on the historical assumption that war was a legitimate instrument of foreign policy, and on the whole the best way of clearing up the knots and tangles that had defied their best statecraft, most diplomats and statesmen of pre-war Europe were unaware that the collective result of their foreign policies and their armaments would all end so tragically. The wars that they had in mind were the limited and localized Prussian-Danish War of 1864, and the Franco-Prussian War of 1870-71. These masterfully planned short wars had both achieved their purpose without excessive sacrifice or the derangement of European society. Whilst the diplomacy previously practiced with its adoption of the ethics of war was well suited to limited localized wars, it became wholly unsuited to the hazards of multi-national alliances that only divided Europe into two hostile groups who became increasingly suspicious of one another, who steadily built up greater armies and navies, and who would all be drawn into the Great War which few wanted. It all came from their system of alliances and therefore remains a question whether this theme of history may be regarded as closed or whether it might be reopened by another oversight of misguided statecraft, and be repeated to an even worse conclusion.

If we search carefully, we can find everywhere underlying the epic struggles of recorded history, (and in those now occurring) an antagonism of some kind between two systems, visions or ideals, which in some particular contention were fundamentally opposed and could not be easily reconciled. State papers and the memoirs of diplomats are apt to confuse the real issues, by setting forth a diary of minor incidents and piquant details, not in their true proportions, but as they appeared at the moment of their occurrence to the awareness of harassed and suspicious officials. For example ; all the archival heritage of the American Civil War upholds the essential fact, that in this conflict a million lives were sacrificed by one of the most intelligent and practical nations of the mid-1800's for no other cause than an irreconcilable difference amongst them ~ for what St. Paul called *the substance of things hoped for*. On one side there was the ideal of Union and a determination to make it prevail. On the Confederate side there was an ideal of Independence and an equal determination to defend it ~ whatever the cost. If war on such basis is possible within a single nation, nurtured in the same traditions and tongue, how futile is any assurance that material considerations can make war impossible between nations. Collisions of interest occur when one Power's interests venture crosswise to the course of another, so the essential question becomes ~ did any nation or Empire have possessions which appeared to obstruct the national ideals of others, and did these others come to believe that alone or in alliance they had the power to redress the perceived balance in triumph?

Cause of the Quarrel
THE GREAT WAR

During the whole or the greater part of the period between the Crimean War and the Great War, there were four principal factors, the interplay of which determined the shape of politics in Europe. First was the fervent hostility of France and Imperial Germany. Second was the desire of Russia to gain full naval access to warm water ~ *to obtain the key of her house* ~ as Tsar Alexander II designated the control of the Straits of the Bosphorus and Dardanelles from the Black Sea through to the Ægean to ensure an outlet into European waters not blocked by winter ice or controlled by Baltic states. Third, the rising contention from 1900 between Britain and Germany over warships, colonies, and the perceived threat of the *Berlin-Baghdad Railway*. Fourth, as the old Ottoman Turkish Empire withered, an escalating alarm in Vienna over the rise of Serbian nationalism and an associated spread of Russian influence through the Balkans, led to a view in the A-H Empire that their very existence was at stake.

As the balance of power changed throughout Europe in military alliances and territorial disputes, new threats took form to set the scene for subsequent conflict. As Austria-Hungary, Germany and Italy established their Triple Alliance, France reached out to Russia to enter into a defensive Éntente with Britain, and became diplomatic partners with their signing of the Anglo-Russian Éntente in 1907. Following the effort of the German group by a succession of shock foreign policy stunts to disintegrate the French-Russian-British Éntente ~ the subsequent focus of both groups on Austro-Russian rivalry in the East became the supreme test of strength for both Alliances. Russian ambition for dominance over the Balkans and Straits spurred southern Slavic dreams for a Greater Serbia to create their erratic Balkan League that during the Balkan Wars all but drove the Ottoman Turkish Empire from Europe, and foresaw the final forcing of the issue between the Alliances, which thereafter loomed, ever nearer.

For years, problems connected with the Near East caused intense international rivalry for future domination over the Balkan Peninsula and Asia Minor ~ and were a continual danger to world peace. These localities, together with the waterways they controlled, formed the great corridor from east to west and from north to south, and constituted the natural thoroughfare from Central Europe to Asia and from Russia to the Mediterranean. Thus, since the advent of the German *Mittel-Europa* scheme, Berlin was just as determined to thrust open the Near-Eastern door via the Balkans to optimize their commercial vitality with the Ottoman Empire - as the Russians were to expand their influence through the same region towards warm water - and the Austrians were to secure their fracturous A-H Empire.

There could hardly be a worse motive for ordering the affairs of a civilized world, but under their Alliance system in a world organized for war, it became irresistible. While many historians seeking the cause of the war waste their ink when they look for it in one root when there are multiple webs, behind every contention were the massed Empires seeing tests of their power in which they could not afford to be vanquished, and within each limited decision matrix, those favouring the neutralist cause failed because there seemed no realization on the side of peace that the danger was most desperate. That they were on the edge of Armageddon… with all psychological momentum on the side of war.

This book examines this history and the condition of Europe with its component nationalities and provides significant new insights on the incidents that occurred before the eruption of the Great War, as it explains the deep-rooted causes of the conflict, the immediate events, and considers the glaring miscalculations that arose out of the myopic misunderstandings between the Chancelleries of Europe. According to German calculations, Britain would stand aside; but Britain took part. Italy would help her allies; but Italy refused. Serbia was a thing of naught; but Serbia shattered several A-H divisions. Belgium would not count; but Belgium by her actions counted for the loss of eight vital days, while by her sufferings she mobilized against the invaders the condemnation of the world.

In the present study, I have attempted to sketch the qualities of the leading actors in this tragedy, to bring forth some clear vision on this torment of world history which society still shudders to recall, to reveal how those who were responsible for the conduct of affairs tried to think in the anxious years before the war and how they applied their conclusions. All through this grand whirlpool of history, the accent is on those individuals caught up in the prelude to war. Accordingly, and having plumbed the depths of archives for vivid detail to bring this cast of characters to life; they can be sensed anew through their words to feel what they felt and knew at that fateful time, for a genuineness of feeling, thought and knowledge ~ uninfluenced by subsequent reaction or recantation. This timeless quality in the perception of events is especially esteemed in crystallizing the sequence of scenes, to help us understand other times, other places, and our shared humanity.

Smoking Lounge and Library

... a wash and a brush up

Interior of the Sleeping Car

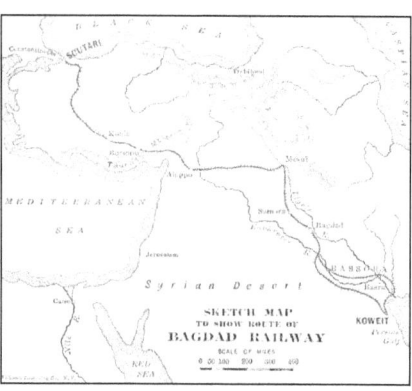

Cause of the Quarrel
THE GREAT WAR

22 MAPS ~ 330 IMAGES ~ 227,000 WORDS

~ SOME PRESS OPINIONS ~

To those who desire a brilliant survey of harrowing history and fine character studies done with precision, we recommend these chapters. MacLean's book on the causes and conditions of the war is by far the best that has yet appeared, in its coherent and lavish survey of the subject. His survey is excellent, and now opens up several new vistas of speculation. ~ *SATURDAY REVIEW*

An enlightening tour of old Europe that delves into long overlooked episodes to grant you complete access to the underlying and immediate origins of the Great War. Imperial foreign policy, nationalism, military expansion, commercial interests and the social conditions within each kingdom are examined in a flowing and entertaining style. This well-researched book reviews the advent of the war from a series of standpoints ~ how Imperial rivalry created a culture of popular imperialism that spurred a culture of popular aggressive nationalism; which in turn generated an enthusiastic expectation of war. The Empires created an aggressive posture of alliances as their best defense and if their defense was triggered - it could not be stopped. Their armies could be mobilized, but not easily stood down. Treaties between Empires were often secret, so there was no clear vision within Germany that their marching through Belgium would cause Great Britain to enter the war, and the inflexibility of their military kept any diplomacy from ever succeeding. ~ *MILITARY GAZETTE*

This new study of the Empires of Europe in their drift into the Great War is deeply researched, highly valuable in its focus on important overlooked episodes, and well-written. It deals with the old Empires of Europe struggling with strong internal nationalism, geopolitics, and commercial interests that detonated the powder-keg of so many underlying ~ and immediate causes for armed conflict. It is a storehouse of diplomatic drama and political thought, set out with a precision and an eloquence which have long been absent from the literature of imperial politics… a rare eloquence and a trove of rarely seen period images. A brilliantly written treatise. Controversial certainly ; but this volume seems to surpass in insight and fairness anything else written on the subject. ~ *MORNING POST*

Already, myriad pages have been published on the Great War and its causes ~ but now at last a sequence of pages grows into a genuine chronicle, and a necessary adventure in difficult truth-telling. With a courage that never for a moment tires, MacLean with his qualities as a good historian ~ his wit and humour, his searching irony, his wide knowledge and unbiased candour ~ have won for him the right to speak for posterity… It has taken us five days to read *Cause of the Quarrel ~ The Great War*, and if all books claimed and merited the same careful study, reviewers would be ruined and the country would be educated. ~ *ETON EXPRESS*

~ BOOK TRAILER ON U-TUBE ~

ARCHIVES
d'Histoire et Militaire

ADELAIDE

29 Alexander Street
LARGS BAY 5016
Australia

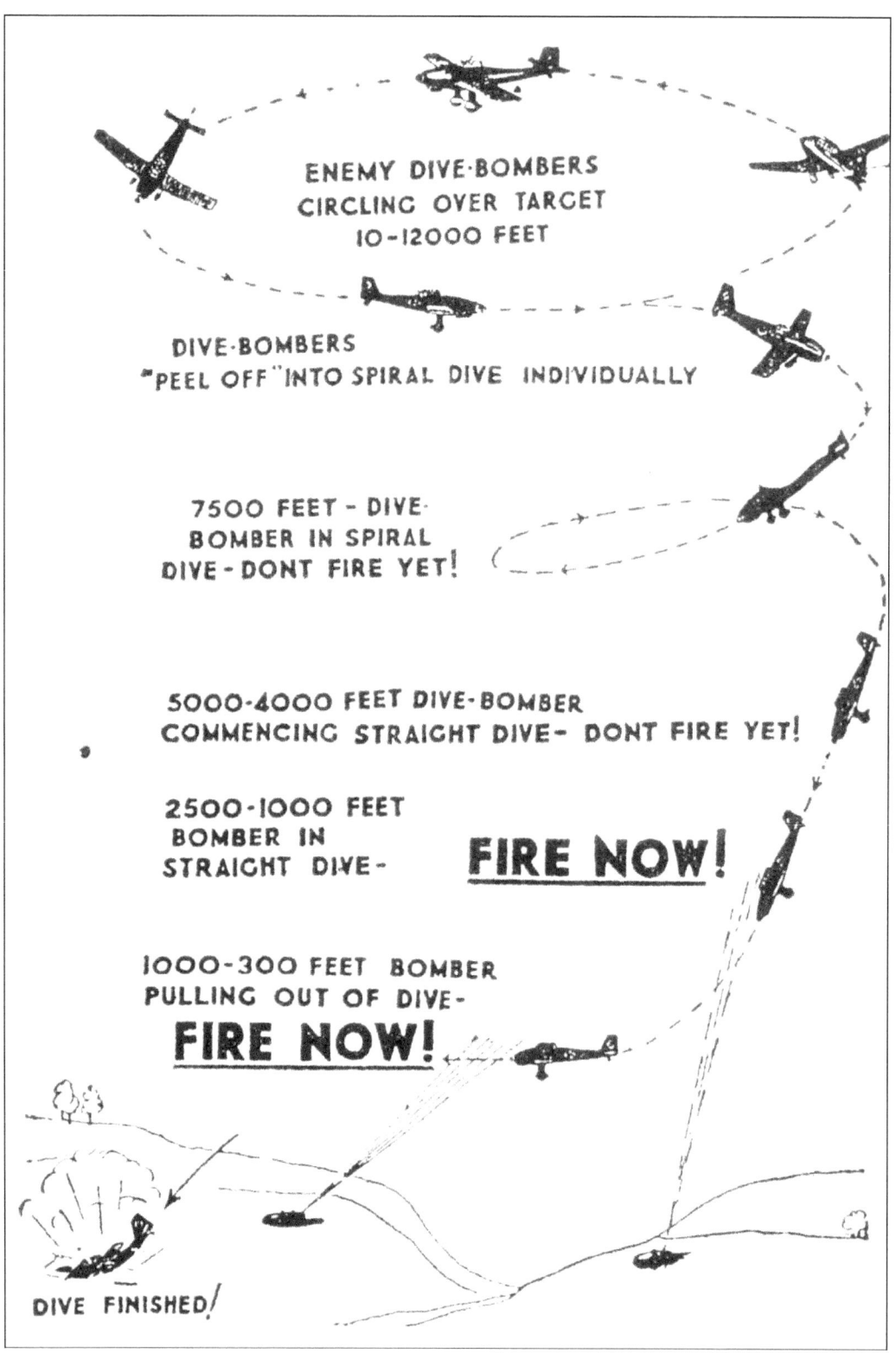

From ~ The Home Guard Training Manual 1941 ~ Edited by John Langdon-Davies, War Correspondent in Spain and Finland.

GLOSSARY OF TERMS & ABBREVIATIONS

AA - Air-to-Air (weapon)
AAA - Anti-aircraft artillery
AAM - Air-to-Air Missile
Acceleration Manoeuver - another term for a low-speed Yo-Yo. (see Yo-Yo, Low-Speed)
Ace - unofficial term for a person with five or more aerial victories over enemy aircraft
ACM - Air Combat Maneouvring
Afterburner - An auxiliary burner attached to the tail pipe of a jet engine for
injecting fuel into the hot exhaust gases and burning it to provide extra thrust
AEW - Airborne Early Warning (or Airborne Electronic Warfare)
AGM - Air-to-ground missile
AGM-45 - Shrike air-to-ground missile, anti-radiation type
AI - Airborne Intercept radar
AIM - Air-intercept missile
AIM-9 - Sidewinder air-to-air missile, various types B to S
ARM - Anti-Radiation (or - Radar Missile)
ASW - Anti-Submarine Warfare
Auto-acquisition - Automatic radar lock-on capability in the cockpit of an aircraft
AWAC - Airborne Warning And Control

BarCAP - Barrier combat air patrol; fighter cover between a strike force and area *of* expected threat
Barrel Roll - rolling maneuver fight path describing a helix about the intended direction of the flight
Bingo (fuel) - Minimum fuel quantity reserve previously established for a given geographical point to permit an aircraft to return safely to its home base, an alternate base, or to an aerial refueling point.
Blip (radar) - A spot *of* light on a radarscope, representing the relative position of a reflecting object such as an aircraft, sometimes called a *pip*.
Bogey - Unidentified aircraft : Bogies - Two or more unidentified aircraft
Bolter - Missed carrier landing - when the hook fails to engage an arresting cable and must go around
Break - Emergency turn of maximum performance to instantly to destroy an enemy tracking solution
Burner - Afterburner

CAM - Catapult Aircraft Merchant
CAP - Combat Air patrol
CarQuals - Carrier Qualification Landings
CAS - Close Air Support
Centerline Tank - A fuel tank carried externally on the centerline of an aircraft
Chaff - confusion reflector of thin narrow metallic strips used to create false signals on radar screens
Close - to decrease separation between aircraft
Closure - relative closing velocity
CO - Commanding Officer
Cool - Use of a gas for cooling the heat-seeker head of the AIM-4D air-to-air missile to prep for firing
Combat Spread - loose formation giving each flight member maximum visual lookout opportunity
Cross-turn - A rapid simultaneous 180 change of heading by the members of an element or flight, in which half of the unit turns toward the other half
Cut-off (tactic) - Employing the shortest route to intercept an enemy airborne target

Deck - A flight altitude just above the surface, as in ;
to hit the deck ~ to fly on the deck ~ to dive toward the deck.
Defensive Spiral - A descending, accelerating dive using high-G
and continuous roll to negate an attack and to gain lateral separation.
Defensive Split - A controlled separation of a target element into different planes,
used in an attempt to force interceptors to commit themselves to one member of the target element.
Defensive Turn - Basic defensive measure to prevent an attacker from achieving a launch or firing position; the intensity of the turn determined by the angle-off, range and closure of the attacking craft.
DF - Direction Finding
Disengage - To break off combat with the enemy
DMZ - Demilitarized zone

GLOSSARY OF TERMS AND ABBREVIATIONS

Echelon – A formation with flight members positioned sequentially on one side of the lead aircraft
ENIGMA -German encryption/decryption machine
ECM – Electronic countermeasures ; the prevention or reduction of effectiveness in enemy equipment and tactics used by electromagnetic radiations; some activities exploit enemy's emission of radiations
ELINT – Electronic Intelligence
Engagement – An encounter, which involves hostile or aggressive action by one or more participants
Envelope - A volume of airspace within which a particular weapon or weapon system must operate, be expended, or be employed in order to achieve maximum effectiveness
EW – Electronic warfare officer
EXOCET, *Aérospatiale* AM 39 anti-ship missile

FAC – Forward air controller
Finger-four/Fingertip – A four-aircraft formation in which the aircraft occupy positions suggested by the four fingertips of either hand, with the fingers being held together in a horizontal plane
Flak German acronym for AAA HE/shrapnel fire
Flak envelope – A varying vertical area of airspace in which a particular type of AAA is effective
Flame(d) out – The extinguishment of the flame in a reaction engine, especially a jet engine
FLIR – Forward Looking Infra-red Radar
Fluid element – The second supporting element in a fluid-four formation, flying high or low
Fluid-four – A tactical formation having the second element spread in both the vertical and horizontal planes **to** enhance maneuverability, mutual support, and lookout ability

G - Unit of acceleration (32.2 ft/sec^2)
G - Unit of force applied to a body at rest equal to the force exerted on it by gravity
Gate – To fly at maximum possible speed or power (full afterburner power)
- also refers to Range Gate, the radar indication of the distance between the interceptor and target
G-load - Theforce exerted upon a pilot (and his aircraft) by gravity or a reaction to acceleration or deceleration as in a change of direction (maneuvering)
Growl - (See Missile tone)

Hangfire – A delay or failure of an article of ordinance after being triggered
Heat – Armament switch setting for using infrared missiles
HEI – High Explosive Incendiary
High-G – Status of having the G-load increased during aircraft maneuvering
Home(d) – Of a missile: to direct itself toward the target by guiding on heat waves, radar, echoes, radio waves or other radiation emanating from the target
Home plate - Nickname for base of origin
Hot Start – A start that exceeds normal starting temperatures

IR missile – an Infra-Red (heat-seeking) missile

JBD – Jet Blast Deflector
Jink (ed) (ing) - Constant maneuvering, both horizontal and vertical to present a difficult target to enemy defenses by spoiling the tracking solution; changes in bank, pitch, and velocity-at random
Joker – A term for fuel planning information: a particular fuel level usually selected to warn that bingo is approaching and further engagements should be avoided
Judy – Term used that the interceptor has contact with the target and is assuming engagement control

Kill – An enemy airplane shot down or otherwise destroyed by military action while in flight
Knot – Kt - one nautical mile per hour

Lead – The lead aircraft in a flight or element, or a reference to a specific lead aircraft or its pilot
Lethal envelope - The envelope within which parameters can be met for successful employment of a munition by a particular weapon system (See Envelope)
Lock-on (lock-up) – To follow a target automatically in one or more dimensions (e.g., range. bearing, elevation) by means of a radar beam
Loose deuce – Fighter tactics in which two to four aircraft maneuver to provide mutual support
Lufbery Circle - A circular tail chase, ascending or descending
Luftwaffe German Air Force

GLOSSARY OF TERMS AND ABBREVIATIONS

Mach – The ratio of the aircraft's velocity to the velocity of sound in the surrounding medium
Maginot Line – French fortifications along the French-German border
Master Bomber ~ Lead bombardier directing air formation bombing
Maximum power – Afterburner power
Maximum turn-rate at which the maximum number of degrees per second is achieved
MIA – Missing in Action
MiG – Name for the Mikoyan-Gurevich series of Soviet jet fighter aircraft
MiGCAP (or MIG cap) – Combat air patrol directed specifically against MiG aircraft (See CAP)
MiGSCREEN – protect a strike force by placing fighters between MiGs and the protected force
Missile tone – Audio signal indicating an AIM-9 Sidewinder is locked–on to an infra-red source
m.g. machine gun
MSL – Mean Sea Level; used as a reference for altitude

NAS – Naval Air Station
Night Owl – Night strike mission(s)
NM (or nm) – Nautical mile: 6,076.1 feet

ONI – Office of Naval Intelligence (U.S.)
Orbit – A circular or elliptical pattern flown by aircraft to remain in a specified area
Overshoot – To pass through the defender's flight path in the plane of symmetry
Overtake velocity – Sudden gain in speed to come up on another aircraft

Padlocked – Term meaning that a crew member has sighted bogies or bandits and has his vision fixed on them; looking away would risk losing visual contact
Panzer – German term for tank
Pendant – The portion of the carrier arresting cable stretching across the deck
Pip – (See Blip)
Pod – Any one of several aerodynamically configured subsystems carried externally on fighter aircraft
POL – Petroleum, oil, and lubricants
Pop-up – A climbing maneuver from low-altitude or other position of concealment, used to gain an advantageous position for weapons delivery; also a maneuver used by enemy aircraft which involved a steep climb from a low- altitude area of concealment to an inbound aircraft or flight of aircraft
POW – Prisoner of War
Powdered - destroyed, caused it to disintegrate (with respect to aircraft)
Pulling lead – Aiming the nose of the aircraft ahead of an enemy aircraft in weapons firing maneuver
Pull-up – pulling up, a pullout, or recovery from a dive ; to bring the nose of an aircraft up sharply
PURPLE : Title of intercepted and decrypted Japanese diplomatic radio traffic

RADAR – Radio Detection and Ranging
Radar signature – Characteristic peculiar to different aircraft distinguishable on a radarscope
RAF Royal Air Force
RAAF Royal Australian Air Force
Ready light - Light indicating a particular avionics/munitions system is operating and ready for use
Recon - Reconnaissance
Refit - repair period for ships
Reticle – Optical sight reticle; a system of lines around a dot (pipper) in the focus of an optical gunsight that provides a reference for aiming and estimating range and distance to the target
RHAW - Radar Homing And Warning; on-board aircraft equipment to warn of active enemy defenses
Ripple fire - Rapid sequential firing of two or more missiles
RFC Royal Flying Corps
R-max – Maximum range
RN Royal Navy of Great Britain
RNAS Royal Naval Air Service
Roger – Term meaning *Message received and understood*
Room 40 Royal Navy HQ for inter-communications intelligence activity
R & R – Rest and Recreation (or Rehabilitation or Relaxation)
RT Radio Telephone
Rudder reversal – A roll reversal using rudder only
- normally used in maximum performance, high angle of attack manoeuvring

GLOSSARY OF TERMS AND ABBREVIATIONS

SAM - Surface-to-Air Missile
Sandwich – Situation wherein an aircraft is positioned between two opposing aircraft
Scissors – A defensive maneuver in which a series of turn reversals are executed in an attempt to achieve the offensive after an overshoot by the attacker
Separation - Distance between interceptor and target aircraft; can be lateral, longitudinal or vertical
Separation maneuver - An energy-gaining maneuver performed with a low angle of attack and maximum thrust, to increase separation (extend) or decrease separation (close)
Shrike - Nickname for the AGM-45 air-to-ground radar-seeking missile
Six - Six (6) o'clock position or area; refers to the rear or aft area of an aircraft
Slice(d) - Maximum performance, hard, descending nose-low turn with more than 90 degrees of bank
Sortie a single air mission by a single aircraft
S maneuver – A weave in a horizontal plane
Snap-roll (ed) - An aerial maneuver in which an aircraft is made to effect a quick, complete roll about its longitudinal axis; the act of putting an aircraft into a snap-roll
Snap-up - Rapid pull-up climb to gain altitude to launch a weapon against a higher enemy aircraft
Speedbrakes - Flaps designed for slowing down an aircraft in flight
Splash – Term meaning that destruction of the target has been verified by visual or radar means
Split-S - 180° rotation about the aircraft's longitudinal axis followed by a 180 change of heading in a vertical plane (a half loop starting from the top)
Strike an aerial attack by numerous aircraft (naval term)
S turn – A turn to one side of a reference heading followed by a turn to the other side; provides a difficult tracking problem for ground radars
Sweep – An offensive mission by fighter aircraft, sometimes accompanied by fighter-bombers over an area of enemy territory to seek out and attack enemy aircraft or targets of opportunity; also the action of flying over an area in malting a search; the path flown in to clear the skies of opposition

Tally-ho - Term meaning that the target has been visually sighted
TARCAP - Target Combat Air Patrol
TF - Task Force
Thud Ridge – Nickname for a mountain range beginning about 20 NM north-northwest of Hanoi and extending about 25 NM northwest, used for navigational and terrain marking.
Tracking - Maintaining the center of the field of view of search radars or airborne sensors on a target
Troll(ed) (ing) - Flying a random pattern by ECM aircraft to detect enemy electronic signals; flying a pattern in a specific area to detect signals of a suspected SAM or AAA site
Trap (noun) - Carrier arrested landing
Trap (verb) – To land aboard carrier
Tuck-under – Tendency of certain aircraft to drop its nose flying at or near its critical mach number

U-boat - German submarine (literally undersea boat)
UHF-DF - Ultra high frequency direction finder
ULTRA - code name given to intelligence material derived from the interception and decryption of WWII German radio transmissions sent using the ENIGMA encoding machine.
USAAC - U.S. Army Air Corps
USAAF - U.S. Army Air Force
USMC – United States Marine Corps
USN – United States Navy
USS - United States Ship (warship)
UXB – Unexploded Bombs

VF U.S. Navy fighter squadron
VMF U.S.M.C. fighter squadron
Vector – A command to direct an aircraft to follow a specific heading

Winchester – Term indicating that all ordinance has been expended
Wingman – Pilot (or aircraft) who flies at the side and to the rear of an element. In an aircraft flight, 02 is wingman to lead 01, and 04 is wingman to 03 Usually, more experienced pilots fly the lead and 03 positions in a flight, and these pilots indicate combat actions while their wingmen fly cover.
WMD – Weapons of Mass Destruction
WSO – Weapons Systems officer; backseater in certain aircraft

BIBLIOGRAPHY

Abe, Saburo *Zero Sen Kaku Tatakaeri!* Zero Fighter Pilots Association 2004
Air Ministry *The Royal Air Force in the Great War* Imperial War Museum, London 1966
Armitage, M.J. *The Royal Air Force ~ An Illustrated History* Arms & Armour, London 1993
Ashmore, Major General E. B. *Air Defence.* London: Longman Green and Co., 1929.
Babcock, Elizabeth *Sidewinder ~ Invention and Early Years* China Lake Museum Foundation 1999
Baring, Maurice *Flying Corps Headquarters 1914-1918* Bell, London 1920
Barnhart, M.A. *Japan Prepares for Total War 1919-1941.* Cornell University Press 1987
Bartlett, C.P.O. *Bomber Pilot, 1916-1918.* Ian Allan, London 1974
Beaver, Paul *The British Aircraft Carrier* Wellingborough : Stephens 1987
Belote, J. & W. *Titans of the Seas ~ Japanese and American Carriers during World War II* 1975
Bergerund, Eric *Fire in the Sky ~ The Air War in the South Pacific* Boulder 2000
Bickers, Richard *The First Great Air War* Hodder and Stoughton, London 1988
Bingham, Henry *An Explorer in the Air Service* Yale Press, New Haven 1920
Bishop, William *Winged Warfare* Grosset and Dunlap 1918
Bleese, Captain F.C. *No Guts, No Glory* USAF 1955
Boelcke, Oswald *An Aviator's Field Book ~*
The field reports of Oswald Boelcke, from August 1st 1914 to October 28th 1916 New York 1919
Bowyer, Chaz *RAF Operations 1918-1938* W. Kimber, London 1988
Boyd, Alexander *The Soviet Air Force Since 1918* Scarborough House, New York 1977
Boyd, Colonel John R. *A Discourse on Winning and Losing* (slide presentation) undated
Boyne, Colonel Walter J *Aces In Command: Fighter Pilots as Combat Leaders* Brassey's 2001
Boyne, Colonel Walter J. *Beyond the Wild Blue: A History of the U.S.A.F.* St. Martin's NY 1997
Boyne, Colonel Walter J. *Clash of Wings: World War II in the Air* Touchstone New York 1994
Brown, Captain Eric N. RN *Duels in the Sky* Naval Institute Press, Annapolis 1988
Bruce, J.M. *The Aeroplanes of the Royal Flying Corps* Putnam, London 1982
Buckley, John *Air Power in the Age of Total War* Indiana University Press 1999
Burge, Gordon. *100th Pioneer Night Bombing Squadron, Independent Force, R.A.F.* London 1919
Cagle, Vice-Admiral M.W. *The Aerial Bombardment of North Vietnam* Naval Institute Press 1986
Castle, H.G. *Fire Over England* Secker & Warburg, London 1982
Chamier, Air Commodore J.A. *The Birth of the Royal Air Force* Sir Isaac Pittmann & Sons, London 1943
Christienne & Lissarague *A History of French Military Aviation* Smithsonian I.P., Washington 1986
Churchill, Randoplh S. & Winston S. *The Six Day War* Houghton Mifflin 1967
Cole, Christopher and E. F. Cheesman *The Air Defence of Britain, 1914-1918* Putnam, London 1984.
Condon, J.P. *Corsairs and Flattops: Marine Carrier Warfare* Annapolis, Naval Institute Press, 1998.
Correlli, B. *Engage the Enemy More Closely ~ The Royal Navy in the Second World War* Norton, NY 1991
Corum & Wray *Airpower in Small Wars: Fighting Insurgents and Terrorists* University Press Kansas 2003
Christienne, Charles, and Pierre Lissarague. *A History of French Military Aviation* Smithsonian 1986
Crowell, Benedict. *America's Munitions, 1917-1918* Government Printing Office, Washington 1919
Cutlack, Frederick, M. *The Australian Flying Corps, Official History* Angus & Robertson, Sydney 1933
Creed, Roscoe *PBY: The Catalina Flying Boat* Annapolis Naval Institute Press 1985.
Cresswell, RN. Commander, John *Naval Warfare* Sampson Low, Marsten & Co. London
Cunningham, Andrew Admiral A.B. Cunningham, *A Sailor's Odyssey* Hutchinson & Co, London 1951
Cynk, Jerzy B. *History of the Polish Air Force 1918-1968* Beachcomber, Marana 1972
Doolittle, James H., and Glines, Carroll V. 1995. *I Could Never Be So Lucky Again.* Schiffer Publishing.
Douhet, Guilio *The Command of the Air* Air Force and History Museum, Washington 1983
Eshel, Lt. Col. David *The Israeli Air Force ~ Born in Battle* Dramit, Tel Aviv 1978
Ethell, Jeffrey & Price, Alfred *The German Jets in Combat* Janes, London 1979
Evans, D.C. & Peattie, M.R. *Kaigun: Strategy, Tactics, and Technology in the Imperial Japanese Navy* 1997
Ferte, Air Chief Marshal Sir Philip Joubert de la *The Fated Sky* Hutchinson, London 1952
Finney, R.T. *History of the Air Corps Tactical School 1920-1940* Washington D.C. 1992
Fitch, Willis S. *Wings in the Night* Marshall Jones, Boston 1938
Fredette, Raymond H. *The Sky on Fire ~ the first Battle of Britain 1917-18* Holt, Rinehart & Winston 1966
Freidman, Norman *Desert Victory: The War for Kuwait* Naval Institute Press, Annapolis 1991
Fonck, René *Ace of Aces* Doubleday 1967
Francillon, R.J. *Japanese aircraft of the Pacific War, Mitsubishi A6M Zero* Naval Institute Press 1995.
Fredette, R.H. *The Sky on Fire: The First Battle of Britain 1917-1918, and the Birth of the RAF* NY 1976
Fuchida & Okumiya. *Midway: The battle that doomed Japan.* United States Naval Institute 1958
Furtrell, Frank *Ideas, Concepts, Doctrine: Basic Thinking in the Air Force, 1907-1960 ~ Volume I*
Maxwell Air Force Base, Alabama: Air University Press, 1989.
Furtrell, Frank *The United States Air Force in Korea* Office of Air Force History, Washington 1983

BIBLIOGRAPHY

Galland, Adolf *The First and the Last* Henry Holt 1954
Garros, Roland *Memories* Hachette Paris 1966
Gill, G.H. *Royal Australian Navy, 1942-45* Australian War Memorial, Canberra 1968.
Glaisher, James & Camile Flammarion *Travels in the Air* Bentley, London 1871
Goss, H.P. *Civil Morale Under Aerial Bombardment 1914-1939* Maxwell AFB 1948
Grinnel-Milne, Duncan *Wind in the Wires* Hurst and Blackett, London 1926
Gross, Charles J. *American Military Aviation: The Indispensable Arm* Texas University Press 2002
Grossnick, R.A. & Armstrong, W.J. *United States Naval Aviation, 1910-1995* Naval Historical Center
Gurney, Gene *Vietnam: The War in the Air* Crown, New York 1985
Hallion, R.P. *The Naval Air War in Korea* Nautical & Aviation 1986
Halsey, William F. *Admiral Halsey's Story* Da Capo Press 1976
Hastings, Max *Bomber Command* Doubleday, New York 1979
Hastings, Max and Jenkins, Simon *The Battle for the Falklands* Norton, New York 1983
Hennessy, Juliette A. *The United States Army Air Arm, April 1861 to April 1917.* Washington 1958
Herzog, Chaim *The Arab-Israeli Wars ~ War and Peace in the Middle East* Random House 1982
Hezlet, Vice Admiral Sir Arthur *Aircraft and Seapower* Stein & Day, New York 1970
Hooton, E.R. *Phoenix Triumphant ~ The Rise and Fall of the Luftwaffe* Brockhampton, London 1999
Horsley, Terence *Find Fix and Strike ~the Work of the Fleet Air Arm*
Hough, Richard *The Great War at Sea 1914-1918* Oxford University Press Incorporated 1984
Hough, Richard *The Longest Battle ~ The War at Sea 1939-1945* Harper Collins NY 1987
Hough, Richard & Richards, D. *The Battle of Britain* Hodder & Stoughton, London 1989
Hoyt, Edwin P. *Japan's War ~ The Great Pacific Conflict* McGraw-Hill, Blacklick 1986
Hughes, Wayne P. *Fleet tactics ~ Theory and Practice* Naval Institute Press, Annapolis 1986
Jackson, Robert *RAF in Action ~ From Flanders to the Falklands* Blandford, Poole 1985
Jackson, Robert *The Red Falcons ~ The Soviet Air Force in Action* Clifton, Brighton 1970
Johnson, Colonel E.C. *Marine Corps Aviation* U.S.M.C. History HQ, Washington 1977
Johnson, J.E. *Full Circle: The Story of Air Fighting* Chatto and Windrus, London 1964
Jones, Ira *An Air Fighter's Scrapbook* Nicholson & Watson, London 1938
Kennett, Lee *The First Air War 1914-1918* Free Press-Macmillan, New York 1991
Killen, John *A History of the Luftwaffe* Bantam Books, NY & Westminster 1986
Knauer, John H. *Flying the A.E.G. G.IV.* Cross and Cockade Journal Summer 1974
Knox, MacGregor *Hitler's Italian Allies ~ and the War of 1940-1943* Cambridge U.P. 2000
Layton & Pineau *I Was There ~ Pearl Harbour and Midway ~ Breaking the Secrets* NY 1985
Lazzarabal, Jesus Salas *Air War Over Spain* (English edition) Ian Allen Ltd. London 1974
Lee, Asher *The Soviet Air Force* Mystic : Lawrence Verry Incorporated 1962
Le May, Curtiss E. *Superfortress ~ The Story of the B-19 and American Air Power* 1988
Liddle, Peter H. *The Airman's War 1914-1918* The Blandford Press, New York 1987
Lehmann, Ernst A. *The Development of the Airship* G.P. Putnam & Sons, London 1927
Loomis, R.D. *Great American fighter pilots of World War II* Random House, New York 1961
Lowe, Thaddeus *My Balloons in Peace and War ~ Memories of Thaddeus S.C. Lowe*
Undated & Unpublished. Manuscript Division, Library of Congress, AIAA Files: Box 81
Lundstrom, J.B. *The First Team ~ Pacific Naval Air Combat from Pearl Harbour to Midway* 1984
Lundstrom, J.B. *The First Team ~ and the Guadalcanal Campaign:*
Naval Fighter Combat from August to November 1942 Naval Institute Press, Annapolis 1994
McCarthy, D.J. *MiG Killers, A Chronology of U.S. Air Victories in Vietnam* Specialty Press 2009
McCudden, James *Five Years in the Royal Flying Corps* The Aeroplane London 1918
McFarlane, Stephen L. *America's Pursuit of Precision Bombing 1910-1945* Washington 1997
McMinnies W.G. Flight-Commander R.N. *Practical Flying* Temple Press London 1918
Marolda, E.J. *Grand Delusion : U.S. Strategy and the Tonkin Gulf Incident, Naval History* 2014
Mead, P. *The Eye in the Air: History of Air Observation for the Army 1785-1945* H.M.S.O. 1983
Mersky. Peter B. *U.S. Marine Corps Aviation ~ 1912 to the Present* Nautical & Aviation Inc. 1986
Michel, M.L. *Clashes: Air Combat Over North Vietnam, 1965-1972.* US Naval Institute Press 1997
Middlebrook, Martin & Everett *Bomber Command War Diaries ~ An Operational Reference Book* 1986
Miller, E.S. *War Plan Orange : The U.S. Strategy to Defeat Japan* Naval Institute Press Annapolis 1991
Miller, Marshal *The Soviet Air Force View of the Bekaa Valley Debacle* Armed Forces Journal, June 1989
Montserrat *Angel on the Yardarm ~ Fleet Radar Defence & the Kamikaze Threat* Naval War College Press
Moore, Admiral Thomas H. *ECM In The Falklands Proves Its Point The Hard Way* Military ECMs 1982
Mitchell, William. *Memoirs of World War I: From Start to Finish of Our Greatest War* Random NY 1960
Musicano, Walter A. *Warbirds of the Seas ~ A History of Aircraft Carriers and Carrier-based Aircraft* 1994

BIBLIOGRAPHY

Neillands, Robin *The Bomber War ~ Arthur Harris and the Allied Bomber Offensive* Overlook NY 2001
Neumann, Major G.P. *The German Air Force in the Great War* Hodder and Stoughton, London 1920
Norris, Geoffrey *The Royal Flying Corps: a History* F. Muller, London 1965
N.Y. Times, *Bomber Gets U-Boat, this time for certain. Probably sank another recently.* 26th May 1943
Okumiya, Horikoshi and Caidin. *Zero: the story of the Japanese Navy Air Force* Cassell, London 1957
Olds, Robin *Fighter Pilot: The Memoirs of Legendary Ace Robin Olds* St. Martin's Press 2010
Peatie, M.R. *Sunburst ~ The Rise of Japanese Naval Air Power 1909-1941* Naval Institute Press 2001
Prange, G.W. *At Dawn we Slept: The Untold Story of Pearl Harbor.* McGraw-Hill. New York 1981
Proctor, Raymond *Hitler's Luftwaffe in the Spanish Civil War* Greenwood Publishing Group Inc. 1983
RAF Historical Air Branch *The Rise and Fall of the German air Force 1939-1945* HMSO London 1983
Raleigh, Walter & Jones *Official History of the War ~ Vol. 1-3* The Clarendon Press, Oxford 1922-1937
Rawlinson, Alfred *Defender of London 1915-1918* Andrew Melrose Ltd, London 1923
Rendall, Ivan *Rolling Thunder ~ Jet Combat From WWII to the Gulf War* The Free Press 1997
Richards, Denis, Sanders & St. George *Royal Air Force 1939-1945 Vol. II The Fight Avails* HMSO 1954
Richthofen, Manfred von *The Red Battle Flyer* Robert McBride, 1918
Rimell, Raymond L. *Zeppelin! ~ A Battle for Supremacy in World War I* Conway Maritime Press 1984
Robinson, Douglas, H. *The Zeppelin in Combat, a History of the German Naval Airship* London 1962
Rolt, L.T.C. *The Aeronauts: A History of Ballooning 1783-1903* Walker and Company, New York 1966
Roskill, Capt. S.W. (Editor) *The Naval Air Service Volume I* Spottiswoode, London 1969
Royal Air Force Historical Society Journal 30: *The RAF in the Falklands Campaign* RAF Hendon 2003
Rumpf, Hans *The Bombing of Germany* Frederick Muller Ltd. London 1963
Sakai, Saburo. Martin Caidan and Fred Saito. *Winged Samurai!* Bantam 1985
Sakaida, H. *Imperial Japanese Naval Aces 1937-1945* Osprey Publishing, London 1999
Sakaida, H. *Japanese Army Air Force Aces 1937-1945* Osprey Aerospace, Oxford 1997
Samson, C.R. *Fights and Flights* Ernest Benn, London 1930
Schaffer, Ronald *Wings of Judgement ~ American Bombing in World War II* Oxford U.P. 1985
Sharp, U.S.G. Admiral *Strategy for Defeat : Vietnam in Retrospect* Presidio, San Rafael 1978
Sherrod, Robert *History of Marine Corps Aviation in World War II* Nautical & Aviation 1987
Sekigawa, Eiichiro *Pictorial history of Japanese Aviation* Allan, London 1974
Scott, Admiral Sir Percy *Fifty Years in the Royal Navy* John Murray, London 1919
Shaw, Robert L. *Fighter Combat: Tactics and Manoeuvring* Naval Institute Press, Annapolis 1985
Sherry M.S. *The Rise of American Air power ~ The Creation of Armageddon* Yale U.P. 1987
Shores, Christopher *Regia Aeronautica ~ History of the Italian Air Force* Signal Publications 1976
Slessor, Sir John *The Central Blue, Recollections and Reflections* Cassel, London 1956
Smith, Gordon *Battles of the Falklands War, by Land Sea and Air* Ian Allan, London 1989
Smith, P.C. *Midway: Dauntless Victory* Pen and Sword Maritime, Barnsley 2007
Snowden Gamble, C. F. *The Air Weapon: Being Some Account of the Growth of British Military Aeronautics From the Beginnings in 1783 until the end of the Year 1929.* Vol. 1. Oxford University Press, London 1931
Spick, Mike *Aces of the Reich: The Making of a Luftwaffe Fighter Pilot* Greenhill Books, London 2006
Spick, Mike *Fighter Pilot Tactics* Patrick Stephens, Cambridge 1983
Steel and Hart *Tumult in the Clouds: War in the Air 1914-1918* Hodder & Stoughton, London 1997
Strange, Lieutenant-Colonel Louis DSO MC DFC *Recollections of an Airman* John Hamilton 1933
Technical Aviation Brief #3, Performance trials: Japanese Zero, Aviation Intelligence, U.S. Navy 1942
Tennant, J.E. *In the Clouds over Baghdad* Cecil Palmer, London 1920
Thayer, L.H. *America's First Eagles: The Official History of the U.S. Air Service, A.E.F.* Mesa, 1983
The United States Strategic Bombing Survey Summary Report ~ Pacific War Washington D.C. 1946
Thetford, Owen *British Naval Aircraft since 1912* (4th Edition) Putnam, London 1978
Tillmann, B. *Clash of the Carriers. The true story of the Marianas Turkey Shoot* Penguin, NY 2005
Udet, Ernst *Ace of the Iron Cross* Doubleday 1970
U.S.A.F. Historical Studies No. 175 : The Russian Air Force in the Eyes of German Commanders Air University, USAF Historical Division, Maxwell AFB, Alabama 1960
Utley, J.G. *Going to War with Japan 1937-1941* New York University Press 1985
Voices in the Storm: the road to Basra Frontline, WGBH PBS online 2011
Voisin, General. *La Doctrine de l'Aviation Française aise de Combat Au Cours de la Guerre* Paris 1932.
Ward RN, Commander Nigel *Sharkey Sea Harrier Over the Falklands* Orion, London 1993
Warneford, Reginald *Combat Report 8th June 1915* ~ quoted in Mary Gibson, *Warneford*
Watts, D.B. *Six Decades of Guided Munitions*, Precision Strike Association, January 2006,
Westrum, Ron *Sidewinder ~ Creative missile development at China Lake* Naval Institute Press 1999
White, C.M. *Gotha Summer: The German daylight air raids on England, May to August 1917* London 1986

~ INDEX ~

INDEX

AAA (anti-aircraft artillery) 10 11 21 (42) 43-50 60 (61) 64 74 80 88 91 103 115 118 285-87 (318)
- *Flaming Onions* 73,
- Japanese multi-coloured AAA 211
ADLERTAG ~ Operation Eagle Day 130
AERIAL ENGAGEMENTS (dogfights)
- Great War 7 11 14-27 34-55 63-69 72-80
- Spanish Civil War 109-16
- World War II : European Theatre 119-36 152-66
- World War II : Pacific Theatre 178-79 184-89 192-94 203-11 218
- Korean War 222-236
- Vietnam War 240-60
- Middle East 261-73
- Falklands 275-84
AFTERBURNER 241 253
AILERONS 159 206
AIR COMBAT MANOUEVRES : *Yo-Yo's* (High & Low) 225-26 230 : *Defensive Spiral* 225 (227) 232 : *Barrel Rolls* (Over-the-Top & Below) (226) (251) : *Split-S* 202 : *Displacement Rolls* : *Lag Rolls* (227) : *Scissors* (227) : *Vertical Rolling Scissors* 250 (250)
AIR CURRENTS & Turbulence 97 : Pacific Ocean : Japan to North America 214 : Korea 224 :
AIR-TO-AIR MISSILES : French Le Prieur rockets 49 51 : see also *Atoll*, Sidewinders, and Sparrow
AIR-TO-SURFACE MISSILES : Hellfire 285 : Hydra 285
AIR SUPPORT GROUND OPERATIONS 2 7 10-11 22-23 34 73-75 109-16 119-21 154-57 263-271 283-91
- Palestine Jordan WWI 100-02 : Crete 148-50 : Pacific 189 202-03 : Korea 221-25 : Vietnam 240 262
AERODROMES of the Greek Isles 92
AIR SUPREMACY 20 261
AIRCRAFT CARRIER development 87
AIR DEFENCE
- of Germany 120 158-166
- of London 35 43-55 119-36 (318)
- of Paris 43 45 56 (58) & France in WWII 120-21
AIR-SEARCH RADAR 126 130 136 140 158
AIR-TO-AIR ROCKETS - Le Prieur 51 (51)
AKAGI, IJN carrier (184)
AKUTAN ZERO (192-92)
ALLENBY, General Edmund 100-02
- offensive in Palestine in September 1918 : 100-02
ALBATROS AIRCRAFT 113
- D.I 17 19 23 26
- D.II 72
- D.III 72 76
ALLEPO 102
ALSACE-LORRAINE 1
AMERICAN CIVIL WAR balloons
ANTI-SUBMARINE WARFARE
- Great War 38 87 91 93 94 97
- World War II 137 141 143-44 171
ARGENTINE AIR FORCE - *Fuerza Aérea Argentina* FAA 275-84
ARMISTICE 102 - *The Heavens are devoid of Huns* 104
ATLANTIC OCEAN - Naval Aviation 86-99 - WWII 137-44
ATOLL, Soviet Vrympel K-13 air-to-air missiles 236 241 260 266 271
AUSTRALIA 170 (178) (181) 221 260
- Japanese plan to invade 182 (183) 184 189
AUSTRALIA, HMAS 184 209
AUSTRALIAN CODE-BREAKERS 184
AUSTRALIAN FLYING CORPS (AFC) 100-01
AUSTRALIAN troops 80 148 192 257 260
AVENGERS (Grumman TBFs & TBMs) (197)
AVIATIK AIRCRAFT 7 24
AWAC (Airborne Warning and Control System) 273 286

INDEX

BALL, Captain Albert (8) 66 73 - S.E.5a & his cockpit 66 67 - his advice 66 - five-flight team attacks 67
B.E.2c ROYAL AIRCRAFT FACTORY 16 17 44 50 (52) 74 91
B.E.2e ROYAL AIRCRAFT FACTORY (52)
B-17 *Flying Fortress* (163) 164 (167) 171 184 (191) 194
B-24 *Liberator* (161) 213 : *Tidewater Tillie* 143-44 : *Cow Town's Revenge* 194
B-25 Mitchell : Doolittle raiders 180 (181) : skip-bombing (194)
B-26 *Marauder* 166
B-29 *Superfortress* : targets in Japan 214 (218-19) 221-22
B-52 *Stratofortress* 252 258-60 288
BADER, Douglas 127 (118)
- *Rules are for the obedience of fools and the guidance of wise men.* 135
BAGHDAD
- 1981 271
- 1990 285-86
BAIT AND SWITCH ruse 241
BARBAROSSA Operation 154-55
BATTLE OF BRITAIN 50 117 121 (128-29) 144
BATTLE OF THE ATLANTIC 137-44
BATTLE OF THE BISMARCK SEA 194-95 (194-95)
BATTLE OF THE PHILIPPINE SEA & LEYTE GULF 209-11
BEKAA VALLEY, Lebanon 272-73
BELGIUM 1-2 120-21
BISHOP, William *Billy* (8) 67 73 104
BISMARCK, Prince Otto von 1
Betty, Japanese G4M2 bomber 178 (179)
BLACK SHEEP SQUADRON 200-02
BLEESE, Major Frederick C. *Boots* (220) 230 - *No Guts, No Glory* 230-36
BLITZKRIEG Lightning War : Spain 109 : Norway 120 : Poland 117 : France 147 : Russia 154-55
BLITZ, 1915 London - Zeppelin & Gotha bomber raids 35-55 - *Blitz* 1940-41 136
BLOODY APRIL 72
BLUE MAX, see *Pour le Mérite*
BOELCKE, *Hauptman* Oswald (12) 14 (15) 16 (18)
- *Dikta Boelcke* 16 17 19 66 113
BOLT, John (174) 201 (220)
BOMBARDMENT by balloon of Venice in 1849
BOMBER, doctrine, operational policy & tactics 60 136 158-59;
- aerial navigation 136
- *Krokodil* 136
- *Luftwaffe* abandons daylight bombing 136
- Master Bomber technique 158
- Pacific theatre bomber escorts 205
- precision bombing at night 158
BOMBING VULNERABILITY of flying unescorted in daylight 136
BOMBSIGHTS : 35 74
BONG, Major Richard *Dick* (174-75)
BOSTON topmast bombing diagrams (193)
BOUGANVILLE Solomons (198)
BOYD, Colonel John (292) 293
BOYINGTON, Major Gregory (175) (198)
BRISTOL AIRCRAFT : F2a Fighter 72 - F.2b Fighter 55 (62) 100-02
BRITISH CARRIERS / Escort Carriers 141 210 222
BRITISH EXPEDITIONARY FORCE 1 2 121 120-21
BRITISH NAVAL BLOCKADE during the Great War 87
BRITISH PACIFIC FLEET 210
BUCCANEER, Blackburn 287
BUCKINGHAM MkVII phosphorous bullet (anti-airship) 49
BUDANOVA, Katya (118)
BUTTLAR-BRANDENFELS, Horst Treusch von - *St Elmo's Fire* 40
BULLETS 202 : explosive 51 - tracer 21 284 - Brock & Pommeroy incendiary-tracer 50

INDEX

CADENA or The Chain 115
CAM SHIPS 137 (138-39)
CAP - Combat Air Patrol, British Pacific Fleet 210-11 : Falklands 277
CAROLINE ISLANDS Truk Atoll 189 211 : Param Island (199)
CATALINA, Consolidated PBY-5 144 (190) 206 : RAAF *Black Cats* 196
CATAPULTS, aircraft launching 87 91 (138-39)
CENTRAL INTELLIGENCE AGENCY (CIA) 239 289
CHURCHILL, Sir Winston 131 148
CLEANUP TRIO (174-75) 185-86 (186) 189
COASTAL AIRSHIPS 90-92
COASTAL COMMAND 144 : HQ Operations Room (142)
COCKPIT : B-17 *Flying Fortress* (191) : B.E.2c 49-51 : Catalina (190) : Do-17 (125) : FW-189A 123 : Heinkel III (122) (129) : Henri Farman 1 : Me II0 (125)
CODES : Tannenberg 7 : German Imperial Navy Cipher book 50 : *Enigma Code* 137 Station X & ULTRA 137 146 148 152-53 : *Triton* 144 : Japanese JN-25 184 : 186 209
CONDOR LEGION 109-16
CORAL SEA 178 184 189
CORSAIRS (Chance Vought F4Us) (198-99) 222
COUTELLE Jean Marie-Joseph 298-99
CRETE 145 146-50
CUNNINGHAM, Admiral Andrew 145-47
CUNNINGHAM, LT. Randall (252) 253
CURTISS, Glen Hammond : P-40 aircraft (173)

DAMASCUS, Syria 1 102 264 267 272
DARDANELLES 92 93
DARWIN, Australia (James Morehead 174) 176 180-81
DAUNTLESSES (Douglas SBDs) 182-83 189 (196-97) 203
DECEPTION : 158 170 264 276 : *Moked* 261
DOGFIGHTS (see Aerial Engagements)

DEFLECTION SHOOTING 63 ~ Colonel William Barker 104
De HAVILLAND : McCudden's DH-2 65 - D.H.4 Day Bomber 50 51 55 72 (62) (83) - DH-9 (81) 100
DEULLIN, Lieutenant Albert Louis - *Pursuit Work in a Single Seater* 28-29
DEVASTATORS (Douglas TBDs) 186 203
DIVE BOMBING 74 117-20 147-49
Dogfights (see Aerial Engagements)
DOOLITLE, Lieutenant Colonel James *Jimmy* 176 180 (181)
DORNIER : Do-17 *Flying Pencil* 113 (125) 130 : Do 215 154 :
DOUGLAS, Lieutenant Sholto 21
DRISCOLL, Lieutenant William 252
DROP TANKS (fuel) 115 120 - Stuka Ju 87R 115 120 : Sabre 225 236 : IAF 237
DUNNING, Squadron Commander Ernest 97
DUNKIRK, France 11 38-39 94 97 121

EARLY WARNING 20 60 117 ; *Disco & Red Crown* 240-41 252 : AWAC 271 : Falklands 284
ELECTRONIC COUNTER MEASURES (ECM) 264
ENGLISH CHANNEL 1 93 104 136
Escadrilles de Chase 22
ESCORT CARRIERS 141
EXOCET, *Aérospatiale* AM-39 air-to-surface anti-ship missile 278-81 284

FARMAN AIRCRAFT 1 F.E.2b 17 & F.E.2d (3) 24 104
F-14 *Tomcat* 271 286
F-15 *Eagle* 271-72 285-86 288
F-16 *Fighting Falcon* 271-73 286 291
F-111F 288
F-117A *Nighthawk* 285-86
F-22 *Raptor* 286

INDEX

FERTÉ RFC, Captain Joubert de la 1 2 7
FIAT AIRCRAFT (see Italian Air Force)
FIGHTER COMMAND, RAF 120 130 131 (128) (134) 136
FINGER FOUR, Douglas Bader RAF develops the, 127
FIRESTORM, Hamburg 158
FLAT TURN 25
FLETCHER, Admiral Frank Jack 184
FLEET AIR ARM 140 146-47 224 277
FLUID FOUR 230 251 (254) 254-55
FLYING-BOATS 94 (95) ; see Catalina
FLYING CIRCUS - The Red Baron's 76
FOCKE WULF AIRCRAFT :
- FW 189A tactical recon (123)
- *Condor* 137 141
- FW 190 155 159
FOKKER AIRCRAFT
- *Eindekker* 14 (14) (15) 16 20
- DR.I Triplane 64 76-77 80 (77-79) (104)
- D.III (78)
- D. VII 104
FORMATIONS 111 Gotha box or diamond 55 : Happe's *Three-V* 24
FORMATION FLYING : British 63-64 67 - Soviet 116
FONCK, *Capitaine* René (5) 22-23 104
FORWARD AIR CONTROLLER (FAC)
FRANCE 1 17 25 67 109 120-21 147 (161) 221 239 261 271
FRANCO-PRUSSIAN WAR
FRANTZ, *Sergent* Joseph 7
FRIEDRICHSHAFEN FF-49 Seaplane 86
FRENCH AIR FORCE, *see L'Aviation Militaire*
FRENCH REVOLUTION, balloons in 292-95 (292-93)
FROZEN COMPASS 41
FUCHIDA, Commander Mitsuo 170-73 (174) 178
FU-GO BALLOONS 214-17 (216-17)

GABRESKI, Francis (118)
GALLAND, General Adolf 113 116 (117) (129) 158 166
GARROS, Rolland (4) 11 38
GAZA STRIP 261-63 267 273
GENDA, Lieutenant Commander Minoru reports Taranto 146
- Pearl Harbour 170 (172)
GERMAN HIGH SEAS FLEET 87-88
GERMAN IMPERIAL AIR SERVICE 7 11 14 16-17 19-20 55 72 76-80 104
GERMAN SPRING OFFENSIVE 72
GIFFARD, Henri 302 (302)
GORING, Hermann 104 109-10 121 130 136 166
GOLAN HEIGHTS 262 269
GOTHA AIRCRAFT 55-56 (56-57) 74 - diamond & box formations 55
GREAT WAR ~ Lessons of the, 115 127
GREIM, Robert Ritter von (13) 166
GROUND SUPPORT,
- see also AIR SUPPORT
- *Bomb Carpet* Spanish Civil War 109
- *Carpet Bomb* Vietnam 240 260 & Iraq 288
GUADALCANAL (Solomons) 187 275
GUAM (Marianas) 184 221 (258)
GUDERIAN, General Heinz 121
GUERNICA 110 (111)
GULF, Persian Wars 285-89
GUYNEMER, *Capitaine* Georges - *The Winged Sword of France* (4) 22 66-67 73

INDEX

HALSEY, Admiral William F. (Jr.) 173 180 208-11
HANDLEY PAGE 0/400 *The Bloody Pulverizer* (52) 55 (57) 74 100 102
HANOI, North Vietnam 240 252-53 260
HARRIER, RAF (272) tactics/viffing (Vector in Forward Flight) 277
- see also, Sea Harrier
HARTMANN, Eric (117) 155 166
- *See-Decide-Attack-Break* 155 291
HAWAII *Operation Z* 170 182
HAWKER HURRICANE 113 120 121 130 131 152
HAWKER, Major Lanoe 10 16
HEAD-ON-ATTACKS 73 136
HEAT : turbulence 40 : air-to-air missile heat discrimination problem 237 :
HEI : High Explosive Incendiary 238
HEINKEL :
- HE 51 109-10 115-16 :
- He 59 Seaplane (142)
- He-111 bomber 110 113 117 (122) (129) 131 154
HELLCATS (Grumman F6Fs) (203) 206-07 (208) 211
HEMMINGWAY, Ernest - *For Whom the Bell Tolls* 109
HERCULES C-130 transport aircraft 239 275-76 281
HINDENBERG, Paul von 7
HIROSHIMA and Nagasaki, Japan 215 218
HITLER, Adolf 108 109 113 116 147 166
HO CHI MINH 239 (244) - Ho Chi Minh Trail 240
HONCHOS - Soviet airmen in Korea 225-26
HOPE, Bob - *Colonel Robin Olds is the largest distributor of MiG parts in South East Asia!* 247
HOTCHKISS machine guns 7 11 QF aerial cannon 24
Hun in the Sun & glare in the air 16 17 65 69 (132) 152

IMMELMANN, *Oberleutnant* Max (12) 14 - *Eagle of Lille* 17
- *Immelmann Turn* 14 266
INTELLIGENCE : 130 170-71 182-84 186 199 221 267 269 272-73: Rolls Royce *Nene* jet engine 222
INTELLIGENCE : *Disco & Red Crown* 240-41 252 : *Moked Project* 261 : *Rooster 53* 264 : ELINT 273
IRAQ 270 285-89
ISLAND HOPPING 187
ITALIAN AIR FORCE - *Regia Aeronautica* 109 152
- Fiat CR-32 fighters 81 113 116
- SM-79 Savoia Marchetti 97 & 81 bombers (112) 113 152
ITALY : invasion of Greece 145 : 148 152

JABARA, Major James (220) 225
Jagdstaffeln or *Jasta* (Hunting Echelons) 17
- *Jasta 1* ~ Richthofen's Flying Circus 104
- *Jagdstaffeln 2* or *Jasta Boelcke* 17 19 20 76
- *Jagdstaffeln 10* or *Jasta Voss* 19
JAPAN 146 169 180 182 (217)
JAPANESE AIR TACTICS : 206 : *Head-on-Attacks* 206 : *Head-to-Tail* 185 : fighter pilots discuss (206)
- atrocities 214-15 218
- carriers : *Akagi* 173 178 187 (188) - *Chitose* 209 - *Chiyoda* 209 - *Kaga* 173 178 187 - *Hiryu* 173 187-88 (188) *Shoho* 184 - *Shokaku* 173 178 184 207 - *Soryu* 173 187 - *Taiho* 207 - *Zuikaku* 173 178 184 - *Zuiho* 209
- espionage 170-73 182-83 (183)
JELLICOE, Admiral Sir John & The Grand Fleet 31 88 91
JOFFRE, General Joseph 7 34
JORDAN 100 102 ~ Jordan River 260 - King Hussein bin Talal 261-62 267 269 271
JUNKERS
- Ju-52 transports 109-10 148-49
- Ju-87A *Anton* - *Stuka* (112) 115 117 118 136
- Ju-87R *Reichweite* (long range anti-shipping version - with a pair of drop tanks) 115 120
- Ju 88 130 154 : night fighter 159

INDEX

KAMIKAZES 209-11 218
KENNEDY, John, F. 240
KIMPO AIRFIELD, South Korea 224-25
KITTEL, Otto (117) 155
KLEEMAN, Commander Henry 270-71 (270-71)
KOKODA TRACK 184

Lafayette Escadrille N. 124 24-26 (25)
L'Aviation Militaire 7 11 24 87 90
LANCASTER bombers 159 (160)
Lawrence of Arabia, Colonel T.E. 100 102
- *Tulip Bombs* 101
L'Entreprenant (*The Enterprise*, French military balloon) 295
Le MAY, Major General Curtiss 218 240
Le Prieur rockets 51 (51)
LEWIS machine guns 1 2 16 21 34 49 50 91 137 - *double drum* 10
LEYTE (Philippines) 208-10
LIGHTNING : Lockheed P-38 (176-77)
LITVYAK, Lillia 154 (118)
LONDON 170 - *see* AIR DEFENCE of,
LOOSE-DEUCE (naval fighter) flight formation 251
LUFBERY, Major Gervais Raoul (5) 24 26 (104) *The Lufbery Circle* 25 240 245 (245)
Luftwaffe 104 109-13 117 121 129-31 148-49 153 164 166 289 - Formations *Rotte & Schwarm* 113
- fighter tactics 153 164 : *freie jagd* or free hunt 130 : roller coaster attacks 165
- redirected against London 291
LUNÉVILLE - Zeppelin Z4 stranded 31-33 (32-33)

MacARTHUR, General Douglas 208
McCLUSKY, Lieutenant Commander Clarence Wade (176) 187
McCONNELL, Capt. Joseph (220) (228-29) 230
McCUDDEN VC, Major James (9) 65
MALAN, Adolph *Sailor* (118) - *Ten of my Rules for Air Fighting* (135)
MALAYAN EMERGENCY 257
MALTA 145 152-53 - Convoy (151) - Aircraft radius of action from Malta (151)
MANHATTAN PROJECT 215 218 221
MANNOCK VC, Major Edward *Mick* (8) - advice & 15 rules 63-64
MARIANA ISLANDS 207 : Turkey Shoot 207 218
MARNE 7
MATA HARI (German agent H-21) 24
MAUBERGE, France 1 299
MEDITERRANEAN SEA 11 92 115 145-47 152-53 261
MESSERSCHMITT : Me 109 113 (114) 115 130 131 (133) 136 147 155 :
- fuel limitation & lack of drop tanks 115
Me 110 120 (125) 136 147 149 154 Me 110G night-fighter 159 :
Me163 & Me 163B *Komet* 164 : Me 262 155 165-66
MIDWAY Island & Atoll 186-87 (188) 189
MiG formations : HI-LO Pairs & Stacked Three (246)
MiG-15 222-23 230 236 (231-35) 252 : MiG 15 bis 225 230 : MiG-17 *Fresco* 236 241 (244-45) 251-53 (259) (266) MiG-19 241 MiG-21 238 (244) 247 (252) 253 264 266 : MiG 21F (263) 272 : MiG-23 (270) 272 :
MiG-25P *Foxbat* (268) : MiG-29 *Fulcrum* (cover image) 286
MiGCAP (MiG Combat Air Patrol, see Vietnam War)
MIRAGE 238 271 : delta wing drag to lose speed in a tight turn 262 : IIIC 261-62 264 266 276-77 279 286
MITCHELL Brigadier-General William *Billy* 105 (105)
Mölders, *Major* Werner 113 (114) (117) (129)
MONS 1 2
MORANE-SAULNIER AIRCRAFT 11 - *Parasol* 38-39 (39)
MUSSOLINI, Benito 109 147
MUSTANG 159 222
MYSTÉRE IV 261 - SUPER MYSTÉRE 261

INDEX

NAGUMO, Vice-Admiral Chuichi 173 178 186-87
NAGASAKI, Japan 218-19
NAPALM 221 240
- Spanish Civil War *Flambos* 116
NAPOLEON Bonaparte and balloons
NAVARRE, Jean - *la Sentinelle de Verdun* (5) 22 24 76
NEW GUINEA 178 182 184 189 194 : Tanahmerah Bay (200)
NIEUPORT : II 23 25 66 - XI *Le Bébé* 16 19 22 (27) - XVII *Superbébé* 23 73 76 - XXVIII 26 - Delange 109
NIGHT FIGHTERS 24 49 136 159
NIGHT WITCHES 154 (156-57)
NIMITZ, Admiral Chester W. 184 186-87 211 - USS *Nimitz* 271
NISHIZAWA, Hiroyoshi (174-75) 185 (186)
NIXON, Richard 260 264
NORTH AFRICA 11 40 100-02 108 148-54 261-71 284-89
NOWOTNY, Major Walter (117) 155 165
Nulli Secundus (Second to None, British Army dirigible) 309
NUNGESSER, Charles *l'Hussar de Mor - Knight of Death* (4) 23 (30)

OBSERVERS & observation 10 21 73 87 88 92 100 130 - German *Drachen* balloons 21
O'HARE, Lieutenant Edward *Butch* (176) 178
O'HARE, T. *Butch* (176) (187)
OHTA, Toshio (174) 185
OKINAWA, Japan 189 210 218 221 (258)
OLDS, Colonel Robin & Operation Bolo 241 (242) 249 260
OODA LOOP 292
OPERATIONS : FS 182-83 (183) : Bolo 241 247 *Linebacker I & II* 258 *Rolling Thunder* 240-42 *Rooster* 53 264
Rimon 20 266 : *Badr* 267 : *Nickel Grass* 269 : *Desert Shield* 285-86 : *Desert Storm* 285 : *Night Camel* 288
OSCAR, Nakajima Ki-43 fighter (219)
OTTOMAN TURKISH EMPIRE 18 92 93 100

PACIFIC OCEAN : WWII 169-218
PALESTINE 100-02 272 285
PANZERS : in Poland 117 - France 120-21 - Crete 150 - Russia 154
PARATROOPS and Crete 147-49
PARIS, France 7 11 24 39 42 121 260 308
PATHFINDER FORCE (No. 83 Bomber Squadron) 158
PATRIOT air-defence missile system 285
PEARL HARBOUR, Hawaii 93 141 146 (168) 169-73 180 184-85 192 209 278
PÉGOUD, Adolphe (5)
PHANTOM F-4 (240) 241 (242-43) (245) 246-47 (246) (248-52) 251-57 260 266-69 272 283
PHILIPPINES 208-09 218
PICASSO, Pablo - *Guernica* 110
PILOT CLOTHING & accessories 1 22 49 50 80
POKRYSHKIN, Alexandr - *Speed-Altitude-Manoeuvre & Fire* 155
POLAND & POLISH AIR FORCE 117 147 - dispersal of, 117
Pour le Mérite 7 17 19 94 104

RABAUL, New Britain 178 182 189
RADAR : 120 130 136 140 144 146-47 169 176 (177) 180 218 222 271-73 264 (265) 266 276 280-81 283 285
RADIO : silence 140 201 260 : NAVIGATION AIDS : X-Great 136
RAM ATTACKS : Soviet *Taran* 154 : FW-190 vs B-17 (162-63)
REAGAN, President Ronald 275
RECONNAISANCE, aerial 20 - Arab-Israeli conflicts - photography 19
RICHTHOFEN, Lothar von (77) 67
RICHTHOFEN, Manfred von (12) 17 19 65 72 76-77 80 (77) 104
- *Everything beneath me is lost* 76
RICHTHOFEN, *Leutnant General* Wolfram von 110 116 121
Rotte & Schwarm German fighter formations 113
RUSES : IAF *let him pass* 262

INDEX

RICKENBACKER, Captain Eddie 26
RECCONAISSANCE 2 10 140
RITCHIE, Capt. Richard S. *Steve* (243)
ROBINSON VC, William Leefe 49-50 (59)
ROECKEL, *Capitaine* René MF7 (4) 7
ROYAL FLYING CORPS (RFC) 1 2 10 20 24 50 63 66 74 100
- anti-balloon instructions 21
ROSE, Major Tricornet de (Stork CO) 22
ROYAL AIR FORCE (RAF) 74 100 152 275-77
- bombing objectives in Europe (160)
- Single Weaving Section 121
ROYAL AUSTRALIAN AIR FORCE (RAAF) 104
ROYAL NAVAL AIR SERVICE (RNAS) 10 26 38 43 74 87 91-94 - Submarine Scout Airships 90 91
RUDEL, Hans-Ulrich - *The Eagle of the Eastern Front* (117) 154
RHYS-DAVIDS, Arthur (9)

SABRE F-86A : 224-25 (227-28) (232-35) : F86E - all moving tail-plane 224 (225) : F-86F 225 238-39
SAKAI, Saburo (174-75) 185 189
SAMs : SA-2 240 (258) 260 264 264-67 (266) 269 : SA-3 264 267 269 : SA-6 269 272 : SA-7 *Strella* 269 273
SAMs : 272-73 : *SAM Song* 264 : *ripple fired* 264 : IAF weaving 269 : Iraqi 286 289
SAMSON, Lieutenant Charles Rumney 92
SASAI, Junichi (174-75) 185 189
SATELLITES : *Cosmos* #596 #597 #598 #600 266
SCHNAUFFER, Heinz (117) 159
SCUD, Soviet SS-1B missiles & launchers 287 289 - *Scud Box* 287
Schwarm - *Luftwaffe* attack formation
S.E.5A British Fighter 66-67 72 74 100
SEA HARRIER RN (272) 282-83
SEAPLANE CARRIERS 88 92 93
SEAPLANE TACTICS 87 94 (197)
SEARCHLIGHTS 40 43 44 49 50 145-47 209
SHOOTING STAR F-80C 221-22 (223)
SHORT BROTHERS SEAPLANES : Type 184 - 87-88 93 (98) : Type 320 - 97 : Sunderland (212-13)
SHRIKE AGM-45A anti-radiation missiles 264 272 276
SIDEWINDER : AIM-9B 236-38 (243) 245 247 251-53 261 266 269 : AIM-9L *Lima* 271-73 275 279-83
SIDEWINDER : AIM-9M with minimal smoke exhaust 286
SINAI PENINSULA 261-62 266-67
SKIP-BOMBING 194-95 (195)
SKYHAWK 264 267 279 282
SLOPING-V 94
SMART BOMBS 271
SOLOMONS : 185 189 194 (191) 211 ~ B-17F *Flying Fortress,* Gizo Island ~ Black Sheep 202 (198-200)
SOPWITH Camel (21) (72) 55 74 76-77 93 97 : RNAS *Tripes* 76 87 : 1½ Strutter 86 : Cuckoo (95) 97 98
SOUND DETECTORS 145-46
SPAD : VII 22 - XII *moteur-cannon* 22 25 (27) - XIII 26
SPANISH CIVIL WAR 109-115
SPARROW AIM-7F air-to-air missiles 266 271 273 286
SPIDER'S WEB patrol pattern 94
SPRAGUE, Rear-Admiral Clinton, *Taffy 3*: 208-11
SPITFIRE (Supermarine aircraft) 113 121 130 131 (128) 140 152 159
STRADDLING technique for depth charge water hammer effect 144
STRAFING 205-06 261
STRANGE, Louis 1 2 (8) 104
STRASSER, *Zeppelin Flotten Kapitan* Peter (13) 46 51
SUBMARINES - German 38 87 89-91 93 94 137 141 143-44 153 : Japanese 173 (173) : Argentine 277
SUKHOI AIRCRAFT : Su-7 238 : Su-20 272 : Su-22 (270) 271
SURFACE-TO-AIR MISSILES, British S.A.S. fired Stinger : see also, SAMs
SWORDFISH (Fairey) planes 140 145
SYRIA 102 261-63 272-73 : Syrian MiGs 272-73

INDEX

TANNENBERG 7
TARANTO, Italy 93 145-47 170
TAUBE AIRCRAFT (6) 7
TEMPEST, Lieutenant Wulstan 51 (59)
THACH, Lt-Com. John S. *Jimmy* (176) 179 186 (187)
- *Thach Weave* 189 251
- *Big Blue Blanket* 211
THUNDERCHIEF F-105 *Thud* 240-41
Thunderjet F-84 (223) - F-84G 236
TILLEY, Captain Reade - *Hints on Hun Hunting* 152-53
TIRPITZ, Grand-Admiral Alfred's ~ proposed *Fire Plan* against London 34-35
TOMAHAWK cruise missile 286 : *Kit-2* carbon filament wire warhead 286
TOP GUN, Mirimar 247 253 286
TORPEDO ATTACKS, aerial 93 (98-99) (140-41) 145-47 152 170-73
- Japanese 168-73 178 *anvil* tactic 184
TRENCHARD, Air Chief Marshall Sir Hugh Montague *Boom* 20
TRUK ATOLL (Caroline Islands) : Param Island 189 (201)
TULAGI (Solomon Islands) 184 189

UDET, Ernst 76
ULTRA, see Codes & Cryptography
UNION ARMY Balloon Corps 298-300
UNITED STATES AIR SERVICE 25-30
UNITED STATES OF AMERICA 24-30 94 141-44 152-53 158-234 (215) 239-60 269-74 284-91
U.S. MARINE CORPS 187 (198-99) 210 275 287
U.S. Navy - Carriers : *Hornet* 178-81 (181) 186 (198) - *Enterprise* 170-71 178 186-87 - *Lexington* 170-71 178 184 (197) 207 - *Ranger* 170 - *Saratoga* 170 186 189 - *Wasp* 170 178 - *Yorktown* 178 184 186-87 (194-95) (203)
USSR - Soviet Union 109 154 166 221
- invasion of Poland in 1939 119
- invasion of, in June 1941 150 154
UXB - unexploded bombs, Falklands 282-83

VERDUN 22 25 76
VICTORIA CROSS 16 39 49 64 65
VOSS, Werner (12) 19 72 (79)
VYRMPEL K-13 *Atoll* 241
VULCAN 276

WAGON WHEEL FORMATION, NVAF 245 (245)
WAKE ISLAND 186 (196)
WARNEFORD VC, Sub-Lieutenant Reginald A.J. (38) 38-39
WATER HAMMER effect 143-44
WILDCAT 185 186-87 (187) 189
WILHELM II, Kaiser 1 7 17 31 46 72 76 91 104
WINDOW (metal foil strips) 158
WINGMAN 14 16 63 189 225 230 233
WING SLIPS 29
WMD weapons of mass destruction 289
WRIGHT, brothers 304 - Flying Boat (52) (86)

YALU RIVER, North Korea 222 224 230
YAMAMOTO, Admiral Isoruko 146 170-71 180 185-87
YOM KIPPUR - OCTOBER WAR (Middle East) 267-69

ZIGZAGS 54 141 - *and in a minefield that is fatal* 93
ZEPPELIN, Count Ferdinand von 31
ZEPPELIN 1 6 13 31-59 (33-38) (40-41) (44) 91 - *The Afrika Zeppelin* 40 - tactics 46
ZERO A6M fighter 171 178 185-86 (186) 189 192 194 199 206 209 211 (219)
- Akutan Zero (192-93)

NOTES

Do not fill in here anything that should be secret from the potential enemy.

NOTES

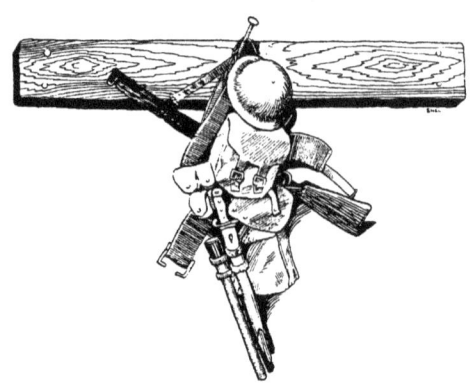

ARCHIVES
d'Histoire et Militaire

ADELAIDE

29 Alexander Street
LARGS BAY 5016
Australia